MICROBIAL DIVERSITY

"If I could do it all over again, and relive my vision in the twenty-first century, I would be a microbial ecologist. Ten billion bacteria live in a gram of ordinary soil, a mere pinch held between thumb and forefinger. They represent thousands of species almost none of which are known to science. Into that world I would go with the aid of modern microscopy and molecular analysis. I would cut my way through clonal forests sprawled across grains of sand, travel in an imagined submarine through drops of water proportionately the size of lakes, and track predators and prey in order to discover new life ways and alien food webs. All this, and I need venture no farther than ten paces outside my laboratory building. The jaguars, ants, and orchids would still occupy distant forests in all their splendor, but now they would be joined by an even stranger and vastly more complex living world virtually without end. For one more turn around I would keep alive the little boy of Paradise Beach who found wonder in a scyphozoan jellyfish and a barely glimpsed monster of the deep."

E.O. Wilson on "The Diversity of Life" in *Naturalist* (1994).

MICROBIAL DIVERSITY

Form and Function in Prokaryotes

Oladele Ogunseitan

Blackwell Publishing

BLACKWELL PUBLISHING
350 Main Street, Malden, MA 02148-5020, USA
108 Cowley Road, Oxford OX4 1JF, UK
550 Swanston Street, Carlton, Victoria 3053, Australia

First published 2005 by Blackwell Science Ltd

Library of Congress Cataloging-in-Publication Data

Ogunseitan, Oladele.
Microbial diversity / Oladele Ogunseitan.
p. ; cm
Includes bibliographical references and index.
ISBN 0-632-04708-9 (pbk. : alk. paper)
1. Microbial diversity. 2. Microbial ecology.
[DNLM: 1. Biodiversity. 2. Microbiology. QW 4 035m 2005] I. Title.

QR73.O486 2005
579′.17–dc22

2004003077

A catalogue record for this title is available from the British Library.
Set in 9½/12 pt Minion
by SNP Best-Set Typesetter Ltd., Hong Kong
Printed and bound in the United Kingdom
by William Clowes, Beccles, Suffolk

The publisher's policy is to use permanent paper from mills that operate a sustainable forestry policy, and which has been manufactured from pulp processed using acid-free and elementary chlorine-free practices. Furthermore, the publisher ensures that the text paper and cover board used have met acceptable environmental accreditation standards.

For further information on
Blackwell Publishing, visit our website:
www.blackwellpublishing.com

CONTENTS

FOREWORD

When we contemplate "evolution of life on Earth" we tend to imagine changes in animals and plants. We picture little ape-men who yelp at their hairy wives or small running Paleocene mammals who run on the third ("middle finger") digits of their fore and hind legs that enlarge and harden to become hooves. We see forests of ancient scaly *Lepidodendron* trees descend to become the little club mosses (also called ground pine or Christmas fern). Mostly the word "evolution" conjures the grunting caveman to singing cave painter transition in northern Spain and southern France. Although we know that no *Eohippus* ever awoke one fine and sunny spring morning, to stretch his legs and watch his toes transform to horny hooves, nor did any bone-splitting, marrow-chomping hairy Neanderthal survey the snowscape to return inside the limestone cavern to outline horned antelope before the fire, such exaggerated evolutionary images enchant and attract us. What is seldom conjured up by the phrase "evolution of life on Earth" is bacteria.

A thorough read of *Microbial Diversity* will alter our worldview. We moderns are grossly biased in our perspective; we are far too preoccupied with far too few forms of life. Of most concern to us are vertebrates that live on land (ourselves, our pets, our draft animals); flowering grasses that do most of the production upon which we depend for sustenance (i.e., barley, corn, rye, wheat); and the fungi as mushrooms, the yeast of bread and beer, or as agents of ringworm, athlete's foot, or allergens. Reminiscent of the five-year old, our anthropocentric view toward the natural world is one of "what's in it for me?" When, in *The Progress of the Soul* (c.1610), John Donne wrote, "Nature's great masterpiece, an Elephant . . .", he was only partly right. He could not have guessed the truth that Professor Ogunseitan builds in this splendid text and, I paraphrase, "Nature's great masterpiece, the bacterium". In *Microbial Diversity* Ogunseitan has written well about the fundamental units of life, the bacteria. His work is surprisingly comprehensive and up to date. By implication and even in explicit reference, he explains their spectacular evolution. But he has not even opened the expansive landscape to their most gifted and crucial descendants: the eukaryotic microorganisms. These larger beings, are the coevolved and integrated communities of bacteria, the immediate kin from which the larger forms of life on Earth evolved. What are the denizens of the glorious microbial world, the microcosmos, omitted from detailed treatment here? The filamentous fungi and the protists, microscopic eukaryotic organisms, refractory to simple classification and to short accurate description are underrepresented.

The fundamental lesson of this book turns our cultural myths inside out. The abundance and diversity of life on Earth has come not from fossil horses or club mosses but from the flourishing of the oldest, most omnipresent life forms, the bacteria. For all intents and purposes the bacteria invented everything of importance: growth, metabolism and reproduction, swimming and chemical sensitivities, oxygen respiration and desiccation-resistant propagules. Some perfected predatory behavior and the kill. They are masters of efficiency and recycling of waste. They invented sex and indulge in it with abandon. They cover the mountaintops, the prairie, and the plains with their offspring. They swim with no thought of sleep. They fashion fuel like methane and ethanol from far less energetic forms of carbon such as CO_2. The prodigious bacteria have created sexual communication and gender, genetic recombination, and consortial living. Some thrive exposed to ferocious winds and blinding sunlight on open cliffs, others burrow into hard limestone rock and photosynthesize right through their chalky covers. As metal workers, bacteria have no peers: some precipitate gold and others mine iron; some manufacture metallic sheens of manganese and others work copper or etch glass. In Ogunseitan's learned tome the crucial importance of bacterial life to our environment is laid bare at a sobering level of scholarship. He pays heed to the recent literature. He does not overstate or overconclude, rather he gives the advanced student access to the professional literature on its own terms.

Microbes, by consent, are live beings too small to be seen as individuals with the unaided eye. Nature's energetic gyration has generated two vast groups, easily distinguished by direct microscopic inspection of their cells. Because no single life form intermediate between these groups has ever been found, all microbial life is unambiguously classified into the Eukarya, organisms with nuclei or the bacteria in the broad sense (the Prokarya). Apparently small and simple when visualized by light microscopy, bacteria are amazingly complex and diverse when studied by more devious means. Chemical, metabolic, macromolecular, and other indirect analyses have revealed the world of diversity in prokaryotes that is the subject of this book. Prokaryotes are single or multicellular organisms in which each cell is of the bacterial kind. Since they were discovered by Antoni van Leeuwenhoek in the late-17th century and analyzed by Louis Pasteur in the late-19th century, bacteria have been studied by chemists, oil scientists, food industry advisers, and especially by physicians. Lately prokaryotes have been the focus of attention of sewage engineers, space scientists, and environmental analysts. Biologists, whether zoologist, botanist, or cell biologist, have tended to exclude prokaryotes from their foci of study. The activities of these wily, insinuating hordes impinge so heavily on human lives that a new vocational term was coined to describe the scientist whose profession it is to study the greater bacteria: he is the microbiologist. By tradition and practice, the microbiologist studies almost all the prokaryotes and one group of eukaryotes, the smallest fungi (the yeast). He systematically, by tradition, excludes the eukaryotic microbes. Microbes composed of cells that contain nuclei, an estimated 250,000 species alive today, form another world. They tend to be studied by zoologists (the protozoa), botanists (the algae), and mycologists (the fungi including water/molds and slime molds). Botanists traditionally claim the cyanobacteria as plants. Ogunseitan rests squarely in the microbiologist's traditions but he is aware of its deficiencies.

His subtitle admits that this is a book about the bacteria as units of life itself. It deals with all organisms made of cells that lack membrane-bounded nuclei. All prokaryotes are still members of the bacterial world, whether in ribosomal composition eubacterial or archaebacterial, or in cell wall structure (two-membraned gram-negative, single-membraned gram-positive, gram-variable, or the aphragmatic that lack cell walls entirely). Bacteria lack chromosomes. In spite of the widespread terminology, "bacterial chromosomes", in my opinion, do not exist. The naked DNA structures of the bacterial genophores are chromonemes, not chromosomes. The histone protein-draped chromatin of animals, plants, protoctists, and fungi provide the material basis of meiotic-fertilization forms of sex. Mitotic spindle movements of real chromosomes assure alternate production of haploids (e.g., plant spores and germ cells) and diploids (plant sporophytes and animal body cells). Such elaborate sexuality in the Eukarya, which requires the breach of the individual haploid cell and the acceptance in toto of a "foreign" nucleus or cell within a common membrane, is a feature essential to eukaryotes and their behavior. The cell-level "emboitement" (known in many guises: fertilization, pinocytosis, phagocytosis, cell fusion, karyogamy, invasion, endocytosis, incorporation) marks as unique all eukaryotes relative to Ogunseitan's prokaryotes. The ability to evolve "a genome at a swallow" is entirely lacking at the prokaryotic level of cell organization.

The paucity of endosymbionts and absence of cyclical cell fusion is what makes Carl Woese's currently preferred term "Archaea" an anathema, as it implies that these microorganisms somehow are not bacteria. Whereas Woese's original concept of "Archaebacteria" led to an immense contribution to the literature of the analysis of microbial life, "Archaea" is a misnomer. Archaebacteria (usually methanogens, halobacters, or sulfoacidophils) are, after all, bacteria in every sense. Like all their prokaryotic brethren they are homogenomic, have small ribosomes and chromonemic organization, transfer small pieces of DNA, and are not products of symbiotic fusions. The great group of bacteria, including archaebacteria, with its incredible diversity, may be characterized by many criteria in addition to the ribosomal RNA DNA genes or a chromonemal gene sequence for a single protein. Unique physiological pathways and environmental distribution are two examples. A singular strength of Ogunseitan's text is its exhaustive, measured, and fair comparison of ways to measure and handle the unruliness of the prodigious diversity of bacteria. He knowingly slights the eukaryotes although he realizes they are composites of bacteria. Clearly, a detailed treatment of them would bring him into a far different realm. Because eukaryotes are all merged complexes of a limited set of prokaryotes another entire book would be required to do them justice. The ecological, behavioral, developmental, metabolic, and genetic magnitude of prokaryotic diversity is staggering enough. Beside it the diversity of all eukaryotes pales. Whether microbial (small fungus or protoctist) or visible with the unaided eye (large fungus and protoctist, animal or plant), the uniformity of the sexual and metabolic biology of eukaryotes is striking relative to the evolutionary diversity of bacteria. Only in morphological splendor, easily delineated species and the production of certain specific metabolites such as plant and fungal hallucinogens, toad and dinomastigote nerve toxins, plant skin irritants, Chinese and English speech, phosphate–nitrate mountains and cities, do the eukaryotes have any claim to unique and original diversity. All important features of life on this planet evolved in bacteria. Enter here and see for yourself.

Lynn Margulis
Distinguished University Professor
Department of Geosciences, University of Massachusetts

PREFACE

Conservation biology is a relatively new academic discipline that is currently enjoying a remarkable popularity among students, researchers, and the public. Its success is striking because it is primarily an exploratory science with few reproducible theories or generalizable concepts. Habitat preservation and species protection programs implemented through government regulations such as the U.S. Government's Endangered Species Act of 1973 have been modestly successful experiments to conserve biological diversity. However, to microbiologists, these landmark steps toward popular appreciation of ecological biology must seem partial and inadequate. Years dedicated to learning that microorganisms sustain the global ecosystem have not reduced the gap between the appreciation of large animal and plant organisms (macrobiodiversity) and the widely acknowledged ignorance of the diversity of microorganisms (viruses, bacteria, protists, and small fungi—microbiodiversity).

In nature, the large population densities and rapid growth rates inferred from working with many microorganisms in the laboratory support the view that prokaryotes cannot be in danger of extinction (Staley, 1997). Microbial pathogens that humans have deliberately tried to extinguish with antibiotics and antiseptics for centuries are still very much with us. However, occasions do arise that raise the prospect of microbial extinction, at least locally. For example, the debate at the end of the twentieth century about whether to terminate the last remaining stocks of the smallpox virus only slightly included discussions about microbial diversity. Instead, the debate was framed by concerns for future bioterrorism and access to vaccines.

This book presents a comprehensive analysis of the concepts and methods that have facilitated recent advances in the understanding of microbial diversity, and a clear linkage to the importance of this understanding for global physicochemical and biological processes that are easily recognizable to investigators in the natural sciences. Microbial diversity is presented here as highly relevant to processes that are proximal to human affairs, a treatment that will be of interest to investigators in other disciplines.

Taxonomy and systematics are unquestionably important to concepts of microbial diversity, but they are treated here only as means to the goal of understanding the diversity of microbial function in nature. Therefore, in certain contexts, it is reasonable to discuss the diversity of specific microbial processes with limited scientific data on the exact number and names of microbes that are responsible for the processes at a particular time and place. The inability to identify and quantify particular microbial species involved in a given ecological process has long been recognized as a major impediment in microbial ecology and environmental microbiology, but the shortcomings are treated here as an opportunity for scientific advancement. The advancement will undoubtedly ameliorate the hindrance that methodological limitations have placed on research programs aiming to understand the contributions of microbial diversity to defined states of ecological and human health.

The book is organized in two major sections with chapters representing major themes within each section. Part I focuses on conceptual and methodological issues, whereas Part

II focuses on principles, applications, and opportunities for research and the advancement of knowledge. Chapter 1 introduces the theoretical and empirical difficulties entailed in defining a microbial "species" as a fundamental unit of microbial diversity, and on the methods for identifying varieties in microbial populations. Chapters 2 through 5 present critical analyses of the major techniques employed for investigating microbial diversity including microscopic, culture, molecular, and phylogenetic systematic methods.

Chapter 6 lays the foundation for integrating microbial diversity with Earth system science through the historical evidence for the influence of the origin of microbial life on the early Earth environment. The sojourn into environmental evolution places contemporary concerns about human impacts on the global ecosystem into the robust context of the coevolution of life and environment. The constant flow of novel and controversial theories about the geological roots of microbial life on Earth makes this topic also of great interest for astrobiology (e.g., see Gold, 1999). Chapters 7 and 8 focus on the relevance of microbial diversity to biogeochemical cycles. This is a well-researched area of microbial ecology, but rather than focusing on the activities of well-known microbial species or specific biochemical pathways, the approach used in these chapters focuses on the interconnectedness of microbial communities and their role in maintaining balanced biogeochemical pathways. In addition, the discussion deals with the consequences to microbial communities, of geochemical disequilibria attributed to human-dominated ecosystems. Chapters 9 and 10 focus on the relationship of microbial diversity to biotic interactions. Many of the commercial applications of microbial diversity fall within this topic area.

The axenic culture conditions that have been very useful in elucidating the characteristics of microorganisms are unfortunately not the state in which microorganisms exist in nature, therefore, it is clearly not appropriate to draw far-reaching conclusions about the driving forces of microbial speciation and diversity from axenic laboratory experiments. Advances in microcosm design have facilitated research on natural microbial communities, but it is difficult to reproduce how these communities behave under natural environmental conditions. As the progenitors of eukaryotic life forms, most microorganisms that have been studied engage in complex multifaceted interactions with plants, animals, and with other microorganisms. The inevitable human-value system prejudges these interactions as either beneficial such as in nitrogen fixation or detrimental such as in the case of pathogens, but there is no question that these biotic interactions and all their ramifications are essential for the sustainability of natural ecosystems. The approach used in these chapters does not underestimate the significance of the categorization of microbial biotic interactions, but the emphasis is on the importance of microbial diversity to such categorizations, and how human impacts on existing diversity may reinforce or modify the biotic interactions in consequential ways.

The final chapter deals with the relevance of microbial diversity to global environmental problems that have been exacerbated recently by inefficient industrial ecology, population growth and urbanization. In many ways, the recognition of dynamic environmental phenomena as intractable problems is rooted in impatience with, and poor understanding of, biogeochemical cycles, some of which are discussed in Part II. An optimistic view is that solutions to these global environmental "problems" may depend in part on better understanding and scientific management of microbial diversity for "correcting" the balance of chemical fluxes in various environmental compartments.

For students majoring in microbiology, this book will be useful for one of the quartet of core courses recommended by the American Society for Microbiology, namely "Introduction to Microbiology", "Microbial Physiology", "Microbial Genetics", and "Microbial Diversity" (Baker, 2001). The order in which this sequence of courses is taken should not adversely affect comprehension of the materials presented in this book, although it is recommended that a solid foundation in general microbiology should come first. The book is presented in a format that can also serve as an accessible resource for a graduate-level, one-semester course in departments where knowledge of applied microbiology is required to sustain specific research programs. The book is also designed for active researchers to reveal the current state of knowledge, including conceptual inconsistencies and existing analytical gaps. All biological diversity can be traced back in time to prokaryotic diversity. In this sense, the Earth is a planet **of** prokaryotes, and the treatment of microbial diversity here leans heavily on prokaryotic forms and functions. Excellent treatments of eukaryotic microbial diversity already exist (notably Margulis *et al.*, 2000). These texts and handbooks should be indispensable supplements to this book for a complete view of microbial diversity.

Special features of the book include:

- Extensive illustrations including charts, maps, graphs, and diagrams.
- Four-color plate inserts with high resolution images of microorganisms, maps, and satellite images.
- Text and image boxes presenting extended discussions of critical concepts and case studies, for those interested in a deeper exploration of the material.
- Challenging questions for further investigation in each chapter offer opportunities for direct field investigation.
- A list of suggested readings in each chapter and an extensive list of references at the end of the book provide up-to-date sources for further exploration.
- An accompanying CD-ROM containing the high resolution images and art used in the text, provided in JPEG format for classroom presentations.
- A dedicated website (**www.blackwellpublishing.com/ogunseitan**) for further consultation, access to special features, and updates.

Many people deserve plenty of gratitude for their unflinching support throughout the evolution of this book project. Without the encouragement of Nancy Whilton at Blackwell, this book could have remained indefinitely as an idea in my mind. Nathan Brown, Elizabeth Wald, and Rosie Hayden also at Blackwell, provided excellent feedback and professional support. Tessa Hanford, in Surrey, England, performed excellent copy editing for which I am very grateful. The quality of the book owes much to the critiques and guidance provided by several peer reviewers at various universities in the United States and Europe. The following reviewers (in alphabetical order) deserve particular gratitude: Mary Ann Burns at Penn State University; James Brown at North Carolina State University; Jocelyne DiRuggiero at the University of Maryland; Stjepko Golubic at Boston University; Lidija Halda-Alija at the University of Mississippi; Lynn Margulis at the University of Massachusetts, Amherst; James Prosser at the University of Aberdeen in Scotland; and Peter Sheridan at Idaho State University. By definition, a scientific book is a "work in progress", and I take full responsibility for any shortcomings that may remain in this edition. A sizable portion of the book was drafted, revised, and completed on my home computer, and without the steadfast support and understanding of Alison, Coryna, and Sofya, this would not have been possible. I thank the Macy Foundation and the administrative faculty of the Marine Biological Laboratory at Woods Hole, Massachusetts for making it possible for me to spend two glorious summers incubating the idea for this book in and around the laboratory at Cape Cod. The perspective of microbial diversity as a fundamental science of the Earth system matured in my mind during the interim period that I spent between these summers as a global environmental assessment fellow at Harvard University, and I am grateful to William Clark and colleagues for providing that opportunity. Finally, I thank my colleagues at the University of California, Irvine, and several investigators around the world who patiently shared the results of their research.

Oladele A. Ogunseitan

CONCEPTS AND METHODS

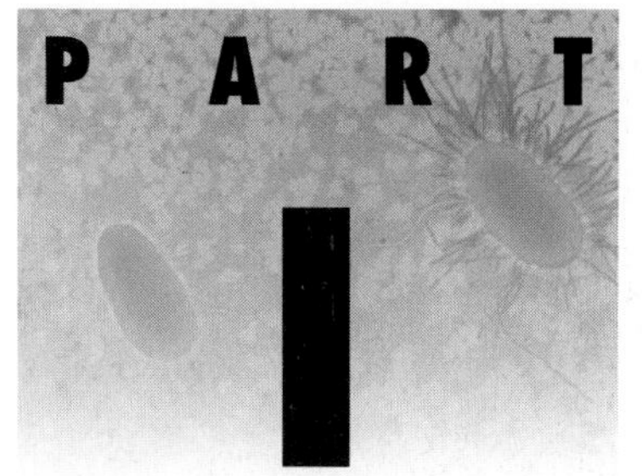

THE CONCEPT OF MICROBIAL SPECIES

INTRODUCTION

What distinguishes microbiology from other disciplines of biology? This question no longer has a straightforward answer that can satisfy all biologists. The traditional answer focused on the extremely small size of organisms under investigation; however, this leaves little room for distinction on the basis of taxonomy because practically all organisms have a microscopic stage during their life cycles. Some organisms that are physiologically closely related to large macroscopic organisms spend their entire life span as microscopic organisms. Nevertheless, physical size remains a dominant conceptual framework for most practicing microbiologists, and most of the discussion in this book is presented from this perspective. Other responses have focused on unicellularity (as opposed to multicellularity) as the defining characteristic of microorganisms, however, viruses are acellular, and many investigators have argued that the so-called unicellular stage of bacteria, for example, is not a naturally occurring phenomenon. Some investigators have advanced the cellularity argument by invoking differentiation as the separating principle, but many "unicellular" organisms also go through developmental differentiation as in the case of sporulation and fruiting bodies. Finally, many investigators focused on the internal anatomy of organisms to identify unique characteristics of microorganisms that are not shared by macroorganisms. This response has been organized around the concept of karyology (pertaining to the organization of genetic material), with prokaryotes and eukaryotes as the main divisions of biological diversity. However, current understanding of endosymbiotic interactions with respect to the emergence of organelles that are usually attributed to the eukaryotes suggests that this response may not be stable over evolutionary time frames. The difficulty of defining the subjective aspects of microbiology is further compounded by the cosmopolitan adoption of incisive tools presented by molecular biology. Molecular analysis has been extremely influential in exposing major unifying concepts in biology, but at the same time, such analysis has revealed a remarkable level of diversity that has proven difficult to organize in discrete packets of information that are consistent with previously held concepts of diversity.

The concept of biological diversity implies consensus on the discrete nature of independent species and on the mechanisms that generate speciation. The recognition of differences and similarities among the discrete features of microorganisms is more challenging and less well understood than for large multicellular organisms. A thorough comprehension of the complexity inherent in the concept of microbial species is fundamental to the appreciation of microbial diversity, and to the understanding of processes that generate differences despite the influence of other ecological pressures that tend to produce similarity in a given environment. The balance of these two seemingly opposing processes (liberation of diversity and conservation of ecological function) has resulted in the emergence of homeostatic conditions that support perpetual phylogenetic lineages. The recognition of independent microbial species is based on an assumption of long-term stability of these homeostatic con-

Chapter contents

ditions. This chapter presents a discussion about the empirical foundation of the microbial species concept, including a balanced view of current controversies that have entangled the interpretation of phylogenetic categories according to recent data on phenotypes and molecular characteristics. The topics are selected to introduce important milestones and alternative perspectives that have been developed to explain the concept of microbial species. The following main points are explored:

1 The differences between the various species concepts; contemporary theories about speciation; and the relevance of these concepts and theories to microorganisms.
2 Comparative assessment of alternative models for speciation, with an explanation of how these models accommodate or ignore special genetic properties of microorganisms.
3 Linkage of the understanding of species concepts and mechanisms of speciation to the theoretical and methodological advances in the assessment of microbial diversity, and the relevance of such advances to the emerging understanding of Earth system processes.

OLD AND NEW CHALLENGES FOR ASSESSING MICROBIAL DIVERSITY

There is a plethora of convincing pieces of evidence that the Earth is a "microbial planet" in the sense that microorganisms predate other life forms, they are the most abundant—both in terms of numbers and distribution. In addition, microbial activities have profound influence on the integrity and functioning of global ecosystems. Despite the widely acknowledged importance of microorganisms on Earth, scientific knowledge of microbial diversity and function is scantier than for physically larger and scarcer organisms (Staley, 1997; Wilson, 1994). The term "diversity" in a biological context presumes a multiplicity of forms that may not necessarily be apparent without sophisticated observation by means of specialized tools. In no other discipline of biological sciences is this truer than in microbiology. To the untrained observer, microorganisms have much more in common than they can possibly have in differences. Therefore, the recognition of microbial diversity has always depended on the methods used for analysis. Early investigators such as Antoni van Leeuwenhoek (1632–1723; see Box 1.1) were limited to gross morphological differences in microbial cell shapes and colonies. It is now possible to recognize substantial differences between organisms at the molecular level, but this scale of analytical power has not yet provided a coherent solution to the persistent questions surrounding the concept of microbial speciation and diversity. These questions fall into three separate but interrelated categories, namely:

1 **Incomplete information on the number of existing microbial species**. The quantitative estimate of the number of microbial species has been limited by the inadequacy of techniques used for recovering, isolating, and cultivating microorganisms present in various ecosystems.
2 **Non-operational definition of "microbial niche"**. The biological concept of niche, as developed for macroorganisms, has not been very useful in microbiology. The niche refers to the multidimensional space where the coordinates are defined by parameters representing the conditions of existence of a given species. Niche is also used in reference to the ecological role of a species in a community. In microbiology the application of the niche concept has been limited by the difficulty of explaining the wide geographical, geological, and ecological ranges in which specific groups of microorganisms occur. In addition, correlations between microbial species diversity and ecosystem functions are very complex and difficult to reduce to any form of numerical modeling.
3 **Loose definition of strains and species**. The occurrence of intra-species and inter-species genetic exchange among microorganisms is recognized as a major driver of evolutionary innovation. The frequency of genetic exchange and the promiscuity of gene-transfer mechanisms have led to a questioning of taxonomic boundaries in microbiology.

BOX 1.1

(a)

(a) The inventor of the microscope, Antoni van Leeuwenhoek (1632–1723), was the first to recognize microbial diversity, although his observations were mostly of eukaryotic organisms. Improvements on the simple microscope led to an expansion of the realm of microbiology to include many species of prokaryotic bacteria and viruses by the early twentieth century. Image by courtesy of Dr. Warnar Moll's private collection, Amsterdam, The Netherlands.

(b)

(b) Carolus Linnaeus (Carl von Linné; 1707–78) invented the basic structure of the system of phylogenetic classification that is still in use today. The Linnaean system was designed to classify large eukaryotic macroorganisms based on morphological differences and reproductive exclusion. Contemporary debates regarding phylogenetic classification have focused on the meaning of "species" as the fundamental unit of biological diversity. Early microbiologists adopted the use of morphological criteria (such as varieties of coccus and bacillus cell shapes) for species classification because of limited knowledge of microbial diversity. New molecular approaches for investigating microbial genetic structure and function have clearly demonstrated the limitation of using morphology as the basis for species categorization in the investigation of prokaryotes. The appreciation of Linnaeus's contributions to natural science is demonstrated by his image on Swedish currency. Image by courtesy of Thomas Hunt.

(c)

(c) Charles Darwin's (1809–82) theories of speciation revolutionized the understanding of the relationship between different "kinds" of organisms, or "species". Modern molecular biology produced an elegant biochemical mechanism that supports the Darwinian theory of evolution by natural selection, which applies to all phylogenetic categories, including microorganisms. However, the actual causes of speciation remain controversial inasmuch as the definition of species remains unresolved. Darwin's work illuminated two different but related conceptual pillars of biological evolution, one dealing with morphological transformation within lineages, and the other with the principle of diversity – leading to an increase in the absolute number of lineages. These concepts, although much better understood since their initial articulation, remain fundamental challenges for most accounts of microbial diversity and function. Image by courtesy of Henry Huntington Library, San Marino, California.

(d)

(d) Darwinian biologist, Ernst Mayr (1904–), developed the salient biological species concept that applies to most eukaryotic species. However, the relevance of the concept to microorganisms is doubtful. Nevertheless, most species concepts dedicated to prokaryotes are more or less variations on the themes that combine genetic and ecological criteria, which were developed for eukaryotes. Ernst Mayr rejected the three-domain universal phylogenetic tree based on comparative assessment of rRNA as proposed by microbiologists led by Carl Woese. In his view, the Archaea and Bacteria belong in the same prokaryote "empire", which is clearly distinguished from the only other domain, namely the eukaryotes. The apparent similarity of structural and morphological characteristics of Archaea and Bacteria cells outweighs the molecular similarity between Archaea and the eukaryotes (Woese, 1998b). Image by courtesy of Ernst Mayr Library, Museum of Comparative Zoology, Harvard University.

The exploration of these categorical questions has produced an expansion of the methodological basis for analyzing microbial diversity. Innovative methods have been used in environmental microbiology and microbial ecology to resolve practical questions while contributing to our understanding of evolutionary systematics. The major lines of methodological advances, namely microscopy, culture, molecular analysis, and phylogenetic bioinformatics are considered in subsequent chapters in this section of the book. These advances have contributed to a widening of the scope of microbiological research, but their applications for solving contemporary problems facing global biodiversity require a comprehensive understanding of what we mean by "species", how they are created, and how they become extinct.

TRADITIONAL CONCEPTS OF SPECIES

The invention of a systematic scheme for classifying organisms was necessitated by the recognition of an expansive biological diversity. Carolus Linnaeus (Swedish name: Carl von Linné; 1707–78) is credited with the first widely accepted hierarchical scheme, which consists today of seven categories, namely kingdom, phylum, class, order, family, genus, and species as the fundamental unit. For Linnaeus, and most of his contemporaries, this taxonomic scheme was immutable because the prevalent doctrine of creation at the time did not include evolution, and the scheme appeared to work well for animals and plants that then dominated the study of biology. Even so, Linnaeus did not explicitly define "species" and modern concepts of species probably include a mixture of attributes that Linnaeus separated into "genus" and "variety".

Charles Darwin (1809–82) argued in *The Origin of Species* that species are not real entities in nature. Since Darwin's time, two major schools of thought have emerged which define the concept of species in ways that are consistent with the emergent synthesis theory of evolution and contemporary understanding of molecular genetics. The **realists** have affirmed that species are real and are the actual units that evolve. The **nominalists** reject this tangible definition of species. One group of nominalists proposed instead that species are breeding populations, also known as **demes**, and are the evolving units. Another group of nominalists acknowledges the usefulness of defining a theoretical concept of species, but does not accept the existence of discrete species units in nature. Four conceptual frameworks have emerged in modern biology to resolve the challenges facing attempts to make the species concept useful. Briefly, these frameworks represent ways of classifying species concepts on the basis of whether they are:

1 **Prospective** with respect to consideration for the future evolution of populations, or **retrospective** by considering species as "dead-end" products of evolution.
2 **Mechanistic** with respect to an ongoing process of speciation, or **historical** through focusing on the outcome of the process.
3 **Trait-based** by focusing on observable characteristic defining traits of organisms with no reference to the inferred lineage of those organisms, or **genealogy based** by focusing on the historical relationships among organisms.
4 **Intrinsic** by invoking "self-imposed" barriers on species mixing through specifically evolved limitations on genetic exchange, or **extrinsic** through focusing on the external or flexible barriers to genetic exchange.

In view of these organizing frameworks, four major species concepts can be compared and contrasted, namely the **typological species concept**, the **morphological species concept**, the **biological species concept**, and the **evolutionary species concept**. There are several variations on these four concepts, and some of these are discussed in the context of their similarities to one or more of the major concepts. Mallet (1995) and Mayden (1997) provide a more detailed coverage of the overlap and trade-offs among the various species concepts.

Typological species concept

The typological species concept predates the Darwinian theory of evolution, and it is not consistent with evolutionist thinking. The concept defines species on the characteristics of a "type specimen". The concept is based on the Platonic and Aristotelian philosophical arguments of the existence of an organismal "archetype" (Plato referred to this as *eidos*). The observed diversity within an archetypal species represents the manifestation of imperfections in an eternal experimental strive to reproduce an immutable perfect state. In this context, species represents a static, non-variable assemblage of organisms that conform to a common morphological plan (Lincoln *et al.*, 1982). The explanatory power of the typological species concept has been seriously challenged by improved understanding of the patterns of variation within populations. Several characteristics are known to vary among members of a single population of interbreeding individuals. Since there are no generally accepted means of specifying the exclusive properties possessed by each and every member of a particular species, it is not possible to generate a complete catalog of "type" characteristics, and the true identity of the "type" species cannot be known. Additional problems with the typological species concept include the observation that apparently different organisms may share the same morphological traits during various segments of their life cycles. Given that the typological species concept is the oldest and simplest of the species concepts, it is perhaps not surprising that many investigators have difficulty accepting it as an operable concept. The shortcomings of the typological species concept led directly to a proliferation of other concepts aiming to improve upon it. However, no specific concept has yet successfully accounted for all the questions that can be raised on how to reconcile the logical products of evolutionary forces acting to maintain species integrity while generating diversity.

Morphological species concept

The morphological species concept considers anatomical (morphological) characteristics to be the primary discriminant function associated with species. It is a derivative of the typological species concept that is preferred by some plant taxonomists and investigators working on organisms that do not reproduce sexually. The concept has been usefully applied to the classification of large groups of undescribed species such as a collection of fossils. The problems associated with interpreting the morphological species concept result from the sometimes arbitrary nature of the evidence used for classification, which in many cases relies on expert opinion about morphological differences. The concept also cannot explain the occurrence of sympatric species that exist in the same habitat, look morphologically identical, but are reproductively isolated from one another. For example, several bacillary bacteria exist in soils and exhibit similar colony morphologies upon culturing, but are clearly physiologically and genetically unrelated when subjected to molecular analyses.

Additional concerns include the inability of the morphological species concept to distinguish among cryptomorphic sibling species, where substantial changes have been introduced into the genome of one species but the changes are expressed in ways other than morphological differences. The concept also does not address the observation of sexual dimorphism, where the male and female versions of the same species look different. In bacteriology, the production of conjugation plasmid-encoded sex pili cannot be explained by the morphological species concept. Finally, the morphological species concept cannot account for genetic polymorphisms that are not directly expressed as distinct morphological characteristics. In view of the fact that speciation at the level of genetic divergence necessarily precedes the expression of morphological differences, the concept focuses perhaps too much on the outcome of evolution at the expense of accommodating the mechanisms that underlie speciation. Overall, the strengths and weaknesses of the morphological species concept are better appreciated when compared to other concepts such as the biological species concept, which accounts for the mechanisms of speciation, including the potential for sexual reproduction.

Biological species concept

Ernst Mayr (1904–) is credited with developing the biological species concept, which defines species as "groups of actually or potentially interbreeding natural populations that are reproductively isolated from other such groups" (Mayr, 1963). The biological species concept considers a species as the fundamental ecological and genetic unit, where consequential interactions occur only between species regardless of the fate of individual members. The species is genetically identifiable only through the population gene pool, which is in constant flux because of adaptive genetic exchange mechanisms. The biological species concept captures the significance of reproductive communities where specialized features prevent the dilution of the species gene pool through intra-specific genetic exchange. However, the phenomenon of natural transformation in which cells uptake "free" genetic materials from the environment is commonly observed in microbial communities. Therefore, species concepts that depend on a permanent reproductive isolation of a population do not fit prokaryotes. The biological species concept also fails to account for certain observations in populations of eukaryotic microorganisms. For example, in reviewing the taxonomy of free-living ciliated protozoa, Finlay and colleagues (1996) confirmed that these organisms have traditionally been identified on the basis of the extant morphological diversity, which is closely related to the natural functions of each of the 3,000 defined species. On this basis, the investigators rejected the biological species concept as inappropriate and impractical. Instead, they favor the morphological species concept for this group of microorganisms as more pragmatic.

Evolutionary species concept

The biological species concept does not accommodate organisms for which asexual modes of reproduction produce clonal species. To address this inadequacy, paleontologist George Gaylord Simpson (1902–84) proposed the evolutionary species concept, where the species is defined as "a single lineage of ancestor–descendant populations which maintains its identity from other such lineages and which has its own evolutionary tendencies and historical fate" (Simpson, 1951). This concept explicitly includes the evolutionary history or lineage of organisms as opposed to a focus on the recognition of current species. Therefore, it has been used extensively in the analysis of the fossil record, particularly in zoology (Ereshefsky, 1992). The evolutionary species concept has been formalized in various renditions of species concepts that focus on phylogenetic lineage, including the phylogenetic species concept in which individual members of a species are considered to be part of a monophyletic group haven descended from a single ancestral taxon (Wheeler and Meier, 2000). The concept does not fully account for genomic hybrids, where genes have passed from one taxon to another, and the genetic make-up of individuals can be traced to different phylogenies or genealogies. Assessments of completely sequenced microbial genomes have demonstrated that such hybrids are common. For example, 5% to 15% of bacterial species' genomes can be attributed to acquisition from other species (Ochman *et al.*, 2000). This makes the evolutionary species concept, as originally proposed, practically irrelevant to prokaryotes.

Other concepts

At least 20 more species concepts have been described, primarily for eukaryotic organisms, but most are variations on themes explored by the four concepts described above. Among the newer concepts, two are particularly noteworthy, namely the **phylogenetic species concept** and the **ecological species concept**, because they are relevant to current attempts to formulate a species concept that is operational for prokaryotic organisms.

Willi Hennig (1913–76), a leader of the phylogenetic school of systematics, championed the phylogenetic species concept. The complex terminology used initially by adherents of this concept makes comparisons with previous concepts tedious; in short, phylogeneticists proposed that relationships among species should be interpreted strictly on the basis of genealogy as **clade** relations. Therefore, a species is defined as "a group of individuals, also known as character-bearing **semaphoronts**, which are interconnected by **tokogenetic** relationships that are strictly defined by the phenomenon of reproduction" (Hennig, 1966). The semaphoront (an individual at a specific period in its life cycle) subconcept was needed to broaden the meaning of an individual by emphasizing the importance of distinct developmental stages, including both phenetic and cladistic differences during the life cycle. Therefore, under this concept, the microbial spore and vegetative cell count as separate semaphoronts. A simpler version defines species as a complex of spatially distributed reproductive communities. This simplification highlights the closeness of the phylogenetic species concept to the biological species concept as defined by Ernst Mayr (Mayr, 1987). Proponents of the phylogenetic species concept, notably Joel Cracaft, Niels Eldredge, and Mary McKitrick considered that an evolutionary view of the Linnaean hierarchy would inevitably produce a nested set of clusters that are linked by shared derived characters (or **synapomorphous** traits). In this sense, a species represents an indivisible cluster of organisms at the base of the Linnaean hierarchy. Following this argument, a definition of species was presented as "a diagnosable cluster of individuals within which there is a parental pattern of ancestry and descent, beyond which there is not, and which exhibits a pattern of phylogenetic ancestry and descent among units of like kind" (Cracaft, 1983). Cracaft (1989) further refined the definition as "the smallest diagnosable cluster of individual organisms within which there is a parental pattern of ancestry and descent." To deal with the problem of genealogical hybrids, the genealogical species concept was proposed by Baum and colleagues as a variant of the phylogenetic species concept by including a consensus of many estimated genealogies of different genes (Baum and Donoghue, 1995). Finally, the inclusion of ecological characteristics and reproductive isolation was added to the basic foundation of the phylogenetic species concept to produce a cohesion species concept (Templeton, 1989).

The ecological species concept is not very well connected to the intrinsic properties of organisms, such as molecular genetic characteristics. Instead it focuses on the occupation of adaptive zones by particular species. The adaptive zones are defined by resource distribution and the biotic and abiotic characteristics of specific habitats. Therefore, the ecological species concept is tied to the concept of ecological niche. The strength of the ecological species concept is that it attempts to categorize organisms by capturing the essence of phenotype as an expression of genomic information and environmental influences. However, it is difficult to consistently recognize ecological species because many organisms can occupy different ecological niches due to adaptation or developmental changes during the life course. The ecological species concept also precludes consideration of directionality in evolution, and it is not consistent with the hierarchical view of species diversity (Ereshefsky, 1992).

None of the traditional species concepts encompasses all groups of organisms. Furthermore, most of the concepts were developed without much thought given to usefulness in organizing the systematics of prokaryotic organisms. Table 1.1 presents a comparative assess-

Table 1.1 Comparative summary of major species concepts, and their relevance to prokaryotes.

Species concepts	Evaluation criteria								Does the concept apply to prokaryotes?
	Implications for the past or future status of populations		Conceptual grasp of process versus outcome of evolution		Basis for evaluating new organisms prior to classification		Barriers to genetic exchange		
	Retrospective	Prospective	Mechanistic	Historical	Traits	Genealogy	Intrinsic	Extrinsic	
Typological	*			*	*		*		No
Morphological	*			*	*		*		Not always
Biological		*	*		*		*		No
Evolutionary	*					*		*	Not always
Phylogenetic	*			*	*			*	Not always
Ecological		*	*		*			*	No

ment of six species concepts according to the four frameworks generally accepted as evaluation criteria for describing the coverage of these concepts.

The scarcity of a fossil record for prokaryotes, coupled with the recognition that most naturally occurring prokaryotes have not been described, posed intractable challenges for the application of species concepts developed for plants and animals to microbial systematics. These problems have led investigators toward proposals of species concepts that specifically consider peculiar characteristics of prokaryotic genetics and ecology. The main features of these "microbial species concepts" are described in the following section.

SPECIES CONCEPTS FOR PROKARYOTES

The Global Biodiversity Assessment program suspects that more than one million species of prokaryotic organisms exist in nature, but not more than 5,000 of them have been described (Rossello-Mora and Amann, 2001). Furthermore, there is currently no official definition of species for microorganisms, although several concepts have been proposed (Colwell *et al.*, 1995; Krawiec, 1985; Rossello-Mora and Amann, 2001; Ward, 1998). A number of criteria have been used to circumscribe existing microbial species categories, depending on the period of discovery and on the objectives of taxonomists working in different disciplines including medical, environmental, or industrial microbiology. The lack of agreement on a microbial species concept has led to an artificial amplification of the number of recognized microbial species because a single species can be identified with different names in different subdisciplines (Rossello-Mora and Amann, 2001).

The conceptual understanding of microbial species has traditionally relied on criteria similar to those used to formulate species concepts for eukaryotic organisms. For example, Ravin (1960) struggled to apply the biological species concept to bacteria by defining the phenotypic clusters of mainstream bacterial systematics as "taxospecies". The taxospecies concept is based on numerical taxonomic methods and defines species as a group of organisms, including strains and isolates, with mutually high phenotypic similarity that forms an independent phenotypic cluster. This concept is analogous to the morphological species concept, but in addition to anatomical features, it includes consideration of physiological characteristics. Ravin also proposed a "genospecies" concept to define groups of bacteria that can exchange genes, but there was very little correlation between the groups of organisms described by the taxospecies concept, and those described as genospecies (Ravin, 1963). This incongruence provoked further dissatisfaction of microbiologists holding traditional species concepts. More recent attempts to make the biological species concept applicable to prokaryotes can be traced to the work of Dykhuizen and Green (1991) who proposed bacterial species as "groups of strains that recombine with one another but not with strains from other such groups." The recognition of historic events of genetic recombination was built into this definition because it had become feasible to reconstruct phylogenetic relationships according to molecular sequence data, which presumably can be used to delineate groups according to the genetic exchange criterion. However, this approach can be called to question on the basis of several observations that many bacteria are capable of exchanging genes both within and between the groups currently nominated as species.

Fred Cohan has argued persuasively that there is a fundamental misconception which limits the success of attempts to develop species concepts for prokaryotes through extrapolation from eukaryotic species concepts. For most eukaryotes, the species represents "a group of organisms whose divergence is capped by a force of cohesion; divergence between different species is irreversible; and different species are ecologically distinct" (Cohan, 2002). Whereas for bacteria, the characteristics of named species do not capture these universal properties, but instead bacterial "ecotypes" fit the definition of eukaryotic species. Ecotypes are defined as "populations of organisms occupying the same ecological niche, whose divergence is purged recurrently by natural selection." Bacterial ecotypes can be recognized by their molecular sequence signatures, and comparative assessments of these signatures have

BOX 1.2

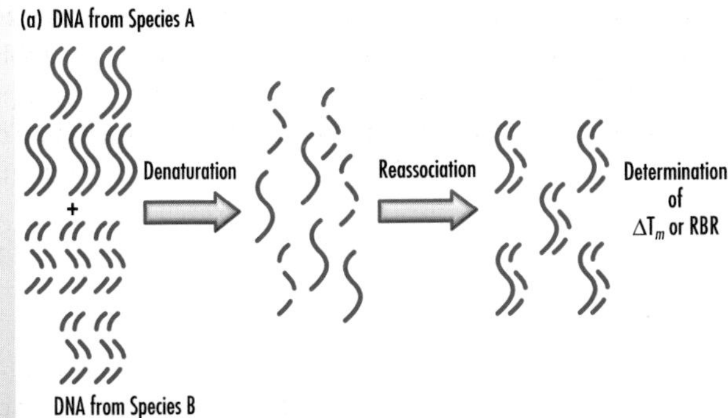

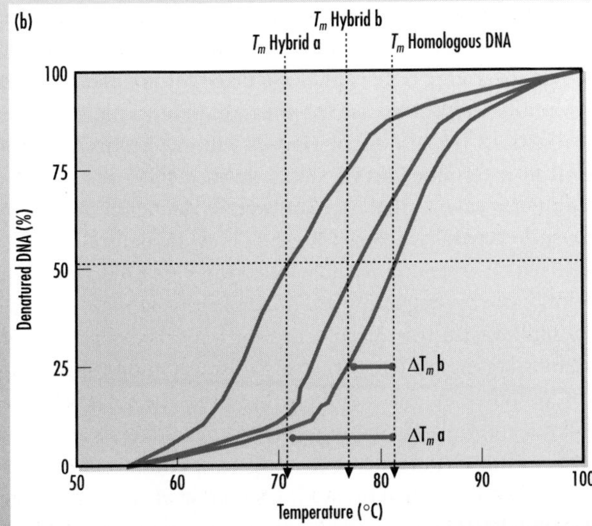

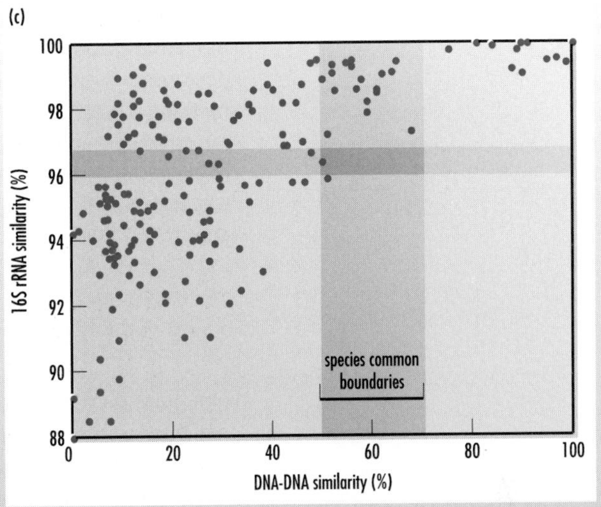

Schematic representation of procedures for quantifying the degree of relatedness between two microbial isolates.

(a) DNA–DNA reassociation kinetics yields a thermal reassociation midpoint (ΔT_m) or the relative binding ratio (RBR). Either RBR or ΔT_m can be used to circumscribe species.

(b) The curves describing the denaturation kinetics of double-stranded DNA. At temperature T_m 50% of DNA in an aqueous solution becomes single stranded. Hybrid double-stranded DNA from two different species (heteroduplex) is less stable than homoduplex DNA from identical organisms. Therefore the T_m of hybrid DNA is lower than that of DNA from single species. The difference between homoduplex DNA and heteroduplex DNA is the ΔT_m. For example, in the graph, the T_m of homologous DNA from a single species is 82°C. The T_m for heterogeneous DNA hybrid "a" is 72°C. Therefore, the ΔT_m for heterogeneous hybrid "a" is 10°C. Similarly, the T_m for hybrid "b" is 76°C, and the ΔT_m for hybrid "b" is 6°C. According to the international committee on the reconciliation of approaches to bacterial systematics, species identities should be circumscribed by a range of ΔT_m lower than 5°C; or by a relative binding ratio of greater than 70% (for example, shaded area in (c)).

(c) The chart compares DNA–DNA and 16S rRNA based on a dataset of 180 values from 27 independent assessments for members of *Proteobacteria*, *Cytophaga-Flavobacterium-Bacteroides*, and Gram-positive bacteria containing a high proportion of guanine plus cytosine base pairs.

Diagrams are modified with permission from Raman Rossello-Mora and Rudolf Amann (see Rossello-Mora and Amann, 2001).

so far indicated that the most named bacterial species in fact contain many ecotypes, each of which exhibits complete attributes of the eukaryotic version of species. According to this line of argument, the groups currently recognized as individual bacterial species are more consistent with the genus level of classification in eukaryote systematics (Cohan, 2001). A version of the ecotype approach was formalized to a certain extent by David Ward who proposed a "natural species concept" for prokaryotes that is similar to the ecological species concept described for eukaryotes (Ward, 1998).

Modern renditions of Ravin's genospecies concept have emerged under the title of "genomic species concept", which defines a species as a group showing high DNA–DNA hybridization values (approximately 70% or greater DNA–DNA relatedness, and 5°C or less ΔT_m; see Box 1.2). In an attempt to anchor the genomic species concept within an ecological and evolutionary framework, Rossello-Mora and Amann (2001) proposed a "phylo-phenetic species concept" where species are defined as "a monophyletic and genomically coherent cluster of individual organisms that show a high degree of overall similarity in many independent characteristics, and is diagnosable by a discriminative phenotypic property." This definition integrates character-based concepts that emphasize the presence of an apparent organismal attribute with history-based concepts that emphasize the degree of relatedness of a new isolate to previously characterized organisms. Furthermore, Sicheritz-Ponten and Andersson (2001) extended the application of the phylo-phenetic concept by developing a "phylogenomic" approach to microbial evolution where phylogenetic information is linked to the flow of biochemical pathways within and among microbial species. By constructing a complete set of phylogenetic trees derived from proteome databases, the phylogenomic approach facilitates the inclusion of horizontal genetic exchange events in the identification of microbial species (Ogunseitan, 2002; Sicheritz-Ponten and Andersson, 2001).

THEORETICAL MECHANISMS OF SPECIATION

Microbial diversity is a dynamic phenomenon, varying across both spatial and temporal dimensions in response to changes in the biotic and abiotic components of ecosystems. The emergence and extinction of species over long time frames are taken for granted according to the overarching theory of evolution. Unlike the case for eukaryotes, however, there is no direct evidence of microbial extinctions because fossil records cannot capture microbial characteristics in sufficient detail to determine species status. In fact, there are almost as many theories about the formation of species as there are concepts to define species (White, 1978). In view of the long time scale generally invoked for evolution by natural selection, most theories of speciation developed for large multicellular eukaryotes are necessarily based on the interpretation of historical evidence regarding the relationships among groups of organisms. To a limited extent, additional pieces of evidence for patterned speciation are based on the analysis of interactions among contemporary organisms and on the extent of similarity between independent genomes. In microbiology, however, the relatively short length of organism generation times and the availability of powerful techniques for genetic analyses and bioinformatics, are making it possible to investigate speciation and natural selection by means of empirical and computational methods (Arnold *et al.*, 2001; Elena and Lenski, 2003; Lenski *et al.*, 2003a and 2003b).

Speciation is the precursor of macroevolution, which leads to the generation of major taxonomic groups. Therefore, plausible theories about the mechanism of speciation need sufficient robustness to explain the outcome of both small and large changes in the parameters that modify phylogenetic lineages. As in the case of species definition and the development of a universal concept of species, most theoretical proposals on the mechanism of speciation are based on observations of eukaryotic organisms. The fossil record provides a particularly rich source of data enabling the discovery of change in eukaryotic lineages over long periods of time. These accessories have revealed two basic patterns that describe modifications in phylogenetic lineages. These patterns are formally known as theoretical frameworks for speciation, namely anagenesis and cladogenesis (Mayr, 2001). It is important first to explore the implications of these two basic patterns of speciation for theories that are directed more clearly toward macroevolutionary processes.

Anagenesis

Anagenesis is among the prominent speciation theories which hold that higher levels of specialization and/or organization are generated from primordial lineages through progressive evolution. As a pattern of evolutionary change, anagenesis involves the transformation of subpopulations to such extents that these subpopulations are sufficiently different from the ancestral population to warrant recognition as a separate species. There are no pre-established criteria for how much change must occur before a new species designation can be conferred. Therefore, the outcome of anagenesis is somewhat arbitrary. Anagenesis is sometimes referred to as phyletic speciation, implying that species are transformed along a phylogenetic lineage from a single progenitor without producing branches from the main trunk of the phylogenetic tree. This means that the total number of existing species remains conserved, excepting extinctions (Mayr, 2001).

Cladogenesis

Cladogenesis is a pattern of speciation that requires the branching of phylogenetic trees through the formation of species that are recognizably different from the parental lineage, which also continues existing. Unlike anagenesis, where species are transformed along a continuous lineage, cladogenesis actually increases the number of existing species, and as such, it leads to an expansion of biological diversity. It is possible to deduce from the fossil record that the cladogenesis pattern of speciation predominates, because the number of species within a taxon typically increases over long periods of time. The underlying mechanism of speciation in cladogenesis is either abrupt or gradual. Abrupt speciation, or "saltation", is the discontinuity in a lineage that occurs through genetic mutations or chromosomal aberrations causing reproductively isolated individuals to establish a new species population. In proposing the theory of "punctuated equilibrium", Stephen Gould and Niels Eldredge built upon the concept of "hopeful monsters", first defined by Goldschmidt in the mid-twentieth century in support of abrupt speciation (Ayala, 1982; Eldredge and Gould, 1988; Gould, 2002).

The evidence for punctuated equilibrium is largely circumstantial in the fossil record. It has not been investigated among the prokaryotes, although laboratory experiments with *Escherichia coli* have provided valuable insights into the emergence of variants in microbial populations (Elena *et al.*, 1996; Rosenzweig *et al.*, 1994). In eukaryotes, polyploidy, the multiplication of chromosome sets, is a well-established mode of abrupt speciation, and by some estimates up to a third of all plant species resulted from polyploidy (Baum and Donoghue, 1995). Karyotypic fission, symbiogenesis, and lateral gene transfer are additional possible mechanisms of abrupt speciation. In contrast to abrupt speciation, gradual speciation is easier to infer from Darwin's theories of evolution by natural selection. Random mutations, most of which have no impact on fitness, produce cumulative genetic divergence until reproductive isolation occurs to separate two or more distinct species (Mayden, 1997).

Macroevolution theories

Most biologists will agree that there is a qualitative difference between speciation as described by anagenesis or cladogenesis and major changes in organism forms and functions that lead to the emergence of completely different phylogenetic lineages. Such drastic change has been invoked in the emergence of the three domains, *Archaea*, *Bacteria*, and *Eukarya*. According to anagenesis and cladogenesis, members of a lineage can be traced back to ancestral organisms either through the fossil record or through structural and functional similarities maintained among a cohort of related species. However, it is more difficult to explain, for example, the emergence of complex structures that differentiate lineages as distant as bacteria are from plants or animals. The origins of multicellularity and complex organismal traits have long challenged evolution theorists. The traditional strategies for addressing these challenges have relied in part on arguments rooted in the relationship between ontogeny and phylogeny. That is, by comparing the size and function of features expressed during the course of growth and the development of an individual from gestation to maturity across phylogenetic groups, it is possible to make generalizations about the trajectory of anagenesis from the rudiments of complex structures.

More recently, researchers in the field of cellular automata—or computer-aided studies in evolution—have written computer software programs that simulate interactions among biological cell analogs to explore plausible outcomes of alternative scenarios in the emergence of multicellularity and the evolution of complex organs. Based on this approach, Pfeiffer and Bonhoeffer (2003) have argued that the benefit of clustering in populations of unicellular organisms includes the reduction of potential interactions with non-cooperative individuals. Furthermore, clustering can evolve as a biological, heritable trait for cells that cooperate to use external energy resources. Along the same lines, Lenski and colleagues (2003b) used digital organisms—computer programs that self-replicate, mutate, compete, and evolve—to argue for the relative simplicity of evolution toward the ability to perform complex logic functions that require the coordinated execution of many genomic instructions. In these quasi-experiments, complex functions evolved by building on simpler functions that had evolved earlier, as long as these functions were also favored by natural selection. Furthermore, no intermediate stages were required for the evolution of complex functions, and seminal genotypes with the ability to perform complex functions differed from their "wild-type" parents by just a few mutations. However, the genetic difference between these seminal genotypes and their distant ancestors was characterized by several mutations, suggesting that it may not be possible to track the lineage of biological complexity through the analysis of molecular sequences. In some cases, mutations that were deleterious when they first appeared served crucial functions in the evolution of complex features.

These results demonstrate that complex functions can indeed originate by random mutation and natural selection. However, it is important to explore organic versions of these experiments before firm conclusions can be reached about their implications for natural evolution. To place the experiments in the context of macroevolution, it is important to discuss earlier attempts to formulate theories that have been proposed to account for the discrete nature of speciation and gaps in phylogenetic lineages. Twelve major mechanisms of speciation are recognized under three theoretical categories, namely phyletic speciation, species fusion, and gradual speciation through populations. Not all of these mechanisms are relevant to speciation in prokaryotic populations but they are all presented to demonstrate the difficulty inherent in deriving a generalized theory of speciation that covers all categories of organisms (Table 1.2). Phyletic speciation or the transformation of existing species along a phylogenetic lineage is probably the best supported through evidence available from analysis of prokaryotes. Phyletic speciation occurs through anagenetic and cladogenetic pathways as discussed above, but there are no well-articulated ways through which these mechanisms can lead to the major transformations required to explain macroevolutionary events. However, species fusion and mechanisms of species multiplication can, in principle, explain major changes in species lineages. These theories remain controversial in microbiology because of the absence of concrete evidence suggesting that the theoretical mechanisms have actually contributed to the apparent diversity of microbial species.

Species fusion theory

The "species fusion theory" describes the formation of entirely new hybrid species from two or more pre-existing species where reproductive isolation mechanisms cease to exist as barriers. This can happen in cases where the reproductive isolation mechanisms are due to ecological factors as opposed to structural, physiological, or genetic factors. It is reasonable to argue in such cases that reproductive isolation was not strict, therefore the species preceding the hybridization are not independent, and instead are variants or strains. Predominance of species fusion over other mechanisms of speciation will result in reticulate evolution where repeated intercrossing between lineages produces a network of relationships in a series of related species. In this case, the phylogenetic map will resemble a net instead of a tree with branches (Hilario and Gogarten, 1993). The endosymbiotic theory as elaborated by Lynn Margulis and her colleagues is a representation of species fusion. Endosymbiosis among prokaryotic cells is the leading uncontested theory that is usually cited to explain

Table 1.2 Major speciation theories and underlying mechanisms.

Major categories of speciation theories	Subcategories of speciation theories	Mechanisms of speciation		Could the theory apply to prokaryotes?
Phyletic speciation (transformation of existing species)	Autogenous transformation (anagenesis and cladogenesis)	Random mutation and natural selection		Yes
	Allogenous transformation	Introgression from other species		Yes
Species fusion	Reduction in the number of species	Extinction of parental lineage due to competition with emergent (newly fused) species		Possible origin of Archaea
	Increase in the number of species	Both parental species coexist with newly formed (fused) species		Not documented
Species multiplication	Instantaneous speciation	Genetic	Single mutations	Yes
			Macrogenesis	No
		Cytological	Chromosomal mutation (e.g. transposition; translocation)	Possible
			Autopolyploidy	No
			Amphiploidy	No
	Gradual speciation	Sympatric speciation		Not documented
		Semi-geographic speciation		Not documented
		Geographic speciation		Not documented

the origin of organelles such as mitochondria and chloroplasts. The contribution of endosymbiosis to the origin of the nucleus and, by default, the origin of the eukaryotic lineage, is more controversial. Nevertheless, endosymbiosis is at least as plausible as any other proposed mechanism used to explain the separation of the Eukarya branch from prokaryotic Archaea and Bacteria progenitors (see Box 1.3; Gray *et al.*, 1999; Margulis, 1996).

The endosymbiosis theory posits that the mitochondria found in all eukaryotic cells evolved from rickettsiae-like aerobic members of the α-*Proteobacteria* living within ancestral host cells (Emelyanov, 2001). In addition, algal and plant chloroplasts are presumed to have evolved from endosymbiotic cyanobacteria (Box 1.3(a)); whereas the cilia of eukaryotes evolved from endosymbiotic spirochetes (Box 1.3(b)). Furthermore, mitosis may have been invented through the creation of the mitotic spindle by basal bodies from which kinetosomes develop. These propositions of the endosymbiosis theory are supported by several lines of evidence, including observations that mitochondria and chloroplasts are not synthesized *de novo* by eukaryotic cells but instead they can only arise from pre-existing mitochondria and chloroplasts. These organelles also maintain independent genomes that bear remarkable semblance to prokaryotic genomes in the sense that they are covalently closed circular DNA molecules with no histones aiding in their packaging. The transcription and translation of mitochondrial and chloroplast genomes are also more related to prokaryotic rather than eukaryotic processes. Finally, the 16S ribosomal RNA (rRNA) signatures of these two organelles clearly demonstrate their affiliation with prokaryotes (Gray *et al.*, 1999; Margulis *et al.*, 1998). However, there are some contradictory pieces of

BOX 1.3

(a)

The endosymbiosis theory of speciation, as articulated by Lynn Margulis (a), is supported by substantial evidence in the prokaryotic origin of two classes of eukaryotic organelles: Mitochondria and the photosynthetic plastids. The relevance of the mechanism to other essential components of eukaryosis is less well understood, although there are indications that endosymbiosis resulted in the acquisition of prokaryotic features resulting in eukaryotic cell motility. Endosymbiosis also provides a plausible mechanism for the origin of viruses and plasmids, although the competing hypothesis based on excision of autonomously replicating units from microbial genomes is equally plausible. Like Mayr, Lynn Margulis rejects the three-domain phylogenetic tree proposed by Carl Woese in favor of a two-domain, five-kingdom system with the eukaryotes emerging from symbiotic interaction between the prokaryotic Bacteria and Archaea (see Margulis and Schwartz, 1998). The picture of Lynn Margulis is by courtesy of Jerry Bauer.

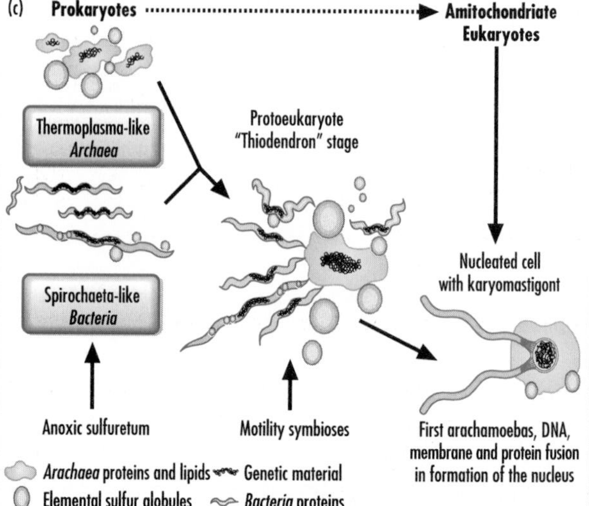

(c) The postulated pathway by which archaeal and bacterial cells enter stable endosymbiotic relationships is based on cell motility under environmental conditions that support sulfur metabolism. The illustration depicts a scenario where Thermoplasma-like Archaea species fuse with Spirochaeta-like Bacteria species in an anaerobic sulfur environment to create a motility symbiosis Proto-Eukarya species, which ultimately emerges as nucleated organisms with features similar to those associated with contemporary members of the Eukarya lineage. Circles represent genomes; triangles represent Archaea proteins and lipids; squares represent Bacteria protein and lipids; rods represent elemental sulfur globules. The diagram is modified by courtesy of Kathryn Delisle from Margulis *et al.* (1998).

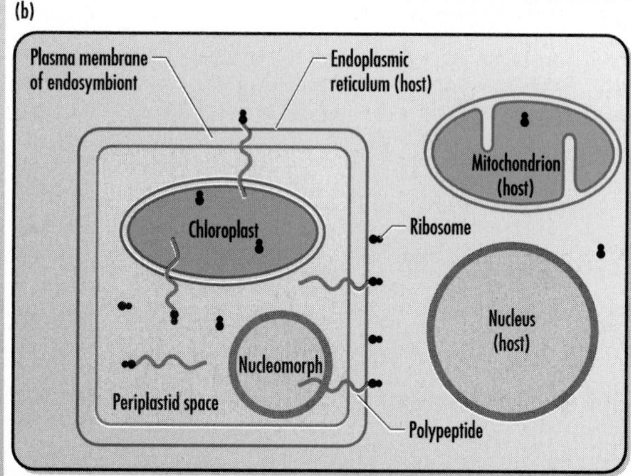

(b) Some of the autonomy maintained by mature endosymbionts such as the chloroplast, including protein synthesis machinery, and the vestigial nucleus (nucleomorph found in a limited number of protists groups). The diagram is by courtesy of John Kimball.

(d)

(d) Carl Woese, a pioneer of molecular phylogenetic studies, focused on the analysis of microbial 16S rRNA to propose that the old kingdom "monera" should be recognized as consisting of two distinct phylogenetic lineages, the Archaea and the Bacteria. The cellular structure of the Archaea resembles the Bacteria, but aspects of their nucleic acid profiles are more like the Eukarya. The speciation process that splits the prokaryotic lineage into two major branches remains an unresolved research question. The Archaea and Bacteria are frequently observed to coinhabit similar ecological niches, and recent evidence of genetic exchange between the two branches may imply prehistoric interactions. Indeed, Radhey Gupta (1998a and 1998b) has shown through comparative analysis of protein sequences called indels that the fundamental phylogenetic distinction among prokaryotes is not the Archaea/Bacteria division because the Archaea show remarkable similarity to Gram-positive bacteria. Rather, prokaryotes may be delineated according to whether cells have a monoderm structure (surrounded by a single membrane, including members of the Archaea and Gram-positive bacteria) or whether cells have a diderm cell structure (surrounded by an inner cytoplasmic membrane and an outer membrane, including all true Gram-negative bacteria). The picture of Carl Woese is by courtesy of Bill Weigand and the University of Illinois at Urbana-Champaign.

evidence against the tempting generalization that all such organelles derive from prokaryotic endosymbionts. Martin (1999), while agreeing with the explanatory power of the endosymbiosis theory in accounting for the origin of mitochondria and hydrogenosomes, has argued that the endosymbiotic origin of the nucleus is not plausible, although the evidence appears incontrovertible for the endosymbiotic origin of the mitotic apparatus found in nuclei, the karyomastigont and other spindles (Margulis *et al.*, 2000). Also, the chloroplasts of some algae may have evolved through a different kind of endosymbiotic process (secondary endosymbiosis) involving the engulfment of pre-existing photosynthetic eukaryotes. In these cases, the organelle retains some eukaryotic properties such as a pseudo-nucleus (nucleomorph) (Archibald and Keeling, 2002). The endosymbiotic theory and its significance to the origin of eukaryotes are discussed further in Chapter 6.

Gradual speciation

The divergence of two or more populations of a single species to extents that they reach distinct status as independent species is called "gradual speciation". The population divergence has been postulated to occur either through geographical separation where gene flow is prevented by extrinsic factors (geographic speciation) or through the colonization of different ecological niches within the same geographical zone and the subsequent prevention of gene flow by intrinsic factors (sympatric speciation). Geographic and sympatric versions of gradual speciation agree on the significance of ecological factors, however, they disagree on the process by which ecological factors constitute a barrier to gene-pool mixing (Nixon and Wheeler, 1990; Templeton, 1989). In his searing criticism of sympatric speciation, Ernst Mayr evaluated several cases that have been proposed as examples in zoology, and reached the conclusion that the hypothesis is neither necessary nor supported by irrefutable facts (Mayr, 1970). Although there are no records of explicit tests of the sympatric speciation hypothesis in microbial populations, its marginalization of genetic dispersal, exchange, and recombination suggests that the hypothesis will be of little use in understanding microbial evolution and diversity. In contrast to sympatric speciation, and after periods of vigorous contestation, many zoologists and botanists now consider geographic speciation to be a universal mode of speciation. However, similarly vigorous explorations of the topic have not taken place in environmental microbiology (Zavarzin, 1994), although the recent development of powerful molecular tools for analyzing microbial populations and the research initiatives to establish microbial observatories may facilitate hypothesis testing in this direction (Petursdottir *et al.*, 2000).

Geographic speciation refers to an elaborate three-step mechanism by which new species are derived from established populations. The first step involves a niche division or spatial isolation caused by geophysical or chemical factors acting non-uniformly upon a habitat. Secondly, the environmental variation produces different selective pressures that favor certain mutations which would otherwise be neutral in the population. Thirdly, the resulting genetic and phenotypic differences are reinforced by reproductive isolation caused by geographic barriers. Geographic isolation can have three possible outcomes. **Character displacement** occurs when genetic differences between species reduce competition for limiting resources, and coexistence is maintained. **Competitive exclusion** may result from insufficient divergence in a recent speciation event, leading to the survival of a dominant species at the expense of a less fit species. The third alternative is the establishment of **hybrid zone** due to incomplete reproductive isolation of geographically adjacent species populations. A population of hybrids is thus established between the geographical zones occupied by the two species (Hull, 1997). Parapatric speciation has been postulated as a variation of geographic speciation in cases of sessile organisms, where physical geographical isolation is unnecessary to produce population isolation. When mutants in a population are able to exist in the same geographical space, but exploit different limiting resources, a contiguous niche is created that can further reinforce the speciation process.

One of the major challenges encountered in attempts to verify the importance of geographic speciation theory in microbial populations is the difficulty of accommodating the continuum of differences across populations and among species. For example, it is widely accepted to discuss morphological diversity as a crucial differentiating factor of microbial species. But so also is evolutionary history, biochemical diversity, behavioral diversity, or genetic diversity. Increasingly, genetic distance as measured by molecular sequence comparisons is emerging as the gold standard of microbial speciation. For example, Petursdottir and colleagues (2000) used allelic variation of 13 genes which code for polymorphic enzymes to explore the genetic diversity of 81 *Rhodothermus* isolates from different geothermal environments in Iceland. Their results revealed 71 distinctive multilocus genotypes with a mean genetic diversity per locus of 0.586 (on a scale of 0 to 1 where 1 represents maximum diversity). They concluded that the relatively high genetic variance observed within *Rhodothermus* isolates from different locations is most likely the result of genetic changes occurring independently in the locations studied. Furthermore, partial or whole sequencing of the 16S rRNA genes of the isolates confirmed that all the isolates belonged to the species *Rhodothermus marinus.* The results strongly suggest that despite phylogenetic and phenotypic similarity, genetic diversity within microbial species at different geographic locations may be very high.

MICROBIAL SPECIATION

One of the enduring debates in microbial evolution centers on the nature of the more important driving forces behind speciation: random mutation or genome acquisition? In an attempt to reinforce the endosymbiosis theory, Margulis and Sagan (2002) presented an argument against the dominant role of random mutation, partly through the recognition that the term "species" does not apply easily to prokaryotes because of the high frequency and extreme promiscuity that characterize genetic exchange in these organisms. They admit that random mutations are important for generating metabolic and genetic diversity in prokaryotes, but since there are no species *per se,* these mutations are irrelevant for speciation. In this view, the real question of speciation can only be posed to the eukaryotes—the animals, plants, protoctists, and fungi where the most secure pieces of evidence for speciation are linked, albeit controversially, to symbiotic genome relationships with microorganisms.

A comprehensive theory about the mechanisms of microbial speciation will likely await a more thorough assessment of microbial diversity with respect to the interactions between genetic and ecological factors that produce and maintain reproductive isolation. Speciation is driven by genetic variation, and in addition to random mutations that occur because of nucleic acid replication, prokaryotic organisms have evolved sophisticated ways for acquiring and losing genetic material. In prokaryotes, three major mechanisms of genetic exchange are well understood, and have been documented to exist in nature with major ramifications for the acquisition of new traits and speciation (Aravind *et al.*, 1998; Cohan, 1994; Ochman *et al.*, 2000; Ogunseitan, 1995). These mechanisms are discussed briefly in the following paragraphs, but a more detailed discussion is presented in Chapter 9.

Conjugation, the direct cell-to-cell transfer of DNA, is the closest process to eukaryotic sex-mediated genetic exchange. Conjugation is capable of moving large sizes of genetic material, including chromosomal genes and entire plasmids from one cell to another, but it is constrained within species. Therefore, it is likely to be very important in the creation of strains, but its role in actual speciation is not presently clear. Small segments of genetic material can also be transferred from one cell to another through the phenomenon of **transduction**, which is mediated by viruses. Virus infection is typically host specific, and the development of resistance is rampant. Therefore, the role of transduction in microbial speciation is also questionable. **Transformation**, the uptake of genetic material directly from the environment, is potentially an efficient mechanism for speciation because the source of transforming DNA can be from a variety of species. Therefore, transformation can introduce novel genes into a microbial population ultimately leading to niche specialization and speciation. However, the occurrence of microbial nucleases suggests that microorganisms have evolved protective measures against transformation, at least at high frequencies.

Given these effective mechanisms of genetic exchange among prokaryotes, it is pertinent to ask how much of microbial speciation can be explained by genome acquisition. Levin and Bergstrom (2000) considered that the genome acquisition theory of adaptive evolution in bacteria is a simple extension of theories developed for sexually reproducing eukaryotes, and that the modes of genetic recombination described above for microorganisms, especially bacteria, are quantitatively and qualitatively different from that of organisms for which recombination is an integral part of the reproduction process. These differences have substantial consequences for the evolution of accessory elements and their role in the adaptive evolution of bacteria as the major driving force for speciation. Lawrence (2001) attempted to correlate the rate of successful acquisition and integration of genetic material obtained through intra-specific lateral genetic transfer with the amount of **genetic headroom** present in recipient species. He defined genetic headroom as the codon usage bias and codon context bias that can be transiently sacrificed to allow a species population to experiment with functions introduced by gene transfer. In this context, Noble *et al.* (1998) also used a chaos-game representation scheme to identify tetranucleotide frequencies in microbial genomes as an index for the residence time of contiguous genomic segments in prokaryotes.

Similar observations regarding biases in codon utilization have led to a robust hypothesis on the differentiation of viral groups according to host range, the equivalence of prokaryotic speciation. By evaluating the relationships between the nucleotide composition of retroviral genomes, the amino acid composition of retroviral proteins, and evolutionary strategies used by retroviruses, Bronson and Anderson (1994) demonstrated that the genome of each viral lineage has a characteristic base composition and that the variations between groups are related to retroviral phylogeny. In experiments conducted with cultures of *E. coli*, several investigators have reported the development of genetic polymorphism arising from a single genetically pure strain (Finkel and Kolter, 1999; Souza *et al.*, 1997; Rosenzweig *et al.*, 1994). Such polymorphisms, when they are not silent (i.e. when they are expressed and confer a selective advantage), are consistent with the parapatric mode of speciation where polymorphic clones arise in a population, but coexist because of differences in niche specialization or resource utilization.

CONCLUSION: EMERGING CONCEPTS AND APPLICATIONS OF MICROBIAL DIVERSITY

Species and speciation are, respectively, the fundamental units and the determinants of microbial diversity. Firm understanding of the scientific bases for species concepts and proposed mechanisms of speciation will achieve more than simple provision of an internally consistent language for taxonomy and systematics. If achieved with confidence, the understanding of microbial speciation will facilitate the formulation and testing of hypotheses on the most important questions facing biology today. An introduction to the relevance of microbial diversity concepts to these questions is presented below. The research paradigms are changing rapidly on many fronts, and salient aspects of the changes are discussed in more detail in subsequent chapters of this book.

It is difficult to overstate the significance of the discovery of prokaryotic Archaea proposed as a third major branch of the universal phylogenetic tree, separate from prokaryotic Bacteria and eukaryotic Eukarya. The new phylogenetic tree is a major revision of the previous five-kingdom phylogenetic tree including plants, animals, fungi, protists, and the monera consisting of all prokaryotes. Many scientists contributed to this discovery, but Carl Woese was particularly influential in facilitating the comparative assessment of rRNA nucleotide sequence information for phylogenetic analysis (DeLong and Pace, 2001; Pace,

BOX 1.4

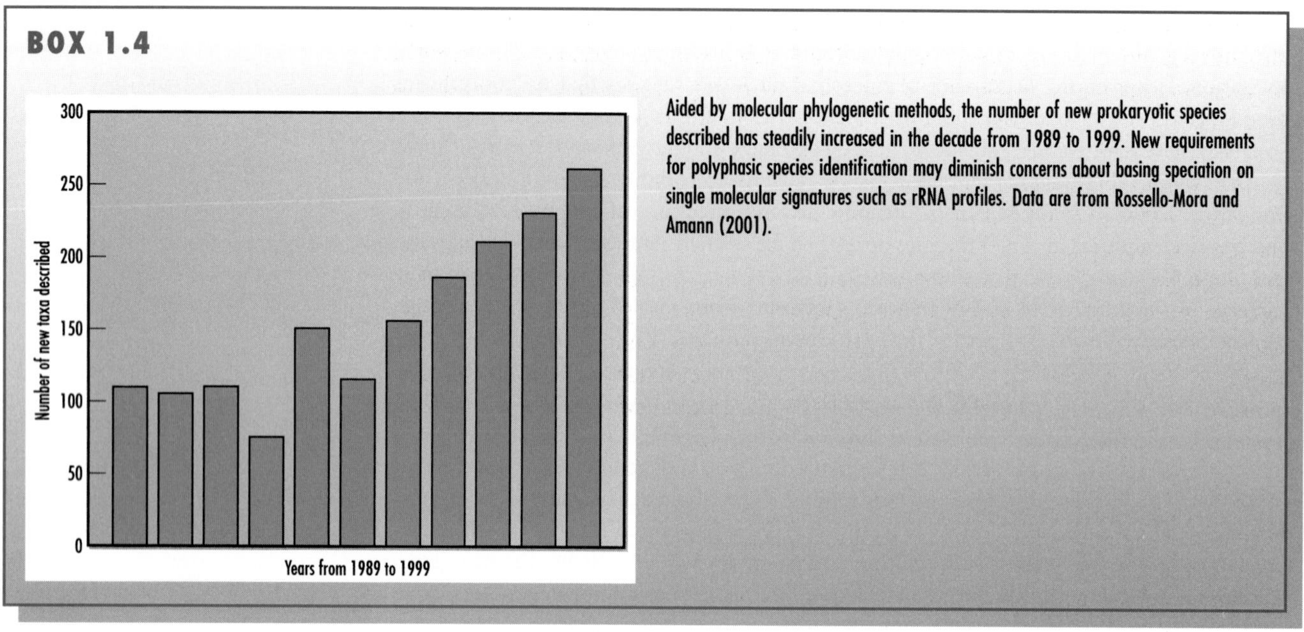

Aided by molecular phylogenetic methods, the number of new prokaryotic species described has steadily increased in the decade from 1989 to 1999. New requirements for polyphasic species identification may diminish concerns about basing speciation on single molecular signatures such as rRNA profiles. Data are from Rossello-Mora and Amann (2001).

1997; Woese, 1998a; Woese *et al.*, 1990). In addition to the proposal to regroup phylogenetic relationships among microorganisms, the use of phylogenetically conserved molecular sequence information allowed microbiologists to venture into the realm of identifying non-culturable organisms (Embley and Stackebrandt, 1997; Ward *et al.*, 1990). The extraction and analysis of nucleic acids from environmental samples contributed to the remarkable increase in the number of microbial genus and species discovered in the past decade (see Box 1.4). These molecular tools have raised new and important questions about the ecological context of microbial diversity. However, new doubts have arisen about the correlation between molecular identity and speciation, particularly with respect to the major domains of the universal phylogenetic tree (Brinkmann and Philippe, 1999; Gupta, 1998a and 1998b; Margulis *et al.*, 2000; Mayr, 1998; Palleroni, 1997; Woese, 1998a). Clearly, there is tremendous structural diversity and resilience in the microbial community. What is less clear is the correlation between the structural diversity and functional diversity. The problem is exacerbated by the impracticality of exhaustive inventories of microbial communities. Regardless of the technique used for defining species, an indulgence in statistical inference is essential for quantitative assessments of microbial diversity (Hughes *et al.*, 2001).

The prevalence of organisms belonging to the Archaea branch in extreme environments such as hot water springs, acidic lakes, and hyper-saline environments suggests that niche specialization plays a crucial role in microbial speciation (Stetter, 1996; DeLong and Pace, 2001). Conversely, the global distribution of the most described microbial species is not limited by geographical boundaries, indicating that geographic separation is not a major contributing factor of microbial speciation (Stoner *et al.*, 2001; Zavarzin, 1994). The deepest branches in the phylogenetic trees of the Archaea and those of the Bacteria (e.g. *Thermotogales* species and *Aquifex* species) are among thermophilic organisms, strongly suggesting that the progenitor of the phylogenetic lineage evolved in a hot environment and under conditions postulated to describe the early Earth environment (Doolittle, 1999; Koonin *et al.*, 1997; Stetter, 1996; Woese, 1998a). Therefore, investigations of microbial speciation and diversity are directly relevant to research on the origin of planetary life, and on how the physiological characteristics of prehistoric microorganisms contributed to the emergent signature of a co-evolving life and environment (Brown *et al.*, 2001; Martin and Müller, 1998; Rasmussen, 2000).

Improved understanding of microbial speciation and diversity can also shed light on the path dependence of the trends associated with biogeochemical cycles that are sensitive to industrial ecological problems such as global warming, toxic pollutants, and disease epidemics (Martin and Müller, 1998). For example, Lake and colleagues (1985) proposed a new group of photocytes as the progenitors of photosynthetic organisms. Photosynthetic prokaryotes are interspersed though most major phylogenetic lineages of Bacteria and Archaea, and they demonstrate a remarkable physiological diversity. An assessment of the diversity of photosynthetic light-gathering arrays in the microbial world can yield biotechnologically adaptive systems that are potentially relevant to the global energy crisis. Similarly, strong cases can be made for investigating the speciation processes underlying the ancient lineage of hydrogen-oxidizing bacteria such as *Aquifex*, and their contribution to models of early atmospheric composition on Earth. These and other cases of the contributions of microbial diversity to the understanding of contemporary global environmental problems are discussed in the final chapter of this book.

QUESTIONS FOR FURTHER INVESTIGATION

1 Find one example each, in the microbial world, of exceptions to the traditional species concepts discussed in this chapter.
2 Find four examples in the scientific literature of evidence to support microbial speciation through (a) genetic exchange mechanisms, (b) niche specialization, (c) geographical isolation, and (d) genetic drift.
3 The proposal to reconstruct the phylogenetic tree of life on Earth to include two main lines of prokaryotes and one line for eukaryotic organisms represented a paradigm shift in the understanding of speciation. Acceptance of this paradigm depends on the definition of species, and the driving forces of speciation, which differ across biological disciplines. All paradigm shifts in science are rigorously contested. Find one example each in the botany, zoology, mycology, and bacteriology literature where Carl Woese's revised phylogenetic tree is contested. Compare and contrast the rationale for each dissenting opinion.

SUGGESTED READINGS

Arnold, F.H., P.L. Wintrode, K. Miyazaki, and A. Gershenson. 2001. How enzymes adapt: Lessons from directed evolution. *Trends in Biochemical Sciences*, **26**: 100–6.

Brown, J.R., C.J. Douady, M.J. Italia, W.E. Marshall, and M.J. Stanhope. 2001. Universal trees based on large combined protein sequence data sets. *Nature Genetics*, **28**: 281–5.

Cohan, F.M. 2002. What are bacterial species? *Annual Review of Microbiology*, **56**: 457–87.

Colwell, R.R., R.A. Clayton, B.A. Ortiz-Condel, D. Jacobs, and E. Russek-Cohen. 1995. The microbial species concept and biodiversity. In D. Allsopp, R.R. Colwell, and D.L. Hawksworth (eds.) *Microbial Diversity and Ecosystem Function*, pp. 3–15. Oxford: CAB International.

Elena, S.F. and R.E. Lenski. 2003. Evolution experiments with microorganisms: The dynamics and genetic bases of adaptation. *Nature Reviews Genetics*, **4**: 457–69.

Gould, S.J. 2002. *The Structure of Evolutionary Theory*, pp. 745–1022. Cambridge, MA: Belknap-Harvard.

Gupta, R. 1998b. Protein phylogenies and signature sequences: A reappraisal of evolutionary relationships among Archaebacteria, Eubacteria, and Eukaryotes. *Microbiology and Molecular Biology Reviews*, **62**: 1435–91.

Hull, D.L. 1997. The ideal species concept—and why we can't get it. In M.F. Claridge, H.A. Dawah, and M.R. Wilson (eds.) *Species: The Units of Biodiversity*, pp. 357–80. London: Chapman and Hall.

Krawiec, S. 1985. Concept of bacterial species. *International Journal of Systematic Bacteriology*, **35**: 217–20.

Lawrence, J. 2001. Catalyzing bacterial speciation: Correlating lateral transfer with genetic headroom. *Systematic Biology*, **50**: 479–96.

Lenski, R.E., C. Ofria, R.T. Pennock, and C. Adami. 2003a. The evolutionary origin of complex features. *Nature*, **423**: 139–44.

Levin, B.R. and C.T. Bergstrom. 2000. Bacteria are different: Observations, interpretations, speculations, and opinions about the mechanisms of adaptive evolution in bacteria. *Proceedings of the National Academy of Sciences (USA)*, **97**: 6981–5.

Mallet, M. 1995. A species definition for the modern synthesis. *Trends in Ecology and Evolution*, **10**: 294–9.

Margulis, L., M.F. Dolan, and R. Guerrero. 2000. The chimeric eukaryote: Origin of the nucleus from the karyomastigont in amitochondriate protists. *Proceedings of the National Academy of Sciences (USA)* **97**: 6954–9.

Margulis, L. and K.V. Schwartz. 1998. *Five Kingdoms* (Third Edition). New York: W.H. Freeman & Co.

Margulis, L. and D. Sagan. 2002. *Acquiring Genomes: A Theory of the Origin of Species.* New York: Basic Books.

Mayr, E. 1987. The ontological status of species: Scientific progress and philosophical terminology. *Biology and Philosophy*, **2**: 145–66.

Ochman, H., J.G. Lawrence, and E.A. Groisman. 2000. Lateral gene transfer and the nature of bacterial innovations. *Nature*, **405**: 299–304.

Pace, N.R. 1997. A molecular view of microbial diversity and the biosphere. *Science*, **276**: 734–40.

Rossello-Mora, R. and R. Amann. 2001. The species concept for prokaryotes. *FEMS Microbiology Reviews*, **25**: 39–67.

Staley, J.T. 1997. Biodiversity: Are microbial species threatened? *Current Opinion in Biotechnology*, **8**: 340–5.

Ward, D.M. 1998. A natural species concept for prokaryotes. *Current Opinion in Microbiology*, **1**: 271–7.

Woese, C.R. 1998a. The universal ancestor. *Proceedings of the National Academy of Sciences (USA)*, **95**: 6854–9.

Woese, C.R. 1998b. Default taxonomy: Ernst Mayr's view of the microbial world. *Proceedings of the National Academy of Sciences (USA)*, **95**: 11043–6.

MICROSCOPIC METHODS FOR ASSESSING MICROBIAL DIVERSITY

Chapter contents

INTRODUCTION

The documentation of microbial diversity is only as good as the methods used to observe and differentiate among microorganisms. Human beings would be completely ignorant about the physical reality of microbes without the aid of magnifying lenses. Microscopes are the traditional instruments used for assessing microbial diversity, and they remain indispensable tools for exploring the morphological, physiological, and genetic diversity present in microbial communities. This chapter discusses microscopy from the perspectives of:

1 Advances in instrumentation and methodology for assessing microbial diversity.
2 The breadth of morphological diversity in microbial communities as revealed by microscopic analysis.
3 Limitations of microscopy for investigating morphology as the fundamental basis of species classification.
4 Suitability of microscopic analysis for assessing microbial cell structures, colony architecture, and community organization.
5 Other approaches that are compatible with microscopy for assessing diversity in various microbial ecosystems, including aquatic, terrestrial, aerobic, and biological samples.

ADVANCES IN INSTRUMENTATION AND METHODOLOGY

The **near point**, or the closest distance at which a human eye can focus clearly, is 25 cm. At this distance, the maximum resolution of human eye acuity is 5×10^{-4} radian. This means that we can discern with minimum clarity individual objects approximately 75–100 μm in diameter (Giancoli, 1991). In general, prokaryotic microorganisms are 5–100 times smaller than the perceptive limits of the human eye. Therefore, the invention of the microscope as a technological aid to human vision was indispensable for the discovery of microbial diversity. The original microscope was basically a magnifying lens with relatively low improvement in resolution, however, contemporary instruments have undergone dramatic technological developments that draw from innovations in optical science, computers, engineering, and specimen preparation (see Box 2.1). In the assessment of microbial diversity, microscopy is usually combined with different analytical techniques, and conclusive inferences about the structure and function of microbial cells, populations, and communities are possible only after integrated analysis of convergent datasets (Braja and Hill, 2001; Casamayor *et al.*, 2000 and 2002; Junge *et al.*, 2002; Upton *et al.*, 2000).

Basic light microscopy

Microscopes cannot resolve details of objects smaller than the wavelength of the electromagnetic radiation spectrum in which the specimen is viewed. The average wavelength for white light is 550 nm. Therefore, the size limit of specimens that can be resolved with microscopes that depend on the visible segment of the electromagnetic spectrum is approximately 200 nm when oil immersion is used, which reduces the wavelength of light by a factor of 1.5 to 1.8. Compound microscopes consisting of both objective and ocular lenses instead of a single magnifying lens are essential for microbiological investigations. Useful magnification for light microscopes is 500 times (Murphy, 2001). However, the doubling of magnifying power to 1000 times is used to minimize eyestrain, beyond that the image is marred by diffraction patterns produced by microscope objective lenses (Lacey, 1999). Light microscopes can be modified in several ways to improve their use in capturing specific features of microorganisms, although the modifications are usually accompanied by trade-offs in magnifying power or other basic features. For example, "dark-field" microscopy is used to visualize transparent cells by reducing the background lighting around those cells. Cells visualized by dark-field microscopy appear bright against a dark background. The enhanced contrast is good for visualizing cells that have not been stained or otherwise prepared in ways that compromise the ability of cells to demonstrate characteristics such as motility (Murray and Robinow, 1994). "Phase-contrast" microscopy is another modification of basic light microscopy, which enhances the visualization of internal cell components. The variations in the density of organelles influence their interaction with incident light. This results in different levels of contrast between the organelles and the surrounding medium. Phase-contrast microscopy uses the differences in contrast levels to produce a sharper image of cellular components than would be possible by regular light microscopy. The recognition of the theoretical and practical limits of the light microscope led to the invention of more sophisticated microscopes that use shorter wavelengths. The scanning, transmission, or tunneling electron microscopes, confocal laser microscopes, and atomic force microscopes are among the most widely used instruments providing higher resolution than the light microscope.

Electron microscopy

Max Knoll and Ernst Ruska built the first practical electron microscope in 1931 at the Technical University of Berlin (Ruska, 1980). **Transmission electron microscopes** (TEM) operate in a manner similar to transmission light microscopes producing two-dimensional morphological information at the scale of atomic diameters. In addition, TEMs facilitate the observation of crystallographic and other biochemical details within a few square-nanometer areas. Specimens must be prepared in extremely thin sections because a beam of electrons is transmitted through the specimen to produce images on a phosphor screen (Glauert and Lewis, 1998). Darker areas of the image represent thicker areas of the specimen where fewer electrons were transmitted, and conversely, light regions of the image represent thin areas on the specimen. The maximum practical resolution of TEM is 0.2–0.5 nm, which is a 1000 times improvement over the resolution attainable with light microscopes. The useful magnification of objects viewed with a TEM ranges from 10^4 to 10^6 (Mayer and Hoppert 2001).

BOX 2.1

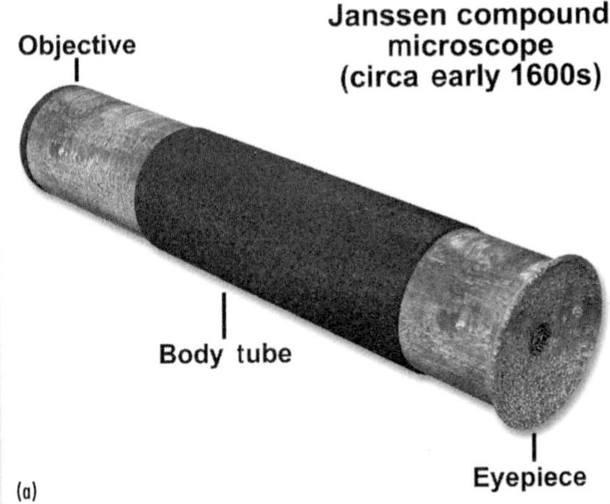

(a)

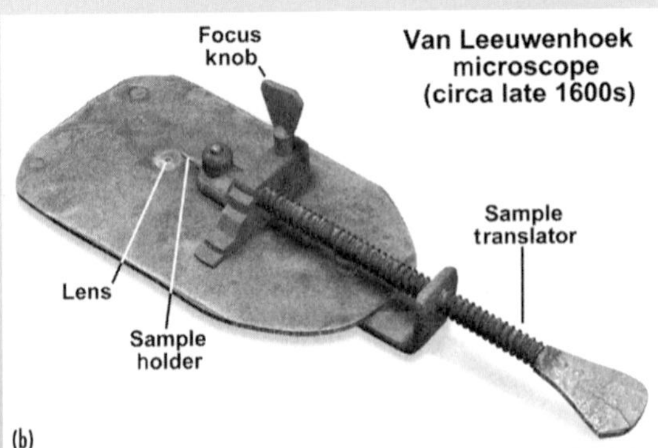

(b)

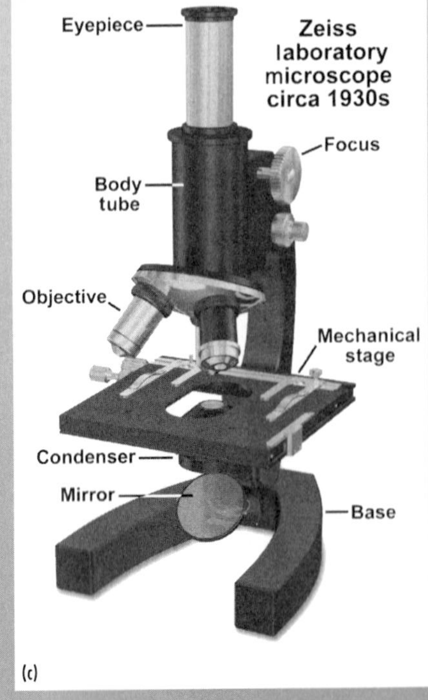

(c)

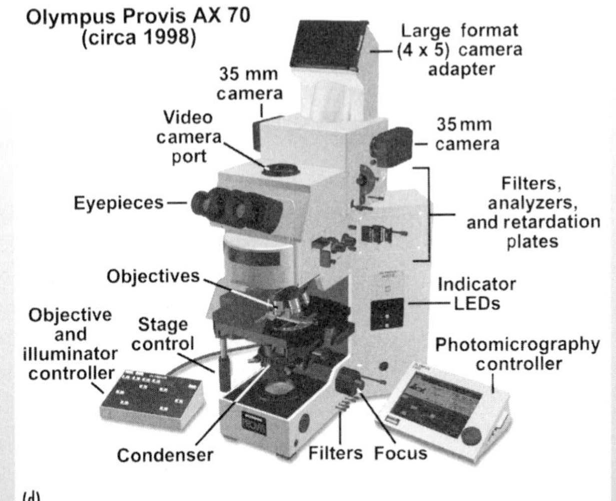

(d)

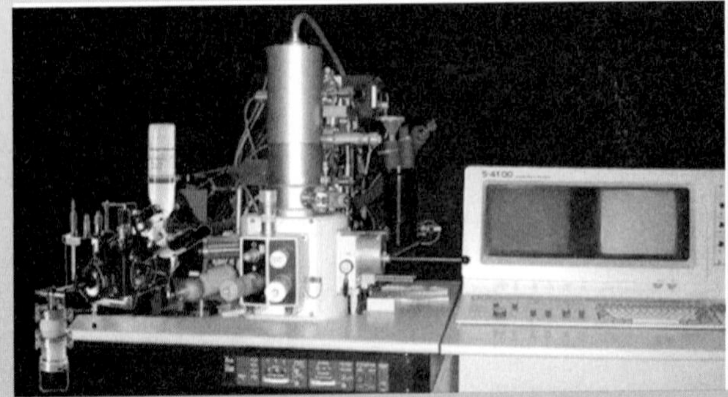
(e)

Technical evolution of the microscope has followed paths of increasing sophistication and size from the simple two-lens cylindrical microscope (a). Credit for the first microscope is usually given to Zacharias Janssen of Middleburg, Holland, circa 1595. The specimen shown is an example of the Royal Janssen microscope, discovered in the seventeenth century. It now resides in the Middleburg museum in the Netherlands (Jones, 1997). It has a maximum magnification power of 6X.

Antoni van Leeuwenhoek's simple, single-lens microscope (b) was a major improvement on the magnification power and resolution of compound microscopes, because of his use of relatively high quality glass. His surprising revolutionary discovery of the microbial world was reported to the Royal Society of London in 1673. Leeuwenhoek's microscopes were approximately two inches long and one inch across. The microscopes consisted of two flat and thin brass plates. The bi-convex lens achieved magnifications between 50X and 200X (Jones, 1997).

Modern monocular (1930s; (c)) and compound triocular microscope (d) with oil immersion can achieve magnifications up to 500X, whereas the electron microscope (e) is capable of magnifying images up to 2 million times. The compound microscope shown is a Grau–Hall scientific accuscope model 3018. The electron microscope is a Hitachi S-4100 field emission transmission electron microscope.

Images in panels (a)–(d) by permission of Michael W. Davidson and Eric Clark, Florida State University; image in (e) by courtesy of the U.S. Department of Agriculture.

Three-dimensional magnified images are produced by the **scanning electron microscope** (SEM) with the aid of electrons instead of light waves. The use of electrons requires that specimens must be able to conduct electricity. This requirement necessitates elaborate specimen preparation, including drying to withstand the vacuum inside the microscope and coating with a conductive metal such as gold. A beam of high-energy electrons focused by magnetic lenses is used to scan the coated specimen, which in turn emits secondary electrons that are detected and converted into an amplified image produced from the number of electrons emitted from each spot on the specimen. The wavelength of electrons, measured at 100 kV is approximately 0.004 nm, and this is the theoretical limit of resolution of SEM, however, aberrations in the magnetic lenses limit the resolution to 5–10 nm. Scanning electron microscopes have been influential in the discovery of microbial cell morphological types and multicellular organization in microbial colonies (Burja and Hill, 2001; Casamayor *et al.*, 2000). In addition, the use of SEM has facilitated the observation of episodic cell–cell interactions such as bacterial conjugation (see Plate 2.1), and population dynamics of microbial host–virus interactions (Hoppert and Holzenburg, 1998; Ogunseitan *et al.*, 1990).

The resolution of electron microscopy was greatly improved by the development of the **scanning tunneling microscope** (STM), where an atom-wide tip is used to scan the specimen surface from a height of approximately 1 nm. Voltage applied between the specimen and the probe mobilizes electrons by tunneling through a vacuum space. The vertical movement of the probe responds to the electron current density at each scanned position on the specimen thereby generating a three-dimensional topographic image of the specimen. The resolution exceeds 0.01 nm vertically and 0.1 nm horizontally. Scanning tunneling microscopy is useful for investigating subcellular components that are important for classifying microbial strains and species (Hayat, 2000).

The **atomic force microscope** (AFM) is a variation of the scanning tunneling microscope (Morris and Kirby, 1999). In contact-mode AFM, a semiconductor tip is used to scan the surface of a specimen, and a laser that is reflected obliquely to a recording device captures the magnitude of tip deflection by the surface (Bhanu and Hörber, 2002). Other variations of the STM and AFM are collectively referred to as **scanning probe microscopes** (SPM). For example, the **chemical force microscope** can be used to probe differences in chemical forces across a specimen surface at the molecular level. In general, these microscopes are too powerful to be widely applicable for investigating morphological differences among various microbial cells but their capabilities can facilitate the investigation of physiological diversity in relation to how environmental factors mediate the expression of genetic potential (Binnig *et al.*, 1986; Hoh and Ansmah, 1992).

Specialized light microscopy

The **confocal laser microscope** (CLM) has been used extensively to investigate microbial phylogenetic diversity (e.g. Upton *et al.*, 2000). The CLM optically dissects artificially stained fluorescent specimens by using light sources of different wavelengths (Paddock, 1999). Popular fluorescent stains used in the assessment of microbial diversity include digoxigenin, 4′-6′-diamidino-2-phenylindole 2 HCl (DAPI), and 5-cyano-2, 3-ditolyl tetrazolium chloride (CTC) (Junge *et al.*, 2002; Porter and Feig, 1980). The typical CLM is equipped with laser sources that emit at different wavelengths, such as argon (458/488 nm) or helium/neon (543 nm or 633 nm) (Diaspro, 2002). When focused on fluorescent specimens, the lasers provoke the emission of high-energy photons that are detected as visible light. The use of a pinhole to exclude light originating beyond the selected specimen area enhances the resolution of CLM. When combined with fluorescence-labeled oligonucleotide probes for DNA or RNA hybridizations (fluorescent *in situ* hybridization; FISH), CLM can provide detailed information on the spatial distribution of microbial cells in their native habitats. In addition, CLM can be used to explore the phylogenetic diversity of natural microbial communities where the majority of species are non-culturable (Amann *et al.*, 1995 and 1996; MacNaughton *et al.*, 1996; Morel and Cavalier, 2001).

Microscopic video image analysis

The diversity of microorganisms in ecosystems is frequently sensitive to temporal and spatial variation in biotic and abiotic environmental factors. The characteristically short generation times associated with microbial growth provide opportunities for "real-time" assessments of the dynamic nature of species and strain diversity as a function of environmental factors (Lenski and Travisano, 1994). Cultural techniques are generally not adequate for capturing real-time changes in the species composition of microbial communities. Similarly, the direct "snapshot" capability provided by typical microscopic analysis cannot capture the variations in the phenotypic expression of changes that are now relatively easily documented at the genetic level. To some extent, the development of video image analysis has facilitated the documentation of natural microbial growth and differentiation with respect to changes in cell structure, population dynamics, colony organization, and community functions. For example, Lewis and coworkers (1994) used digitized video microscopy in combination with fluorogenically labeled enzyme substrates to detect the expression of specific genes in a diverse group of bacteria. This approach has been used to demonstrate that gene expression could be induced and monitored in non-culturable starved cells of *Salmonella enteritidis.* The integration of enzyme-specific fluorescent staining with nucleic acid-specific dyes facilitated the production of a video image documentary on the morphological changes that accompany sporulation in *Bacillus subtilis.*

In a different but equally innovative use of high resolution video microscopy, Matz and colleagues (2002) investigated the influence of bacterial phenotypic characteristics, including cell size, shape, and capsule formation, on the selective feeding of *Spumella* sp., a freshwater heterotrophic nanoflagellate. The use of video microscopy made it possible to empirically distinguish the effect of bacterial swimming speed from the effects attributable to cell size and surface charge on nanoflagellate grazing behavior. The two studies described above represent attempts to capture a more continuous assessment of microbial diversity than is possible by means of traditional light and electron microscopy. Applications of video microscopy have been limited by cost and time considerations, but the development of digital video recording technology is improving rapidly. These improvements have led to the increasing sophistication of research questions that impinge upon the assessment of microbial diversity (for example, see Wilkinson and Schut, 1999).

OBJECTIVES FOR MICROSCOPIC ANALYSIS IN MICROBIAL DIVERSITY ASSESSMENT

Technological improvements in microscopic instrumentation have been the major driving forces behind strategies to discover solutions to long-standing questions in microbiology. For example, on the question of the determinants of cell morphological types, microscopic analysis is indispensable for comparative assessment of variations in cell shape at different points in a colony's life cycle, and variations among different species. Despite the extensive morphological diversity in the microbial world, there is considerable overlap of forms that masks a high level of physiological and genetic diversity in natural microbial communities. Therefore, microscopy is frequently combined with other forms of analysis to provide comprehensive information on the relative abundance of species and how they interact with one another in complex microbial communities. A few recognizable themes permeate the applications of microscopic analyses in the context of microbial diversity assessment. These themes are circumscribed by the available level of technological capability and the sensitivity of different microorganisms to the protocols demanded by specific techniques. Nine thematic objectives for integrating microscopic analysis into assessments of microbial diversity are described in the following sections.

Microbial cell morphological types

Early in the history of microscopic analysis, cell morphology was a useful way of categorizing bacterial species until the realization that many physiologically distinct bacteria could look identical (see Plate 2.1). The relics of an era in which the typological species concept dominated bacteriology are still present in taxonomic appellations such as *Coccus*, *Bacillus*, and *Spirillum*, which denoted spherical, rod-like, and spiral, respectively. In many ways, the taxonomy of viruses infecting prokaryotic cells remains dependent to a large extent on viral capsid morphology.

Spherical and cylindrical shapes dominated descriptions of bacterial cell morphological characteristics until 1980 when square-shaped halophilic microorganisms belonging to the genus *Haloarcula* ("salt-box") were discovered in hypersaline environments in the Sinai Peninsula region of Egypt (Horikoshi *et al.*, 1993; Kessel *et al.*, 1985; Oren *et al.*, 1999). Members of this cosmopolitan group of organisms can grow under conditions of salt concentration up to 4 M Na^+, and under alkali conditions of pH 9 (Takashina *et al.*, 1994). *Haloarcula quadrata* and other members of this group were discovered fortuitously by investigators using microscopes to examine the structure and function of intracellular vacuoles in microbes present in buoyant aquatic systems.

The morphological novelty of *Haloarcula* species has been subjected to intense microscopic scrutiny. The group now includes members such as *H. japonica* with triangular cell morphology (Nishiyama *et al.*, 1995; see Fig. 2.1). Although microscopic analysis was the exclusive primer for the discovery and preliminary classification of these organisms, subsequent molecular analyses, including polar lipid composition, 16S ribosomal RNA profiles, and the guanine + cytosine base pair composition of DNA were used to identify new members of the genus and to place *Haloarcula* in the phylogenetic domain of Archaea (Oren *et al.*, 1999; Takashina *et al.*, 1994).

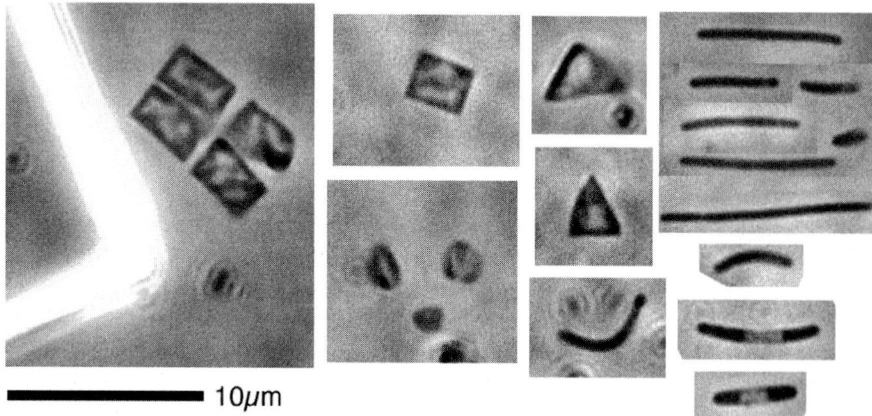

Fig. 2.1 Extremely halophilic archaeal species inhabit natural salty lakes, artificial salt-evaporating ponds, and salty geological formations such as gypsum halite evaporite. A collage of images produced by oil immersion lens (100× objective), phase contrast microscopy showing different cell morphologies. Note the squares and triangles (typical of *Haloarchaea*). The left-most picture has a salt crystal next to the four square cells. The cup-shape discs (lower middle) are like *Haloferax* cells. Other unusual microbial morphology represented by the triangular cellular shapes are characteristic of *Haloarcula* species. Images by courtesy of Dr. Dyall-Smith, University of Melbourne, Australia.

In experiments conducted under the "Biopan" program of the European Space Agency, microscopic analysis was indispensable for investigating the survival of non-sporulating *Haloarcula* and *Synechococcus* species in the vacuum and high ultraviolet exposure conditions of extraterrestrial space. The microorganisms, which were originally isolated from the tidal flats along the coast of Baja California Sur, endured long-term exposure to extraterrestrial conditions on Russian Foton rockets (Mancinelli *et al.*, 1995).

Multicellular organization in microbial colonies

Microbial cells are typically organized into structured arrangements within colonies. These arrangements are presumed to promote access to nutrients while reducing exposure to potentially toxic waste products. Multicellular organization may be as simple as clusters resulting from undifferentiated cell division, for example in the genus *Staphylococcus*. More complex secondary structures are produced by certain microbes only during specific periods of the life cycle, or only in response to specific environmental cues. Based on microscopic investigation the genus names *Streptococcus* and *Staphylococcus* were originally used to distinguish chain from cluster organization of spherical (coccoid) cells, respectively (see Plate 2.1). However, for some microorganisms, the apparent multicellular arrangements may be artifacts generated by cultivation under laboratory conditions. Microbial biofilms present opportunities to investigate multicellular organizations under natural conditions. Multicellular organization in biofilms has been aided by innovative microscopic techniques such as confocal laser microscopes. With this technique, thick specimens can be penetrated optically to produce three-dimensional maps of the organization of cells within biofilms. The combination of electron microscopy and confocal laser microscopy has been used extensively to investigate natural microbial biofilms to advance the understanding of **quorum sensing** (the ability of individual cells in a colony to estimate and respond to cell density in a colony; see Chapter 9) and the architectural organization of cells in ecological niches as diverse as the rhizosphere, epilithon, potable water distribution systems, oral cavity, and extraterrestrial micro-gravity conditions (see Chapter 9; Araujo *et al.*, 2000; Bloemberg *et al.*, 2000; Christensen *et al.*, 2002; Lawrence *et al.*, 2001; Li *et al.*, 2002; McLean *et al.*, 2001). Microscopic images of microbial biofilms are defined by structural characteristics that are categorized as either textural or spatial. The texture refers to the heterogeneity of the biofilm, whereas the spatial category describes the morphological relationship between the size and shape of biofilm surface features. In an impressive application of microscopy to the analysis of natural microbial colony structures, Yang and colleagues (2000) developed methods for extracting morphological features from biofilm images. These methods facilitate the quantitative assessment of the relationship between the physical properties and biochemical characteristics such as nutrient concentration and distribution, and species diversity that support biofilm stability (Fig. 2.2).

Relative abundance of species in a community

Despite the metabolic diversity of many natural microbial communities, it is not unusual to discover that a given physiological process is conducted by unique species because successful colonization of environments where physical and chemical conditions fluctuate requires niche diversification. Consequently, it is not surprising that processes such as nitrogen fixation, methanogenesis, transformation of metallic ions, and biodegradation of certain toxic organic chemicals are usually carried out by a narrow spectrum of microbial species. It is important to quantify the abundance of these kinds of species for the purpose of developing models for characterizing the potential impacts of anthropogenic influences on ecosystems. It is also important to develop methods for monitoring spatial and temporal changes in the distribution of microbial diversity in a manner that facilitates the linkage of

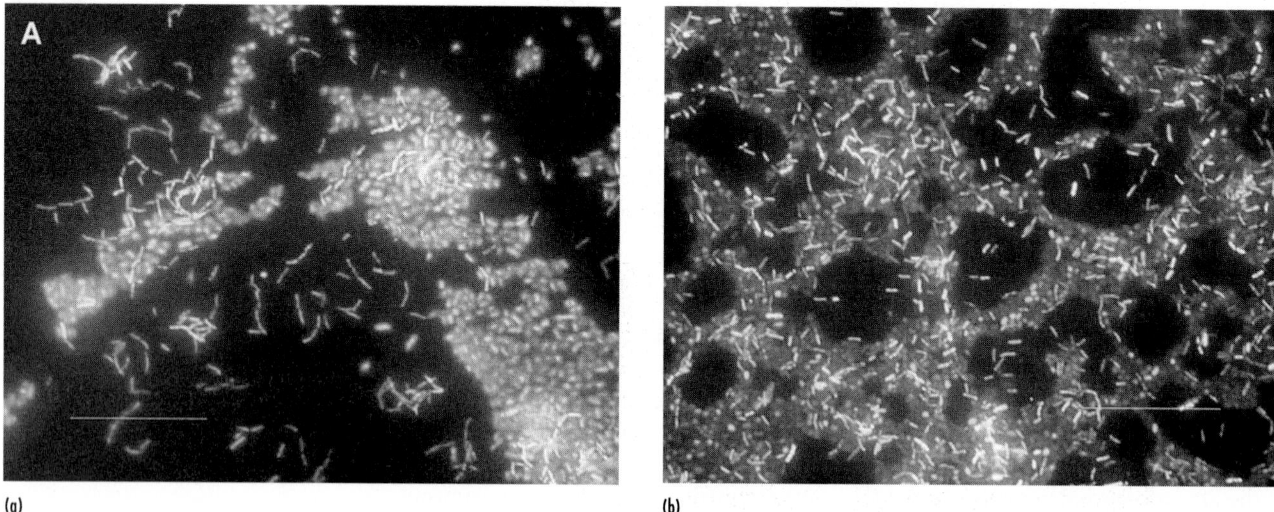

(a) (b)

Fig. 2.2 The patchiness of microbial biofilms can be illustrated by light microscopic images, which underscore the importance of structural heterogeneity in the stability and persistence of natural microbial colonies. Patchiness is influenced by both external factors such as nutrient supply and by factors intrinsic to the biofilm microbial community such as species diversity and the dynamics of nutrient flow through channels in the biofilm (Yang *et al.*, 2000). (a) shows polymicrobic biofilms (containing many different species) grown on stainless steel surfaces in a laboratory potable water biofilm reactor for seven days, then stained with 4,6-diamidino-2-phenylindole (DAPI) and examined by epifluorescence microscopy. The bar is equivalent to 20 μm. (b) shows a similar 14-day-old biofilm. Photographs are by permission of Ricardo Murga and Rodney Donlan, Center for Disease Control (Donlan, 2002).

genetic potential to ecological functions (Ogunseitan, 1998 and 2000; Ogunseitan *et al.*, 2001).

The combination of microscopy and species-specific oligonucleotide hybridization probes has become the preferred method for tracking relative species abundance in microbial communities. For example, Upton and colleagues used a combination of nucleic acid hybridization probes and confocal laser scanning microscopy to map the spatial distribution of methanogenic bacterial populations in an extensive peat bog ecosystem (Upton *et al.*, 2000). On a smaller scale, Rocheleau and colleagues (1999) used similar techniques for mapping the distribution of methanogens in the anaerobic sludge component of sewage treatment. These applications of microscopic analyses are gaining importance partly because of the increasing realization that microbial process at local levels can contribute substantially to global environmental change. Methane is 20 times more effective as a greenhouse gas than carbon dioxide, and its concentration in the atmosphere has been increasing at an annual rate of 1%. Attempts to control methane emissions require detailed understanding of the physiological and ecological diversity of methanogenic bacteria, which collectively account for approximately 70% of the methane released into the atmosphere (Richie *et al.*, 1999). Confocal laser scanning microscopy and fluorescent *in situ* hybridization (FISH) have provided valuable insight into the diversity and distribution of major methanogenic bacterial species, including *Methanosarcina barkeri*, which is considered one of the most versatile mesophilic methanogens. This organism is capable of producing methane from hydrogen, carbon dioxide, methanol, methylamines, and acetate (Lloyd *et al.*, 1998; Raskin *et al.*, 1994; also see Fig. 2.3).

Cell–cell interactions

It has long been recognized that microorganisms communicate using specialized processes mediated by the release and detection of biochemical molecules. Such communication

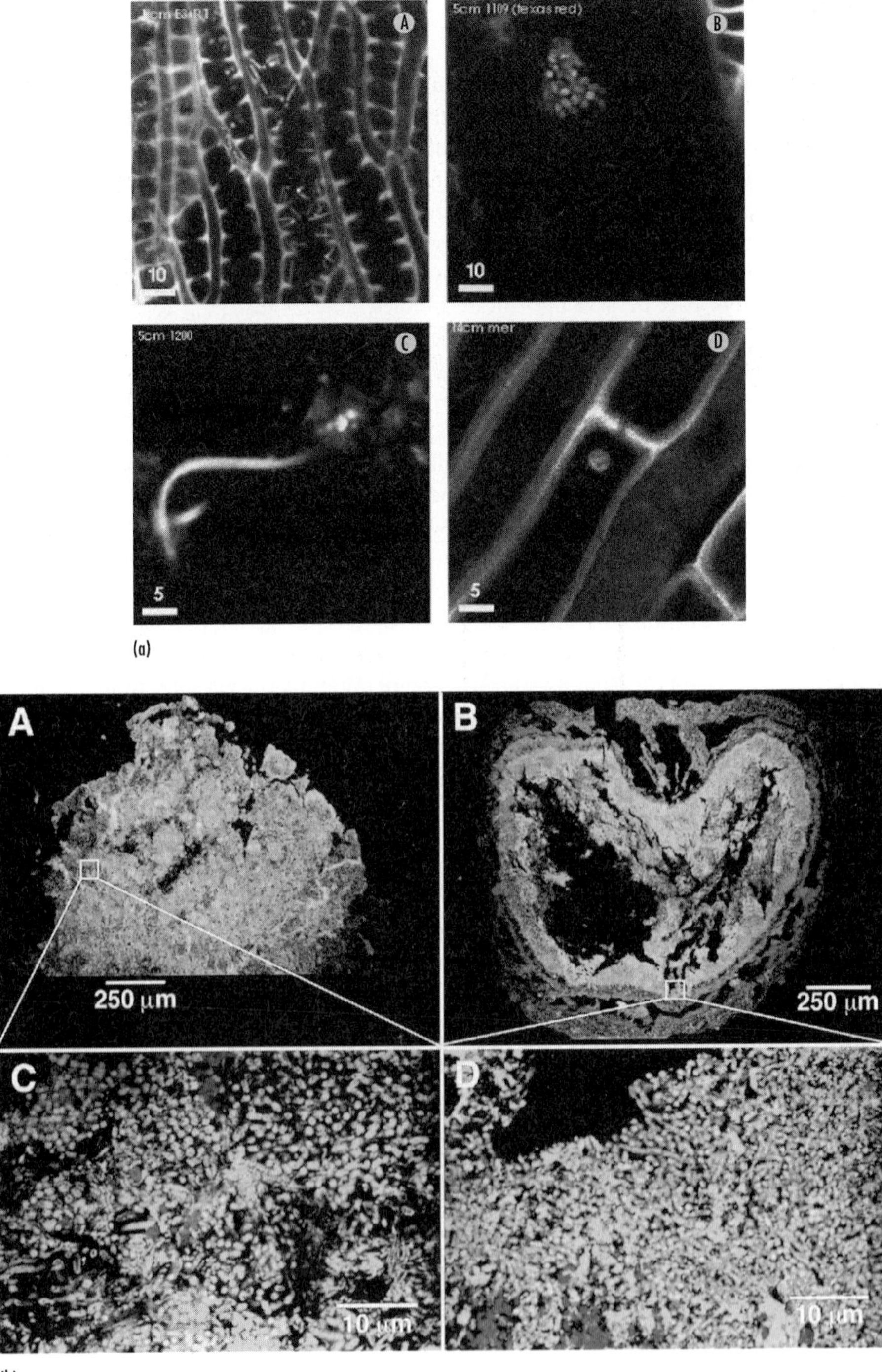

Fig. 2.3 Confocal laser scanning microscopy (CLSM) is used to reveal information on the three-dimensional structure and functional characteristics of natural microbial communities. (a) Upton and colleagues used CLSM to probe core samples of peat bog from a region that represents the source of 40–60% of the global methane output. Phylogenetically conserved (16S rRNA sequences) and functional (gene encoding for methyl-coM-reductase) fluorescence-labeled nucleic acid hybridization probes were used in conjunction with CLSM images to investigate the distribution of methanogenic bacteria at depths of 5 cm and 14 cm in panels (A) and (B), *Methanosarcina* species in panel (B), and *Methanomicrobiales* species in panel (C). The spiked structures in panel A are tissues of a Sphagnum leaf to which the methanogens are attached. Images are with permission of Clive Edwards (Upton *et al.*, 2000). (b) In a more domestic application of microbial methanogenesis, Rocheleau and colleagues (1999) used CLSM and species-specific oligonucleotide probes to differentiate the most versatile mesophilic methanogen, *Methanosarcina barkeri* from other methanogenic bacteria in anaerobic granular sludge samples. The combination of fluorescent *in situ* hybridization (FISH) and CLSM facilitated the construction of a topographic model of sludge granules with a multi-layered structure (panels A and B) where *M. barkeri* grows in the outermost layer, followed by the strict *acetoclastic* bacterium, *Methanosaeta concili*, and an inner core consisting of other bacteria (the brightest cells in panels C and D).

strategies are considered to be the mechanisms that underpin numerous microbial characteristics including colony architecture, responses to variable environmental factors that influence cellular differentiation, and population dynamics in host–parasite interactions. Research advances in microbial cell–cell communications have produced new understanding of **quorum sensing** and its relevance to chemotaxis, swarming behavior, biofilm stability, production of secondary structures, and the biological control of microbial pests (Bassler, 2002). The evidence for microbial communication through biochemical sensing was gathered in large part by assaying culture filtrates, and by matching cellular or colony responses to varying concentrations of chemicals (Li *et al.*, 2002).

In addition to the biochemical pathways for cell–cell communication, physical contact is a common mechanism of cell–cell interactions. The demonstration of intra- and interspecific physical contact communication required extensive microscopic analysis. Such analysis was essential for elucidating the genetic exchange process of conjugation, which requires the production of a conjugation tube, or pilus, through which genetic material is physically transferred from a donor cell to the recipient cell (Clewell, 1993). The image of two *E. coli* cells attached by a conjugation tube is depicted in Plate 2.1 (the first panel of (a)). The length of conjugation tubes can be up to five times the length of a single cell, but their thinness makes them fragile. For this reason, disrupted conjugation provided an excellent strategy for mapping linked genes on bacterial chromosomes. In *E. coli*, approximately 100 minutes is required for the replication and transfer of the entire chromosome from the donor cell to the recipient. Therefore, the positions of genes on the *E. coli* chromosomes are identified by minutes-denoted contiguous segments (Clewell, 1993). The fragility of conjugation tubes also makes them difficult to visualize in natural environmental systems, and it is doubtful that conjugation can be effective in highly dynamic environments experienced by planktonic organisms in aquatic ecosystems. Confocal laser scanning microscopy is a promising technique for capturing conjugation in naturally occurring sessile microbial communities. Microscopy is essential for investigating conjugation in natural ecosystems because it can be difficult to distinguish the outcome of conjugation (exconjugants) from the outcome of other genetic exchange mechanisms such as transformation and transduction, unless there are experimental controls to exclude these processes.

Viability and metabolic activities

One of the most challenging difficulties of investigating microbial diversity is the inadequacy of methods used for isolating and cultivating microorganisms under laboratory conditions. Fewer than 10% of microbial species occurring in nature can be cultivated. Molecular methods based on 16S rRNA and *in situ* microscopy have confirmed the existence of the suspected large numbers of non-culturable microorganisms in natural ecosystems (Colwell and Grimes, 2000). However, there is considerable debate about the viability of organisms detected by means of microscopy or FISH to phylogenetic probes. Methods that cannot be used to determine cell viability are limited in their usefulness because active participation in metabolic reactions at the level of individual cells is the fundamental basis of microbiological contributions to ecosystem function.

Several techniques have been developed to identify metabolically active cells in natural environments without actual cultivation. The most widely used techniques include **flow cytometry** or **cell sorting**, confocal laser scanning microscopy, and epifluorescence microscopy. Flow cytometry has limited use because it only applies to cells existing in aqueous media (Urbach and Chisolm, 1998; Wallner *et al.*, 1997). The simplest and most useful of these techniques is microscopic examination of samples that have been amended with specific dyes which change color after metabolic transformation (Betts *et al.*, 1989; McFetters *et al.*, 1991). Popular dyes include propidium iodide and SYTOX Green^{TM} for assessing bacterial membrane integrity, 5-cyano-2, 3-ditolyl chloride (CTC) for detecting microbial redox activity, fluorescein diacetate (FDA) for assessing intracellular enzyme activity, and rhodamine 123 or bis-(1,3-dibutylbarbiturate) trimethine oxonol (DiBAC™) for measuring cell transmembrane electrochemical potential (Smit *et al.*, 2000). Not all microorganisms respond to these chemical agents for assessing viability, and the choice of a specific agent for a particular microbial community must be determined through experimentation.

Several applications of microscopy for determining viability in natural microbial communities are documented in the literature. For example, Schaule and colleagues (1993) used CTC and microscopy for quantifying planktonic and sessile respiring bacteria in drinking water, where microbial regrowth in distribution pipelines can cause public health problems under conditions that support low concentrations of residual chlorine disinfectant. Smit and colleagues (2000) used DiBAC as a vital microscopic stain to assess the viability of microbial biofilms subjected to stressful environmental conditions including abrasion and starvation.

Cell components

The formation of inclusion bodies in microbial cells has several implications for the biogeochemical cycle of elements. The desire to analyze the cellular location and elemental composition of inclusion bodies has necessitated the integration of microscopic techniques with other methods such as spectrophotometric assays and X-ray fluorescence. For example, the accumulation of phosphorus as polyphosphate granules in bacteria has been used as a strategy for tertiary treatment of wastewater to eliminate excess phosphorus from domestic sources (Pauli and Kaitala, 1997). Jeon and colleagues (2000) used electron microscopy to investigate the morphological diversity of microorganisms in sludge that has been acculturated for enhanced biological phosphorus removal from the aqueous phase in sequencing batch reactors. After approximately two years of microscopic observation, they noticed a decline in morphological diversity in favor of dominance by perpendicular cuboidal bacterial cells arranged in an octameric formation. The cells produced lactic acid and they existed in a nutritional symbiotic relationship with a coccus-shaped bacterium that accumulated phosphorus granules.

In addition to iron oxides, magnetotactic bacteria (see Plate 2.1, (g), second panel) can accumulate other metals and phosphorus. Lins and Farina (1999) and Keim and colleagues (2001) ingeniously employed a combination of electron microscopy and energy-dispersive X-ray analysis to demonstrate that uncultured magnetotactic bacteria, which typically contain iron oxide magnetite (Fe_3O_4) granules, are also capable of accumulating phosphorus in addition to metals such as zinc, manganese, strontium, cadmium, aluminum, and chromium (Fig. 2.4). The analysis provided support for the idea that phosphorus accumulation in microorganisms serves other environmentally sensitive functions in addition to the regulation of nutritional status (Rees *et al.*, 1993).

Microscopy has been used for investigating other cell components in addition to inclusion bodies. For example, current understanding of the structure and function of ribosomes in protein synthesis would not have been possible without the aid of microscopes. Perhaps more importantly, the lack of a membrane-bound nucleus is one of the most important factors that distinguish prokaryotes from eukaryotes. In cases of newly isolated microorganisms, microscopic evaluation is crucial for making a judgment about genome organization, and consequently about phylogenetic classification, although the data are usually supplemented with rRNA profiles.

Predation and parasitism that regulate populations

The ultramicroscopic size (usually less than 100 nm) of viruses renders them observable only through the high level of magnifications achievable by electron microscopy. The simplicity

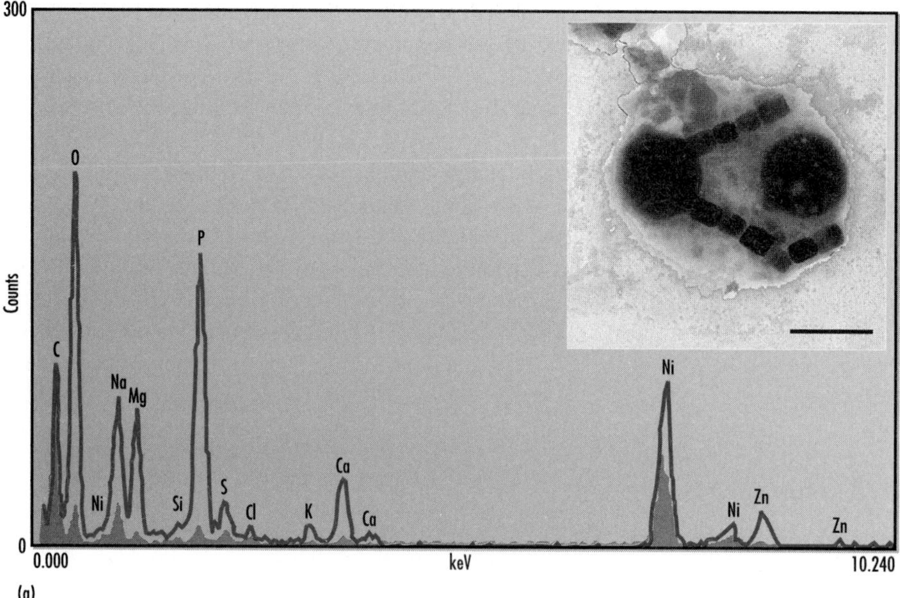

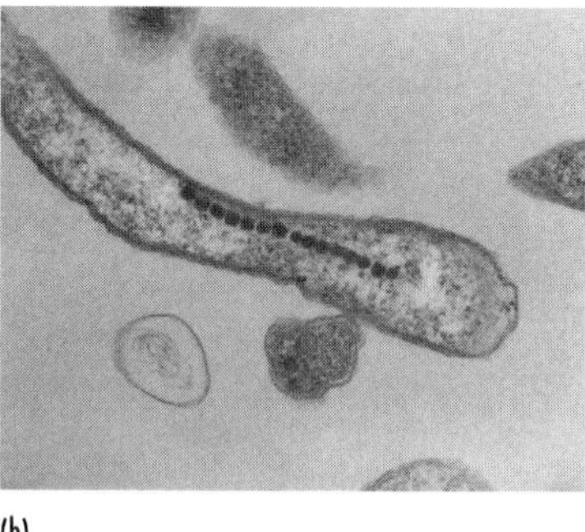

Fig. 2.4 Photomicrograph and element abundance curves showing the accumulation of phosphorus and metals in uncultured magnetotactic bacteria as revealed by microscopic and energy-dispersive X-ray analyses. Note that the peak for nickel is attributed to the supporting grid, an artifact of the experimental technique. Bar represents 0.5 μm. Image (a) is by courtesy of Marcos Farina (Keim *et al.*, 2001). Image (b) shows the details of the internal components of a magnetotactic bacterial cell (*Aquaspirillum magnetotacticum*) revealed by a transmission electron microscope. The magnetic granules are aligned to a magnetic field. Image (b) by courtesy of Dennis Kunkel.

of viral components, consisting of one or two nucleic acid molecules and typically fewer than 10 different kinds of protein molecules, masks the morphological diversity of viruses as revealed by electron microscopy (Fig. 2.5(a)). In addition to the investigations of viral architecture and morphological diversity in natural environments, microscopic analysis has been extremely useful in understanding various aspects of host–virus interactions. In some cases, microscopy is used in combination with other techniques to reveal the spatial and temporal patterns of virus infections. For example, Nakamura and colleagues used biotinylation (biotin-labeling) and confocal laser scanning microscopy to visualize filamentous bacteriophage particles that infect *E. coli.* The sensitivity of microscopic observation of biotinylated bacteriophage particles can be increased substantially because of the high affinity of biotin to fluorescein-conjugated avidin (Nakamura *et al.*, 2001).

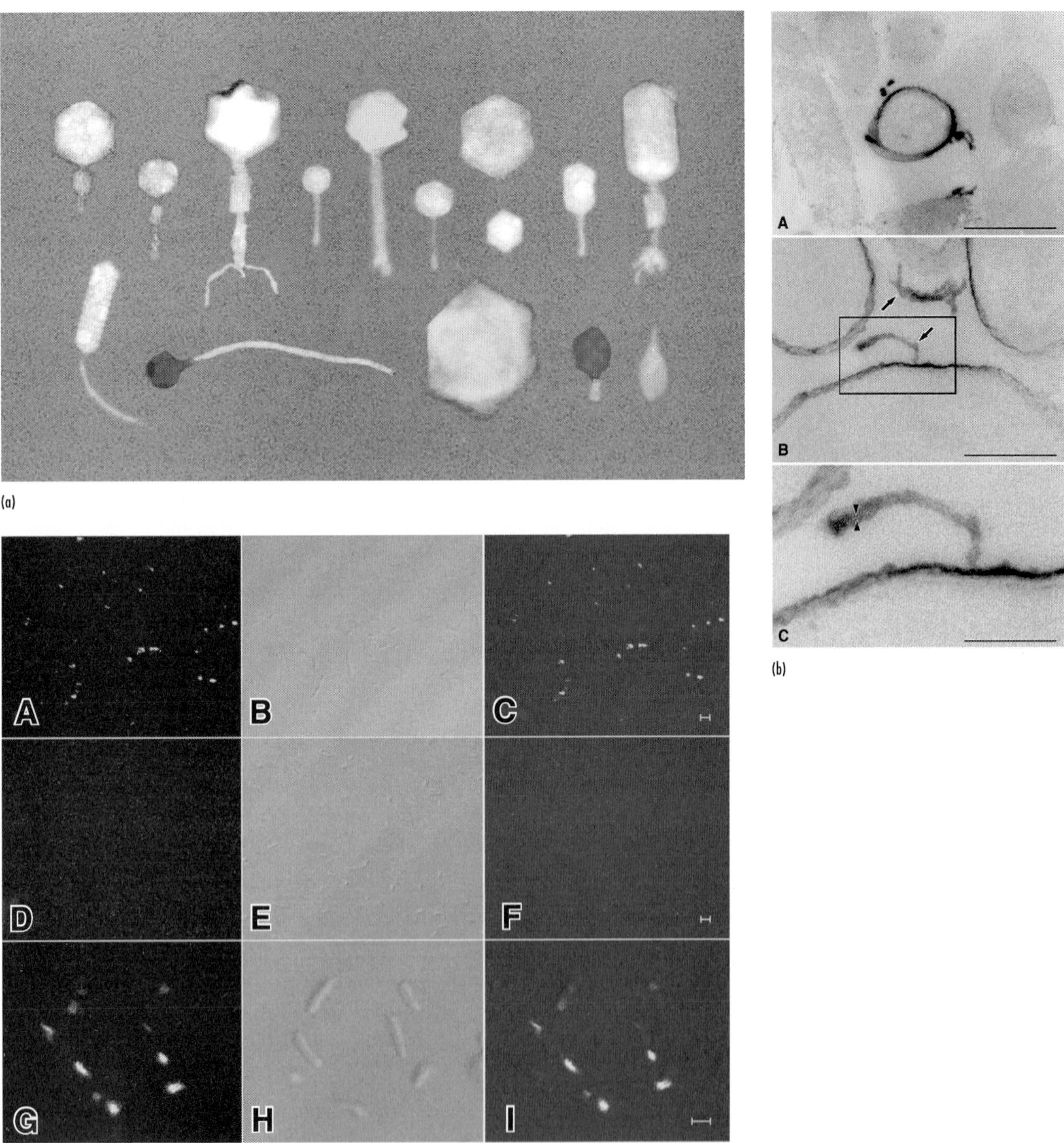

Fig. 2.5 (a) High level of bacteriophage diversity in a saline wetland environment where the multiplicity of infection (ratio of virus to bacteria) exceeds 100 because of the rareness of eukaryotic grazers. Bacteriophage morphology is used for classification according to size and geometrical shape of the head, presence of tail, and whether the tail is sheathed. Images were acquired through electron microscopy and reproduced by permission from David Bird and Richard Robarts. Bar is 10 nm. (b) Electron micrograph of biotinylated filamentous bacteriophage infecting *E. coli* JM109. Intact bacteriophage particles and protein derivative are seen as dark products of peroxidase reaction. Frame A shows the dark intact labeled bacteriophage filaments. Frame B shows the attachment of labeled phage to a bacterial cell and the intracellular location of bacteriophage protein molecules. Frame C is an enlarged box from frame B. (c) Confocal laser scanning microscope images of avidin-fluorescence conjugated filamentous bacteriophage infecting *E. coli* JM109. Microscopic techniques used are fluorescence (A, D, and G) and phase contrast (B, E, and H). Frames C, F, and I contain digital superimpositions of images in A and B; D and E; and G and H. Images in (b) and (c) are from Nakamura *et al.* (2001). Original magnifications in (b): A = 312,000; B and C = 324,000. Bars: A = 1 mm; B = 0.5 mm; C = 0.125 mm. In (c), bars: A–F = 4 mm; G–H = 2 mm.

Differentiation and life cycles

The production of inclusion bodies in microbial cells was discussed under the section entitled "Cell components". Certain transformations of microbial cell morphology also occur externally, such as the production of "stalks" that facilitate a sessile lifestyle of *Caulobacter* species (Ong *et al.*, 1990; Smit *et al.*, 2000). *Caulobacter crescentus* cells are capable of differentiating from a mobile (**swarming**) behavior to stationary cells attached to a substratum with the aid of stalks. Stalked cells attach to surfaces through a polysaccharide adhesive called "holdfast" at the tip of the stalks. The holdfast results from the elongation of the stalk at a site previously occupied by another cellular appendage, the flagellum, during differentiation (Janakiraman and Brun, 1999). Morphological studies on stalk production in *Caulobacter* require sophisticated microscopic analysis. Janakiraman and Brun (1999) used an innovative approach for understanding the spatial and temporal regulation of holdfast expression by labeling stalks with fluorescein-conjugated lectin followed by visualization of cells with combined fluorescence and phase-contrast microscopy (Fig. 2.6).

Microscopy is also essential for investigating spore formation by some microorganisms. Most fungi produce spores or "fruiting bodies" that allow them to withstand harsh environmental conditions during periods of nutrient deficiency or dehydration. Spore formation also contributes to the dispersal of microorganisms around the planet. Among bacteria, many Gram-positive microorganisms form spores. For example, the spores of *Bacillus anthracis* are notorious because of their potential use as biological weapons. However, it is the spores of the related bacterium, *B. subtilis* that have been studied most intensely. *Bacillus subtilis* produces spores with thick outer cell walls, which makes them resistant to ultraviolet radiation, heat, and disinfection. Bacterial spore hardiness has made them prime subjects for exobiology research (Horneck, 1998). Horneck and colleagues (2001) exposed spores of *B. subtilis* to extraterrestrial space conditions aboard the Earth-orbiting FOTON satellite for two weeks. Under these conditions, the survival rate of the spores remained at 100% when mixed with soil, but the survival rate was close to 0% when the spores were exposed as dry layers. The tolerance of *B. subtilis* spores to high levels of ultraviolet radiation is attributed to a protective cortex and the presence of both inner and outer spore coat layers. The structure of these layers was revealed by transmission electron microscopy (Horneck *et al.*, 2001).

Fossil microorganisms

One of the most famous meteorites in recent scientific history has the cryptic name of ALH84001. The fame of ALH84001 stems from the publication by McKay and colleagues in 1996 that microscopic analysis of uncontaminated sections of the Martian meteorite suggested that it harbored relics of early biological activity (McKay *et al.*, 1996). There are profound implications for any announcement suggesting that the characteristics of a Martian meteorite match at least five of the six postulated qualities of biologically produced magnetite crystals, namely size range and width/length ratio, chemical purity, crystallographic perfection, linear chain arrangement of crystals, unusual crystal morphology, and elongation of crystals in the (111) direction (for a color-coded image of magnetotactic aquatic bacterium, *Aquaspirillum magnetotacticum*, see second panel of (g) in Plate 2.1). The announcement by McKay and coworkers was controversial in part because of disparate interpretations of the electron micrograph images generated from the meteorite sample. Exobiologists considered the absence of magnetite crystal chains from ALH84001 as the missing evidence that disqualifies the argument of a biological origin for the meteoritic structural components. In addition, many microbiologists considered the 100 nm size ranges of the ovoid shapes identified in ALH84001 as too small to be related to terrestrial magnetotactic bacteria. Despite the intense disagreements regarding the interpretation of microscopic data, it is indisputable that the physical, chemical, and potentially biological characteristics of

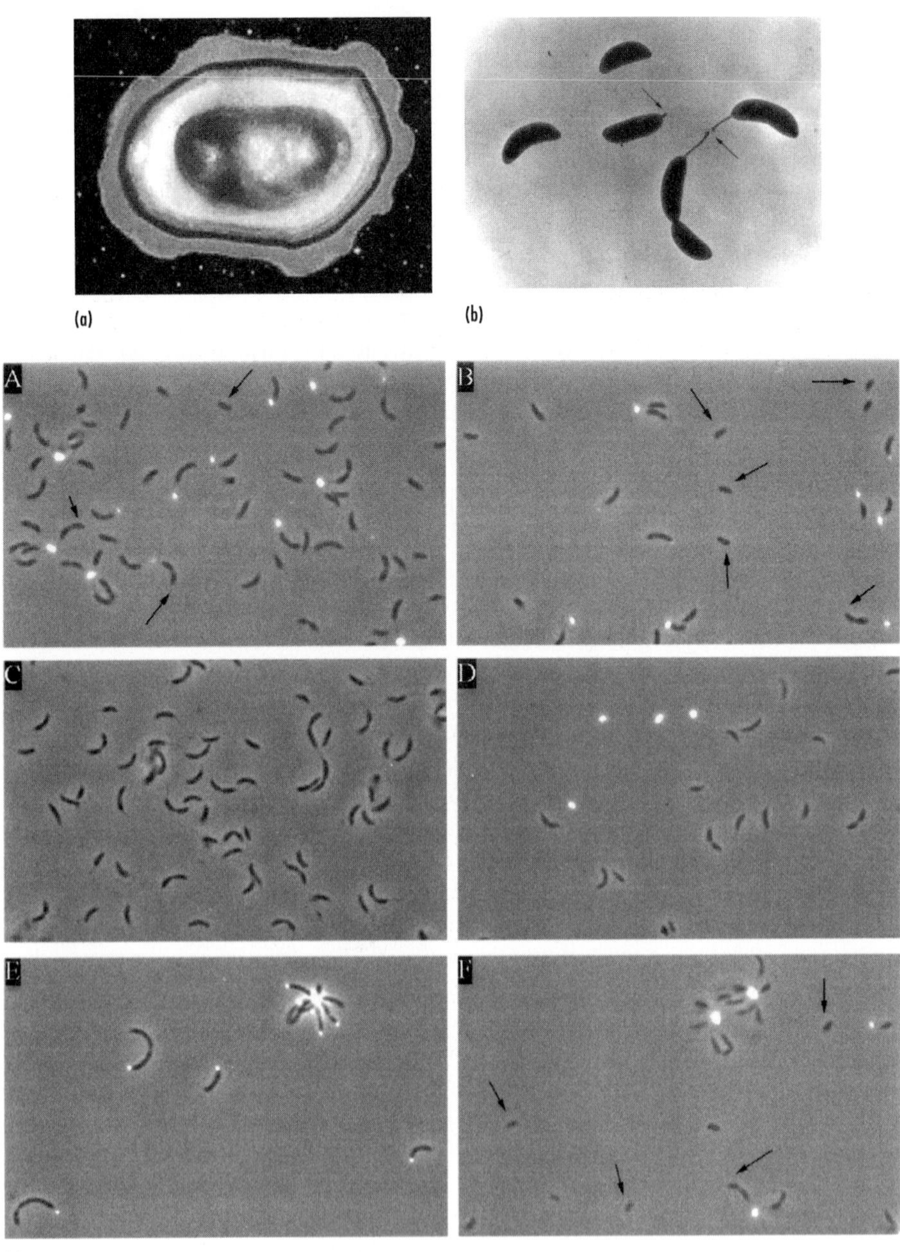

Fig. 2.6 (a) Electron micrograph of a cross-section of *B. subtilis* spore showing the genome preserved inside the core, which is surrounded by a protective cortex and inner and outer spore coats. The spore is 1.2 μm and approximately 0.25 μm². Image is from Horneck *et al.* (2001). (b) Transmission electron micrograph showing the production of stalks by *Caulobacter crescentus.* (c) Visualization of lectin-labeled holdfast at the tip of *Caulobacter crescentus* stalks. A combination of fluorescence microscopy and phase-contrast microscopy was used for visualizing the wild-type strain produced images in frames A and B. Frame C is an image of a mutant that does not produce holdfast. The images in frames D, E, and F are of mutants that shed the holdfast. Arrows in frames A, B, and F point to swarmer cells that have not differentiated through stalk production. Epifluorescence photomicroscopy was performed on a Nikon Eclipse E800 light microscope equipped with a Nikon B-2E fluorescein isothiocyanate (FITC) filter cube for FITC and a 100× Plan Apo oil objective. For transmission electron microscopy, cells were stained with 1% uranyl acetate for 30 seconds and washed four times with water. The cells were visualized with a Philips model 300 electron microscope at 60 kV. Images in (b) and (c) are by permission of Yves Brun (Janakiraman and Brun, 1999).

ALH84001 provide a fascinating sample for investigating the prospects of microbiological survival in extraterrestrial materials.

Friedmann and colleagues used high-power backscattered scanning electron microscopy (SEM-BSE) to find and record intact magnetite crystal chains in the Martian meteorite ALH84001 (Friedmann *et al.*, 2001). The usual method to demonstrate magnetite crystals enclosed in rocks involves acid cleaning which destroys such chains. While standard scanning electron microscopy produces images of surface structures, SEM-BSE records chemical compositions. Structures below the surface composed of heavier elements appear brighter than those of lighter ones. (In this respect, SEM-BSE images resemble X-ray images). The technique depends on backscattered electrons from beneath the surface of a specimen (400–1000 nm below). The research led the investigators to conclude that ALH84001 does in fact contain magnetite crystal chains which show all the six characteristics of such chains produced by terrestrial magnetotactic bacteria.

A wide array of microorganisms is entombed in amber deposits, some of which are older than 50 million years (Greenblatt *et al.*, 1999; Poinar *et al.*, 1993). There is a possibility that some of the entombed microorganisms can be revived to shed light on prehistoric microbial diversity and on certain aspects of microbial evolution. This possibility has generated a considerable body of research based on microscopic analysis in combination with molecular assessment and culture methods to document the diversity of microbial communities in ancient materials. For example, Cano and colleagues (1995) reported the isolation of spore-forming *Bacillus sphaericus* species from the abdomens of bees entombed in amber. As in the case of the Martian meteorite, microscopic analysis of cell morphology is not sufficient to unequivocally determine the presence of small microorganisms in amber, although there is generally no dispute about the presence of large eukaryotic microorganisms such as fungi. Inferences about microbial evolution are intimately linked to the techniques used to determine the age of amber deposits, which are also controversial. For example, Schmidt and colleagues determined that the Mesozoic amber of Schliersee in southern Germany is approximately 130 million years younger than previously proposed (Schmidt *et al.*, 2001). The revision of the temporal scale of amber specimen origin had consequences for the interpretation of phylogenetic and evolutionary change in *Palaeodikaryomyces baueri*, an archaic fungus that has now been reclassified as abundant during the Paleozoic as opposed to previously believed Triassic geological period (Schmidt *et al.*, 2001; also see Fig. 2.7).

Finally, the record for the longest period of time that bacteria can survive in stable geological formations is held by the spore-forming halobacterium *Bacillus* species 2-9-3 which was discovered in a brine inclusion within a 250 million-year-old salt crystal from the Permian Salado Formation in New Mexico (Vreeland *et al.*, 2000). Comparative assessment of the 16S rDNA from *Bacillus* species 2-9-3 showed that it is related to the lineage of *Bacillus pantothenticus* (97.5% similarity) and *Bacillus morismortui* (99% similarity). The nucleotide sequence similarity between the 250-million-year old *Bacillus* and *B. morismortui* is remarkable, and it suggests a very high level of genetic conservation, at least in the 16S rDNA locus. Alternately, skeptical investigators may consider that such observations of extreme genetic conservation demonstrate the difficulty of evaluating fossil evidence while avoiding contamination with contemporary microorganisms. Verification of methodological integrity regarding sample preparation is absolutely essential for acceptance of evidence for ancient microbes. In the case of *Bacillus* 2-9-3, the investigators' use of experimental controls and infallible sterilization techniques support the authenticity of the discovery.

CONCLUSION

Microscopes have provided a unique window through which our view of the microbial world was established and continues to be revised. These instruments have undergone remarkable technological improvements since their invention, with the result that they remain key features of investigative approaches dedicated to the understanding of diversity and all its

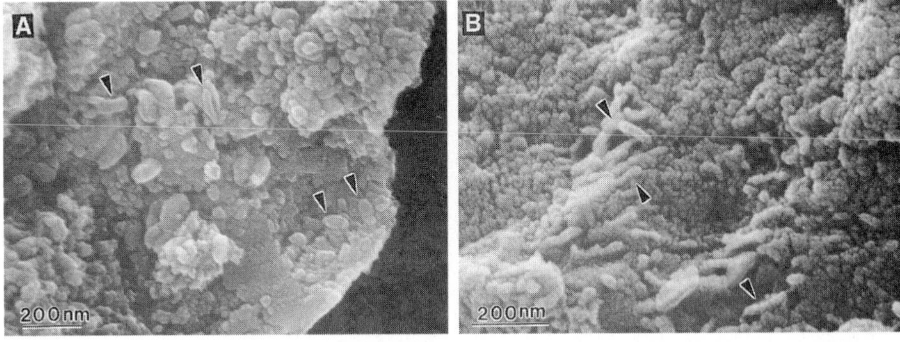

(a)

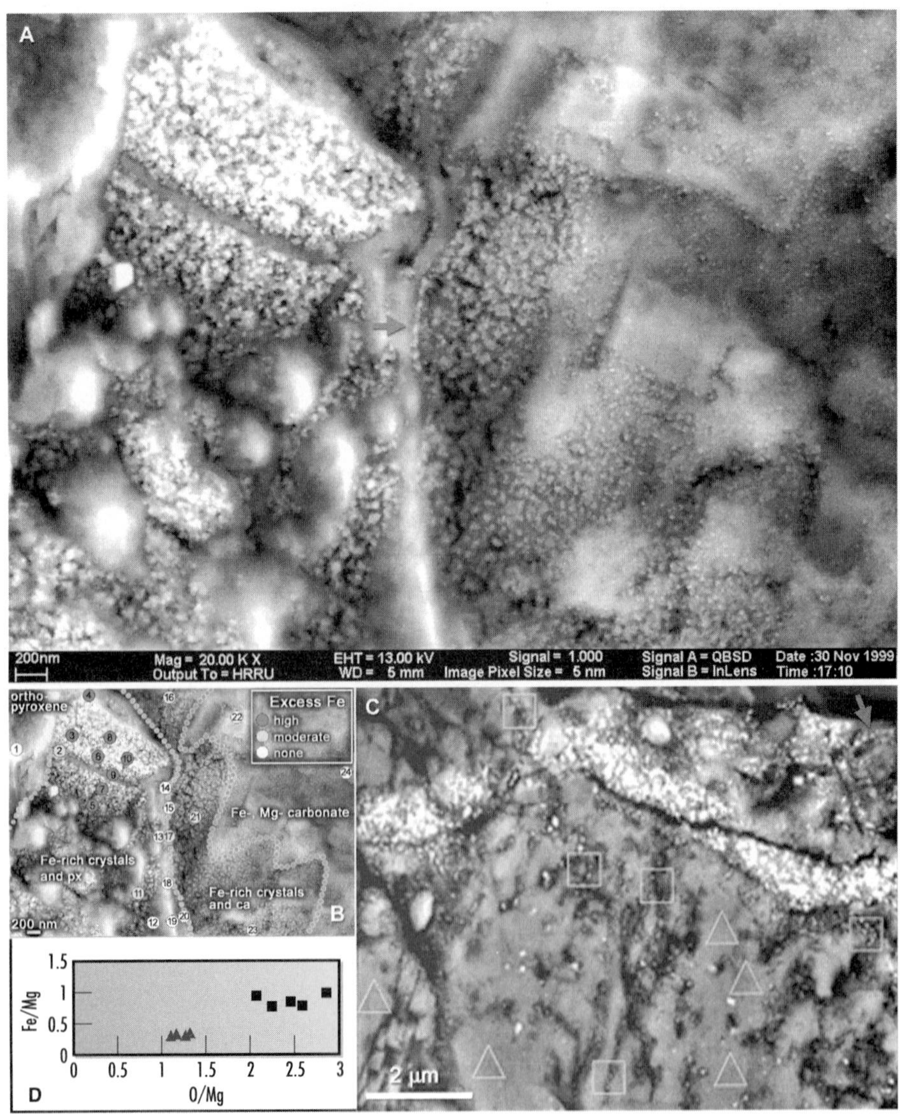

(b)

Fig. 2.7 (a) Details of the Martian meteorite ALH84001 as revealed by scanning electron microscopy showing (arrowed) ovoid and tubular shaped structures originally putatively identified as bacteria fossils, although the consensus is now that these structures are not fossil bacteria because they are so small that they cannot contain even a single strand of DNA. Frame B is a close-up view of the carbon-rich sections of the specimen. Images are by courtesy of NASA (McKay *et al.*, 1996).

(b) Frames A–C: High-power backscattered scanning electron (SEM-BSE) micrographs of magnetite crystals in the rim region of carbonate globules in Martian meteorite ALH84001. Frame A shows the image through the intact surface of a freshly fractured specimen, showing chain of large elongated crystals (arrow). Frame B shows energy-dispersive X-ray spectrometry analyses of elemental composition showing the position of the Fe-rich rim. Frame C shows a similar site in a resin-embedded, sectioned, and polished specimen, one chain marked by an arrow, also the site of Auger electron spectroscopy analyses. Frame D shows the results of analyses of areas outlined in frame C: squares, magnetite crystal chains present; triangles, absent; *ca*, carbonate; *px*, orthopyroxene. Images are by courtesy of Imre Friedmann at NASA Ames Laboratory (see also Friedmann *et al.*, 2001).

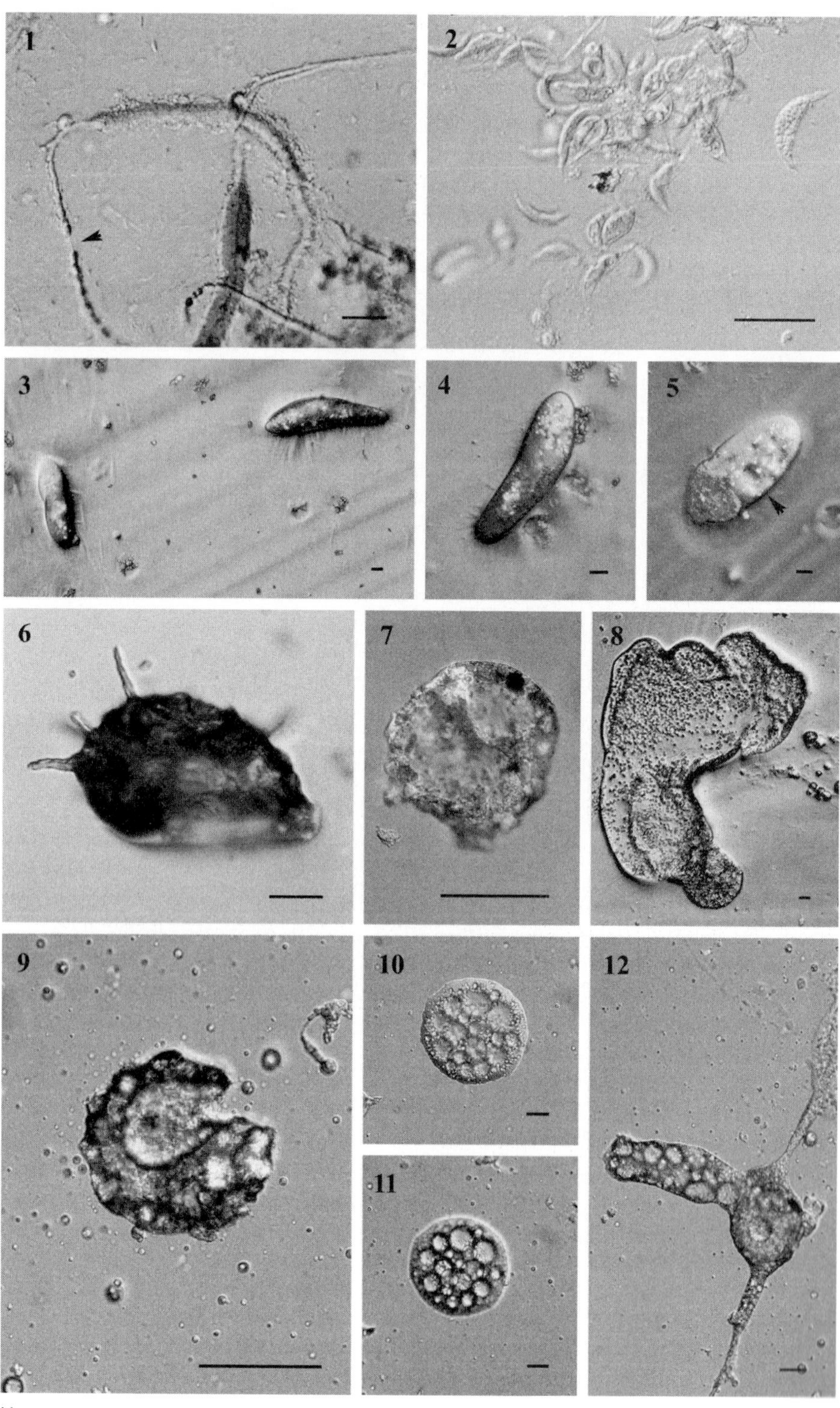

(c)

Fig. 2.7 *Continued* (c) Electron micrographs of amber resin materials from the Schliersee region of southern Germany.
Frames 1–8 represent bacteria, algae and Protozoa:
Frame 1: *Leptothrix discophora* in the resin of *Cycas revoluta.*
Frame 2: *Monoraphidium terrestre* in the resin of *Cycas revoluta.*
Frames 3 and 4: *Paramecium caudatum* in the resin of *Cycas revoluta.* The drainage lines around the cells are suggestive of a particular structure of resin.

ramifications in microbial communities. Increasingly, microscopy is supplemented with molecular techniques to reveal information on relative species abundance and the presence of non-culturable microorganisms in several ecosystems. The assessment of cell viability is of central importance to questions regarding the linkage between microbial diversity and ecosystem function. Although microscopic analysis contributes important answers to these questions, culture methods remain paramount for deeper exploration of microbial physiology. The next chapter focuses on concepts and methods dedicated to the investigation of microbial growth and proliferation.

QUESTIONS FOR FURTHER INVESTIGATION

1 Microscopy has been used most extensively for characterizing microbial cell morphology. However, the available database on microbial morphological types may be biased by the relative ease of cultivation of various microbes to prepare them for microscopic analysis. Search electronic and published databases for microscopic images of microorganisms belonging to at least three major phylogenetic categories. For the first 100 images you recover, construct a table with four columns, namely phylogenetic group, morphological type, cultivation conditions (if available), and natural habitat. Determine whether there are obvious correlations among these four categories, and write a brief essay to explain your inferences.

2 Imagine four different environmental samples recently recovered from (a) deep sea hydrothermal vents, (b) sand from the Sahara desert, (c) the stratosphere, and (d) a meteorite. For each sample, develop a protocol for initial assessment of microbial diversity based on microscopic methods.

SUGGESTED READINGS

Amann, R.I., W. Ludwig, and K.H. Schleifer. 1995. Phylogenetic identification and *in situ* detection of individual microbial cells without cultivation. *Microbiology Reviews*, **59**: 143–69.

American Society for Microbiology. 2002. Microbe Library. http://www.microbelibrary.org/.

Bhanu P.J. and J.K.H. Hörber (eds.). 2002. *Atomic Force Microscopy in Cell Biology*. San Diego: Academic Press.

Diaspro, A. (ed.) 2002. *Confocal and Two-Photon Microscopy: Foundations, Applications, and Advances*. New York: Wiley-Liss.

Glauert, A.M. and P.R. Lewis. 1998. *Biological Specimen Preparation for Transmission Electron Microscopy*. Princeton, NJ: University Press.

Hayat, M.A. 2000. *Principles and Techniques of Electron Microscopy: Biological Applications*. Fourth Edition. New York: Cambridge University Press.

Hoppert, M. and A. Holzenburg. 1998. *Electron Microscopy in Microbiology*. New York and Oxford: Springer in association with the Royal Microscopical Society.

Jones, T. 1997. *The History of the Light Microscope*. Third Edition. http://www.utmem.edu/~thjones/hist/hist_mic.htm.

Junge, K., F. Imhoff, T. Staley, and J.W. Deming. 2002. Phylogenetic diversity of numerically important Arctic sea-ice bacteria cultured at sub-zero temperature. *Microbial Ecology*, **43**: 315–28.

Kramer, S. and D. Kunkel. 2001. *Hidden Worlds: Looking Through a Scientist's Microscope*. Boston, MA: Houghton/Mifflin.

Lacey, A.J. (ed.) 1999. *Light Microscopy in Biology: A Practical Approach*. Oxford: Oxford University Press.

Lloyd, D., K.L. Thomas, A. Hayes, B. Hill, B.A. Hales, C. Edwards, J.R. Saunders, D.A. Ritchie, and M. Upton. 1998. Micro-ecology of peat: Minimally invasive analysis using confocal laser scanning

Fig. 2.7 *Continued* Frame 5: *Paramecium caudatum* with recognizable organelle in the resin of *Cycas revoluta*. The arrow points to a vacuole within the cell.
Frame 6: *Centropyxis aculeata* in the resin of pine (*Pinus strobes*).
Frame 7: *Centropyxis sphagnicola* in the resin of *Cycas revoluta*.
Frame 8: *Amoeba proteus* in the resin of *Cycas rumphii*.
Frames 9–12 represent organisms found in the resin of *Pinus strobus*, which resemble those found in the Schliersee resin.
Frame 9: Amoeba-like organisms.
Frames 10 and 11: Disc-forming organisms at two different levels of sharpness.
Frame 12: Irregular shaped organism with internal pock-marked structures.
The bars represent 20 μm. The images are reproduced by courtesy of Alexander Schmidt at Jena and Berlin, Germany (see also Schmidt *et al.*, 2001).

microscopy, membrane inlet mass spectrometry, and PCR amplification of methanogen-specific gene sequences. *FEMS Microbiology and Ecology*, **25**: 179–88.

MacNaughton, S.J., T. Booth, T.M. Embley, and A.G. O'Donnell. 1996. Physical stabilization and confocal microscopy of bacteria on roots using 16S ribosomal RNA-targeted, fluorescent-labeled oligonucleotide probes. *Journal of Microbiological Methods*, **26**: 279–85.

Mayer, F. and M. Hoppert. 2001. *Microscopic Techniques in Biotechnology*. Chichester: Wiley.

Matz, C., J. Boenigk, H. Arndt, and K. Juergens. 2002. Role of bacterial phenotypic traits in selective feeding of the heterotrophic nanoflagellate *Spumella* sp. *Aquatic Microbial Ecology*, **27**: 137–48.

Morel, G. and A. Cavalier. 2001. *In Situ Hybridization in Light Microscopy*. Boca Raton, FL: CRC Press.

Morris, V.J. and A.P. Kirby. 1999. *Gunning Atomic Force Microscopy for Biologists*. London: Imperial College Press.

Murphy, D.B. 2001. *Fundamentals of Light Microscopy and Electronic Imaging*. New York: Wiley-Liss.

Murray, R.G.E. and C.F. Robinow. 1994. Light microscopy. In P. Gerhardt, R.G.E. Murray, W.A. Wood, and N.R. Kreig (eds.) *Methods for General and Molecular Bacteriology*, pp. 8–20. Washington, DC: American Society for Microbiology.

Nishiyama, Y., S. Nakamura, R. Aono, and K. Horikoshi. 1995. Electron microscopy of halophilic *Archaea*. In S. DasSarma and E.M. Fleischmann (eds.). *Archaea: A Laboratory Manual: Halophiles*, pp. 29–33. New York: Cold Spring Harbor Laboratory Press.

Paddock, S.W. (ed.) 1999. *Confocal Microscopy Methods and Protocols*. Totowa, NJ: Humana Press.

Tomb, H. and D. Kunkel. 1993. *Microaliens: Dazzling Journeys with an Electron Microscope*. New York: Farrar, Straus, and Giroux.

Wilkinson, M.H.F. and F. Schut. 1999. *Digital Image Analysis of Microbes: Imaging, Morphometry, Fluorometry, and Motility Techniques and Applications*. New York: Wiley.

Yang, X., H. Beyenal, G. Harkin, and Z. Lewandowski. 2000. Quantifying biofilm structure using image analysis. *Journal of Microbiological Methods*, **39**: 109–19.

CULTURE METHODS

INTRODUCTION

Many microorganisms exhibit short generation times under favorable growth conditions. It is extremely difficult, however, to predict and reproduce "favorable growth condition" for all microorganisms. Therefore, most microbes existing in nature have not formed viable colonies under laboratory conditions. Nevertheless, cultivation remains a powerful method for investigating microbial populations, especially when it is combined with powerful direct methods such as microscopy and molecular analyses. This chapter provides an overview of the major approaches for observing microbial forms and activities after cultivation. These approaches highlight the strategies for resolving the following challenges in the assessment of microbial diversity:

1 The discrepancy between microbial growth in nature and under cultivation in the laboratory.
2 The limitations of microbial cultivation as an independent method for assessing microbial diversity with respect to axenic cultures, enrichment cultures, selective cultures, and community cultures.
3 The phenomenon of viable but non-culturable organisms.

Chapter contents

CULTIVATION AND DIVERSITY ASSESSMENT

In physics it is acceptable to announce the discovery of new elements with no direct evidence of their existence in nature. Thus elements with atomic numbers 104 through 118, created through the experimental fusion of existing nuclei, have extremely short half-lives and are recognized only by inferred chemical signatures. The naming of new chemical elements is frequently subjected to acrimonious debate among physicists despite the fact that these elements may not "physically" exist (Winter, 2002). This tradition is very different from the practice in microbiology where the isolation of organisms from nature and their cultivation in axenic form under laboratory conditions has been the required evidence for naming new species and for inferring phylogenetic relationships to already described organisms (Palleroni, 2001).

Microbial growth through cell proliferation can be decoupled from metabolic activity, but there is a preference for cultivation as the vital indicator of occurrence and activity of microorganisms in natural ecosystems (Kell *et al.*, 1998). The requirement for independent growth is reflected in two of the four key points delineated by the postulates of Robert Koch (1843–1910), who is credited with inventing the enrichment culture technique and for cultivating some of the most potent pathogens (Kell and Young, 2000; see Box 3.1). Several investigators have used molecular signatures including unique 16S rRNA sequences and fatty

BOX 3.1

(a)

Modern research approaches to cultivate microorganisms isolated from natural environments owe much to the pioneering work of **Louis Pasteur** (1822–95; (a)) who is best known for demonstrating that microorganisms are not spontaneously generated in liquid broth cultures. Pasteur's work also illuminated the current models of microbial nutrition by demonstrating the differences in culture byproducts when yeasts are cultivated under various conditions of aeration. Image of Pasteur is reproduced by courtesy of the Chantal Policieux; copyright, Musée Pasteur, Paris.

(b)

The principles of enrichment culture methods developed by **Robert Koch** (1843–1910) are still valuable today (b). **Richard Julius Petri** (1852–1921) invented the ubiquitous Petri dish while working in Koch's laboratory (Petri, 1887). Similarly, Fanny and Walter Hess invented the cultivation of microorganisms on solid agar disc media, while working in Koch's laboratory (Hitchens and Leikind, 1939). Image of Koch is reproduced by courtesy of Beste Grüsse at the Museum der Wissenschaft, Germany.

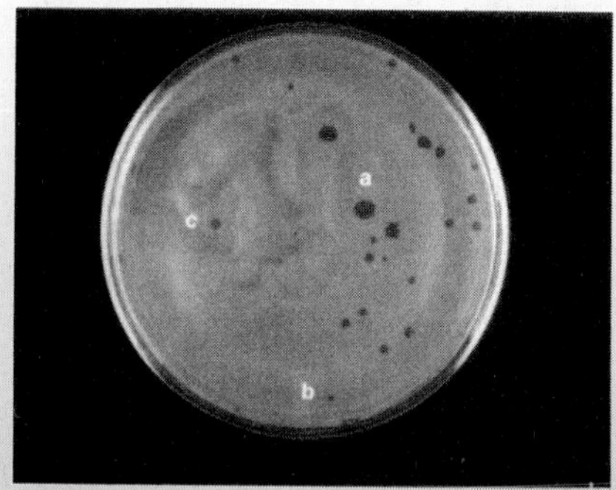

(e)

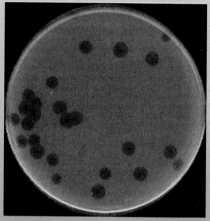

(f)

acid profiles from environmental samples as sufficient evidence for the existence of new microbial species (Amann *et al.*, 1995; Chung *et al.*, 1997; Hugenholtz *et al.*, 1998b). This practice has been controversial, particularly when inferences about ecological functions are made regarding organisms identified only by molecular sequence signatures. Many microbiologists still consider cultivation as the gold standard in assessing functional characteristics of microbial diversity (Palleroni, 1997; Watve and Gangal, 1996). The advocates of cultivation readily admit that the current status of microbial cultivation under laboratory conditions leaves much to be desired with respect to the assessment of species composition and their activities in environmental samples. It is generally assumed that much less than 10% of existing microbial diversity in aquatic, terrestrial, and biological ecosystems can be accounted for by cultivation under laboratory conditions (Barer *et al.*, 1993; Barer and Harwood, 1999; Kell *et al.*, 1998). For organisms that are culturable, there is incontrovertible evidence of considerable differences between their activities in nature and their activities under laboratory conditions where colony formation is encouraged. For example, the **diazotrophic** (capable of using atmospheric nitrogen as the only source of nitrogen for growth) cyanobacterium *Trichodesmium* species is capable of forming colonies under laboratory conditions which significantly enhances the organism's ability to conduct reactions which are dependent on nitrogenase, an oxygen-sensitive enzyme. This organism is

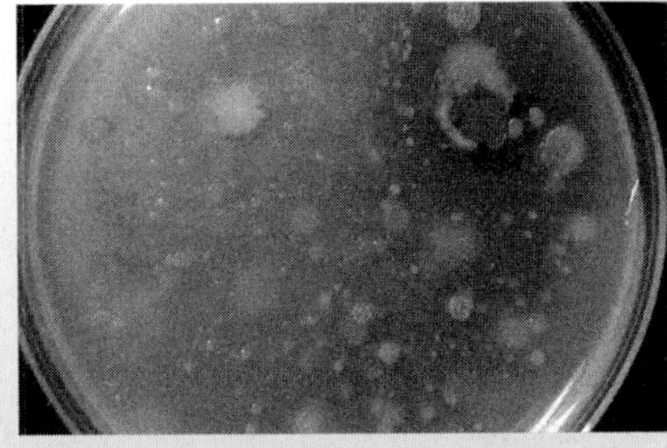
(c)

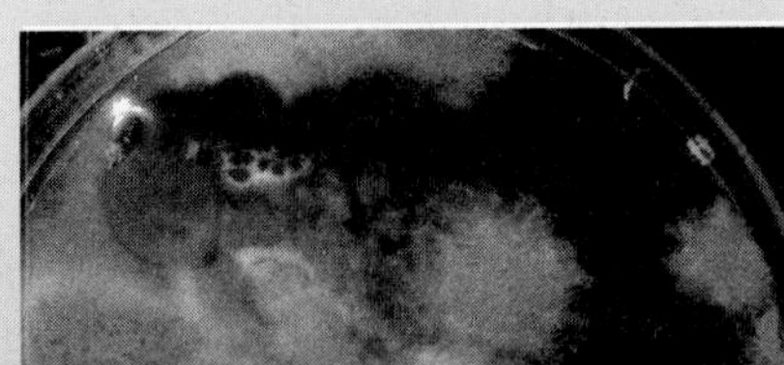
(d)

Colony morphologies for bacteria (c) and filamentous fungi (d) cultivated on agar plates are useful for preliminary assessment of microbial diversity. Similarly, plaque morphologies exhibited by viruses cultivated on uniform lawns of microorganisms are useful for the preliminary assessment of viral diversity (turbid plaques in (e); clear plaques in (f)). Images are from the author's collection ((c), (d), and (e)), and Drew Endy, Molecular Sciences Institute, Berkeley, CA (f).

(g)

Martinus Beijerinck (1851–1931; (g)) pioneered the isolation and cultivation of microorganisms that play central roles in the biogeochemical cycling of elements. For example, he cultivated the first sulfate-reducing bacterium, *Desulfovibrio desulfuricans*. He also used the now popular enrichment culture technique for obtaining a pure culture of the nitrogen-fixing root nodule bacterium, *Rhizobium* species (Beijerinck, 1888). Image of Beijerinck is reproduced with permission from Lesley Robertson, Curator of the Delft School of Microbiology Archive at Delft University of Technology.

(h)

Sergei Winogradsky (1856–1953; (h)) pioneered the studies of microbial cultures in microcosm models that closely simulate natural biotic and abiotic environmental conditions. He invented the "Winogradsky column" as a model for investigating microbial succession in response to nutritional demands in a community of organisms. Winogradsky also established the concept of autotrophy and its consequence for natural geochemical cycles. His investigation of *Beggiatoa* species helped to demonstrate the use of hydrogen sulfide as an energy source, and carbon dioxide as a carbon source (Winogradsky, 1890). Photograph of Winogradsky taken by H. Manuel is reproduced by permission, ©Institut Pasteur, Paris, France.

(i)

Rita Colwell (1934–; (i)) reinvigorated research into the limitations of culture methods for assessing microbial diversity by demonstrating the ability of many routinely culturable microorganisms to enter into a viable but not culturable phase depending on environmental factors. Image of Colwell is reproduced with her permission.

frequently observed individually (as single trichomes) in marine environments. Therefore, laboratory colony-based assessments are likely to overestimate its contribution to nitrogen cycling in nature (Letelier and Karl, 1998).

Many attempts have been made to increase the sensitivity of techniques developed for isolating and cultivating microorganisms from natural environments (Biotol, 1992; Jaffal *et al.*, 1997; Jensen *et al.*, 2001; Tanner, 2002). Most techniques attempt to provide balanced nutritional requirements such as a carbon source and trace elements (Atlas, 1995 and 1997). It is also necessary to optimize environmental parameters that influence microbial growth, including temperature, pH, oxygen tension, and humidity. Even when these nutritional and environmental conditions are optimized for a particular species, it is usually not possible to reproduce the biotic environmental factors which provide an important context for microbial proliferation in climax microbial communities (Balows *et al.*, 1991; Benlloch *et al.*, 1996; Caldwell *et al.*, 2002). Several microbial species are known to require growth factors produced by proximal but different species (Salyers and Whitt, 2001). In addition, the presence of different species in very close proximity may generate highly varied environmental

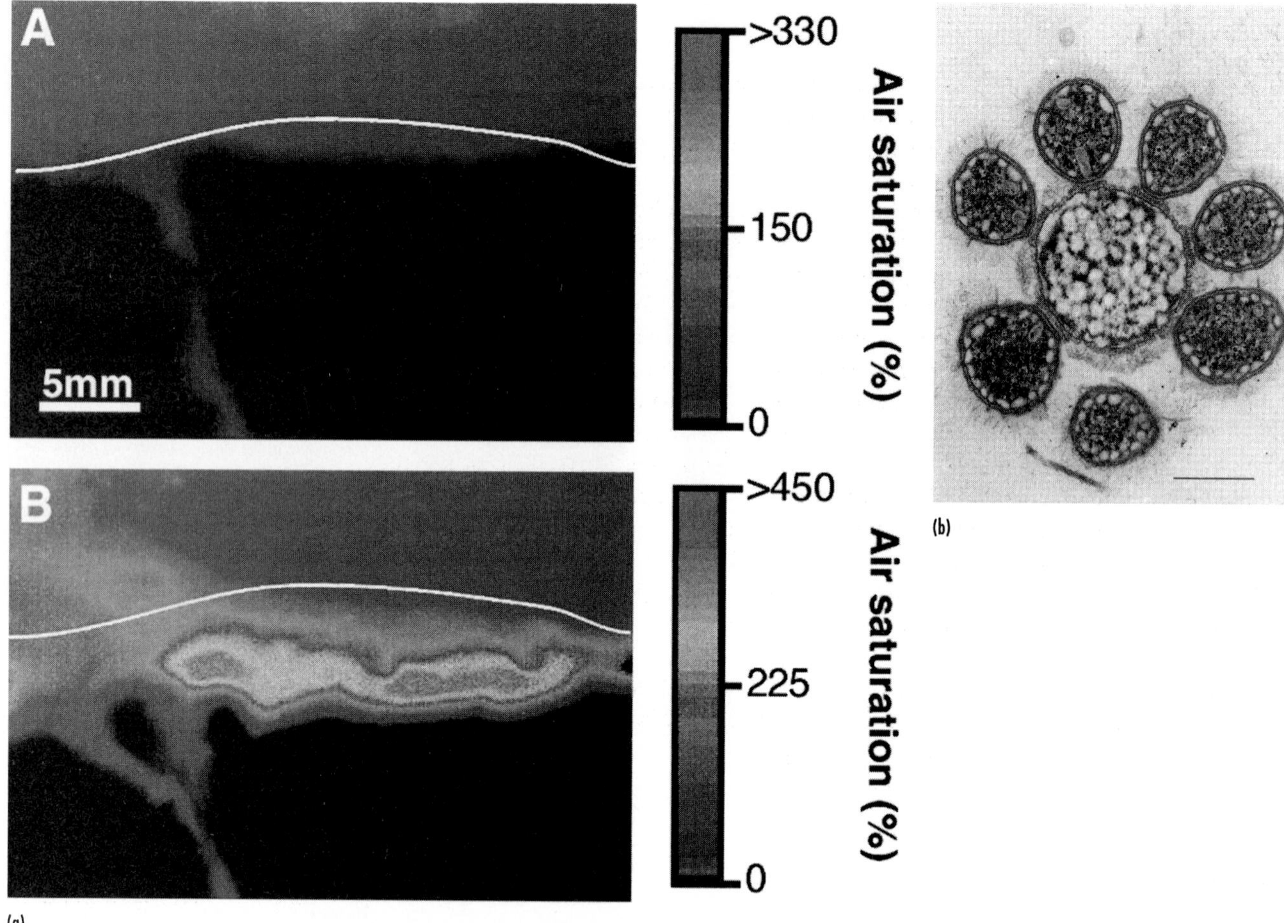

Fig. 3.1 Spatial variation in environmental factors exerts considerable influence on the distribution of microbial diversity across extremely small distances. In (a) Fenchel (2002) demonstrated very steep oxygen gradients ranging from complete lack of oxygen to supersaturation at 300% over a distance of 2 mm in a vertical section of the upper 2 cm of a 1-m deep marine sediment at night (A) and during the day (B). In (b) the importance of cross-species interactions for microbial nutrition is demonstrated by the association between seven photosynthetic green sulfur bacterial cells at the periphery of a single cell of an anaerobic heterotrophic bacterium. The bar is 0.5 μm long. Image is by permission of Tom Fenchel and Ken Clarke with CEH-Windermere, UK (Fenchel, 2002).

conditions at very small spatial scales. These variations may exert considerable influence on the diversity of organisms in different habitats (see Fig. 3.1). The lack of detailed information about the structural and functional components of natural microbial communities creates a difficult challenge for attempts to reproduce biotic and abiotic environmental conditions in a manner that could facilitate the discovery of new species (Christensen *et al.*, 2002). Therefore, inferences about relative species abundance in natural ecosystems should not be made on the basis of data gathered through cultivation methods under laboratory conditions (Boschike and Bley, 1998; Dupin and McCarty, 2000). Martinus Beijerinck (1851–1931) and Sergei Winogradsky (1856–1953), who pioneered culture methods for assessing natural microbial communities, expressed similar reservations. These reservations have been alleviated to some extent by the development of innovative techniques for accessing non-culturable microorganisms (see Box 3.1). In the following sections, overviews of the traditional and innovative techniques are presented from the perspective of their respective contributions to the investigation of microbial diversity.

AXENIC CULTURES

An axenic culture is defined as a homogeneous population of microbial cells deriving from a single species (Cummings and Relman, 2002). Cultivation in axenic or pure cultures provides unique opportunities for investigating microbial physiological and genetic characteristics without the restrictions that would ordinarily be imposed by low population densities in natural environmental samples. However, very little about microbial ecology and diversity has been glimpsed from axenic cultures because heterogeneous multicellular existence predominates microbial life in nature (Shapiro and Dworkin, 1997). Studies on the diversity of microbial forms and functions have revealed important information on colony morphologies, which provide a certain level of confidence in systematic classifications. However, the recovery of microorganisms and the characteristics of colony morphologies, such as shape, texture, color, and size, are known to change according to the composition of growth media and incubation conditions (Stewart *et al.*, 1995). In addition, axenic cultures are of limited use in assessing microbial diversity because of the small proportion of microbial species that have been cultivated, and the extensive variation in cultivability across phylogenetic categories (Fig. 3.2). Nevertheless, axenic cultures are required to produce cells in sufficient numbers for proper characterization of systematic features such as the microscopic analysis of cell-bound staining reactions, which can be used to rapidly distinguish between Gram-positive and Gram-negative bacterial species (Gupta, 2002). Similarly, axenic cultures are required for in-depth analysis of molecular profiles that can reveal much information about phylogenetic relationships among microorganisms (Von Wintzingerode *et al.*, 1997). Advances in understanding microbial nutritional requirements and natural growth conditions are required to improve methods for efficient recovery and cultivation of new and fastidious organisms present in many habitats.

Modeling microbial nutrition

The common assumption that microorganisms adapt well to both stable and fluctuating environmental conditions in their habitats suggests that it might be possible to reproduce physical and chemical aspects of habitats under laboratory conditions to support microbial growth. Therefore, the simulation of natural environments is frequently the first step in the isolation and cultivation of microorganisms. Molecular methods for assessing the impact of nutritional status on microbial biomass and community structure *in situ* have contributed to the development of model growth conditions to support laboratory cultivation (White, 1993). For free-living microorganisms, artificial media, including solid and liquid varieties have been invented to support the cultivation of bacteria and fungi. Obligate intracellular parasites such as viruses must be cultivated on their host cells, and their detection is facilitated by visible production of clear zones or plaques on monolayer lawns made up of host cells. Despite the remarkable feat of reproducing natural growth conditions for thousands of microorganisms, the recognition that there are many more thousands, perhaps millions, which cannot be successfully cultivated, is evidence for lack of comprehensive understanding of microbial nutrition as a complex integration of biotic and abiotic environmental parameters. For viruses, the identification of host cells and characterization of the host cell susceptibility conditions are not always sufficient for reproducing virus proliferation under laboratory conditions because of the difficulty encountered in modeling the interactions between two different kinds of organisms. However, for free-living organisms, including most prokaryotes, it is useful first to analyze the chemical composition of cells in order to model the environmental conditions that can support growth and proliferation.

In their normal vegetative growth state, free-living cells consist mostly of water (70–85% by weight), up to 15% of protein as enzymatic and structural molecules, 7% of polymeric material such as polysaccharides that make up the cell wall, 3% of lipids, 3% of RNA, and

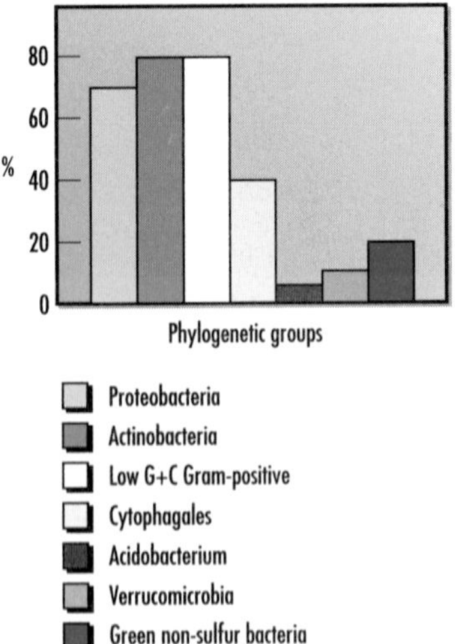

Fig. 3.2 The estimated proportion of bacterial species from various phylogenetic groups that have been cultivated probably reflects a bias in available microbial isolation and culture conservation techniques (Colwell, 1997). The estimates are based on methodological comparisons between the detection of molecular (16s RNA) signatures, DNA reassociation kinetics, and available bacterial cultures in depositories such as the American Type Culture Collection (ATCC). Data are from Hugenholtz *et al.*, 1998a.

1% of DNA. Furthermore, the elemental composition of microbial cells includes 50% carbon, 20% oxygen, 14% nitrogen, 8% hydrogen, 3% phosphorus, 1% sulfur, 1% potassium, 0.5% calcium, 0.5% magnesium, and 0.2% iron. Given this cellular composition, it seems plausible to satisfy growth requirements by careful preparation of culture media to support cell growth and proliferation. In reality, very small concentrations of limiting nutrients and growth factors that are not found in all environments are responsible for controlling the rate of microbial growth, and for selecting the kinds of species that can survive in a given environment.

Microbial trophic systems

Microbial growth is supported by three major categories of nutrients, namely energy sources, carbon sources, and trace elements. Energy sources include materials that generate adenosine triphosphate (ATP) through various forms of respiration or fermentation, depending on the relative availability of oxidizing and reducing agents. Carbon sources are needed for biomass construction, and trace elements are required for specialized biochemical activities to optimize, for example, enzymatic processes. Most growth media developed for cultivating microorganisms contain balanced quantities of water, carbon source, energy source, nitrogen, sulfur, phosphorus, potassium, magnesium, calcium, oxygenated molecules, trace elements and organic growth factors. Trace elements generally include metallic cofactors such as cobalt, copper, iron, manganese, molybdenum, and zinc. Additional trace elements including essential amino acids may be required for certain organisms that are not capable of synthesizing these compounds *de novo*.

Aeration

Prokaryotes have adapted to growth under different concentrations of atmospheric chemicals, and conversely, the consequences of microbial growth contribute to the modification of atmospheric components. Microorganisms continually change the Earth's atmosphere through their involvement in biochemical oxidation and reduction (redox) processes to support the energy-generating electron transport chain. Four categories of microbial interaction with molecular oxygen are recognized. These categories are informative about the diversity of niche specialization in microbial communities with respect to the co-evolution of life and the global environment. At one extreme are the obligate aerobes that require molecular oxygen at the concentrations found in contemporary atmosphere (about 20.9% at sea level). The pervasiveness of atmospheric gases suggests that obligate aerobic microorganisms are geographically widely distributed and metabolically diverse. Although the wide-spread distribution of aerobes has been verified by numerous studies, it is also true that the organisms at the other extreme, the obligate anaerobes for which molecular oxygen is actually toxic, are also widely distributed and metabolically versatile (Tanner, 2002). Obligate anaerobes have been discovered in both artificial and natural environments with limited access to atmospheric gases, or in situations where atmospheric oxygen is rapidly consumed by biological or chemical processes. Anaerobic organisms can also persist in aerated environments if they are not metabolically active, because the toxic byproducts of oxygen metabolism are produced only during active electron transport.

Between the extreme situations of obligately aerobic and anaerobic organisms are microaerophilic and facultatively anaerobic organisms. Microaerophiles require molecular oxygen for growth but exhibit the effects of oxygen toxicity at concentrations approaching contemporary atmospheric oxygen levels. Facultative anaerobes can grow both in the presence and in the absence of molecular oxygen because they are able to use alternative chemicals as electron acceptors in redox reactions (Tanner, 2002). One of the important differences between aerobes and anaerobes is the production of enzymes such as catalase

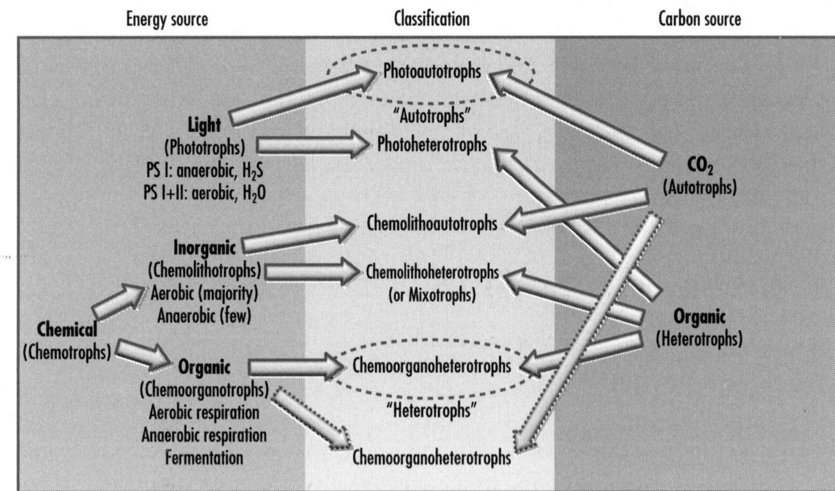

Fig. 3.3 Microorganisms can be classified according to nutritional categories based on preferences for carbon substrate and energy generation. Organisms equipped with molecular light-gathering arrays (Photosystem (PS) I or Photosystem (PS) II) can use sunlight as the energy source. Other microorganisms specialize in using either inorganic (lithothrophic) or organic (organotrophic) compounds. For carbon source, autotrophic organisms use carbon dioxide, and the others are heterotrophic. Diagram is reproduced by courtesy of Joseph Vanillo, Marine Biological Laboratory, Woods Hole, MA.

and superoxide dismutase used by aerobes for detoxifying the byproducts of oxygen metabolism, namely hydrogen peroxide and superoxide, respectively.

Carbon and energy sources

Microorganisms are distinguishable on the basis of their preferences for carbon substrates and energy sources. Autotrophism and heterotrophism refer to organisms' ability to use inorganic or organic forms of carbon, respectively; whereas phototrophism and chemotrophism refer to the ability to use light or chemicals as primary sources of energy, respectively (Zelenev *et al.*, 2000; See Fig. 3.3). Autotrophic microorganisms are at the base of the food chain because of their ability to use atmospheric carbon dioxide for synthesizing carbon compounds that are more complex than the substrate. Microorganisms that use light for generating energy and carbon dioxide for building biomass are "photoautotrophic". In contrast, heterotrophic microorganisms use non-carbon dioxide compounds as the source of carbon to support growth. Organisms that use light and non-carbon dioxide sources of carbon are "photoheterotrophic". Unlike the photoautotrophic microorganisms, photoheterotrophic microorganisms cannot use carbon dioxide as the primary source of carbon, and they cannot generate oxygen. "Chemoautotrophic" organisms use chemical compounds as electron donors for energy production, but they depend on carbon dioxide for building carbon-based biomass. These organisms tend to inhabit environments that are not well lighted, such as deep-sea vents where they constitute the primary producers (Deming and Baross, 2000). Chemoheterotrophic microorganisms use chemical compounds for energy production and organic compounds as the source of biomass-building carbon. For example, several chemoheterotrophic organisms can use glucose for both energy and carbon sources.

Chemotrophic microorganisms can grow by utilizing simple energy-laden organic compounds such as glucose to generate ATP. These organisms can also use more complex compounds such as lipids, proteins, unsaturated aliphatic carbohydrates, and aromatic compounds to support the production of energy needed for cell proliferation. In addition, simple inorganic chemicals such as hydrogen, sulfur, ammonia, and certain metals can also be used for driving ATP-producing reactions. Like plants, phototrophic microorganisms can use light as the source of energy. "Fastidious" organisms have complex nutritional requirements, and it is usually not possible to cultivate fastidious organisms on chemically defined media, where the quantities of all the ingredients are known. Instead, fastidious organisms are cultivated in complex media, which typically contains materials such as "yeast extract" with slightly different compositions depending on the preparation batch.

Selective growth conditions

Selective growth media or selective incubating conditions have been used successfully to narrow the scope of species recovered from a microbial community. Selectivity is based on the recognition that microorganisms adapt to physicochemical factors that are not necessarily nutritional (Wu and Chen, 1999). For example, many microorganisms can proliferate at the neutral pH of 7.0, but extremely "alkaliphilic" microorganisms can be selected by maintaining the pH of growth media at greater than 10.0 (Horikoshi and Akiba, 1982; Matthies *et al.*, 1997). At the other end of the pH tolerance spectrum, "acidophilic" microorganisms have been isolated from acid mine drainage pits and acidic lakes by adjusting the pH of the growth medium to 2.0. Selective microbial growth can also be controlled by adjusting incubation temperature, with the range of tolerance ranging from less than 0°C ("psychrophilic" organisms) to higher than 100°C ("thermophilic" organisms) (Zdanowski and Weglenski, 2001). Atmospheric pressure, salt concentration, antibiotics, and light conditions are other strategies for selecting microbial growth in axenic cultures. In the assessment of microbial diversity, it is not possible to reproduce all the possible environmental conditions to facilitate the recovery of all organisms. This problem, coupled with the dynamic nature of the interactions among various environmental parameters, leads to the general inadequacy of cultivation techniques, used alone, for comprehensive assessment of microbial diversity (Palumbo *et al.*, 1996; Van den Berg, 2001).

"Enrichment culture" methods have been used to create conditions that support the isolation of a narrow diversity of microorganisms in order to increase the likelihood of recovering microorganisms from a sparsely populated ecosystem (Hirsch *et al.*, 1979; Kataoka *et al.*, 1996). For example, Beijerinck (1888) originally isolated *Rhizobium* species from root nodules through enrichment by limiting nitrogenous substrates. Contemporary studies using the enrichment approach have focused on the discovery of species capable of metabolizing xenobiotic or anthropogenic chemical pollutants. These enrichment processes are typically referred to as "bioaugmentation" or "molecular breeding", and several examples have been reported in the literature (Burlage *et al.*, 1998; Cerniglia and Shuttleworth, 2002; Karthikeyan *et al.*, 1999; Ogunseitan, 1994; Vogel and Walter, 2002).

MICROCOSM CULTURES

Microbiological processes that are of consequence to global ecosystem functioning can be represented as balanced chemical reactions, which are catalyzed by enzymes. It is not unusual to isolate specific organisms from the environment in order to inquire whether their role in large-scale ecological processes can be reproduced in axenic cultures. Such experiments typically underestimate or overestimate the contribution of free-living microorganisms because their activities in nature are invariably influenced by the presence of other groups of organisms (Slater *et al.*, 1983; Staley *et al.*, 1997). Many attempts have been made to capture the integral activities of microbial communities through the use of special cultivation techniques (Christensen *et al.*, 2002). Microbial community cultures are different from mixed population cultures, which are produced by combining two or more pure cultures while excluding other organisms by incubating the mixture under aseptic conditions (Caldwell *et al.*, 2002). True community cultures accommodate interactions between variable biotic and abiotic factors that influence the microbial species composition of the culture, and the fate of individual species as the community develops.

Community cultures are typically maintained in microcosm vessels which are constructed for incubating natural samples in the laboratory. Microcosm cultures have enabled investigations on how specific changes in ecosystem processes are accompanied by changes in the diversity of microorganisms (Muller *et al.*, 2001; Radajewski *et al.*, 2000). The results of such investigations have contributed to the development of theoretical models used to describe interactions among different species. For example, the "competition model" stipulates that communities emerge through competition between individuals or groups of organisms. In contrast, the "cooperation model" suggests that communities arise through cooperative interactions that maximize the utilization of available resources and niche specialization. The "proliferation model" suggests that communities result from the growth of nested series of proliferating organisms rather than through specific competition or cooperation (Caldwell *et al.*, 2002). These theoretical models have not been thoroughly tested with respect to how well they describe various microbial communities. A notable exception is the reproducible use of the "Winogradsky column" to demonstrate the development and succession of aquatic sediment microbial communities (see Box 3.2). Practical application of data collected through such microcosms requires advanced knowledge of intrinsic microbial diversity in environmental samples, and detailed information on the environmental conditions that influence the development of the climax microbial communities (Liu *et al.*, 2001). Other microcosm strategies for cultivating microbial communities include the use of chemostats, nutristats, and microstats for continuous propagation of complex microbial communities (Veldkamp, 1976; Willke and Vorlop, 1994; Wimpenny, 1998).

SOMNICELLS AND MICROBIAL DIVERSITY ASSESSMENT

Several investigators have encountered the difficulty of culturing apparently viable microorganisms in various ecosystems. There is a growing realization that such "viable but nonculturable" (VBNC) microorganisms represent a common state of organisms in natural environments (Colwell and Grimes, 2000; Kaprelyants *et al.*, 1993). Other terminologies used to describe VBNC cells include "somnicells" and "dormant non-sporulating" microorganisms (Box 3.3). The existence of such cells is of concern for attempts to quantitatively estimate microbial diversity, but there is some controversy over the appropriate terminology to use, and the correct interpretation of encounters with non-cultivability of cells in natural microbial communities (Barer *et al.*, 1993; Barer and Harwood, 1999; Kell and Young, 2000; Kell *et al.*, 1998). Furthermore, there is a paucity of data on whether the VBNC phenomenon affects the estimates of diversity that have been presented for non-bacterial microorganisms (Azofsky, 2002; Colwell, 1996; Finlay *et al.*, 1996). The phenomenon is further complicated by some confusion about the difference between the VBNC state and the status of microorganisms that survive harsh environments through the induction of developmental changes that facilitate hardiness and dispersal (Kwaasi *et al.*, 1998; Mandeel *et al.*, 1995).

Several hypotheses have been proposed to explain the VBNC state for some, but not all, bacterial species that have been, or are readily, cultivated under different circumstances. Physical or chemical injury to cells, protracted periods of starvation, and extremely slow metabolism are among the leading hypotheses (Colwell and Grimes, 2000; McDougald

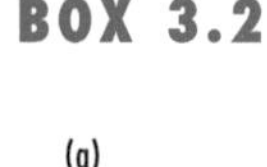

BOX 3.2

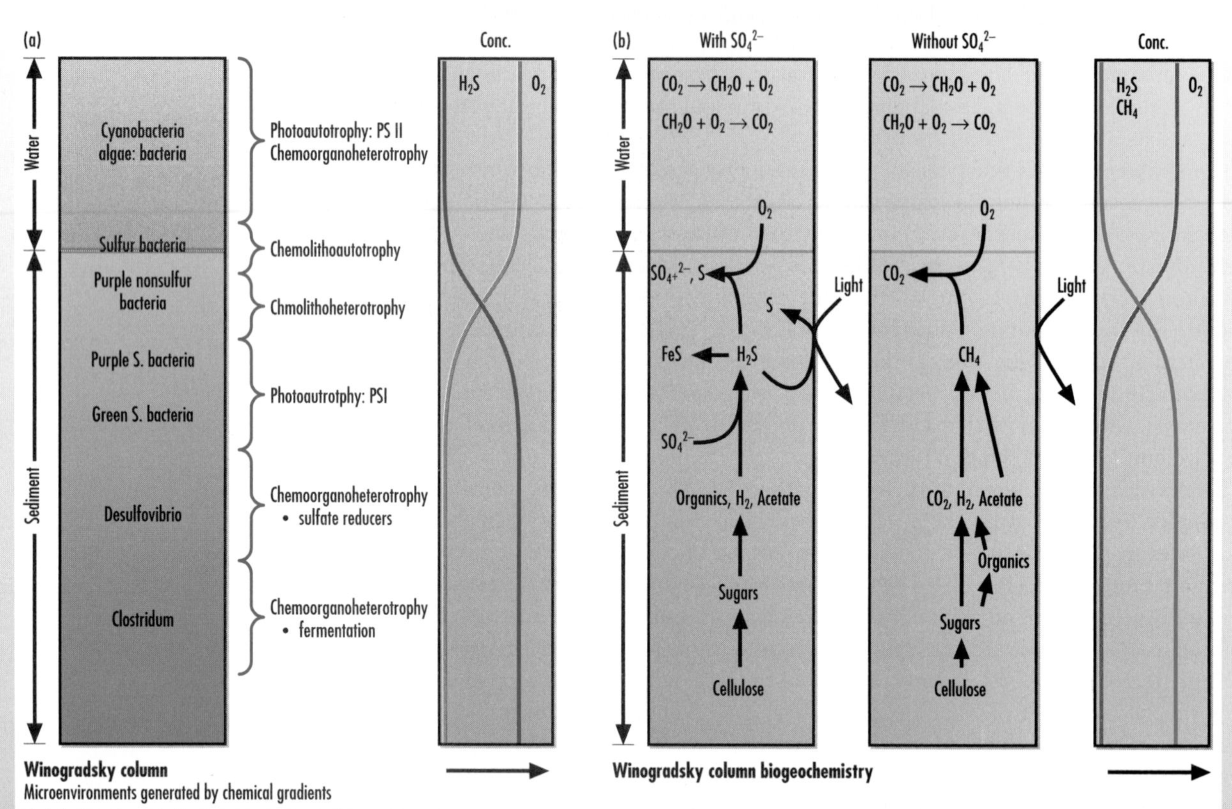

(c)

Microbes and processes in Winogradsky column

Aerobic Environment

- Algae and cyanobacteria (photoautotrophy using PS II)
- Bacteria and eukaryotes respiring (chemoorganoheterotrophy).
- Sulfide oxidizers: $H_2S + O_2 \rightarrow S$ or SO_4^{2-}
  - Some use CO_2 (chemolithoautotrophs),others use organic compounds (chemolithoheterotrophs)
  - Examples, *Thiobacillus* sp. And *Beggiatoa* sp.
- Methanotrophs: $CH_4 + 2O_2 \rightarrow CO_2 + 2H_2O$ (chemolithoheterotrophs)
  - Example, *Ralstonia* sp., *Pseudomonas* sp.

Anaerobic Environment

Fermentors (chemoorganoheterotrophs)

- Break down cellulose, etc. and ferment sugars into:
  - alcohols acetate
  - organic acids hydrogen
- Many bacterial groups can conduct fermentation, but not all of these have the ability to decompose polymeric compounds such as cellulose
- Example, *Clostridium* species

(d)

Anaerobic Environments

Sulfur compounds

- Sulfate reducers: use sulfate, $SO_4^{2-} + e^- \rightarrow S$ or H_2S, to oxidize organic compounds produced by fermentors, (chemoorganoheterotrophs).
  - Many genera of bacteria. Example, *Desulfovibrio* sp.
- Phototrophic bacteria: Use light and H_2S as electron acceptor (PS I) (photoautotrophs).
  - Examples, purple and green sulfur bacteria.

Methanogens and Acetogens

- Methanogens: $CO_2 + 4H_2 \rightarrow CH_4 + 2H_2O$ (chemolithoautotrophs)
  Acetate$^-$ + $H_2O \rightarrow CH_4 + HCO_3^-$ (chemoorganoheterotrophs)
  - Example: *Methanobacterium* (Archaea)
- Acetogens: $2CO_2 + 4H_2 \rightarrow CH_3COOH + 2H_2O$ (chemolithoautotrophs)
  - Example: *Homoacetogens*

(e)

Other possible microbes

Aerobic Environments

Hydrogen

- Hydrogne oxidizers: $H_2 + {}^1/_2O_2 \rightarrow H_2O$ (both chemolithoheterotrophs and chemolithoautotrophs). However, it is unlikely that H_2 will make it to the aerobic interface (it will be used in the anaerobic environment first)
  - Example, *Ralstonia eutrophus*

Iron

- Iron oxidizers: $Fe^{2+} + H^+ + {}^1/_4O_2 \rightarrow Fe^{3+} + {}^1/_2H_2O$ (chemolithoautotrophs)
  Occurs only at low pH (~2)
  - Example: *Thiobacillus ferrooxidans*

Ammonium

- Nitrifiers: $NH_3 + 1{}^1/_2O_2 \rightarrow NO_2^- + H^+ + H_2O$
  $NO_2^- + {}^1/_2O_2 \rightarrow NO_3^-$
  - Example: *Nitrosomonas* and *Nitrobacter*, respectively. Both chemolithoautotrophs

Anaerobic Environments

Nitrate

- Denitrifiers: $NO_3^- + 6H^+ + 5e^- \rightarrow {}^1/_2N_2 + 3H_2O$
  - Reaction combined with oxidation of organic matter.

Iron

- Iron reducers: Many organisms can utilize Fe^{3+} as electron acceptor.

The Winogradsky column has been used extensively to demonstrate prokaryotic metabolic diversity. The self-contained column processes are driven by light energy, and it illustrates the interdependence of different microorganisms in maintaining a balanced cycling of elements on a small scale. The glass columns, approximately 30 cm tall and 5 cm diameter, are filled up to one-third depth with lake or river sediment supplemented with a carbon source (e.g. cellulose paper), sodium sulfate, and calcium carbonate. The column is then filled to the top with environmental water samples. The tube is capped and placed near a window with supplementary light. Microbial community cultures develop over a period of 2–3 months of incubation during which cells originally present in low numbers proliferate, and niche specialization is observed through the development of distinct pigmentation zones. Within the column, it is possible to isolate microorganisms that represent each of the four basic forms of carbon and energy metabolism: photoautotrophs, chemoheterotrophs, photoheterotrophs, and chemoautotrophs (a). Figures (b)–(e) represent the biogeochemical reactions that take place in the column as conducted by the four major categories of metabolic activities. PS I and PS II represent photosystems I and II, respectively. Schematic diagrams are reproduced by courtesy of Joseph Vanillo, Marine Biological Laboratory, Woods Hole, MA.

BOX 3.3

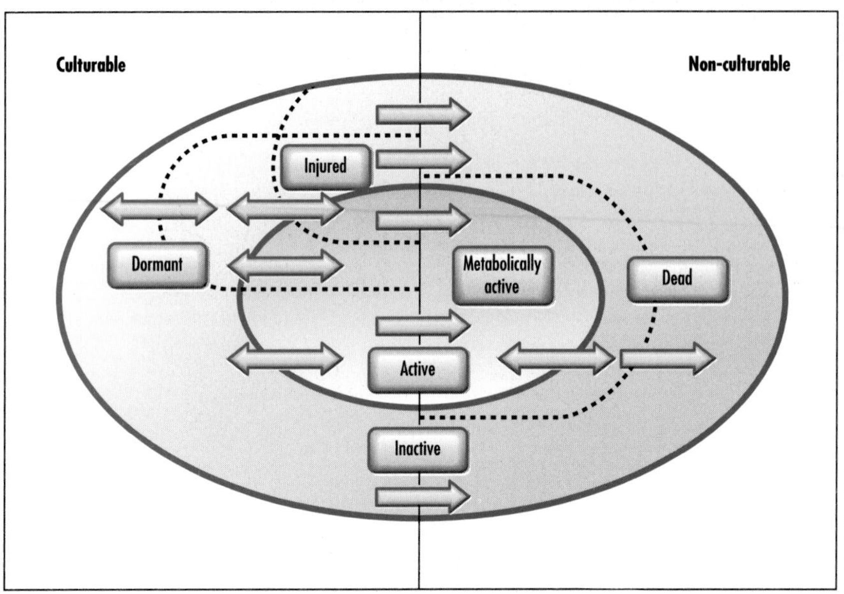

In nature, microbial cells can exist in one of four possible physiological states, which influence their ability to form visible colonies upon isolation and attempts at cultivation. Three of the four physiological states represent viable and culturable organisms which can exist in either **normal vegetative**, **dormant**, or **injured** states. Presumably, organisms in a normal vegetative state are metabolically active and proliferate in the environment. These organisms may even form colonies as observed under laboratory cultivation. The isolation, enumeration, or identification of organisms in the normal vegetative state simply requires the provision of adequate growth conditions, including nutrients, physical-chemical environmental parameters, and necessary ecological interactions. In the dormant state, which can be provoked by lack of nutrients or extreme environmental conditions, microorganisms can be resuscitated by returning them to normal environmental and nutritional conditions that support growth. In the injured state, cells are damaged by exposure to toxic chemicals or harsh environmental conditions, but they can be resuscitated in artificial growth medium after a protracted period of repair under supportive environmental conditions. The fourth state in which microbes exist in nature represents non-culturable microorganisms, which are demonstrably metabolically active. These organisms cannot be cultured either because the artificial growth medium is inadequate, or because a microbial consortium is necessary for growth. Non-culturable organisms can also be metabolically inactive and non-culturable, a state that is difficult to distinguish from being dead.

The concepts of VBNC microorganisms were developed from the perspectives of individual cells, and as such, the impact of the non-culturable state on the enumeration of the population density of known organisms is considerable. This is because in a given population of cells belonging to a single species, it is highly probable that a fraction of the total number of cells will not form colonies upon being subjected to culture methods. However, very little information has been contributed by the VBNC concepts to the recovery of new species that have never been cultured because every single cell belonging to that putative species is in a non-culturable state. In those cases, the most likely reason for the relatively small fraction of microbial species recovered by culturing, compared to the information available on genetic diversity, is that the culture methods are inadequate or that some microbial cells exist in a natural state of symbiotic interaction with cells belonging to other species such that individual cells cannot form independent colonies. This phenomenon may partially explain the inability of enrichment culture techniques pioneered by Martinus Beijerinck (1851–1931) to demonstrate a geographical dimension of microbial diversity (Azofsky, 2002; Finlay, 2002). Diagram is based on a model described by Kell and Young, 2000.

et al., 1998 and 1999; Relman, 1999). Comparative analysis of cultivation and molecular methods (e.g. 16S RNA sequence profiles) is usually necessary to distinguish between organisms that cannot be cultivated because of injury as opposed to organisms that have never been cultivated (Feldman and Harris, 2000; Dunbar *et al.*, 1999). In addition, flow cytometry and cell sorting can be used to determine the proportion of a microbial population that has lost the ability to grow on artificial media, although this technique is generally feasible only for relatively simple environmental samples with low organism population density (Vesey *et al.*, 1994). Research on techniques for "awakening" culturable organisms is important for improving the understanding of microbial physiology under natural environmental conditions. To expand the knowledge base for microbial diversity, it is essential

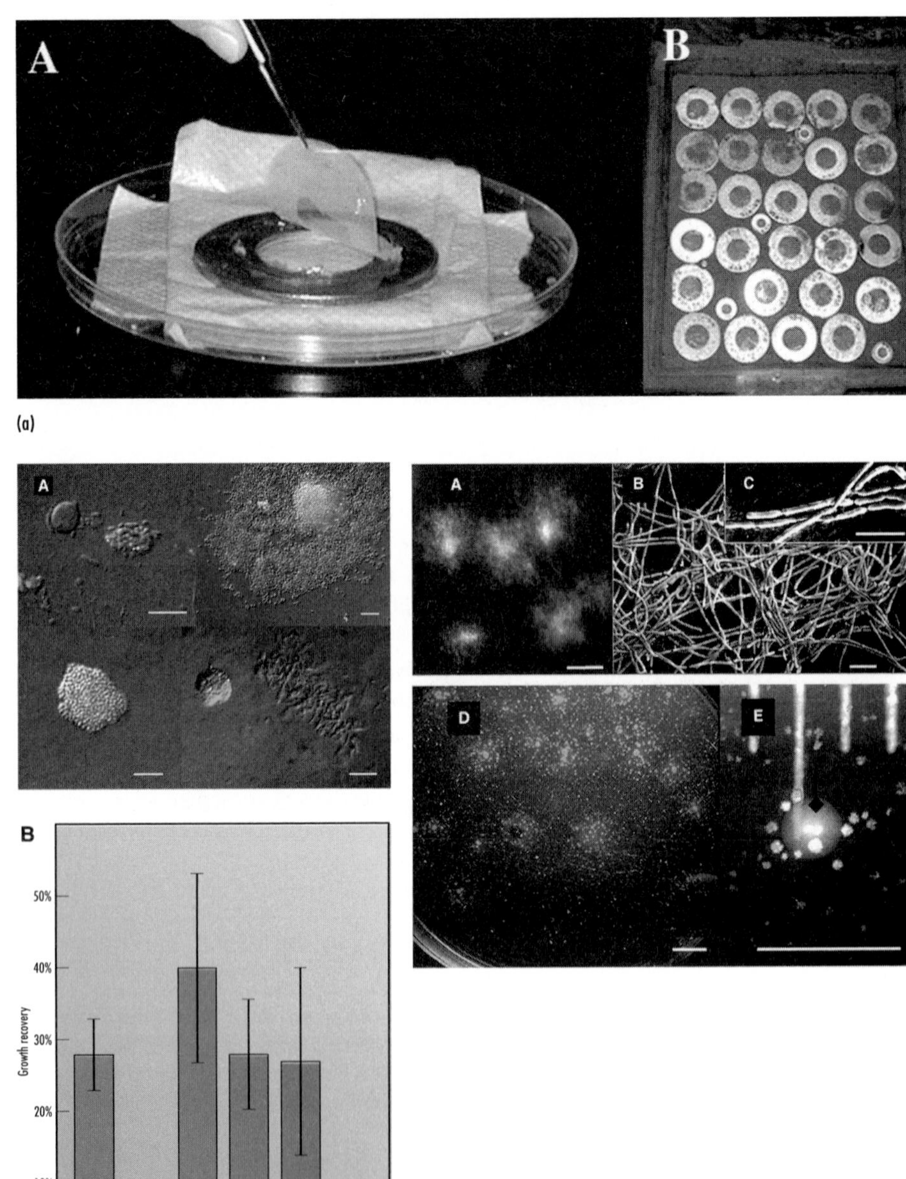

Fig. 3.4 The use of diffusion growth chambers for *in situ* cultivation of environmental microorganisms promises to reduce the number of instances where the "viable but non-culturable" explanation is used for the poor characterization of microbial diversity. The diffusion chamber is formed by a washer that is placed between two polycarbonate filters with pore size of 0.03 micrometers (frame A in (a)). Several diffusion growth chambers can be incubated in an array format directly in the environment to recover microbial colonies on the filters (frame B in (a)). The diffusion chamber approach has been used to recover new microbial species from marine sediments (Kaeberlin *et al.*, 2002). Microscopic observation of representative colonies that developed on the filters shows considerable diversity of colony morphology (frame A, (b), right). The VBNC phenomenon in this particular case may be influenced by seasonal variation as demonstrated by growth recovery in the diffusion chambers over a period of several months (frame B, (b), right). Elongated cells in the colony of new marine sediment isolate identified as MSC1 is shown in frame C. MSC1 colonies formed on casein agar only in the presence of another species, MSC2 (frame D, magnified view in frame E). Images are reproduced by courtesy of Slava Epstein and permission of the American Association for the Advancement of Science.

to develop techniques for recovering microorganisms that have never been cultivated in the laboratory.

An innovative approach for "awakening" previously uncultured species is the use of diffusion chambers that were invented for simulating natural nutritional and environmental conditions in the laboratory. When inoculated with whole environmental samples, the diffusion chamber can overcome some of the limitations due to the inability to replicate natural growth conditions that support colony formation in the laboratory (Kaeberlein *et al.*, 2002; Fig. 3.4). Most species recovered by means of the diffusion chamber cannot exist as pure cultures on artificial media, but they are capable of forming colonies in the presence of other organisms. This observation strongly suggests that the VBNC phenomenon can be explained in part by the requirement for biologically produced growth factors that are not available in most synthetic growth media.

CONCLUSION

The growth of microorganisms from invisible single cells to visible macro-colonies requires the coordination of several events, including genetic expression, numerous enzymatic activities, and various ecological interactions with neighboring cells and responses to environmental cues. The achievement of microbial cultivation under laboratory conditions is a tribute to the depth of available knowledge on microbial physiology and ecology. However, only a small fraction of existing microorganisms have been successfully cultured in the laboratory. Therefore, culture methods must be used in combination with direct techniques including the analysis of nucleic acids and proteins to provide comprehensive information on the diversity of microorganisms in natural ecosystems. The next chapter deals primarily with new developments in the molecular level assessment of microbial diversity.

QUESTIONS FOR FURTHER INVESTIGATION

1 It is generally assumed that greater than 90% of existing microbial species have never been cultivated in the laboratory, and presumably, members of the uncultivated majority grow to reach population densities that make them visible as colonies under natural conditions. Write an essay in which two lines of argument are developed, in support of, and against, the relative importance of uncultivated microorganisms for global ecosystem functions such as the biogeochemical cycling of elements.

2 Imagine a research project aimed at exhaustive characterization of culturable microbial diversity in (a) 1 gram of soil, (b) 1 liter of freshwater, and (c) 1 cubic meter of urban aerosol. Based on a compendium of microbiological growth media (e.g. Atlas, 1995 and 1997), tabulate the kinds of selective, non-selective, and enrichment media that will be necessary to produce the best estimate of diversity in each environmental sample. Does your analysis reflect a bias in the development of culture techniques?

SUGGESTED READINGS

Atlas, R.M. 1995. *Handbook of Media for Environmental Microbiology.* Boca Raton, FL: CRC Press.

Barer, M.R. and C.R. Harwood. 1999. Bacterial viability and culturability. *Advances in Microbial Physiology*, **41**: 93–137.

Caldwell, D.E., G.M. Wolfaardt, D.E. Korber, S. Karthikeyan, J.R. Lawrence, and D. Brannan. 2002. Cultivation of microbial consortia and communities. In C.J. Hurst, R.L. Crawford, G.R. Knudsen, M.J. McInerney, and L.D. Stetzenbach. *Manual of Environmental Microbiology*, pp. 92–100. Washington, DC: American Society for Microbiology Press.

Colwell, R.R. 1997. Microbial diversity: The importance of exploration and conservation. *Journal of Industrial Microbiology and Biotechnology*, **18**: 302–7.

Colwell, R.R. and D.J. Grimes. (eds.) 2000. *Nonculturable Microorganisms in the Environment.* Washington, DC: American Society for Microbiology.

Fenchel, T. 2002. Microbial behavior in a heterogeneous world. *Science*, **296**: 1068–71.

Hugenholtz, P., B.M. Goebel, and N.R. Pace. 1998. Impact of culture-independent studies on the emerging phylogenetic view of bacterial diversity. *Journal of Bacteriology*, **180**: 4765–74.

Kaeberlein, T., K. Lewis, and S.S. Epstein. 2002. Isolating "uncultivable" microorganisms in pure culture in a simulated natural environment. Science, **296**: 1127–9.

Kataoka, N., Y. Tokiwa, Y. Tanaka, K. Takeda, and T. Suzuki. 1996. Enrichment culture and isolation of slow-growing bacteria. *Applied Microbiology and Biotechnology*, **45**: 771–7.

Kell, D.B., A.S. Kaprelyants, D.H. Weichart, C.R. Harwood, and M.R. Barer. 1998. Viability and activity in readily culturable bacteria: A review and discussion of practical issues. *Antonie van Leeuwenhoek*, **73**: 169–87.

McDougald, D., S.A. Rice, D. Weichardt, and S. Kjellerberg. 1998. Non-culturability: Adaptation or debilitation? *FEMS Microbial Ecology*, **25**: 1–9.

Palleroni, N.J. 1997. Prokaryotic diversity and the importance of culturing. *Antonie van Leeuwenhoek*, **72**: 3–19.

Stewart, S.L., S.A. Grinshpun, K. Willeke, S. Terzieva, V. Ulevicius, and J. Donnelly. 1995. Effect of impact stress on microbial recovery on an agar surface. *Applied and Environmental Microbiology*, **61**: 1232–9.

Tanner, R.S. 2002. Cultivation of bacteria and fungi. In C.J. Hurst, R.L. Crawford, G.R. Knudsen, M.J. McInerney, and L.D. Stetzenbach (eds.) *Manual of Environmental Microbiology*, pp. 62–70. Washington, DC: American Society for Microbiology Press.

Van den Berg, H.A. 2001. How microbes can achieve balanced growth in a fluctuating environment. *Acta Biotheoretica*, **49**: 1–21.

Watve, M.G. and R.M. Gangal. 1996. Problems in measuring bacterial diversity and a possible solution. *Applied and Environmental Microbiology*, **62**: 4299–301.

CHAPTER 4 MOLECULAR AND GENOMIC METHODS

Chapter contents

INTRODUCTION

The ultimate determinants of microbial diversity are the molecules which, by virtue of their fixed structures and variable functions, can generate and maintain differences between lineages of microorganisms. Therefore, methods designed to evaluate the relationships among these molecules are directly relevant to the recognition of diversity and the classification of newly discovered species. This chapter presents an overview of new molecular approaches for investigating microbial communities from the perspective of diversity assessments. The strengths and limitations of these methods are explored through comparisons with microscopic and cultivation methods. Ideally, molecular methods are best used to complement other methods. The discussion focuses specifically on:

1 The relationship between the molecular composition of microbial cells and phenotypic features that are used in traditional systematics.
2 The concept of nucleic acid-based approaches for assessing microbial diversity, and the interpretation of data generated by these methods.
3 The concept of fatty acid-based approaches for assessing microbial diversity.
4 The concept of protein-based techniques for assessing microbial diversity.

THE MOLECULAR CONTEXT OF MICROBIAL DIVERSITY

Unlike other branches of ecology, the theoretical discord between the fundamental approaches of reductionism and holism has not produced a seemingly irreconcilable schism in the discipline of microbial ecology. The substantive issues that divide these scientific worldviews do exist in microbiology; however, as the primary domain of molecular biology, there is a near unanimous acceptance of the tenet that meaningful answers to important questions can be gained through exclusive analyses of the molecular components of microbial cells. The molecular approach has been enormously powerful for approaching key questions in microbial physiology and genetics at the single-species level. The analysis of multi-species microbial communities required further innovation in methodological tools to facilitate the proper application of molecular techniques. For example, the variable nature of microbial sensitivity to cell lyses techniques, differences in the tenacity of cellular adsorption to abiotic particles, and the disproportionate representation of individual species in microbial community gene pools, all posed specific but surmountable challenges for the reproducible application of routine molecular techniques to natural microbial communities (Young and Levin, 1992).

The advent within the past three decades of molecular tools for investigating microbial

BOX 4.1

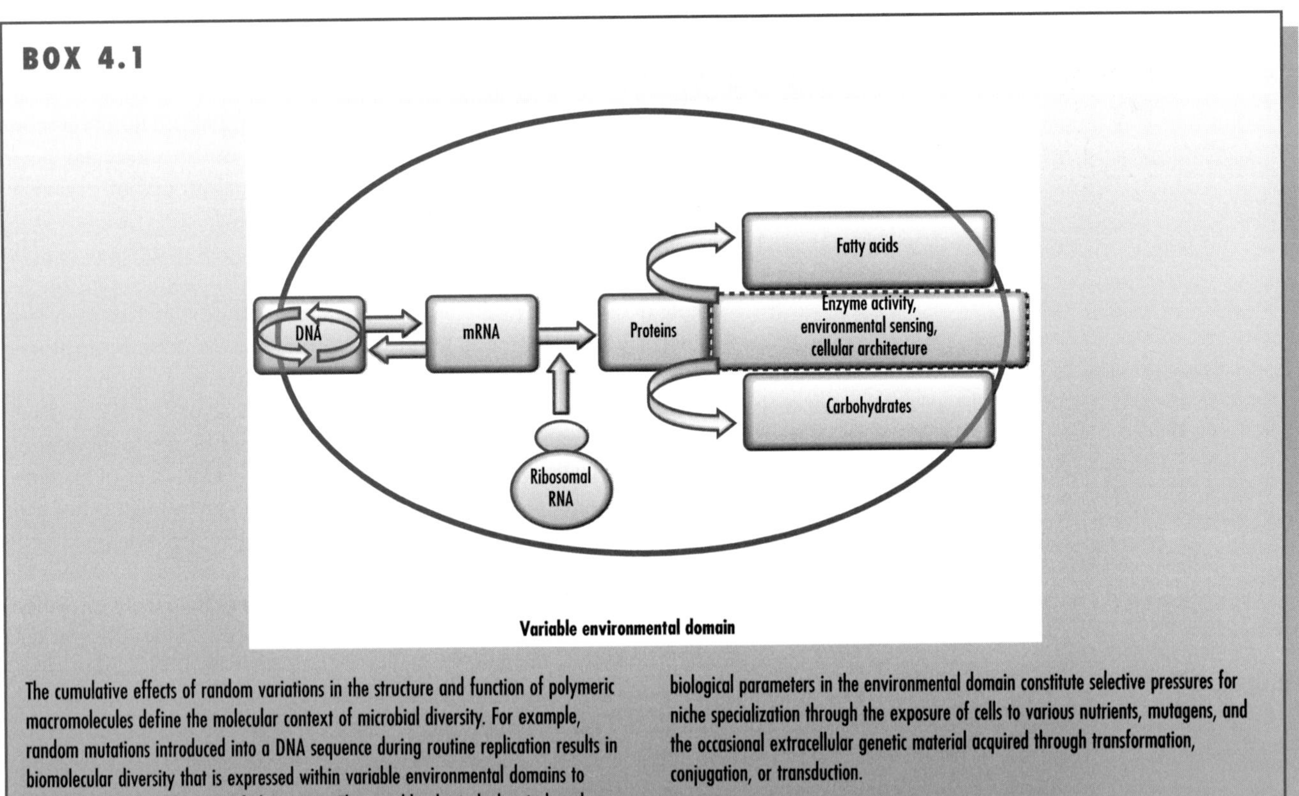

The cumulative effects of random variations in the structure and function of polymeric macromolecules define the molecular context of microbial diversity. For example, random mutations introduced into a DNA sequence during routine replication results in biomolecular diversity that is expressed within variable environmental domains to produce the observable array of phenotypes. The variable physical, chemical, and biological parameters in the environmental domain constitute selective pressures for niche specialization through the exposure of cells to various nutrients, mutagens, and the occasional extracellular genetic material acquired through transformation, conjugation, or transduction.

diversity has led to a cadre of specialists trained exclusively to analyze nucleic acids, lipids, or proteins. Unfortunately, narrow "molecular" specialization may have slowed the pace at which the relatively slippery microbial species concept may finally be settled (Liu and Stahl, 2002). The elucidation of the major driving processes behind microbial speciation, and the relationship between molecular diversity and functional diversity may have been slowed by the dominance of new and practical but non-comprehensive methodological approaches (Cain and Provine, 1992). The pace may have been slow, but it is difficult to envisage a path different from molecular assessment that could have produced the vast amount of new information on the structure and organization of microbial communities (e.g. Reysenbach and Shock, 2002). Developments in molecular microbial ecology have provoked new questions about microbial diversity and about the appropriate level of analysis within the "central dogma" of molecular biology upon which the concept of microbial speciation can be rooted (Box 4.1). Regardless of the level of analysis and the independence of specialized research approaches, there is general agreement that robust answers to questions regarding various dimensions of microbial diversity will require coincident assessment of different biochemical molecules. Knowledge of the relative abundance of molecules unique to specific phylogenetic groups, and the clarification of exchange mechanisms that regulate their distribution are needed to support inferences about the sensitivity of microbial diversity to environmental pressures and the process of natural selection (Kemp, 1994).

Interpretation of molecular diversity

In a landmark publication, Emile Zukerkandl and Linus Pauling proposed the use of polymeric biomolecules, specifically nucleic acids and proteins, to reconstruct evolutionary distances and phylogenetic relationships among organisms. The proposal was based on the recognition that the cumulative effect of random mutations, which occur as a result of infidelity in the molecular aspects of genetic replication, produces distinct lineages of organisms that may be tracked chronologically (Zukerkandl and Pauling, 1965). This observation led directly to the invention, calibration, and subsequent revisions of the "molecular clock" that has been used extensively for phylogenetic analysis of several groups of organisms including bacteria, fungi, viruses, and microscopic eukaryotes (Ayala, 1997; Moran *et al.*, 1993; Omland, 1997; Otsuka, 1999; Scherer, 1990). The characterization of differences in the sequence of molecules that form polymeric cell components in different organisms has progressed from simple descriptions to the level of predictive assessments based on large collations of phylogenetically conserved and unique sequence databases. It is not surprising, however, that not all seriously proposed molecular sequence information is relevant for phylogenetic analysis. The shortcomings of "molecular clocks" include lateral genetic exchange among distantly related organisms, limited penetration of molecular sequences in different branches of the universal phylogenetic tree, and disproportionately high rates of mutation in certain sequences due to sporadic integration of exogenous fragments associated with mobile genetic elements such as transposon, plasmids, and viral genomes (Ochman *et al.*, 1999; Ogunseitan, 1995; Liu and Stahl, 2002).

To establish the link between molecular diversity and the observable phenotypic differences among species, it is necessary to define limits for each category of molecular sequence data. For example, the redundancy inherent in genetic transcription and translation means that a certain level of difference is tolerated at the nucleotide sequence level among individuals of the same species. Conversely, differences at the level of protein structure and function typically produce differences in ecological fitness among organisms inhabiting similar environments. These molecular differences are, presumably, the fundamental determinants of speciation and niche diversity in prokaryotic populations. Large differences in protein function also relate to differences in the structures and functions of secondary macromolecules such as fatty acids and carbohydrates that have also been compiled and used successfully for assessing diversity in microbial communities. Methods for extracting and analyzing polymeric molecules in microbial communities have been the driving force behind strategies for assessing microbial diversity (e.g. Bintrim *et al.*, 1997). These methods are at different stages of development, and in many cases, they have not been used concurrently for investigating the linkages between microbial diversity and ecosystem function.

NUCLEIC ACID SEQUENCE COMPARISONS

The DNA content of microbial cells exists in various forms including single unenclosed chromosome(s) with or without satellite plasmid(s) in prokaryotes, and nucleated and mitochondrial genomes in microbial eukaryotes. Viruses contribute extracellular "mobile" DNA to both prokaryotic and eukaryotic microorganisms, and in certain cases, the phenomenon of lysogenic conversion resulting from resident viral genomes produces phenotypic differences among otherwise identical host strains. There is no consensus on the contribution of such discrete and "mobile" sources of genotypic and phenotypic differences to the process of speciation in microorganisms (Ogunseitan, 1995).

DNA-based approaches for assessing microbial diversity have relied in part on the complementary nature of the double-helix structure of DNA molecules, and on the efficiency of hybridization between two strands from heterogeneous sources under controlled experimental conditions. There are many variations on the methodological theme of DNA hybridization as a fundamental approach for exploring microbial diversity and systematics (DeLong and Pace, 2001). For determining species identity among culturable organisms, the most straightforward strategy is whole-genome hybridization with results expressed as percent homology between an unidentified organism and one that is well characterized. The proportion of homology between two genomes should be used more as a criterion for including or excluding organisms from a broadly defined taxon rather than for unequivocal assignment into the taxonomic categories of genus or species. For example, a score of 70% or more in whole-genome DNA hybridization homology is considered adequate for including an unknown isolate within a species category. The difference in melting temperatures (ΔT_m) for DNA extracted from closely related species should not be more than 5°C (Liu and Stahl, 2002; also see Chapter 1, Box 1.2). Microbial genomes cover a wide range of sizes. For example, chromosomes can be as small as 0.58 Mb in *Mycoplasma genitalum*, an obligate intracellular parasitic bacterium that lacks a distinct cell wall, or as big as 7.59 Mb in *Mesorhizobium loti*, a nitrogen-fixing endosymbiont inhabiting several *Lotus* plant species. The sequence of nucleotides has been determined for an increasing number of species (see Appendix). The resulting sequence databases have substantially enhanced the performance of rapid comparisons of sequence distribution among microorganisms, with the goal of identifying phylogenetically conserved or unique varieties.

Several interactive programs have been developed to facilitate the comparative analysis of microbial genomes for which complete DNA sequence information is available. One of the most comprehensive programs is entitled "Microbial Genome Database (MBGD) for Comparative Analysis", which is managed by the Japanese National Institute for Basic Biology (http://mbgd.genome.ad.jp/). The aim of MBGD is to facilitate comparative genomics from various points of view such as orthologue identification, paralogue clustering, motif analysis, and gene order comparison (see Box 4.2 and Box 4.3). The program uses "function-level" categories that are based on those assigned by the extremely prolific genome sequencing research organization, The Institute for Genomic Research (TIGR; http://tigrblast.tigr.org/). However, correspondence between the original and reference categories may not align perfectly, especially when different classification policies are used in different organisms.

Specific nucleic acid-based methods

Insight into species diversity and ecological interactions within microbial communities has been acquired through the application of techniques based on the complementary nature of the double-helix structure of DNA (Ogunseitan *et al.*, 1992; Tebbe *et al.*, 2002). The recognition of single-stranded RNA molecules by partially complementary strands of DNA to form RNA–DNA hybrid molecules has also contributed to molecular techniques available for assessing diversity (Dockendorff *et al.*, 1992; Matheson *et al.*, 1997). These hybridization and reassociation reactions have been enhanced by advances in polymerase chain reaction (PCR) and fine-scale sequencing to provide information on the level of relatedness among well-characterized species and on the composite characteristics of microbial community gene pools (Liu and Stahl, 2002).

Single-stranded nucleic acid probes labeled with radioactive or fluorescent substrates have been instrumental in tracking the spatial and temporal distribution of specific microorganisms across geographic gradients in many environments (Amann *et al.*, 1996b; Beja *et al.*, 2002; Brosch *et al.*, 1996; Neilan, 1996). These assessments advanced knowledge on the relative abundance of microbiological niches involved in ecosystem processes such as nutrient cycling, toxic chemical biodegradation, and biological population dynamics (Buchholz-Cleven *et al.*, 1997; Eder *et al.*, 2001). There are several variations on the theme of "nucleic acid sequence analysis and hybridization", which have facilitated the measurement of microbial diversity with implications for the discovery of new taxonomic groups. The most salient of these technical approaches are described in the following paragraphs. A summary of their relative strengths and weaknesses is presented in Table 4.1.

- **DNA reassociation kinetics** was originally developed to assess intra-genomic complexity in eukaryotic organisms and the technique has been adapted for the measurement of diversity in natural microbial communities containing chromosomes of different sizes and gene composition. The result of DNA reassociation kinetics experiments is the length of time it

BOX 4.2

	ape	sso	sto	pai	afu	hal	mth	mja	mac	mma	mka	pab	pfu	pho	tac	tvo
ape	*	71	71	71	60	51	44	44	51	51	45	57	59	56	48	48
sso	42	*	91	48	42	35	32	32	39	37	30	37	40	35	40	41
sto	42	92	*	47	41	34	32	32	38	37	30	36	39	35	40	40
pai	64	73	72	*	60	49	45	45	53	53	45	53	57	50	49	49
afu	45	52	51	49	*	51	54	53	70	68	48	53	56	50	41	42
hal	50	58	57	53	68	*	56	52	71	71	48	56	59	53	53	54
mth	37	47	46	42	62	48	*	67	74	74	63	51	64	48	39	40
mja	39	48	48	44	63	47	71	*	70	69	66	58	60	54	39	40
mac	25	32	32	29	46	35	43	39	*	92	34	34	36	32	28	28
mma	25	31	31	29	46	36	44	39	94	*	35	34	35	32	27	28
mka	46	53	53	51	68	51	78	78	72	72	*	59	61	56	41	42
pab	43	46	46	43	53	42	45	49	52	51	42	*	89	86	38	38
pfu	42	47	46	45	53	42	45	48	52	50	42	85	*	80	38	38
pho	42	45	45	42	52	41	43	46	50	49	41	89	87	*	37	37
tac	43	61	60	49	50	48	41	40	51	49	36	46	48	43	*	92
tvo	43	61	60	48	50	49	42	40	52	49	36	45	48	43	92	*

Domain	Phylum	Order	Species (strain)	Notation
Archaea	Crenarchaeota	Desulforococcales	*Aeropyrum pernix* K1	*ape*
		Sulfobales	*Sulfolobus solfataricus* P2	*sso*
			Sulfolobus tokodaii	*sto*
		Thermoproteales	*Pyrobaculum aerophilum* IM2	*pai*
	Euryarchaeota	Archaeoglobi	*Archaeoglobus fulgidus* DSM4304	*afu*
		Halobacteria	*Halobacterium sp.* NRC-1	*hal*
		Methanobacteria	*Methanobacterium thermoautotrophicum* (delta H)	*mth*
		Methanococci	*Methanococcus jannaschii* (DSM 2661)	*mja*
			Methanosarcina acetivorans (C2A)	*mac*
			Methanosarcina mazei (Goel)	*mma*
		Methanopyri	*Methanopyrus kandleri* (AV19)	*mka*
		Thermococci	*Pyrococcus abyssii* (GE5)	*pab*
			Pyrococcus furiosus (DSM 3638)	*pfu*
			Pyrococcus horikoshii (OT3)	*pPho*
		Thermoplasmata	*Thermoplasma acidophilum*	*tac*
			Thermoplama volcanum (GSS1)	*tvo*

Degree of whole-genome relatedness (% homology) among *Archaea* species with completely sequenced genomes. Such rapid web-based (http://mbgd.genome.ad.jp/) assessment of genetic relatedness facilitates the formulation of hypotheses about prokaryotic diversity.

takes for a mixed pool of single-stranded nucleic acids to hybridize, forming double-stranded molecules. The length of time is proportional to the concentration of complementary sequences in the pool. The method produces a quantitative measure of heterogeneity in a sample of DNA pooled from various sources. The main contribution of this approach has been estimates of the number of genomes per unit of environmental samples such as soil or water (Torsvik *et al.*, 1993).

• **Restriction fragment length polymorphism (RFLP)** refers to the technique where differences in the distribution of DNA sequences recognized by restriction enzymes are displayed through electrophoretic resolution. RFLP patterns have been used extensively for mapping the level of relatedness among isolated organisms and for tracking the influence of genetic exchange mechanisms on the dissemination of "marked" genes in a microbial community (Moyer *et al.*, 1996).

BOX 4.3

Domain	Phylum	Order	Family	Species (strain)	#
Proteobacteria	alpha-proteobacteria	Caulobacteriales	Caulobacteriaceae	*Caulobacter crescentus*	1
		Rhizobiales	Bradyrhizobiaceae	*Bradyrhizobium japonicum*	2
			Brucellaceae	*Brucella melitensis* (16M)	3
				Brucella suis (1330)	4
			Phyllobacteriaceae	*Mesorhizobium loti* (MAFF)	5
			Rhizobiaceae	*Agrobacterium tumefaciens* (C58C)	6
				Agrobacterium tumefaciens (C58D)	7
				Sinorhizobium meliloti (1021)	8
		Rikettsiales	Spotted fever group	*Rikettsia conorii* (Malish 7)	9
			Typhus group	*Rikettsia prowazekii* (Madrid E)	10
	beta-proteobacteria	Burkholderiales		*Ralstonia solanacearum*	11
		Neisseriales		*Neisseria meningitidis* (MC58)	12
				Neisseria meningitidis (Z2491)	13
	gamma-proteobacteria	Alteromonadales		*Shewanella oneidensis* (MR1)	14
		Enterobacteriales	Buchnera	*Buchnera* (*Acrythosiphon pisum*)	15
				Buchnera aphidicola (Bp)	16
				Buchnera aphidicola (Sg)	17
			Escherichia	*Escherichia coli* (K-12)	18
				Escherichia coli (CFT073)	19
				Escherichia coli (O157: H7)	20
				Escherichia coli (EDL933)	21
			Salmonella	*Salmonella enterica* (CT18)	22
				Salmonella typhimurium (LT2)	23
			Shigella	*Shigella flexneri* (2a 301)	24
			Wigglesworthia	*Wigglesworthia brevipalpis*	25
			Yersinia	*Yersinia pestis* (CO-92)	26
				Yersinia pestis (KIM)	27
		Pasteurellales	Haemophilus	*Haemophilus influenzae* (KW20)	28
			Pasteurella	*Pasteurella multocida* (PM70)	29
		Pseudomonadales	Pseudomonas aeruginosa group	*Pseudomonas aeruginosa* (PAO1)	30
			Pseudomonas putida group	*Pseudomonas putida* (KT2440)	31
		Vibrionales	Vibrionaceae	*Vibrio cholera* (N16961)	32
				Vibrio vulnificus (CMCP6)	33
		Xanthomonadales	Xanthomonas	*Xanthomonas axonopodis* (c 306)	34
				Xanthomonas campestris (33913)	35
			Xylella	*Xylella fastidiosa* (9a5c)	36
				Xylella Temecula (1)	37
	delta/epsilon subdivisions		Campilobacteriaceae	*Campilobacter jejuni* (NCTC 11168)	38
			Helicobacteriaceae	*Helicobacter pylori* (26695)	39
				Helicobacter pylori (J99)	40

Continued

1	2	3	4	5	6	7	8	9	10	11	12	13	14	15	16	17	18	19	20	21	22	23	24	25	26	27	28	29	30	31	32	33	34	35	36	37	38	39	40	# of genes	Color-coded gene categories
																																					●	●	●	162	
●	●	●	●	●	●	●	●	●	●	●	●	●	●	●	●	●	●	●	●	●	●	●	●	●	●	●	●	●	●	●	●	●	●	●	●	●	●	●	●	155	
					●	●	●																																	140	
																	●	●	●	●	●	●	●																	130	
																																	●	●	●	●				120	
				●	●	●	●																																	119	
													●																		●	●								110	
	●	●	●	●	●	●	●																																	89	
																	●	●	●	●	●	●	●		●	●														86	
																	●	●	●	●			●																	77	
	●			●			●																																	69	
●	●	●	●	●	●	●	●																																	67	
		●	●	●	●	●	●																																	67	
																	●		●	●																				64	
																	●	●	●	●																				54	
●																																	●	●						54	
	●			●	●	●	●																																	52	
																		●	●	●																				46	
				●	●	●																																		45	
	●				●	●																																		43	
																	●	●	●	●	●	●																		43	
																	●		●	●			●																	41	
	●				●	●	●																																	39	
										●																							●	●						35	
													●				●	●	●	●	●	●	●		●	●	●	●			●	●								35	
Total																																								1942	

Amino acid biosynthesis
Purines, pyrimidines, nucleosides, and nucleotides
Fatty acid, phospholipid and sterol metabolism
Biosynthesis of cofactors, prosthetic groups, and carriers
Central intermediary metabolism
Energy metabolism
Transport and binding proteins
DNA replication, restriction, modification, recombination, and repair
Transcription
Translation
Regulatory functions
Cell envelope
Cellular processes
Other categories
Hypothetical

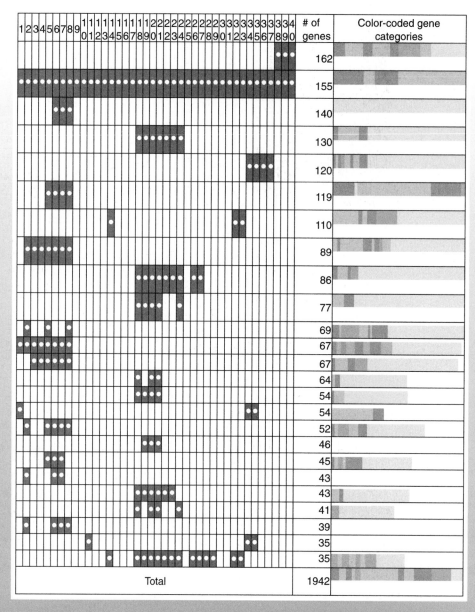

Number of homologous gene clusters among members of the proteobacteria, beta-proteobacteria, gamma-proteobacteria, and delta/epsilon-proteobacteria with completely sequenced genomes. This approach to the comparative assessment of genomes brings the evaluation of diversity to the level of genetic functions. For example, all members of this group share 155 genes (second row), with the largest number of genes relating to the cell envelope. See http://mbgd.genome.ad.jp/ to experiment with different genomes.

Table 4.1 Strengths and limitations of nucleic acid-based methods.

Molecular technique	Strengths	Limitations
DNA reassociation kinetics	Facilitates comparative assessments of diversity in community gene pools. It is a quantitative measure with an easily cross-referenced index ($C_0t_{½}$), which is based on the length of time required for 50% of a given concentration of single-stranded DNA to hybridize. The quantitative measure correlates with the level of heterogeneity (diversity) in the pool of DNA.	The results are only as reliable as the source of DNA used for reassociation experiments. If the DNA mixture is based on a known species, there is very little bias, but if the DNA is from a complex microbial community, bias due to DNA extraction techniques and impurities is reflected in the index. The correlation between reassociation index and species diversity needs to be demonstrated on a case-by-case basis, using supplementary techniques to corroborate the results of reassociation kinetics.
Restriction fragment length polymorphism	Large choice of available restriction enzymes facilitates the production of clear electrophoretic banding patterns for restriction fragments. This clarity leads to unambiguous differentiation among the genetic profile of organisms.	If two unknown isolates exhibit identical restriction fragment banding patterns, it is still not appropriate to regard them as the same species. Typically, many restriction enzymes must be used to detect differences. There is no guarantee that the "right" enzyme will be available.
Terminal restriction fragment length polymorphism	Combines the specificity of ribosomal RNA typing with the flexibility of restriction fragment length polymorphism. Phylogenetically conserved primers to the small subunit RNA can be used first to narrow the scope of diversity assessment, followed by RFLP to estimate the approximate number of different taxonomic groups in the sample. This technique can also be used to estimate relative activities of a specific microbial population by using the correlation between the ratio of rRNA ÷ rDNA and growth rate.	The results may not reflect true diversity of microbial communities because of biases attributable to nucleic acid extraction and polymerase chain reaction. These biases can be minimized by using appropriate controls for several steps during the process.
Denaturing gradient or temperature gradient gel electrophoresis	Allows the identification of dominant phylogenetic groups in a complex microbial community on the basis of 16S rDNA sequences. The method is extremely sensitive, having sufficient resolving power to detect single nucleotide differences in genes.	Biases in the procedures used for extracting DNA from microbial communities are reflected in a skewed estimation of dominant phylogenetic groups. In view of the fact that spacer sequences between rRNA operons vary, individual species can produce multiple results on a gradient gel, thereby introducing false positive estimates.
Single-strand conformation polymorphism	Extremely sensitive to small variations in nucleic acid sequences. Therefore, the technique can be used to differentiate strains belonging to the same species. Products of the reaction can be processed further by sequencing.	Multiple bands may be generated from one species, leading to uncertainty about the interpretation of results with respect to relative species diversity. Steps can be taken to minimize this problem, but the procedure is cumbersome, and it lengthens the processing time (Liu and Stahl, 2002).
PCR-amplicon length heterogeneity	Variation in amplicon lengths resulting from the amplification of coding regions in the small subunit rRNA or intergenic-spacer regions corresponds to phylogenetic differences. This is a relatively straightforward way to compare the species richness of different microbial communities.	The technique uses electrophoresis to separate nucleic acid fragments on the basis of size. This is a limitation because imprecise determination of fragment sizes reduces confidence in the estimates of sequence diversity. The taxonomic implications of small differences in the sequence of target nucleic acids are not yet established.
Low molecular weight (LMW) RNA pattern analysis	Facilitates comprehensive assessment of both population diversity and relative species abundance. The RNA molecules do not vary with microbial growth phase or growth conditions, so they support a consistent interpretation for species composition.	The methods required for separating low molecular weight RNA molecules (electrophoresis and chromatography) are limited in their capacity to reveal the presence of several different species. The upper limit of the resolving power needs to be determined empirically for each method.
Reverse sample genome probing (RSGP)	Reproducibly tracks the occurrence, and possibly, the abundance of specific, well-described microorganisms in a complex environment.	The technique requires extraction of DNA from known species. Since these species have to be cultivated, the technique is limited by the problem that only a small fraction of microbial species has been successfully cultivated under laboratory conditions.

• **Terminal restriction fragment length polymorphism (T-RFLP)** depends on the amplification of a region of the small subunit of the ribosomal RNA (SSU-RNA) gene using primers that are fluorescently labeled at the 5′ end. The PCR products are then subjected to RFLP preferably with a tetrameric restriction enzyme (Noble *et al.*, 1998). If, for example, primers specific for the SSU-RNA genes of major bacterial phylogenetic categories are used,

T-RFLP can provide detailed information on the diversity of bacterial species in DNA samples extracted from microbial communities. T-RFLP has also been used for analyzing the distribution of functional genes in ecosystems. Marsh and colleagues (2000) have published a web-based interactive program for selecting combinations of primers and restriction enzymes that give optimum conditions for microbial community analysis.

• **Denaturing or temperature gradient gel electrophoresis (D/TGGE)** allows the detection of very small differences in the sequence of nucleotides within phylogenetically conserved genes (Kozdroj and van Elsas, 2001). The technique depends on physical conditions that retard the movement of DNA molecules through a gel due to partial melting within discrete regions of double-stranded DNA. The melting is caused by gradients of elevated temperature or increasing concentrations of denaturing agents such as urea and formamide. Typically, fragments larger than 500 base pairs cannot be separated with this technique, a restriction that limits the detail of phylogenetic information generated (Muyzer and Smalla, 2000).

• **Single-strand conformation polymorphism (SSCP)** detection has been applied in conjunction with PCR using primers specific for variable regions of ribosomal RNA genes to fingerprint changes in microbial population structure in response to variable environmental parameters. The technique can separate single-stranded DNA (ssDNA) molecules on the basis of fragment length and conformation because ssDNA folds to form stable structures that are defined by sequence composition. Small differences in these "conformers" can be resolved electrophoretically (Schweiger and Tebbe, 1998).

• **PCR-amplicon length heterogeneity (PCR-ALH)** has been used for comparative assessments of diversity within microbial communities as they change across geographic gradients (Fisher and Triplett, 1999). The technique is based on the amplification of 16S–23S rRNA interspacer regions resulting in DNA fragments of different lengths. The different size fragments are subsequently separated electrophoretically (Delbes *et al.*, 1998; Eder *et al.*, 2001). Fragment length can vary by as much as 1.2 Kb, but the technique is limited in the ability to resolve long fragments that differ in sequence identity by few nucleotide bases.

• **Low molecular weight (LMW) RNA pattern analysis** depends on the size distribution of transfer RNA and 5S ribosomal RNA molecules in species within a heterogeneous microbial community. The composition of LMW RNA is not affected by environmental conditions, and it is therefore a stable measure of species occurrence. The resolution of LMW RNA extracted from a complex natural environmental sample requires sophisticated pattern recognition programs to identify the presence of known species, and the possible discovery of novel species (Hofle, 1988).

• **Reverse sample genome probing (RSGP)** depends on the extraction of bulk DNA from environmental samples, and the use of this community gene pool to probe immobilized DNA that has been extracted from known microbial species (Voordouw *et al.*, 1993). The technique is particularly sensitive to the need to track specific organisms in different ecological contexts, and to identify dominant species within perturbed ecosystems.

SIGNATURE LIPID BIOMARKERS

Lipids, in particular fatty acids, have been used extensively for microbial identification and the assessment of species diversity and physiological status in microbial communities (Pinkart *et al.*, 2002; White *et al.*, 1979; White *et al.*, 1997; Tunlid and White, 1992). Fatty acids are long straight or branched chain hydrocarbon compounds with a terminal carboxylate group. The nomenclature of fatty acids is in the form A:BωC, where "A" represents the number of carbon atoms, "B" is the number of double bonds, and "C" is the count of the number of carbon atoms from the methyl (omega, "ω") terminus of the molecule to the first unsaturated (double) bond. For example, 18:1ω9 represents oleic acid (cis-9-octadecenoic acid) with the following chemical structure:

$CH_3(CH_2)_7CH{=}CH(CH_2)_7COOH$

An example of a saturated fatty acid is stearic acid (octadecanoic acid) represented by the numerical code 18:0 and the chemical structure $CH_3(CH_2)_{16}COOH$. Geometric isomers within the fatty acid structure are referred to by "c" for "cis" and "t" for "trans". References to methyl branching ("me") within the fatty acid structure are made using the prefixes "i" for "iso", "a" for "anteiso". Cyclopropyl rings within fatty acid structures are referenced by "cy" (Navarrete *et al.*, 2000).

Branched chain fatty acids with 9 to 20 carbon atoms are particularly useful in differentiating phylogenetic groups. Among the bacteria, fatty acids are part of the phospholipids in cell membranes and they also occur in cell lipopolysaccharides. The microbial cell membrane contains several different lipids, particularly the short chain length, unsaturated molecules that enhance the fluidity of fatty acids and their derivatives (Fig. 4.1). Membranes

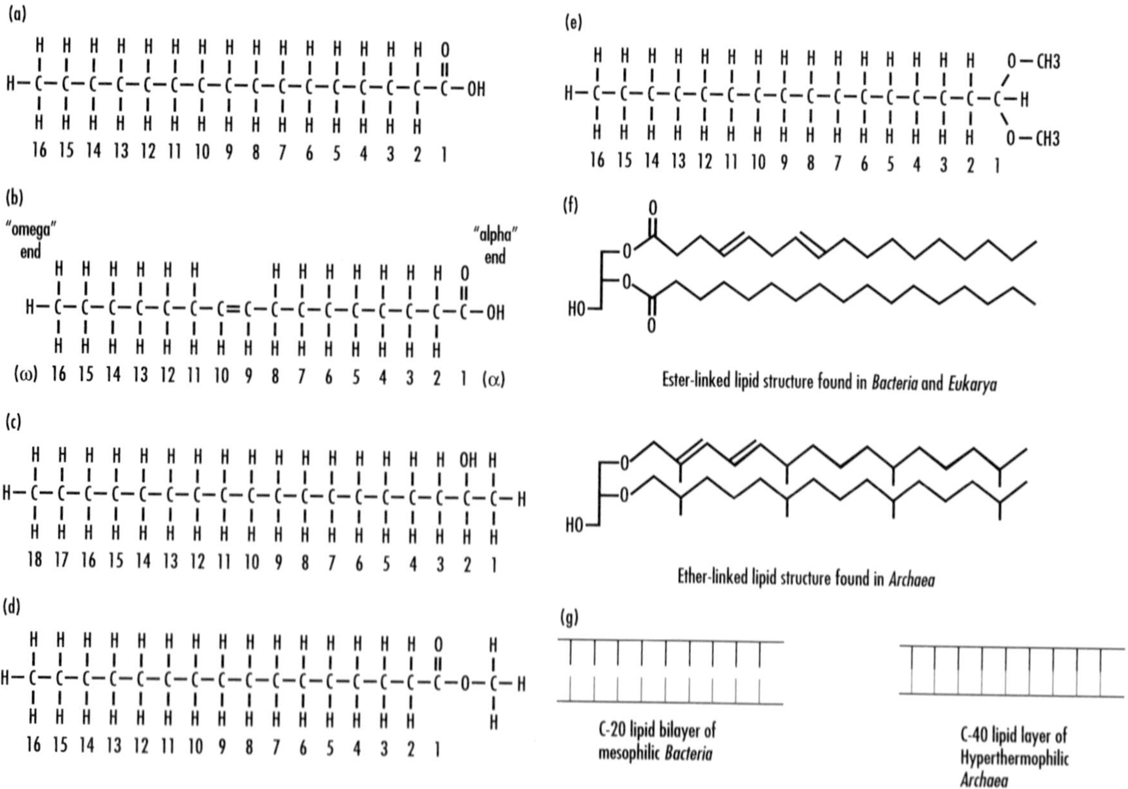

Fig. 4.1 Structure and nomenclature of some long-chain aliphatic compounds useful in assessing microbial diversity. (a) Straight-chain fatty acid, 16:0. (b) Cis-conformation of the monounsaturated fatty acid, 16:1ω7c. (c) 2-octadecanol (shown) and 2-eicosanol (20 carbon atoms long) are typically associated with *Mycobacterium* species. (d) Fatty acid methyl ester, 16:0 FAME. A methyl group has been added to the carboxyl end to add volatility for GC analysis. (e) Dimethyl acetal, 16:0 DMA. These compounds occur as analogs of the fatty acids found in anaerobic bacteria. (f) Archaeal lipids are characterized by ether-linked fatty acids as opposed to the ester-linked varieties found in other phylogenetic domains. (g) The membranes of hyperthermophilic Archaea species are not "bilayers" in that they contain 40-carbon lipids that span the whole membrane. In contrast, the phospholipids of Mesophiles are arranged in a bilayer. It is suspected that the arrangement of lipids in thermophiles reduces excess fluidity of the membrane at extremely high temperatures.

also contain glycolipids, which include carbohydrate molecules that are attached to the primary hydroxyl unit of fatty acids usually with an even number of carbon atoms. Phospholipids derive from glycerol (phosphoglycerides) or sphingosine, fatty acid chains, and phosphorylated alcohol. Sterols are present in most eukaryotes, but not in most prokaryotes (White *et al.*, 1997).

Lipids can be extracted from various ecological samples including aquatic systems, sediments, and soils to assess the distribution and viability of bacteria, fungi, protozoa, and metazoa (White *et al.*, 1996; Zelles *et al.*, 1995). The esterification of fatty acids extracted from microorganisms is usually performed in order to render them more volatile prior to analysis by gas chromatography and mass spectrometry (GC/MS). Hence, fatty acid methyl ester (FAME) analysis has been adopted by several investigators as a powerful technique for chemotaxonomic purposes and for assessing prokaryotic community structures (Guckert *et al.*, 1985; Lechevalier and Lechevalier, 1988). Phospholipids associated with non-viable microorganisms are degraded to diglycerides within a few hours of cell death. Therefore, FAME analysis provides excellent assessment of viable microbial diversity. Data gathered on fatty acid profiles are usually subjected to principal component statistical analysis (PCSA) to facilitate sample diversity groupings across spatial or temporal gradients (Bruggemann *et al.*, 1995; Pankhurst *et al.*, 2001). Quantitative conversion factors are often used to estimate the number of viable cells in a given environmental sample based on the quantities of phospholipid fatty acids (PLFA) extracted. For example, in some experiments with sediment samples, 10^8 picomoles of PLFA per gram of dry weight of sample was determined to be equivalent to 2.5×10^{12} cells per gram of sample dry weight. This provides a conversion factor of 2.5×10^4 cells per picomole of PLFA (Balkwill *et al.*, 1988; Findlay and Dobbs, 1993). However, these conversion factors can produce estimates that vary by as much as one order of magnitude, and appropriate controls using complementary methods must be conducted with each experimental procedure to increase the level of confidence in ecological inferences resulting from PLFA analysis.

The formation of cyclopropane fatty acids from their monoenoic homologs and the ratio of "trans" to "cis" monoenoic phospholipid fatty acids are frequently used to assess the physiological status of microbial communities. The production of trans-monoenoic fatty acids accompanies exposure to toxic environmental conditions, whereas starvation conditions induce the formation of cyclopropane PLFA (Guckert *et al.*, 1986). These physiological status indicators can be used in conjunction with PLFA taxonomic markers to generate a functional diversity "map" of microbial communities under the influence of dynamic environmental conditions (Navarrete *et al.*, 2000). The presence of monoenoic PLFAs is indicative of Gram-negative bacteria and cyanobacteria, whereas terminally branched chain fatty acids (TBCFA) are characteristic of Gram-positive bacteria; however, some anaerobic Gram-negative bacteria have also been found to produce TBCFA. Anaerobic microorganisms are typically associated with branched monounsaturated fatty acids and mid-chain branched saturated fatty acids (Dowling *et al.*, 1986).

Although PLFA profiling is reasonably well established for assessing the occurrence and distribution of bacterial and eukaryotic microorganisms in the environment, the situation is not as well developed for monitoring the occurrence and relative abundance of members of the Archaea. Lipids found in the Archaea represent a distinctive feature of this group of microorganisms, and the lipids are involved in their survival in extreme environments (Konings *et al.*, 2002). Archaeal lipids contain branched hydrocarbons, and hydrocarbons attached to glycerol by ether links rather than the ester links found in the Bacteria. Polar lipids, including phospholipids, sulfolipids, and glycolipids are also notable features of the Archaea. However, up to 30% of membrane lipids are non-polar derivatives of squalene. The membranes of organisms inhabiting extreme environments consist mostly of tetraether monolayers (Eguchi *et al.*, 2003). Phospholipid etherlipid (PLEL) analysis is a recently developed approach specifically for the assessment of Archaea distribution and abundance in the environment. Using this approach, Gattinger and colleagues (2003) demonstrated that monomethyl-branched alkanes were the most dominant lipids, accounting for 43.4% of the total identified ether-linked hydrocarbons in a soil microbial community harboring archaeal species.

PROTEIN PROFILES

Nucleic acid-based approaches for investigating microbial communities rely invariably on the coordinated functions of protein molecules that serve as enzymes for recognizing, replicating, and amplifying specific nucleic acid sequences. In addition to this fundamental linkage between protein function and the assessment of genetic diversity, protein molecules represent the final result of genetic expression, and through their functions as physiological catalysts, structural components, signal transducers, and mediators of intercellular communication, they control key reactions in ecological processes performed by microorganisms in aquatic, terrestrial, and certain artificial environments (Ogunseitan, 2000, 2004; Ogunseitan *et al.*, 2002). For these and other reasons, several investigators, including Giraffa (2001), Gupta (1998b), and Galperin and Koonin (1999) have suggested that the construction of phylogenetic trees by means of protein-coding sequences are likely to produce more robust measurements of diversity than the current predominant reliance on coding sequences that are not translated into functional proteins. The analysis of microbial proteins has traditionally focused on comparative quantitative assessments and functional characteristics of a few phylogenetically conserved molecules. For example, cytochromes, protein elongation factors, and ATPases are among the most widely studied protein molecules in systematics (Liu and Stahl, 2002).

Recent methodological developments have enhanced strategies for conducting whole-cell protein assessments based on the rationale that a complete **proteome** map will facilitate the discovery of unique polypeptides whose production is mediated by rare environmental cues. Although such discoveries are increasing in frequency, there is still an imbalance in the depth of investigation of microbial proteomics within different fields of specialization. For example, comparative analysis of microbial proteins has facilitated the construction of phylogenetic trees and the derivation of a "molecular clock" that has greatly contributed to the development of classification schemes and to the exploration of both quantitative and qualitative dimensions of microbial diversity. However, information on the biochemical ecology of proteins has typically been inferred from axenic cultures (Hantula *et al.*, 1990 and 1991; Jaan *et al.*, 1986; Jackman, 1985; Kersters and De Ley, 1980), and questions persist about the verification of phenomena discovered under such conditions in natural ecosystems. One response to these questions is the emergence of a repertoire of methods and techniques that support the analysis of protein synthesis, diversity, and function in natural heterogeneous microbial communities (Garcia-Cantizano *et al.*, 1994; Hantula *et al.*, 1990 and 1991; Ogunseitan, 1993–2000 and 2002; Ogunseitan *et al.*, 2001 and 2003).

Strategic analyses of microbial proteins to elucidate microbial diversity and ecosystem-level activities in the environment require an appreciation of the complexity inherent in protein structure when compared to other biomolecules such as nucleic acids. Protein complexity increases the richness of information that can be obtained with proper tools of investigation, but few techniques have been developed specifically for resolving proteins extracted from natural environmental samples. The choice of technique depends on the investigator's desire for qualitative or quantitative information, or both. The information contained in protein molecules is based in the primary structure, or the sequence of ~20 possible amino acids making up the polypeptide chain. Procedures for constructing protein profiles, based on molecular size and isoelectric properties through one- or two-dimensional polyacrylamide gel electrophoresis, are readily available for studies focused on temporal and spatial variations in genetic induction of protein synthesis. Comparative amino acid sequence analysis of specific proteins can also address questions of evolutionary diversity. In addition to the primary chain structure, most protein molecules also exist in secondary (folded protein), tertiary (globular protein), or quaternary (several interacting folded polypeptides) forms. These morphological conformations contribute significantly to the ability of proteins to perform crucial functions such as enzymatic catalysis and organelle construction. The abundance and diversity of microbial proteins suggest fairly straightforward extraction methods, but attention must be paid to extraction conditions that preserve protein integrity and function while reducing interference from co-extracted substances such as nucleic acids and humic materials. It is particularly important that studies investigating protein enzymatic functions employ extraction, resolution, and detection techniques that optimize stability of protein conformation while minimizing interference by potential inhibitors such as metal ions and detergents, which act as denaturing agents.

Quantitative methods for direct extraction of amino acids and proteins from natural environments were invented by biogeochemists interested in the nitrogen cycle (Cheng, 1975; Evens *et al.*, 1982; Greenfield *et al.*, 1970). Numerous semi-qualitative methods for protein extraction from natural environmental samples were subsequently developed to investigate specific enzymes important in agriculture and environmental contamination (Wright, 1992). The occurrence of enzyme polymorphisms at the molecular level encoded (in eukaryotes) by different alleles has also provided the opportunity for using the molecular resolution of enzymatic activity or fine-scale amino acid sequence as a tool for measuring microbial diversity (Boerlin, 1997; Reid *et al.*, 2001; Selander *et al.*, 1986; Zhong *et al.*, 2002). Recent interest in elucidating species diversity and metabolic productivity of microbial communities required fine-structure resolution of key molecules such as nucleic acids and fatty acids extracted directly from nature (Paul, 1993). The molecular resolution of microbial community proteins presents a challenging endeavor due to the extremely large number of different proteins synthesized by different species, even in axenic microbial cultures (Blom *et al.*, 1992; Bohlool and Schmidt, 1980). However, the availability of immunological techniques and numerous enzyme assays reduces the need for extensive resolution in cases where attention is focused on a particular polypeptide or enzyme (Nybroe *et al.*, 1990; O'Connor and Coates, 2002; Selander *et al.*, 1986; Wright, 1992).

MOLECULAR MICROARRAY SYSTEMS

Microarray systems for processing large numbers of molecules have the potential to revolutionize the assessment of natural microbial diversity, however, their use remains limited primarily to engineered systems or simple environments, pending more technological developments. DNA and protein microchips present dense arrangements of molecular probes such as oligonucleotides immobilized on solid surfaces that are subsequently challenged with molecules extracted from environmental samples to determine the presence or absence of target organisms or functional genes (Guschin *et al.*, 1997a and 1997b). Although bacterial messenger RNA (mRNA) is short-lived, microarray systems can be designed to assess the production of mRNA as an index of genetic expression under variable environmental conditions. Due to the limitations imposed by the need for protocol standardization, it is likely that information based on nucleic acid microarray systems will have to be complemented with alternative methodologies. These limitations restrict the exploration of how molecular level interactions in microbial ecology are sensitive to small differences in physical and chemical conditions. For these purposes, pattern recognition programs such as neural network computing and other bioinformatics protocols are bound to prove indispensable (Sugawara *et al.*, 1996).

Technologies for proteome analysis are intrinsically more complex than for nucleic acid analysis, partly because of the larger number of amino acids compared to the number of nucleic acid bases, but mostly because of the post-translation modifications that endow each category of polypeptide chain with a unique set of characteristics. Microarray systems combine multiple technologies in a unit platform that facilitates protein immobilization, purification, analysis, and processing from complex biological mixtures. In particular, the development of surface enhanced laser desorption and ionization and time-of-flight

(SELDI-TOF) mass analysis has increased experimental options for differential display analysis of proteomes for organisms cultivated under different ecological conditions. These developments in coupled mass spectrometric analysis of proteomes in microarray formats are typically presented as an approach that is considered a necessary prelude to the exploration of ecological dimensions of microbial proteome assessment. These dimensions include spatio-temporal and quantitative mapping of protein involvement in cell–environment interactions, and inter- or intra-specific cell–cell communications (Ogunseitan *et al.*, 2001 and 2003).

CONCLUSION

The interactions between microorganisms and their environments are as varied and complex as the numerous attempts to define the terms "microbial diversity" and "environment" by means of discrete bits of information. There is an increasing recognition of both microbial diversity and the environment as continuums of interrelated concepts where the emergence of novel entities and relationships is defined by selection and adaptation at the molecular level. In this context, the potency of molecular creativity is embedded in the structure and function of nucleic acids, but protein molecules that have the capacity to sense and respond to environmental change and ecological stimuli mediate its actualization. In turn, fatty acids and other structural molecules accomplish the ecologically sustainable strategies for organism survival and competition. Therefore, the analysis of molecules can provide meaningful information on the diversity of microorganisms at a finer scale than would be possible through investigation of morphological characteristics.

QUESTIONS FOR FURTHER INVESTIGATION

1 According to one view of the Darwinian theory of evolution, speciation and radiation of biological diversity result from changes in the molecular composition of cells. Discuss the origin of these molecular changes, and the pathways through which they ultimately produce the array of diverse organisms.

2 Discuss the advantages and limitations of DNA, RNA, lipids, and proteins as key molecules used in the assessment of microbial diversity.

3 Design an experiment, based on the analysis of an environmental sample, where the use of each of the molecules mentioned in question 2 contributes essential information to the assessment of microbial diversity.

SUGGESTED READINGS

Gattinger, A., A. Guenthner, M. Schloter, and J.C. Munch. 2003. Characterization of *Archaea* in soils by polar lipid analysis. *Acta Biotechnologica*, **23**: 21–8.

Giraffa, G. 2001. Protein coding gene sequences: Alternative phylogenetic markers or possible tools to compare ecological diversity in bacteria? *Current Genomics*, **2**: 243–51.

Gupta, R.S. 1998. Protein phylogenies and signature sequences: A reappraisal of evolutionary relationships among archaebacteria, eubacteria, and eukaryotes. *Microbiology and Molecular Biology Review*, **62**: 1435–91.

Konings, W.N., S.-V. Albers, S. Koning, and A.J.M. Driessen. 2002. The cell membrane plays a crucial role in survival of bacteria and archaea in extreme environments. *Antonie van Leeuwenhoek*, **81**: 61–72.

Liu, W.-T. and D.A. Stahl. 2002. Molecular approaches for the measurement of density, diversity, and phylogeny. In C.J. Hurst, R.L. Crawford, R. Knudsen, M.J. McInerney, and L.D. Stetzenbach (eds.) *Manual of Environmental Microbiology* (Second Edition), pp. 114–34. Washington, DC: ASM Press.

Ogunseitan, O.A. 2004. Assessing microbial proteomes in the environment. In G. Bitton (ed.) *Encyclopedia of Environmental Microbiology*, pp. 305–12. New York: Wiley.

Ogunseitan, O.A., J. LeBlanc, and E. Dalmasso. 2001. Microbial community proteomics. In P.A. Rochelle (ed.) *Environmental Molecular Microbiology*, pp. 125–40. Norfolk, England: Horizon Scientific Press.

Pinkart, H.C., D.B. Ringelberg, Y.M. Piceno, S.J. MacNaughton, and D.C. White. 2002. Biochemical approaches to biomass measurements and community structure analysis. In C.J. Hurst, R.L. Crawford, R. Knudsen, M.J. McInerney, and L.D. Stetzenbach (eds.). *Manual of Environmental Microbiology* (Second Edition), pp. 101–13. Washington, DC: ASM Press.

Sugawara, H., S. Miyazaki, J. Shimura, and Y. Ichiyanagi. 1996. Bioinformatics tools for the study of microbial diversity. *Journal of Industrial Microbiology and Biotechnology*, **17**: 490–7.

White, D.C., J.O. Stair, and D.B. Ringelberg. 1996. Quantitative comparisons of in situ microbial biodiversity by signature biomarker analysis. *Journal of Industrial Microbiology*, **17**: 185–96.

Young, J.P.W. and B.R. Levin. 1992. Adaptation in bacteria: Unanswered ecological and evolutionary questions about well-studied molecules. In T.J. Crawford and G.M. Hewitt (eds.) *Genes in Ecology*, pp. 169–92. Oxford, England: Blackwell Scientific.

Zuckerkandl, E. and L. Pauling. 1965. Molecules as documents of evolutionary history. *Journal of Theoretical Biology*, **8**: 357–66.

PHYLOGENETIC ANALYSIS

INTRODUCTION

Chapter contents

The establishment of quantifiable similarities and differences among various microbial taxonomic groups is one of the most thrilling endeavors encountered in investigations of microbial diversity. Distinctive characteristics are increasingly recognized at the genomic level, and rapid progress in the sequencing of microbial genomes has energized research interests in the comparative analysis of differences and similarities at this level. However, not all differences in genomic sequence contribute to the observable differences in microbial phenotypes that are interpreted to represent diversity. The availability of extensive databases for nucleic acid and amino acid sequences has necessitated the development of tools for "mining" the databases, and representing information on differences and similarities in a way that contributes to the understanding of diversity beyond genetic potential. Phylogenetic analysis requires an equivalent level of skill for the analysis of environmental samples to discover novel organisms and to produce molecular sequence data. Phylogenetics is generally at the center of empirical strategies designed to produce meaningful information on how microbial communities are organized. This chapter presents an overview of methods and applications of phylogenetic analyses, with particular emphasis on:

1 The rationale for creating different kinds of phylogenetic trees and their role in the assessment of microbial diversity.
2 The accessibility of molecular sequences from established databases, and their uses for constructing phylogenetic trees.
3 The interpretation of phylogenetic trees as visual representations of evolutionary divergence among groups of organisms.
4 The implications of phylogenetic relationships for the understanding of genotypic and phenotypic diversity among microorganisms.

THE RATIONALE FOR PHYLOGENETIC TREES

The rapid accumulation of nucleic acid and protein sequence datasets in publicly available formats represents the pinnacle of reductionism as a dominant analytic approach in the biological sciences. The remarkable success of this approach has necessitated the invention of reliable strategies for comparing sequence information generated from various sources. Methods were also needed for integrating such comparative assessments into the analytic frameworks being developed to explore phenotypic diversity and evolutionary convergence of microbial traits that support adaptation in various ecological contexts. Systematists who focus on microorganisms are generally interested in reconstructing phylogenetic relationships among extant taxonomic groups, and occasionally in identifying the position of ancestral species in the lineage of more complex organisms. Phylogenetic trees are designed to reveal these relationships in the simplest way possible without compromising the understanding of the complicated mechanisms underlying the translation of genetic potential into phenotype expression.

Methods for constructing phylogenetic trees emerged before the proliferation of molecular sequencing. The use of morphological and physiological characteristics made it relatively easy to recognize differences among groups of organisms within the major phylogenetic branches. It is more difficult to recognize such differences in phylogenetic trees constructed with molecular sequence information. Powerful computing capabilities are now required and they must still be trained by expert knowledge systems to produce meaningful information from the routine evaluation of multiple databases compiled for numerous phylogenetic groups.

The construction of phylogenetic trees with molecular sequence data is increasingly appealing to microbiologists because of the reputation of this technique as a very powerful way to visually represent fine-scale differences in biological diversity (Berbee and Taylor, 1993; Chao and Carr, 1993; Gibbs *et al.*, 1998; Gillespie, 1984). Nucleic acid and protein sequences are being generated far in advance of useful information about phenotypic characteristics associated with specific molecular signatures. This knowledge gap has led, in some cases, to the attribution of specific functions to protein molecules solely on the basis of sequence homologies (Hall, 2001). The negative consequences of assigning function to nucleic acid and protein sequences are exacerbated by assumptions that the true function of the first protein to be sequenced within a group has been categorically determined. Furthermore, assumptions that similar proteins perform identical functions in different organisms dealing with different ecological constraints are questionable at best. Although the construction of a phylogenetic tree can facilitate the detection of relative homology between newly identified proteins and the sequence of molecules associated with specific phenotypes, post-translational modification and protein folding may obscure the straightforward interpretation of sequence data to predict ecological functions.

Early attempts at quantitative comparisons of molecular sequences involved simple expression of pairwise homologies as percent similarities. This strategy became limited in its application by the expansive database which precluded the publication of exhaustive estimates of homologs within phylogenetic categories. Therefore, multiple alignment programs were invented (Higgins and Sharp, 1988; Thompson *et al.*, 1997). Multiple alignments involve the construction of a "guide tree" from all possible pairwise comparisons; however, such trees do not truly reflect the domains that are homologous across all the sequences used for pairwise comparisons (Hall, 2001). The construction of "true" phylogenetic trees was necessitated in part by the desire to avoid errors in the interpretation of guide trees constructed by pairwise comparisons.

The most challenging – and time-consuming – aspect of producing phylogenetic trees is the sequencing of nucleic acid or protein molecules that will be compared to sequences already existing in available databases. Sequencing is now mostly automated but protein sequences are still routinely inferred from structural DNA sequences, although perfect correspondence should not be assumed. The recognition of bioinformatics, functional genomics, and proteomics as fully fledged sub-disciplines of molecular biology has intensified progress toward more robust approaches for translating genetic potential, as recorded in nucleic acid sequences, into functional diversity in microbial communities (Ogunseitan *et al.*, 2002).

The increasing production of molecular sequences and their storage in electronic databases by numerous investigators necessitated the development of computer-assisted search engines with user-defined search parameters for retrieving sequence data. One of the most widely used computer software programs for exploring and downloading molecular sequences is BLAST (Altschul *et al.*, 1997). BLAST is managed by the United States government through the National Center for Biotechnology Information under the auspices of the National Institutes of Health (http://www.ncbi.nlm.nih.gov/). Nucleic acid ("blastn") or protein ("blastp") sequences may be uploaded into BLAST for comparison with other sequences preexisting in the database. Alternatively, deposited sequences may be recalled to the query form through "Accession" numbers allocated to specific sequences deposited by investigators.

The result of a query submitted to BLAST is a list of sequences in the database, which are judged to be related to the specific target sequence. The judgment of relatedness is identified by the "bit score" and the "E value" attributed to sequences related to the target. For nucleic acid sequences, the bit score is derived through a summation of 1 and 0 values assigned to each match and mismatch, respectively, and through subtracting penalties for gaps in the sequence. In the case of protein sequences, the nature of mismatch is taken into consideration when assigning the bit score. BLAST presents the related sequences in descending order of bit-score values. In general, high bit scores represent sequences that are more closely related to the target sequence. The E value is a probabilistic estimate of the chance occurrence of identified matches in the database. Small E values indicate a high level of confidence in the inference that the similarities between two sequences are due more to common descent than to chance. Programs such as BLAST recover more sequences than are needed for constructing robust phylogenetic trees. Therefore, it is necessary to set a limit on the minimum level of sequence relatedness (bit score) to be included in the analysis. The limit can be set at naturally occurring gaps in the E-value score, below which there is a high possibility of chance occurrence of sequence similarities. Selected sequences can then be downloaded and saved in formats (e.g. "FASTA") that facilitate subsequent multiple sequence alignments.

MULTIPLE SEQUENCE ALIGNMENTS

Molecular sequences that are deposited into public databases are typically byproducts of research projects designed to explore fundamental questions about the relationship between specific genotypes and protein function in broadly defined ecological contexts. Therefore, it is probable that sequence data begins and ends at different locations in different genomes. To compare sequences from different sources and different organisms reliably, it is important to ensure that the sequences are aligned in a realistic manner to avoid spurious correlations that can nullify inferences regarding gene flow and protein function. The quality of sequence alignment is quite likely the most important aspect of constructing reliable phylogenetic trees.

The alignment of molecular sequences from different sources is aided by computer algorithms, although the process usually requires user-specified parameters that influence the rigor of the alignment. Several software programs have been developed to assist in the alignment of molecular sequences. The "Clustal" series is one of the most widely used programs in microbiology, and it is available at http://inn-prot.weizmann.aci.il/software/ClustalX.html. "CodonAlign" is useful for converting protein sequence alignments to DNA alignments (Hall, 2001). Most programs begin with pairwise comparisons among sequences in the database, followed by the production of a guide tree, which is then used to create multiple alignments. The existence and length of gaps between two sequences are "penalized" to optimize the reliability of alignment scores. Ideally, the outcome of multiple alignments should produce meaningful data regarding the relative biochemical significance of motifs or domains in protein sequences that are conserved across phylogenetic groups. The recognition of these motifs can lead to testable hypotheses about protein function.

CONSTRUCTING PHYLOGENETIC TREES FROM ALIGNED SEQUENCES

The construction of phylogenetic trees proceeds according to one of two basic approaches, namely "tree searching" and "algorithms". Popular tree-searching methods, including **maximum parsimony**, **maximum likelihood**, and **Bayesian** analysis, generate several preliminary trees from an aligned sequence database, and then use certain criteria to select the best tree. Algorithmic methods, including **neighbor joining** and the less common **unweighted pair-group method with arithmetic mean** (UPGMA), use an algorithm to construct a single tree from the database. The algorithmic methods are also referred to as "distance methods" because they are based on matrices generated from computed pairwise differences or "distances" between aligned sequences.

Several software programs are available to support the construction of phylogenetic trees, using one or more of the methods described in the paragraph above. Some programs (e.g. ClustalX) designed primarily for producing sequence alignments can also produce phylogenetic trees, although it is generally preferred to import aligned sequences for use by programs designed explicitly for phylogenetic tree construction because they tend to offer more user-specified options for guiding the format of the tree. Among the most commonly used programs are PAUP* (Swofford, 2000; available at http://www.sinauer.com); Mr. Bayes (Mau *et al.*, 1999; available at http://brahms.biology.rochester.edu/software.html); PUZZLE, which uses the maximum-likelihood technique (Schmidt *et al.*, 2002), is available at http://www.tree.puzzle.de; and Phylogeny Inference Package (PHYLIP) is available at http://evolution.genetics.washington.edu/phylip.html. For a thorough discussion of advantages and disadvantages of these methods and tutorials on their uses, consult Hall (2001).

INTERPRETING PHYLOGENETIC TREES

The ideal outcome of the phylogenetic tree construction process is the revelation of branching order and distance from a common ancestor(s) that describes the evolutionary pathway of sequences represented in the database. The branch order is characterized by branches from internal nodes representing the common ancestor of two or more external nodes, also known as clades or phyletic groups. The distance is estimated from the number of molecular substitutions that have occurred along a branch. The reliability of inferences regarding evolutionary distance on the basis of molecular substitutions can be compromised by genetic exchange events that cut across clades. The interpretation of molecular clocks based on sequence divergence has been debated extensively in the literature (Ayala, 1997; Gibbs *et al.*, 1998; Gillespie, 1984; Kimura, 1980; Rannala and Yang, 1993; Rambaut and Bromham, 1998; Robinson and Robinson, 2001; Rodrigo *et al.*, 1994; Schierup and Hein 2000; Thorpe, 1982).

It is not always possible to identify the single common ancestor for all the taxonomic groups represented in a phylogenetic tree. This implies that the "direction" of evolution is difficult to obtain from phylogenetic analysis based on a single gene or protein. There is usually not sufficient information to "root" a phylogenetic tree on the basis of available sequences. However, trees can be rooted if there is independent and complementary evidence from other sources about the evolutionary relationship among the clades under investigation. For example, if other biochemical evidence or the fossil record points to a particular organism as evolutionarily "older" than the other organisms in the database, then it is reasonable to select the oldest organism as the "outgroup", which is used as the root of the tree, but this strategy must be justified rigorously. It is not easy to achieve this justification unequivocally for microorganisms because of the scant fossil record and the possibility of lateral gene flow. The possible number of rooted trees increases considerably as the number of taxonomic groups being compared increases according to the following equation:

$$\text{The number of possible rooted trees} = (2t-3)! \div 2^{t-2}(t-2)!$$

where t is the number of taxonomic groups (Hall, 2001).

According to this equation, approximately 34.5 million rooted trees are possible for a phylogenetic analysis based on sequences from 10 taxonomic groups.

The difficulty of rooting phylogenetic trees is compounded by the questionable reliability of groupings or the assignment of specific sequences to a given clade, as they are presented in the tree. Some investigators are interested primarily in the reliability of the branching order, whereas others are more interested in group associations. The reliability of phylogenetic trees is typically estimated by a randomized sampling technique known as "bootstrapping", where the frequency of occurrence of a given clade with the same component sequences occurring over 100 to 1,000 bootstrap trials is determined. Bootstrap values reflect the level of confidence attributable to branches of the phylogenetic tree, and high values (e.g. >95%) inspire confidence; whereas low values (e.g. <30%) inspire little confidence in the phylogenetic groupings. It is not unusual to discover that bootstrap values are uneven across phylogenetic branches represented in a tree. Branches with unusually low bootstrap values may reflect considerable genetic exchange events within those branches. The occurrence of horizontal genetic exchange events has made it difficult to construct "universal" trees for prokaryotes. But robust trees have been constructed using large protein data sets (see **www.blackwellpublishing.com/ogunseitan** for an example).

Case study of phylogenetic relationships and niche diversity

The relatively easy access of investigators to computer software for downloading and aligning sequence data has expanded the range of questions that can be asked about processes that define natural diversity of microorganisms. The following case study of the phylogenetic analysis of a conserved enzyme illustrates some of the benefits and challenges that are encountered in similar studies.

delta-Aminolevulinic acid dehydratase (ALAD; porphobilinogen synthase; EC: 4.2.1.24) is a 128 kDa metalloenzyme that catalyzes the first common step in the biosynthetic pathway for all tetrapyrroles including the oxygen carrier heme, the light-gathering array component chlorophyll, vitamin B_{12}, and cofactor F_{430} (Box 5.1; Jaffe *et al.*, 2000; Jones *et al.*, 1994; Krishnamoorthy, 2000; Tanaka *et al.*, 1995). ALAD activity has been demonstrated in most organisms that have been investigated across several phylogenetic groups. The enzyme is a homo-octameric protein where each subunit is in the form of the TIM-barrel fold. The TIM-barrel tertiary protein fold is named for triosephosphate isomerase, an enzyme whose interesting structure now typifies the geometrically conservative but functionally diverse group of proteins belonging to at least 15 distinct enzyme families (Wierenga, 2001). These enzymes use this TIM-barrel scaffold (eight alpha and eight beta sheets or strands) for their active site, which is usually at the C-terminal end of the eight parallel beta strands of the barrel. The N-terminal arms of ALAD, consisting of 20–30 amino acid residues, engage in inter-subunit interactions such that the octamer has all of the eight active sites exposed on the protein surface (Box 5.1). In most organisms, each active site of ALAD has two lysine residues at positions 210 and 263. ALAD also requires either zinc, magnesium, or both for optimum functioning. Mammalian ALAD requires zinc whereas plant ALAD typically requires magnesium. Microbial ALAD is very diverse, presumably reflecting the ecological niche to which organisms have adapted. In the proximity of the enzyme active site is the metal-binding domain, which for zinc consists of three cysteines at positions 133, 135, and 143, and a solvent molecule, presumably a hydroxide ion. Asparagine at positions 135 and 143 has been postulated as essential for magnesium-dependent ALAD (Erskine *et al.*, 1999).

Few microbial ALAD systems have been described, and the metallic component of ALAD differs among species (Chauhan *et al.*, 1997; Gudmundsdottir *et al.*, 1999; Mamet *et al.*, 1996; Rhie *et al.*, 1996). For example, zinc is required for ALAD activity in yeasts, whereas magnesium is required in *Bradyrhizobium japonicum* and *Pseudomonas aeruginosa*, and both zinc and magnesium are required by the *Escherichia coli* ALAD, although the requirement for magnesium in *E. coli* is not stringent (Chauhan *et al.*, 1997). Moreover, the sensitivity of ALAD to toxic metals depends on the identity of the metallic cofactor (Tanaka *et al.*, 1995; Chauhan and O'Brian, 1995). A genetic polymorphism at the ALAD locus has been described

BOX 5.1

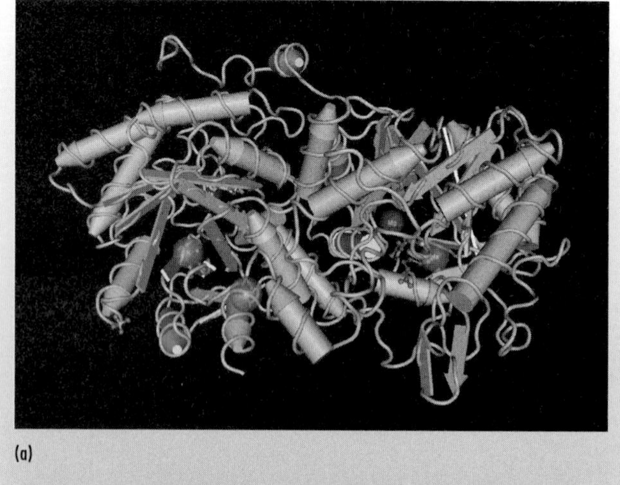

(a)

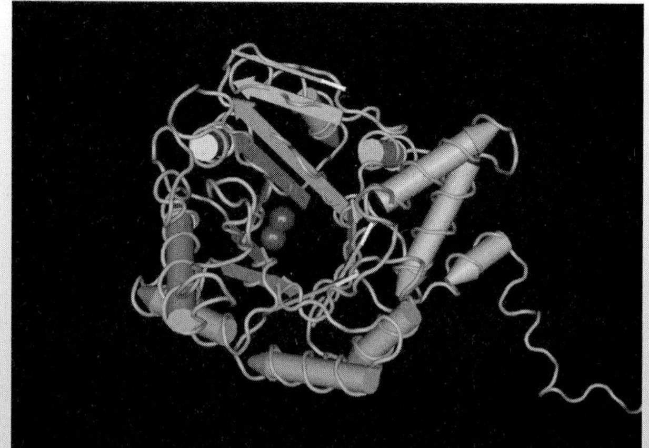

(c)

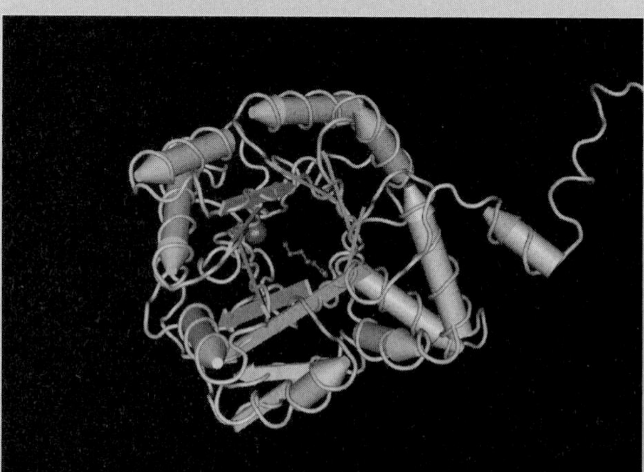

(b)

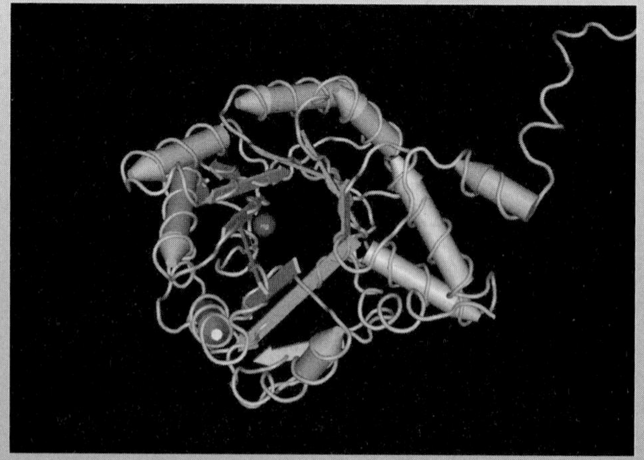

(d)

Porphobilinogen synthase, also known delta-aminolevulinic acid dehydratase (ALAD) catalyzes the reaction between two molecules of aminolevulinic acid (ALA). The ALA molecules engage in a Knorr-type condensation reaction to produce porphobilinogen (a). Across all phylogenetic domains, porphobilinogen and its derivative molecules play central roles in key biochemical reactions, including the trapping of electromagnetic radiation for driving photosynthesis (e.g. chlorophyll), and the transportation of oxygen for respiration (e.g. hemoglobin). Therefore, the enzyme is conserved across phylogenetic groups. However, there are variations in the enzyme's molecular structure and differences in its sensitivity to environmental factors suggesting limited influence of selective pressure on the evolution of ALAD. These differences are exemplified by the requirement for different metallic cofactors in the enzyme active site.

The three-dimensional structure of the ALAD protein showing the metal-binding domains is presented in (a) and (b). (a) The Mg-dependent ALAD of *P. aeruginosa* with the position of Mg shown relative to other metal ions. The N-terminal arms of the *P. aeruginosa* ALAD do not project out of the octamer, and the dimers are asymmetric, unlike in zinc -dependent ALAD found in yeasts, mammals, and Archaea. (b) Phylogenetic analysis of the amino acid sequence of the metal-binding domain of ALAD indicates that the occurrence of various metal ions in ecosystems inhabited by microorganisms exerts a strong selective pressure on the evolution of preference for specific metal ions. For example, the zinc version of ALAD is very sensitive to lead (Pb) and mercury (Hg) when these toxic metals replace zinc at the metal-binding domain of the protein (c and d, respectively). Protein structures were created and manipulated by means of the Cn3D software, version 4.0.

in humans, with implications for susceptibility to lead poisoning (Wetmur, 1994). The evolutionary origin of the ALAD polymorphism has not been established, but it is clear that geographic and strain-specific factors define the distribution of the two recognized ALAD alleles and the evolution of ALAD genotypes (Bishop *et al.*, 1998). The increasing number of genome sequences available for phylogenetic analysis could potentially illuminate some

CD:	COG0113.1, HemB, Query added	PSSM-Id:	9988	Source:	Cog
Description:	Delta-aminolevulinic acid dehydratase [Coenzyme metabolism]				
Taxa:	cellular organisms	Related:	pfam00490		
Status:	Alignment from source	Created:	7-Oct-2002		
Aligned:	53 rows	PSSM:	330 columns	Representative:	Consensus

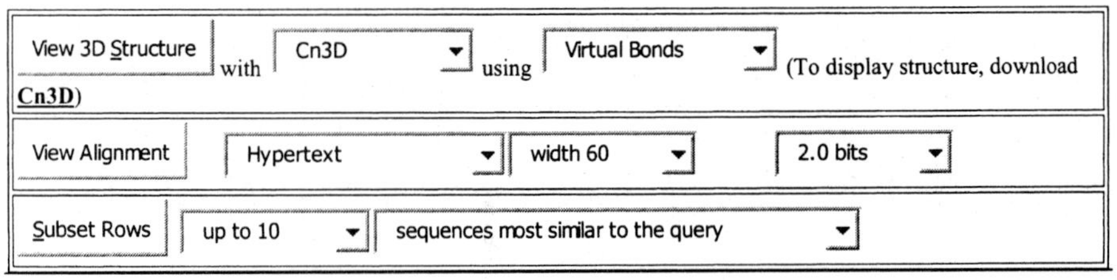

View 3D Structure with Cn3D using Virtual Bonds (To display structure, download Cn3D)

View Alignment Hypertext width 60 2.0 bits

Subset Rows up to 10 sequences most similar to the query

Fig. 5.1 **Partial output of the user interface produced by the BLAST database search engine used to recover sequences that share homology to the aminolevulinic acid dehydratase protein (ALAD encoded by bacterial *hemB* gene). Sequences were viewed and saved in the FASTA format, which was later imported into the ClustalX software for multiple alignments.**

of the key questions regarding ALAD metal requirement and the ecological constraints on protein sequence diversity.

Phylogenetic analysis was performed on the basis of the sequence of 335 amino acid residues in the C-terminal domain, including the metal-binding motif, of the protein. The analysis, representing 97% of the ALAD coding sequence in 55 taxonomic groups, included Bacteria, Archaea, and Eukarya branches. The sequences were recovered through the BLAST search engine (Fig. 5.1). The phylogenetic tree was constructed using the neighbor-joining method employed by ClustalX software (version 1.83). The multiple alignment parameters used were as follows:

Gap opening penalty = 15.00
Gap extension penalty = 0.30
Protein weighting matrix = Gonnet series
Delay divergent sequences = 30%
Residue specific penalties = "on"
Hydrophilic penalties = "on"
Hydrophilic residues = G P S N D Q E K R
Gap separation distance = 4.0
End gap separation = "off"

A partial result of the alignment is presented in Fig. 5.2. The aligned sequences were subsequently used to construct a bootstrapped neighbor-joining phylogenetic tree.

Two forms of the un-rooted phylogenetic tree are presented in Figs. 5.3 and 5.4, respectively. Bootstrapping based on 1,000 randomized trials showed that the ALAD sequences in the Eukarya and Archaea branches were very robust at 96–100% and 94–100% respectively, whereas the Bacteria branches were weakly robust at 30–100%.

The results of the phylogenetic analysis indicate greater diversity in Bacteria ALAD genes than the other phylogenetic groups. The data also support the "ecological affiliation" hypothesis for the evolution of ALAD metal requirement (Ogunseitan *et al.*, 2000). Representatives of Archaea and Bacteria lineages require zinc and/or magnesium as a cofactor depending on ecological affiliation, whereas mammals require zinc and plants require magnesium exclusively. Interesting exceptions to this rule include *Rickettsia prowazekii*, an obligate intracellular parasite that appears to have no requirement for metallic cofactors. Among

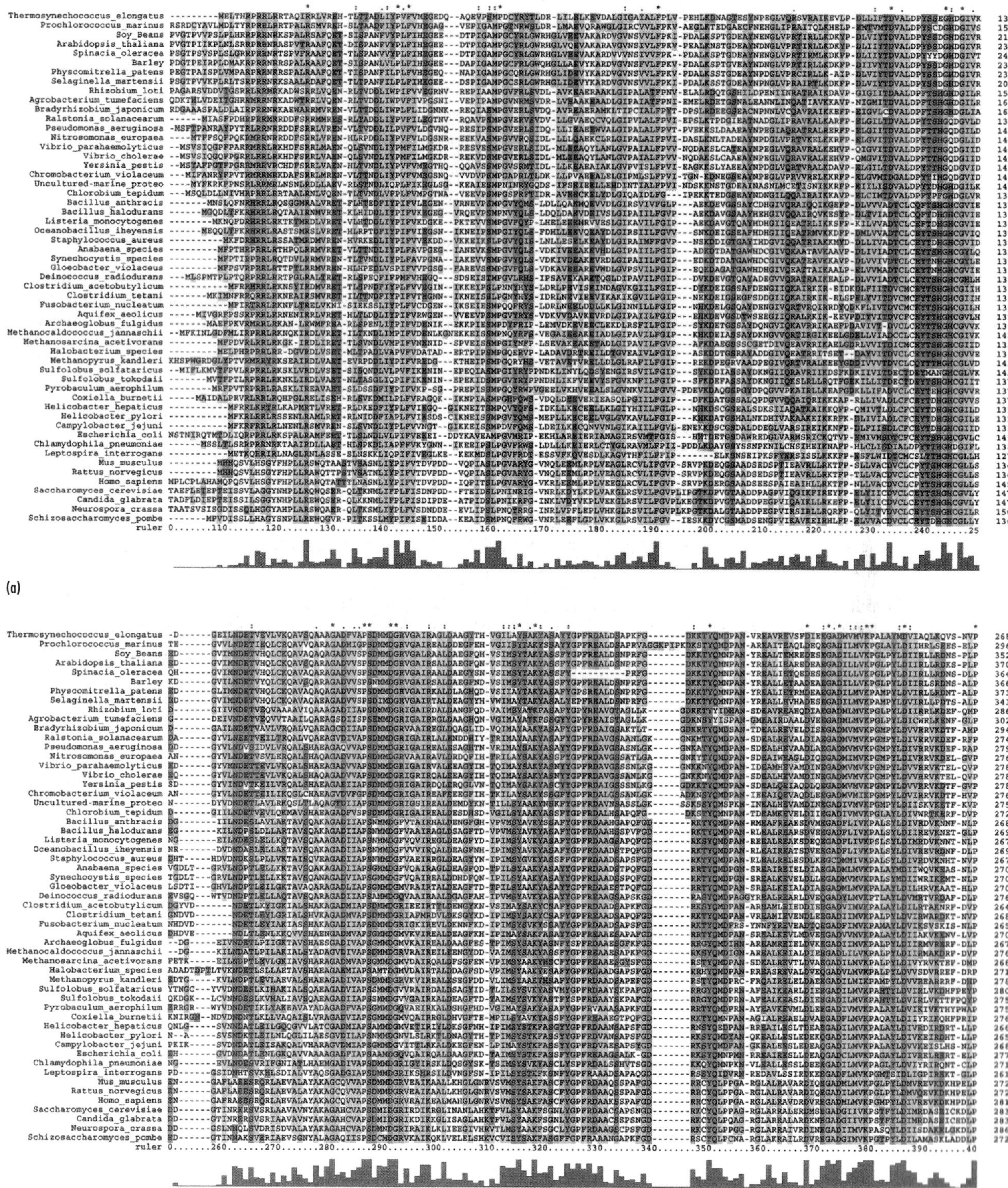

Fig. 5.2 Partial result of the multiple alignments performed with the ClustalX software on the amino acid sequence of ALAD from 54 organisms. * = fully conserved positions; : = one of the "strong" group of amino acids is fully conserved; . = at least one of the "weak" groups of amino acids is fully conserved. The bar heights represent intensity of consensus.

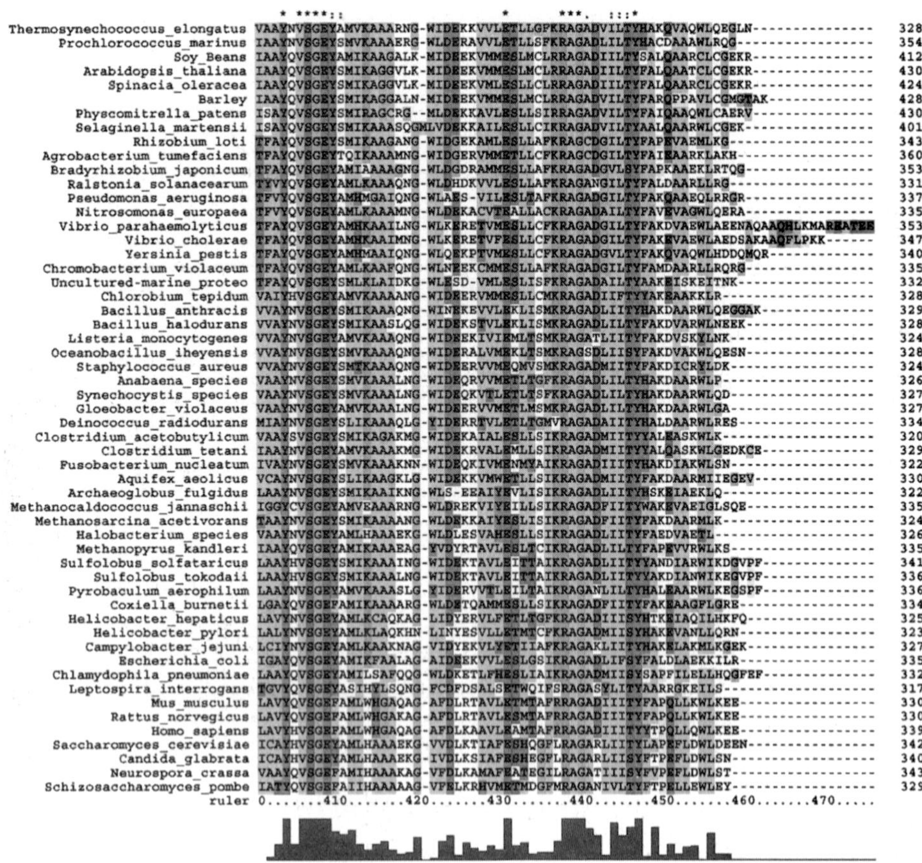

Fig. 5.2 *Continued*

the Bacteria, *Bradyrhizobium japonicum*, a bacterium that exists as a plant symbiont, uses four zinc atoms and eight magnesium atoms per ALAD molecule, and *E. coli* uses eight zinc atoms and eight magnesium atoms. The analysis supports the hypothesis that the distribution of metals in the immediate environment of organisms has contributed strong selective pressure on the molecular evolution of ALAD activity.

One possible reason for the emergence of magnesium-resistance dependent ALAD in photosynthetic organisms is the heavy demand for zinc in the ribulose-5-phosphate bis carboxylase (RubisCO) pathway. In the scarcity of zinc, photosynthesis is repressed. Therefore, the functions of supporting enzymes leading to RubisCO activity are optimized for an alternative metallic cofactor. This also explains the presence of magnesium-dependent ALAD in *Bradyrhizobium japonicum*, a nodule-forming plant symbiont. The thermodynamic equilibrium between Zn-protoporphyrin and Mg-protoporphyrin favors the zinc complex by a factor of 3,000. The thermodynamic factor requires that free Zn^{++} be eliminated from the centers of photosynthetic activity (Krishnamoorthy, 2000).

The example of ALAD is provided primarily to illustrate the three-pronged procedural steps and outcome of phylogenetic analysis and tree construction. The three steps are:

1 Searching and collating related sequences from a sequence depository, in this case, by means of the BLAST search engine.
2 Conducting multiple sequence alignments with ClustalX.
3 Constructing a phylogenetic tree using the neighbor-joining procedure.

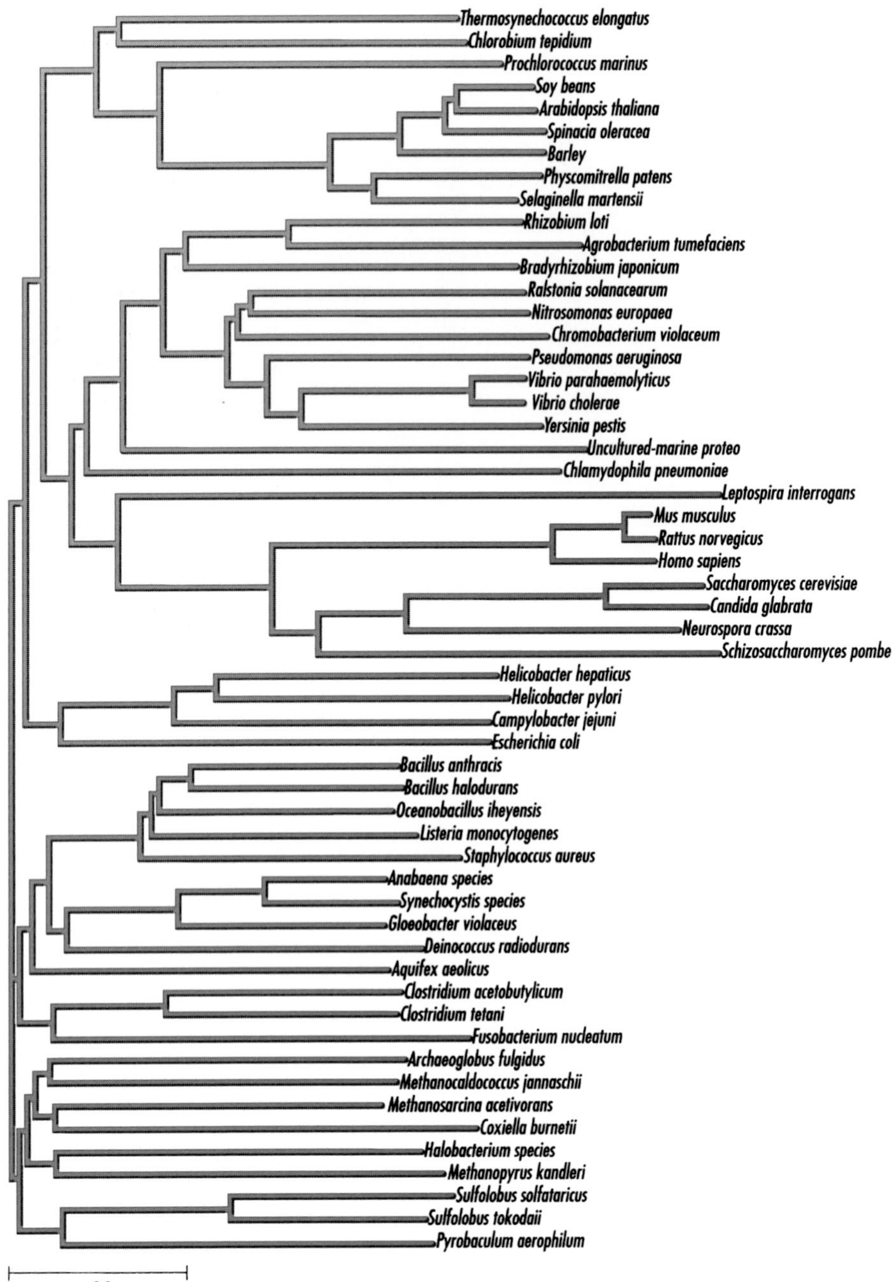

Fig. 5.3 Radial presentation of the phylogenetic tree based on 335 amino acid sequence of ALAD in 54 organisms. The radial presentation is not as easy to read as the slanted cladogram (Fig. 5.4), however, it makes it unequivocally clear that the tree is not rooted.

CONCLUSION

In a little more than two decades, molecular approaches for exploring microbial diversity have developed rapidly, emerging from rudimentary comparisons of gross DNA homologies that were inferred from hybridization signal intensities, to the level of computer-assisted processing of molecular sequences to generate fine-scale phylogenetic trees. The rapid accumulation of molecular sequence data engendered the discipline of bioinformatics, and

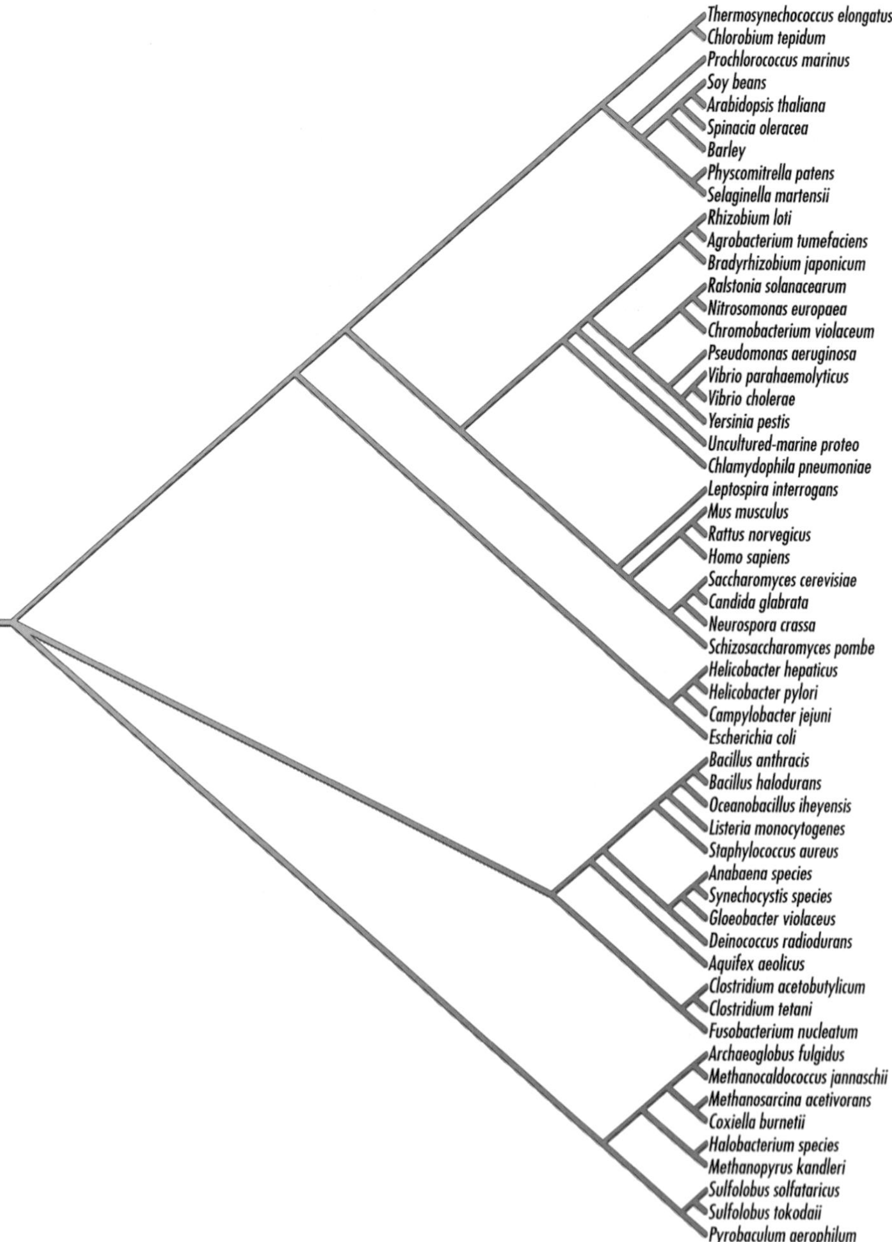

Fig. 5.4 Slanted cladogram of the unrooted phylogenetic presented in Fig. 5.3. This mode of presentation facilitates the visualization of species clusters which are related by common molecular sequences. Near the top are plant-associated clusters; in the middle are animal-associated clusters; and a preponderance of *Archaea* at the bottom. The *Bacteria* are interspersed throughout the tree, suggesting more diversity in this phylogenetic group than in the others.

innovative tools developed in that discipline have become indispensable for rapid analysis of microbial diversity. The reliability of phylogenetic trees depends first on the integrity of the sequence data used to generate the tree. Molecular sequencing is now mostly automated, but there remains a certain level of uncertainty, and caution must be exercised in using sequences from sources other than the investigator's own laboratory. There is no single "best" technique for producing phylogenetic trees. It is to be expected that the proliferation of techniques will continue, but in all likelihood, some convergence will occur to minimize the sources of error in each technique. Finally, the interpretation of phylogenetic trees and their use in generating testable hypotheses is where most of the intellectual effort remains for students of microbial diversity. Microorganisms do not generally leave fossils, and genetic exchange is rampant. Therefore, the reconstruction of microbial phylogenetic relationships based on molecular sequence information can be challenging, however, the strategy provides a rich opportunity for exploring the relationship between genotype and phenotype across groups of organisms with fascinating natural histories shaped by ecological adaptation and speciation over geological time frames.

QUESTIONS FOR FURTHER INVESTIGATION

1 Search for the protein sequence of the ribulose bisphosphate carboxylase large chain (water-stress responsive proteins 1, 2, and 14) using BLAST (see example web page below).

CD:	COG1850.1, RbcL, Query added	PSSM-Id:	11560	Source:	Cog
Description:	Ribulose 1,5-bisphosphate carboxylase, large subunit [Carbohydrate transport and metabolism]				
Taxa:	cellular organisms	References:	2 Pubmed Links	Related:	pfam00016
Status:	Alignment from source	Created:	7-Oct-2002		
Aligned:	13 rows	PSSM:	429 columns	Representative:	Consensus

View Alignment as Hypertext width 60 color at 2.0 bits

Subset Rows up to 10 sequences most similar to the query

2 View and save the sequences belonging to microbial and plant taxonomic groups in the "mFasta" format.
3 Import the sequences into "ClustalX" and conduct multiple sequence alignment. Experiment with the various user-specified parameters to discover how these parameters change the alignment.
4 Use ClustalX or other software to create the phylogenetic tree.
5 Describe the main features of the phylogenetic tree in terms of number of clades, branch lengths, and bootstrap values.
6 What story does the phylogenetic tree tell about the evolution of ribulose bisphosphate carboxylase?

SUGGESTED READINGS

Altschul, S.F., T.L. Madden, A.A. Schaffer, J. Zhang, Z. Zhang, W. Miller, and D.J. Lipman. 1997. Gapped BLAST and PST-BLAST: A new generation of protein database search programs. *Nucleic Acids Research*, **25**: 3389–402.

Hall, B.G. 2001. *Phylogenetic Trees Made Easy. A How-to Manual for Molecular Biologists*. Sunderland, MA: Sinauer.

Rambaut, A. and L. Bromham. 1998. Estimating divergence dates from molecular sequences. *Molecular Biology and Evolution*, **15**: 442–8.

Schierup, M.H. and J. Hein. 2000. Consequences of recombination on traditional phylogenetic analysis. *Genetics*, **156**: 879–91.

Swofford, D.L. 2000. *PAUP*: Phylogenetic Analysis Using Parsimony and Other Methods* (software). Sunderland, MA: Sinauer.

Thompson, J.D., T.J. Gibson, F. Plewniak, F. Jeanmougin, and D.G. Higgins. 1997. The ClustalX–Windows interface: Flexible strategies for multiple sequence alignment aided by quality analysis tools. *Nucleic Acids Research*, **25**: 4876–82.

PRINCIPLES AND APPLICATIONS

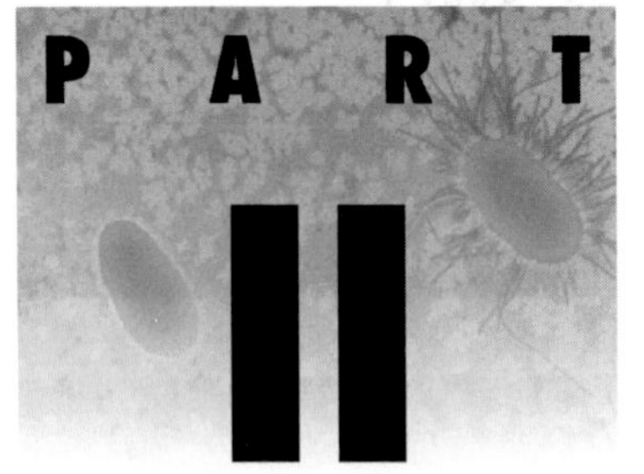

"While Lamarck was wrong to believe that organisms could incorporate the outer world into their heredity, Darwin was wrong in asserting the autonomy of the external world. The environment of an organism is not an independent, preexisting set of problems to which organisms must find solutions, for organisms not only solve problems, they create them in the first place. Just as there is no organism without an environment, there is no environment without an organism. 'Adaptation' is the wrong metaphor and needs to be replaced by a more appropriate metaphor like 'construction.' . . . The proper view of evolution is then of coevolution of organisms and their environments, each change in an organism being both the cause and the effect of changes in the environment. The inside and the outside do indeed interpenetrate and the organism is both the product and location of that interaction."

R.C. Lewontin, on *Genes, Environment, and Organisms* (1995).

ENVIRONMENTAL EVOLUTION

CHAPTER 6

INTRODUCTION

This chapter deals with the reciprocal causative interactions between the evolution of microbial life forms and the physical-chemical environment. The discussion begins with the conditions that may have facilitated the self-organization of molecules into primordial microbes and the sensitive dependence of environmental evolution on the physiological processes that characterized life in the Archaean period. There are two major goals for research on the early environmental conditions of Earth, and on the structure and function of early life forms. The first goal is to reveal the possible existence of extraterrestrial life. The second goal is to elucidate the historical contributions of living systems to the causes of environmental change at the global level. With these goals as primers, the chapter focuses on:

1 Biogenesis theories in the context of early environmental conditions on Earth.
2 The record of microbial diversity in stromatolites.
3 The importance of microbial mats for understanding physiological and ecological diversity of microorganisms.
4 The concept of environmental homeostasis and its implications for the diversity of microorganisms in extreme environments.
5 The endosymbiotic theory and genetic innovation in microbial communities.
6 Applications of the knowledge of interactions between microbial diversity and environmental evolution to global issues with the Earth system.

Chapter contents

BIOGENESIS

The year 1859 marked two seemingly diametrically opposite landmark events in scientific understanding of the origin of living systems and their interactions with the environment. In that year, Charles Darwin (1809–82) published his thesis on "The Origin of Species", which gave us a remarkable framework for understanding evolutionary divergence through spontaneous mutation and natural selection. Darwin also believed that the root of the evolutionary tree emerged spontaneously from an interaction between energy and inorganic matter. He wrote in a letter to botanist Joseph Dalton Hooker (1817–1911):

> . . . if we could conceive in some warm little pond with all sorts of ammonia and phosphoric salts, light, heat, electricity, etc., present. . . . that a protein compound was chemically formed ready to undergo still more complex changes . . . (Darwin Correspondence Project; http://www.lib.cam.ac.uk/Departments/Darwin/)

Also in 1859, Louis Pasteur (1822–95) started the classical experiments which have been interpreted to demonstrate unequivocally that spontaneous generation of living cells does not occur from inorganic and/or organic chemical matter (Robbins, 2002). Darwin's views of chemical evolution and Pasteur's evidence against spontaneous generation have significantly shaped the direction of research in biogenesis, although contemporary versions of this debate tend to focus on whether chemical evolution occurred on Earth or elsewhere in the expanse of galactic space. There is currently no scientifically testable alternative to chemical evolution as the quintessential process that led to biogenesis.

The evidence for co-evolution of life and environment is apparent in comparative assessments of planetary characteristics in the solar system. The non-equilibrium state of the geochemical cycling of elements on Earth has been attributed to ecosystem functions maintained by biological diversity, and conversely, episodic incidences of environmental change have been linked to periods of evolutionary innovation (Hoffmann and Parsons, 1997; Kasting and Seifert, 2002). For example, the atmospheric composition of Earth, when compared to the composition on Mars and Venus, provides strong large-scale sustained impacts of physiology on the physical parameters that define planetary characteristics (Des Marais *et al.*, 2002; Kasting *et al.*, 2001; McElroy, 2000) (Fig. 6.1). There is considerable

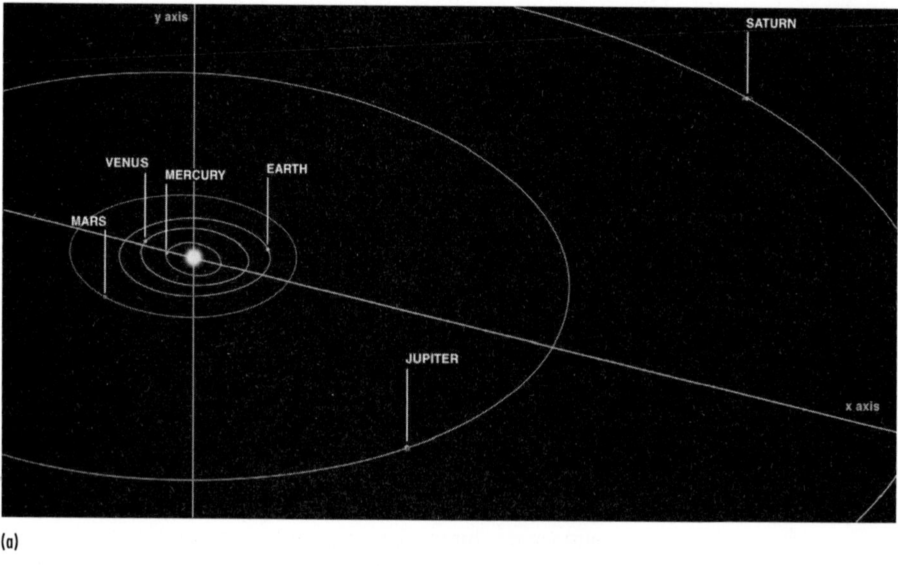

(a)

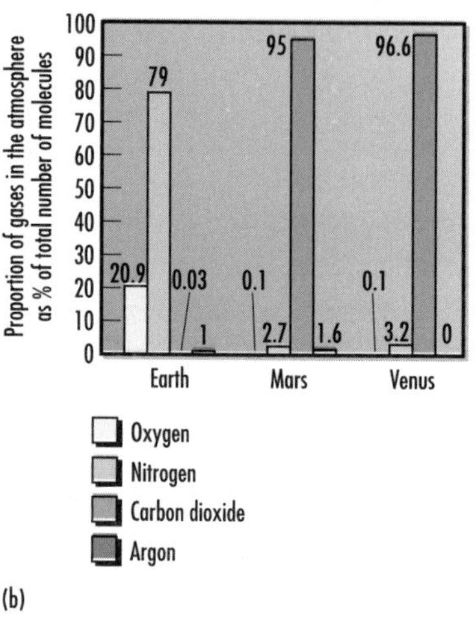

(b)

Fig. 6.1 (a) The inner planets of the solar system, Mercury, Mars, Earth, and Venus were probably created by similar processes, but have taken distinct evolutionary pathways, with Earth being the only planet with environmental conditions to harbor life as we know it. Image reproduced by courtesy of NASA. (b) The evidence for life on Earth can be inferred from the unusual atmospheric composition when compared to the atmospheric composition of Mars and Venus.

interest in predicting the fate of this magnificent co-evolution with respect to recent anthropogenic events that influence global environmental conditions. However, planetary biogeochemistry is clearly not yet a predictive science because the sample size from which inferences can be drawn is too small (von Bloh *et al.*, 2003). To understand the fate of co-evolution of life and environment requires improved assessment of biological signatures of functional diversity in ecosystems, and how these relate to planetary properties. This level of understanding needs to occur before reliable models can be developed to predict the conditions necessary for extraterrestrial life, based on the record of conditions that led to the origin of living systems on Earth.

It is axiomatic that independent unicellular microorganisms are the precursors of complex multicellular life forms (Hartman *et al.*, 1987). There is near scientific consensus on the generation of phylogenetic diversity through spontaneous mutation, endosymbiosis, and natural selection, but there is considerably less agreement on the origin of unicellular life forms, and on the process that led to multicellularity from these primal cells (Maynard Smith, 1999; Miklausen, 1997). Since Louis Pasteur's brilliant experiments, the idea of spontaneous generation of living cells from inanimate matter has been extinguished, leaving a gaping void where pompous hypotheses on the origin of life once engaged serious scientists (Robbins, 2002). The void was soon filled with experiments at the nexus of geochemistry and biology, where the production of organic compounds was recorded after prolonged incubation of inorganic chemicals under conditions that simulated the early Earth environment (Miller, 1953; Honda *et al.*, 1989; Oparin, 1953; Oro *et al.*, 1990; Urey and Miller, 1959). The results of those early experiments in biogenesis were sufficiently encouraging to sustain active research programs, but the real gains toward understanding the origin of living systems were very modest. Among those gains are those arising from the concerted effort to understand the form and function of the genetic material at the molecular level, culminating in the discovery of self-catalyzing types of ribonucleic acids, which are now considered to be among the first generation of molecules with the capacity to serve as the template for evolution (see Box 6.1).

Despite the rapid accumulation of knowledge concerning the ubiquitous nature of the molecular synthetic reactions that sustain living systems, it has been argued that the odds against the formation of such self-catalyzing molecules within the temporal and parametric constraints of prehistoric Earth may be insurmountable. The alternative to Earth-bound biogenesis was supported by a school of dissenting investigators who favor variations on the theme of panspermia, where preformed genetic systems are transported on meteorites and are presumed to have inoculated Earth in the distant past (Kamminga, 1982; Lazcano, 2000). The original versions of panspermia are usually attributed to Svante Arrhenius (1859–1927) and Hermann von Helmholtz (1821–94) among others, although several investigators have promoted the modern version (Cahan, 1994). These investigators include the astrophysicists Fred Hoyle (1915–2001) and Chandra Wickramasinghe, cosmochemist Cyril Ponnamperuma, and molecular biologist Francis Crick (Crick, 1981; Crick and Orgel, 1973; see Box 6.2).

The case for panspermia

Astronomers have long recognized that Space is not empty between visible entities, and the evidence is incontrovertible that interstellar matter consists of dust particles ranging in diameter from 10 to 300 nm (Hoyle and Wickramasinghe, 1990). The chemical composition of interstellar dust particles has been the focus of intense research for more than a century and two major approaches are routinely used to shed light on the question. The first is based on remote sensing of light absorbance by interstellar dust clouds, and this approach has formed the cornerstone of Hoyle and Wickramasinghe's experiments and microbiological inferences. The second approach is the physical analysis of Space materials such as

BOX 6.1

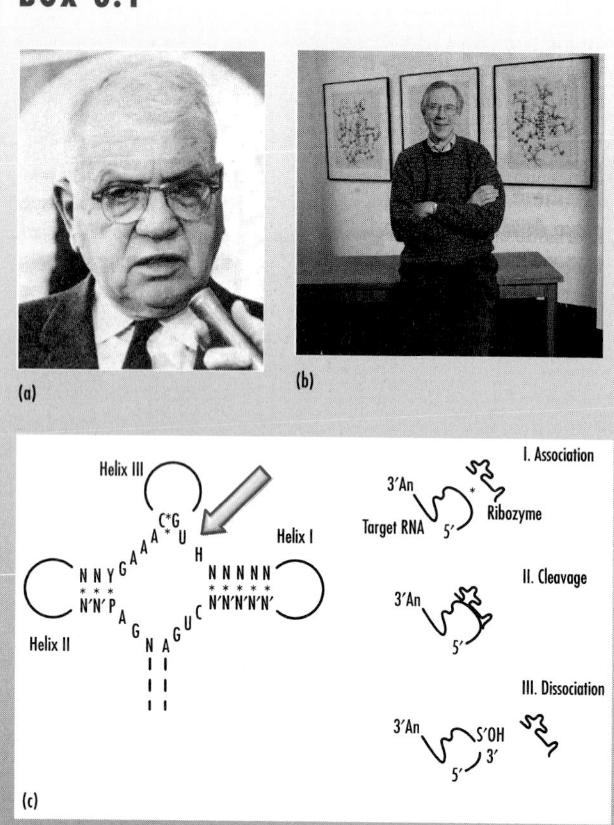

(a) Harold Urey (1893–1981), photograph by courtesy of NASA history archives; (b) Thomas Cech, photograph copyright Paul Fetters for Howard Hughes Medical Institute; (c) The catalytic reaction of ribozymes involves a three-step reaction, namely association, cleavage, and dissociation (Diagram after R.L. Miesfiled, The University of Arizona; see also Scott, 1999; Scott *et al.*, 1996).

Biogenesis theories, which are based on terrestrial mechanisms for producing autocatalytic self-replicating molecules from inorganic matter and energy, were championed by two separate camps of investigators. One camp focused on autotrophic origins, whereas the other camp emphasized heterotrophic origins. Among those who focused on autotrophism is Mexican biologist Alfonso Herrera who proposed plasmogenesis as the process of producing photosynthetic organisms from a mixture of hydrogen cyanide and formaldehyde—compounds thought to be abundant in the early Earth environment (Lazcano, 2000; Oro *et al.*, 1990; Perezgasga and Silva, 2003). Alexander Oparin is one of the best-known advocates of heterotrophism. He proposed that anaerobic bacteria emerged with the capacity to use organic material that was originally synthesized from components of the early Earth atmosphere (Oparin, 1953). To prove that such complex organic materials could have been synthesized abiotically from a mixture of ammonia, methane, and hydrogen, Stanley Miller and Harold Urey designed simple flask microcosm experiments to simulate the primitive atmosphere. The experiments resulted in the detection of amino acids, nucleotides, and lipids—the main molecular building blocks of extant cells (Miller, 1953). It is generally agreed that autopoiesis (self-making) is a quintessential feature of living systems, but there is considerable debate on the identity of the initial molecule(s) (proteins, DNA, or RNA) which endowed living systems with this property (Dillon, 1978). Thomas Cech is credited with the discovery of autocatalytic RNA (ribozymes) having the capacity for genetics and metabolism (Doudna and Cech, 2002; Zaug and Cech, 1986). Many investigators now accept ribozymes as the most likely candidates for the primordial molecule in biogenesis (Joyce, 2002; Zamore, 2002).

chondrites that fall intact to Earth, providing valuable samples for analysis by cosmochemists such as Ponnamperuma.

Early theories regarding the chemical composition of interstellar dust were based on quantitative measurements of dimmed starlight between the wavebands 550 nm and 350 nm. The studies led to the most widely accepted theory, proposed in the 1940s, that interstellar dust consists of frozen inorganic solids including solid ammonia, hydrogen cyanide, methane, and water ice (Matthews, 2000). Although many subsequent measurements supported this theory, it was eventually disqualified because more sophisticated measurements in the 1960s using satellite-based spectrophotometers demonstrated a significant discrepancy between the extinction of light in the ultraviolet range for interstellar dust and known extinction curves for water ice. Following the demise of the water-ice theory, Hoyle and Wickramasinghe, then at Cambridge University, became involved in the subject and were the first to argue, in a series of papers published between 1962 and 1969 in *Nature* and *Monthly Notices of the Royal Astronomical Society*, that solid carbon (graphite), iron, and silicates were plausible constituents of interstellar dust (Hoyle and Wickramsinghe, 1978–97). The experiments were based on comparative analysis of spectral signatures, in particular the discovery of a 10-micrometer spectral feature, of interstellar materials and known elements or compounds.

Despite the success of the early studies, it remained doubtful that inorganic materials alone could account for the quenching of light by interstellar materials. Innovative techniques developed in radioastronomy were instrumental for exploring the presence of organic

BOX 6.2

(a)

(b)

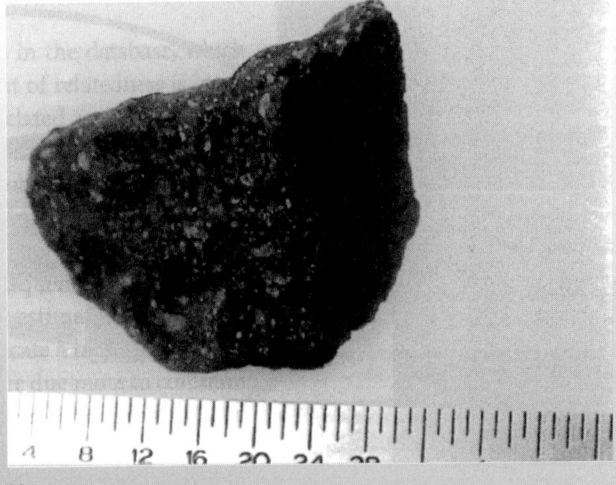

(c)

(a) Svante Arrhenius (1859–1927) is credited with invigorating the panspermia hypothesis (Arrhenius, 1903, 1908). He won the Nobel Prize for Chemistry in 1903 on the basis of his research in electrochemistry. (b) Sir Fred Hoyle (1915–2001) was the first to propose a theory of the origin of heavy elements within stars. He was awarded the 1997 Crafoord Prize for this and for his pioneering work on stellar evolution. Hoyle and Chandra Wickramasinghe worked together on the modern version of panspermia through a proposal that considered interstellar dust particles as biological in origin. Jointly with Hoyle, Wickramasinghe was awarded the International Dag Hammarskjold Gold Medal for Science in 1986. Among the investigators aiming to discover the hard evidence for panspermia, Cyril Ponnamperuma studied the chemical composition of the Murchison meteorite (c), a carbonaceous chondrite which fell in Australia on September 28, 1969. Analysis of the meteorite revealed first evidence of the presence of a large number of non-biological amino acids representing equal amounts of left- and right-handed molecules (Ponnamperuma, 1992).

Photograph credits: (a) © The Nobel Foundation, Sweden (http://www.nobel.se/chemistry/laureates/1903/arrhenius-bio.html); (b) California Institute of Technology—Institute Archives; (c) author's own collection.

materials in interstellar dust. In a groundbreaking publication, Wickramasinghe (1974) proposed polyformaldehyde as a major component of interstellar dust. The foray into microbiology came soon afterwards in 1976–77 with the coincidental discovery of a 220-nm wavelength interstellar absorption feature in carbonaceous chondrites and the Murchison meteorite, followed by the first explicit reference to polysaccharides in reference to the spectra of galactic sources (Hoyle and Wickramasinghe, 1978; Ponnamperuma, 1992). In an attempt to find materials that exactly match the spectrum of interstellar dust, including absorptions in the 2–4 micrometer band, the 8–12 micrometer band, and a peak at 18 micrometers, Hoyle and Wickramasinghe (1978) identified dry cellulose as having the best match.

The introduction of polysaccharides as candidate components of interstellar dust was heavily criticized as noted by Hoyle and Wickramasinghe in their book pointedly titled *From Grains to Bacteria* (1984a):

> The more precise the correspondences we calculated between our models and the observations, the greater was the measure of opposition we received from individuals, from journals and from funding agencies. The introduction of polysaccharides, because of their biological associations apparently, became a signal for papers to be turned down by journals, and/or even the most modest grant applications to be thrown back in our faces. In contrast to this development, which became ever more severe from 1978–1980 onwards, our earlier less successful efforts—less successful in matching the facts that is to say—had always secured support and publication without much effort on our part, and on occasions been afforded modest approbation by science umpires. It was in this period that we encountered the phenomenon of "socially welcomed disproofs".

Hoyle and Wickramasinghe amassed a large body of evidence on bacteria spectral signatures matching those of interstellar dust better than any substance ever examined. With microbiologist Shirwan Al-Mufti at Cardiff University in Wales, the team produced a series of spectrograms demonstrating that the transmittance curve for *Escherichia coli* cells, whether at 20°C or at 350°C, showed invariance over 3.3–3.5 micrometers, clearly matching flux measurements for the cosmic body GC-IRS 7. Inspired by Svante Arrhenius' speculations and Louis Pasteur's experiments, Hoyle and Wickramasinghe also produced data showing nearly perfect correspondence between the infrared spectrum of the Trapezium nebula and diatom cultures, and they published treatises on the optical properties of bacterial grains (Hoyle, 1980; Hoyle and Wickramasinghe, 1978–97). They produced experi-mental results showing the extraordinary diversity of microbial resistance to extreme environmental conditions, including electromagnetic radiation and heat. They also produced the first theoretical analysis of limiting nutrients for bacterial proliferation in cosmic dust, dispelling doubts regarding insufficient phosphorus in Space to sustain nucleic acid-based biological proliferation (Hoyle and Wickramasinghe, 1984b; Westherimer, 1987).

According to Hoyle and Wickramasinghe, all the evidence led inevitably to the improbable conclusion that the galaxy is a microbiological system. In 1983, the Royal Astronomical Society agreed to sponsor a conference entitled "Are Interstellar Grains Bacteria?" At the meeting, Hoyle and Wickramasinghe summarized the evidence for the microbial theory of interstellar dust, and concluded with a proposal about the role of planets as centers for the amplification of biological materials. Hoyle also recalled the century-old argument between Charles Darwin and the equally famous geologist Charles Lyell on the contradiction of "evolution by natural selection" under circumstances of relative environmental stability on Earth. The only plausible resolution of the contradiction, according to Hoyle, is that there is a continuous input of novel genetic material from Space that drives biological evolution on Earth. Presumably, the environmental selective pressure on Earth is not sufficient to support the apparent diversity of species, but the radiation occurs nonetheless due to the intrusion of microorganisms present in interstellar dust (Hoyle and Wickramasinghe, 1986). To advance the implication of their controversial speculations, Hoyle and Wickramasinghe approached some intractable questions in microbial ecology, however, their ideas were generally considered too far-fetched, scientifically dubious, and lacking parsimony. For example, they proposed that the linkage between influenza virus epidemics and seasonal trajectories, and the emergence of new pathogens resulted from the intrusion of extraterrestrial particulate matter carrying bacteria and viruses descending into the Earth's stratosphere during the winter months (Hoyle and Wickramasinghe, 1979; Hoyle *et al.*, 1985a–c).

The controversial idea that meteorites may transport actual biological cells from outer space to Earth first enjoyed mainstream evaluation in 1996 when scientists at NASA published photographs suggesting the presence of "nanobacteria" in the Martian meteorite ALH84001, a 4.5 billion-year-old orthopyroxene cumulate (Fig. 6.2; McKay *et al.*, 1996). Although the biological nature of the cylindrical structures observed in the Martian meteorite is questionable, the discovery energized interest in astrobiology by provoking and expediting NASA's Mission to Mars program (Nishioka, 1998).

To further probe the nature of interstellar dust, NASA launched on February 7, 1999, a solar-powered spacecraft known as *Stardust* with the mission to collect samples of interstellar dust. The interstellar dust samples will be returned to Earth for physical analysis. The *Stardust* spacecraft is equipped with a Dust Flux Monitor Instrument that could reproduce Hoyle and Wickramasinghe's remote spectrometric experiments. The spacecraft also harbors a Cometary and Interstellar Dust Analyzer Instrument (IDAI) that will conduct real-time analysis of dust components. Preliminary data from these mass spectroscopic experiments suggested that interstellar dust particles are made up of three-dimensionally cross-linked polymeric-heterocyclic aromatics. This result is consistent with Hoyle and Wickramasinghe's

(a)

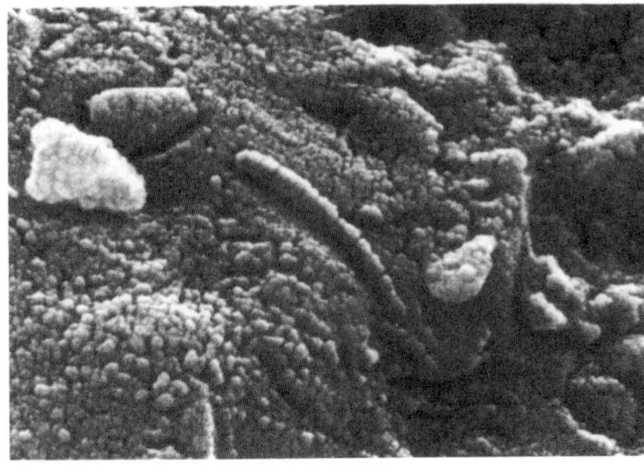

(b)

Fig. 6.2 (a) Martian meteorite ALH84001 before processing. (b) A high-resolution transmission electron microscope image of a chip from the meteorite showing the outline of what were initially thought to be microscopic fossils of bacteria-like organisms. Image reproduced by courtesy of NASA.

model, but the data do not support more prominent models that proposed planar polyaromatic hydrocarbon structures as the main components of interstellar dust (Hoyle and Wickramasinghe, 1990). In a much more "down to Earth" scenario regarding the existence of primordial hydrocarbon compounds, Gold (1999) proposed a provocative theory entitled the "deep hot biosphere" where abiogenic petroleum deposits dating back to the formation of Earth played a vital role in originating and nurturing a deep subterranean microbiosphere that is independent of photosynthesis. According to this view, Earth surface biosphere derived from this subsurface community. Furthermore, such subsurface biospheres are speculated to be present in many other planets of the solar system. The implications of the *Stardust* findings for Gold's theories are not yet clear, but they warrant close investigation.

Stardust is not expected back on Earth until January 2006, but there have already been very exciting technological spin-offs from the *Stardust* project for microbial ecologists and environmental scientists. To collect dust samples from space, NASA scientists invented a new material known as "aerogel", which contains an extremely durable material capable of trapping dust particles moving at a speed of 20–25 km per second relative to the spacecraft (Fig. 6.3). The aerogel is designed to bring the particles to a stop within a distance of less than a centimeter, without compromising their chemical integrity. However, it is not possible to preserve the biological integrity of the samples under such conditions. *Stardust* collected the first set of samples between February and May 2000. The speed with which sample materials are stopped generates 4–6 eV per H atom, which is sufficient to inactivate cellular integrity and carbon–hydrogen bonds in aliphatic side chains. However, cross-linked aromatic structures, polysaccharides, and possibly proteins are expected to withstand such impact. The preliminary data from *Stardust* are consistent with the presence of these biochemical molecules, but the data do not prove conclusively that they are of biological origin. There is a lot riding on the dusty payload of NASA's *Stardust*. With the rapid pace of microbial genome sequencing and proteomics research, knowledge of the molecular determinants of microbial resistance to "unearthly" extreme environmental conditions is expanding. These terrestrial and extraterrestrial explorations are fundamental to the advancement and refinement of the current framework on biogenesis. To follow the progress of *Stardust*, visit the web site at http://stardust.jpl.nasa.gov/index.html.

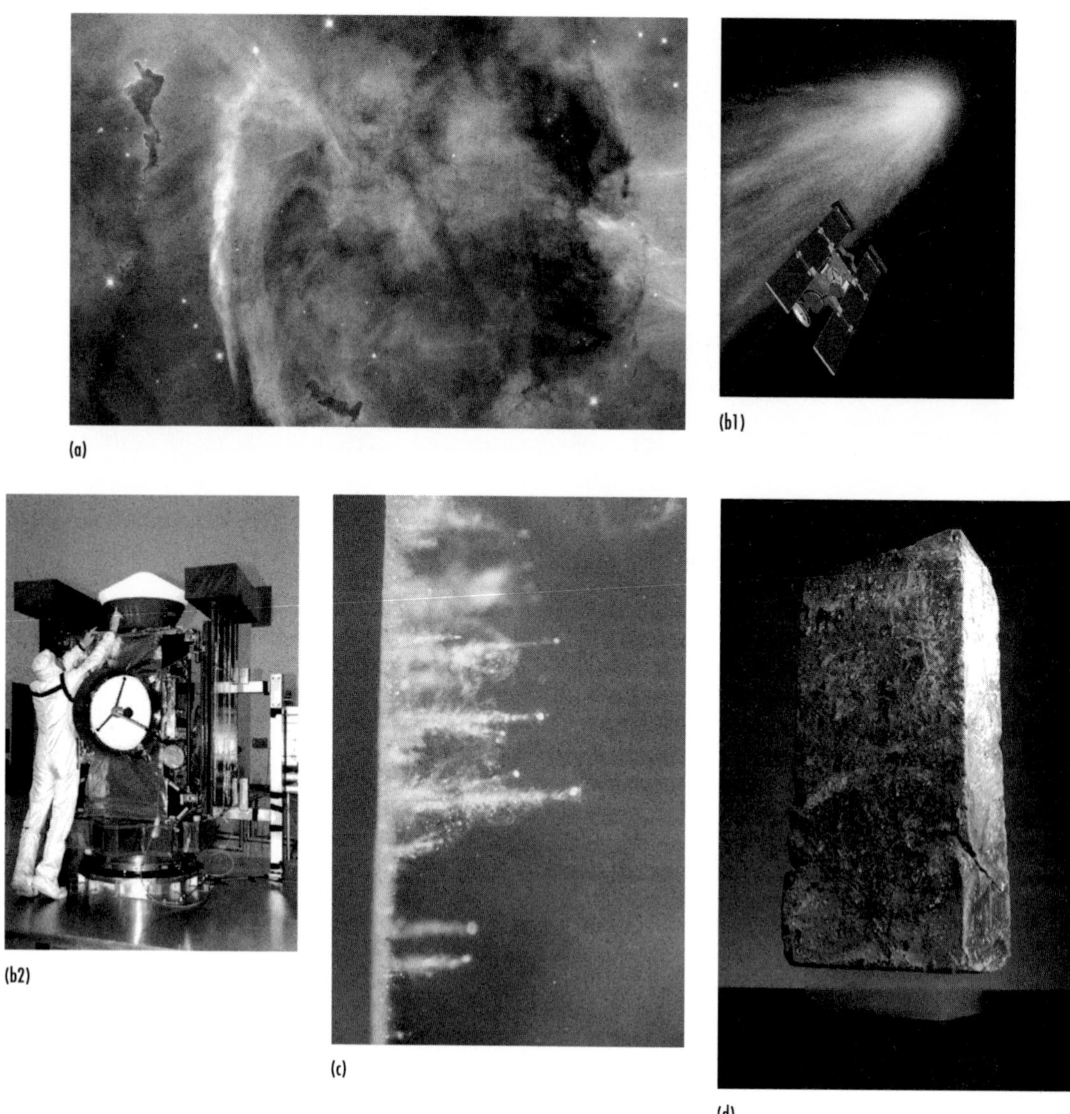

Fig. 6.3 (a) Detail from Hubble Space Telescope's image of the Carina Nebula (NGC 3372), a giant cloud of cosmic dust from an expired star. The image is a montage assembled from different photographs made in April 1999 with Hubble's Wide Field Planetary Camera 2, which used six different color filters. Image credit: Courtesy of NASA Hubble Heritage Team (AURA/STScI). (b1) Artist's conception of the NASA *Stardust* Spacecraft designed to collect interstellar dust particles for analysis. The relative size of the *Stardust* Spacecraft sample collection module is shown here in comparison to a NASA technician (b2). (c) Particles captured in "aerogel", an extraordinary material being used in the *Stardust* Spacecraft to collect interstellar dust samples. Particles shot into aerogel at high velocities leave elongated cone-shaped trails (Nishioka, 1998). (d) The strength of aerogel is demonstrated by the ability of a 2 g piece of the material to support a 2.5 kg brick. Image credit: Courtesy of NASA.

THE HISTORY OF MICROBIAL DIVERSITY IN STROMATOLITES (MICROBIALITES)

The solar system is estimated to be 4.6 billion years old, and the inner planets including Earth were formed soon afterward. The age of Earth is inferred from the oldest known crystalline rock, which has been dated to approximately 4 billion years, and from the oldest known sedimentary rock, the Isua formation at the west coast of Greenland, which has been dated at 3.8 billion years (Emiliani, 1992; Moorbath, 1995). The age of the oldest living systems has been more difficult to establish, although three lines of evidence point to the occurrence of abundant microbial life at least 3.5 billion years ago (Fenchel, 2002a; Lazcano

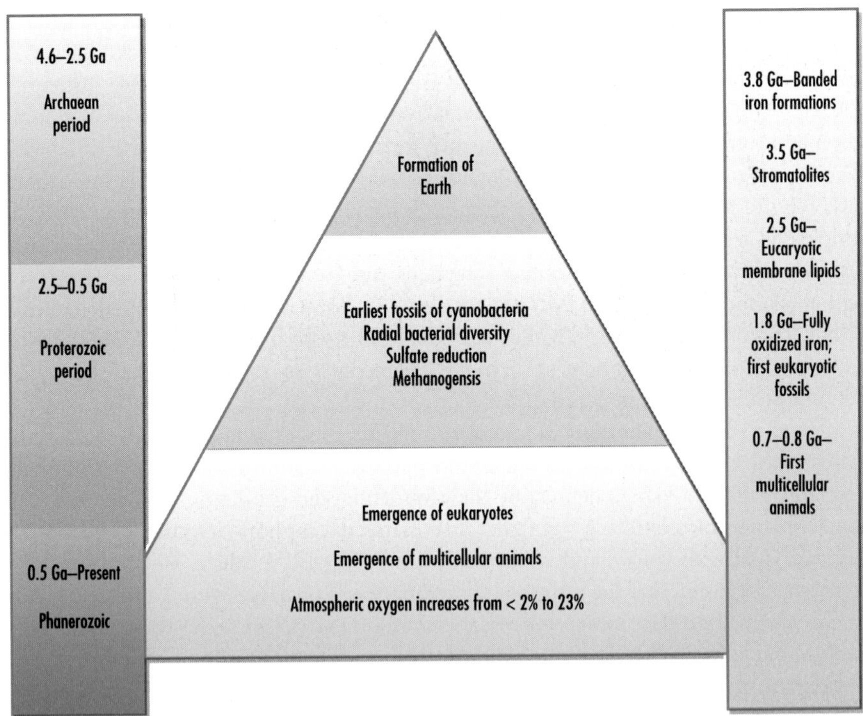

Fig. 6.4 Biogeological time frame and record of microbial evolution on Earth (http://www-curator.jsc.nasa.gov/curator/antmet/marsmets/alh84001/ALH84001-EM4.htm).

and Miller, 1994; Morris *et al.*, 2002; see Fig. 6.4). There is general agreement that 80% of the natural history of life on Earth is exclusively a history of microbial communities (Colwell *et al.*, 1996). The first line of evidence regarding the antiquity of microbial life forms consists of ancient microbial fossils which have been found at almost all places where prehistoric environmental conditions support their preservation (Corsetti *et al.*, 2003). Thus, flint-like siliceous rocks (cherts) may contain fossils of bacteria, particularly of large cyanobacterial cells with specialized secondary structures embedded in silica (Kasting, 2001).

Stromatolites (Fig. 6.5) are defined as fossilized microbial communities that form laminated relief structures typically found in Precambrian carbonate sediments consisting of limestone, dolomite, or silicate (Riding, 1999). There is no doubt that structurally stromatolites are consistent with the theory that their structure and morphology have been influenced by microbial communities dominated by cyanobacteria present in shallow photic zones in aquatic ecosystems during the Precambrian period. However, the implications of the occurrence of microfossils in stromatolites have been debated extensively. Some investigators have suggested an interaction between biotic and abiotic depositional factors being responsible for the texture of stromatolites (Corsetti *et al.*, 2003; Grotzinger and Rothman, 1996; Neilan *et al.*, 2002; Olivier *et al.*, 2003; Sabater, 2000).

Stromatolite domination of the Precambrian aquatic ecosystems has been documented in formations dating from 3.5 billion years (e.g. Onverwacht chert from Southern Africa) to approximately 600 million years ago when preservation conditions became considerably more tenuous (Westall, 1998). The second line of evidence supporting the antiquity of microbial life involves traces of biochemical compounds in old sedimentary rocks. The interpretation of such evidence is subject to innumerable sources of error due to contamination by contemporary sources of organic compounds of biological origin. For example, hydrocarbon derivatives of microbial membrane lipids, porphyrin structures, and kerogen are commonly found components of sedimentary rock, but these are considered reliable fingerprints for the recent occurrence of microbial groups in sedimentary rock (Burner and Moore, 1987; Fenchel, 2002; Grotzinger and James, 2000). The ratios of stable isotopes ^{12}C to ^{13}C or ^{32}S to ^{34}S have also been used extensively to characterize the biological origin of carbon and sulfur compounds in ancient rock formations. The ratios of these isotopes can indicate biological origins because biological systems preferentially use the lighter isotopes which require less energy to involve them in biochemical reactions (Sumner, 2001). For example, the graphite component of the Isua rock formation is enriched with ^{12}C, which is consistent with photosynthetic sources (Stephens and Sumner, 2003; Awramik, 1992). Differential enrichment of carbon isotopes has also been used as evidence for methanogenesis in old sedimentary rock where ^{13}C has been depleted by up to 8% (Catling, 2001; Teske *et al.*, 2003). Up to 4% enrichment of ^{32}S in old sedimentary rocks has also been shown to be associated with the earliest record of sulfate respiration (Visscher *et al.*, 1999).

Inferences about the diversity of microbial communities present in ancient stromatolites can be made on the basis of the dominant morphological conformation of the stromatolite. Stromatolite morphology is defined by the proportion and geometric arrangement of the laminae (planar folds resulting from microbial mats), the support (irregular branching surfaces associated with voids), and voids filled with herringbone calcite. Sumner (1997) and others have suggested that the morphological differences observed among the supports and laminae are due to differences in microbial community structure with respect to metabolic function and tactical behavior. According to this view, support structures were built by microorganisms exhibiting tactic responses to light or nutrients, and the laminae were built by sessile organisms. Additional support for the microbiological interpretation of stromatolite morphology is provided by differences observed in the rate of deposition and total amount of precipitated herringbone calcite, which is higher on the supports than on the laminated mats, suggesting different metabolic preferences by the microbial communities inhabiting the Archaean aquatic ecosystems (Sumner, 1997).

(a)

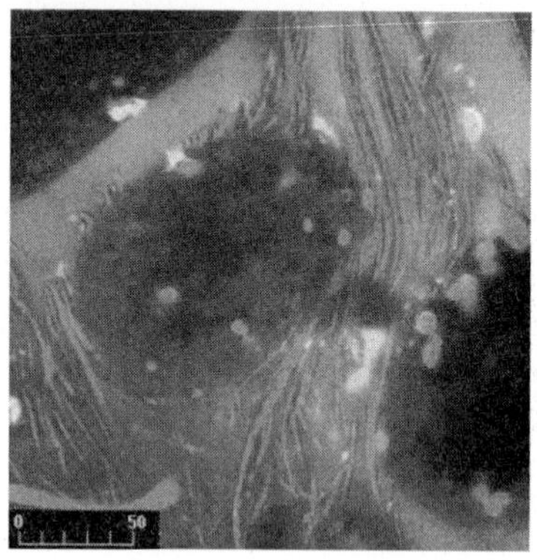

(b)

Fig. 6.5 Stromatolites represent the oldest preserved evidence of microbial life as early as 1 billion years after the formation of the Earth. (a) shows the array of stromatolite formations off the coast of Australia (photography by courtesy of NASA www.resa.net/nasa/mars_life_gifossil.htm). (b) represents a confocal multi-photon image of a dissected stromatolite showing filamentous cyanobacteria-type organisms. Photograph in (b) by permission of Bio-Rad and Tomohiro Kawaguchi and Alan Decho University of South Carolina (Decho and Kawaguchi, 1999).

CONTEMPORARY MICROBIAL MATS

The environmental conditions that supported the development of stromatolitic structures by microbial activities during the Archaean and Proterozoic eons are still relevant to the contemporary development of multilayered mats of microbial communities. These microbial mats have been dissected to reveal important information about extant and fossilized communities. Microbial mats are primarily self-sustaining micro-ecosystems, but their development and stability are sensitive to global ecosystem changes including redox potential, incident electromagnetic radiation, and biodiversity (Golubic, 2000; Guerrero *et al.*, 2002). For example, Des Marais (2001 and 2003) investigated contemporary hypersaline microbial mats to understand the environmental context of early evolution of microbial physiological processes. Molecular and phylogenetic assessment of microbial mat composition can reveal valuable information on the early evolution of photosynthesis, generation of oxygen, and the dynamic nature of biogeochemical gradients (Fig. 6.6).

Photosynthetic microbial mats are defined by an abundance of cyanobacteria close to the surface, where they synthesize carbohydrates (CH_2O in Fig. 6.6) through photosynthetic carbon fixation:

$$6CO_2 + 6H_2O \rightarrow 6<CH_2O> + 6O_2$$

Filamentous and sheathed cyanobacterial populations, including *Spirulina* and *Lyngbya* species constitute the "green layer" of microbial mats (Box 6.3). The primary production sustained by the cyanobacteria is linked directly to the consumption of organic matter by aerobic heterotrophic bacteria through aerobic respiration:

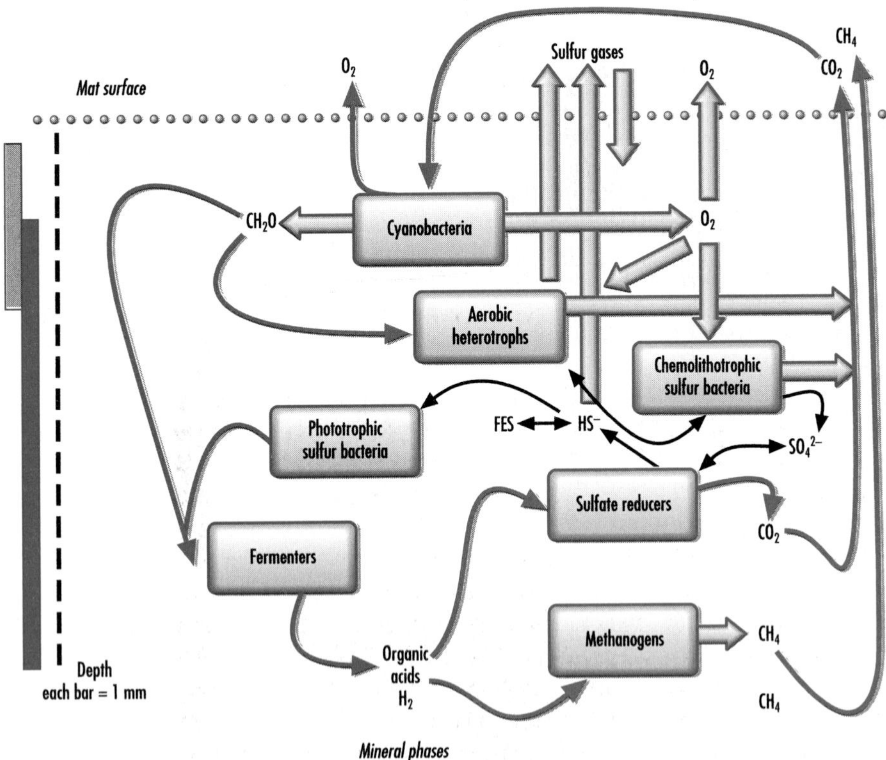

Fig. 6.6 Biogeochemical cycling in a contemporary microbial mat. The major trophic categories of microorganisms are in the seven boxes arranged by location according to depth from the mat surface (hatched bar on the left). The mat "breathes" according to a diurnal cycle with oxygen tension (gray bar on the left) limited to the top 4 mm during the day, and overlapping with the anaerobic conditions defined by hydrogen sulfide tension (solid green bar to the left) increasing from 2 mm to the bottom of the mat around 12 mm. The main movements of the carbon cycle are designated by green arrows, of the oxygen cycle by broad pale green arrows, and of the sulfur cycle by solid black arrows. Scheme is adapted from NASA's program on microbial mats (http://nai.arc.nasa.gov/).

$$6\langle CH_2O\rangle + 6O_2 \rightarrow 6CO_2 + 6H_2O$$

The supply and consumption of oxygen in microbial mats result from complex interactions between physicochemical factors regarding gaseous diffusion and the diversity of organisms in the stratified mat layers. These interactions are influenced by light, pH, and nutritional conditions that support the preferential proliferation of distinct categories of organisms (Jonkers and Abed, 2003).

Glud and colleagues (1999) used planar optodes (optical electrodes) and imaging techniques to demonstrate that vertical and horizontal variations in oxygen tension within a microbial mat are tightly coupled to photic zones within the mat. The stratifications are influenced by moderate oxygen production by diatoms in an upper zone, high oxygen production by *Microcoleus*-like cyanobacteria in a middle zone, and high oxygen consumption by microalgae in a lower zone. These observations point to the possibility that light-stimulated respiration is influenced by enhanced heterotrophic activity, which is supported by organic carbon produced by phototrophic microorganisms. Nuebel and colleagues (2002) investigated the photosynthetic layer of the microbial mat of an alkaline hot spring (Mushroom Spring, Yellowstone National Park, U.S.), where they discovered that *Chloroflexus* and related species dominate the highly oxygenated surface layers using bacteriochlorophyll "a" for photosynthetic activity at 70°C. In addition to photosynthetic activity, these organisms were also observed to assimilate organic carbon (Nuebel *et al.*, 2002).

BOX 6.3

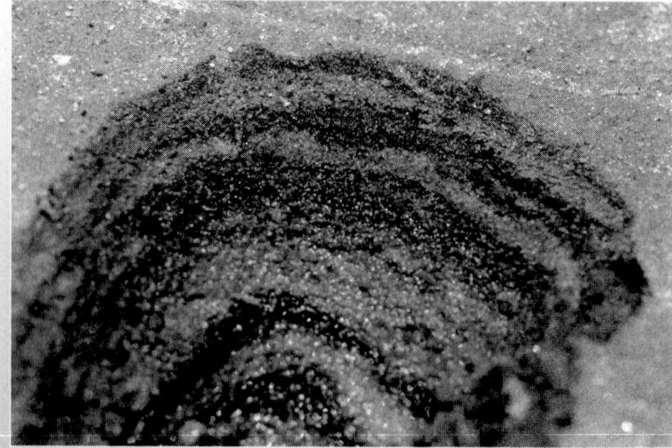
(a)

(a) A cross-section of the microbial mat from the Great Sippewisette Salt Marsh near the Marine Biological Laboratory, in Woods Hole at Falmouth, MA, USA. The sample is approximately 5 cm across, and the different layers are shaded by bacterial pigments. The darkest layers are due to iron sulfide, the lightest layers are due in part to pyrite.

(b) and (c) Filamentous cyanobacteria from the top layer of the microbial mat depicted in (a). The layers are dominated by cyanobacteria and diatoms which are observed in the upper 2 to 5 mm. Under natural light conditions, the presence of carotenoids contributes a pinkish hue to the microbial cell layers. The carotenoids are attributed to pigment formation by phototrophic purple sulfur bacteria. Beneath the surface layers is the black layer colored by iron sulfide precipitated by anaerobic sulfate-reducing bacteria. These bacteria consume the organic matter formed by the cyanobacteria in the top layers. Beneath the microbially stratified colored layers lies a gray mineral zone consisting primarily of pyrite and buried biomarker compounds. (b) Phase contrast microscopic image; (c) the same view as in (b) after special illumination. The autofluorescence demonstrates the presence of phycobiliproteins.

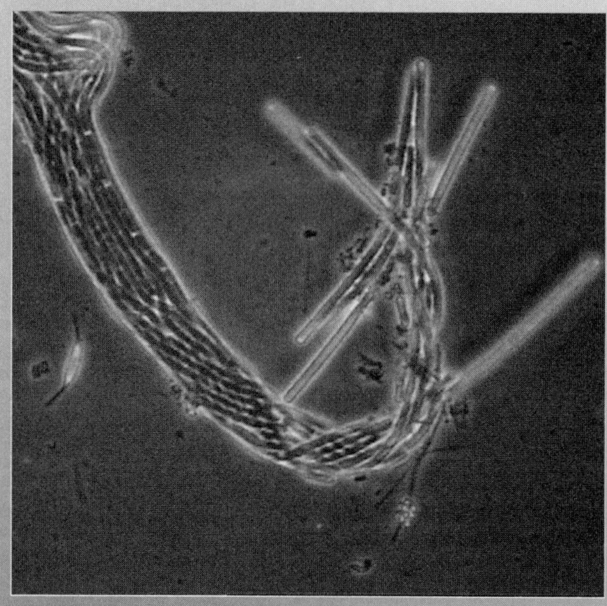
(b)

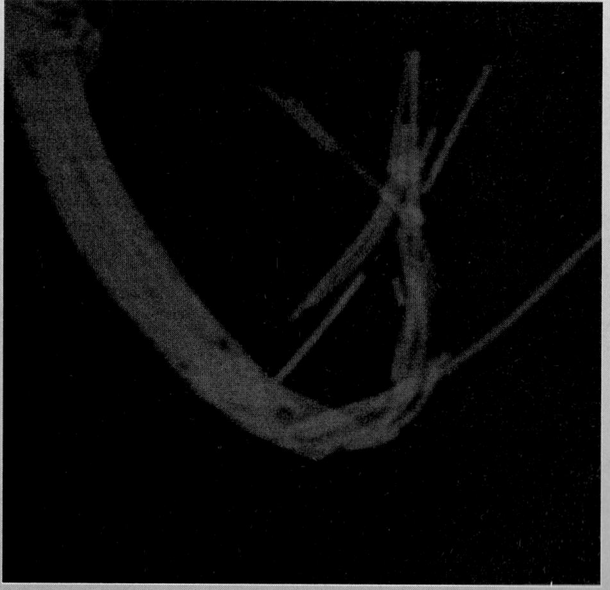
(c)

Continued

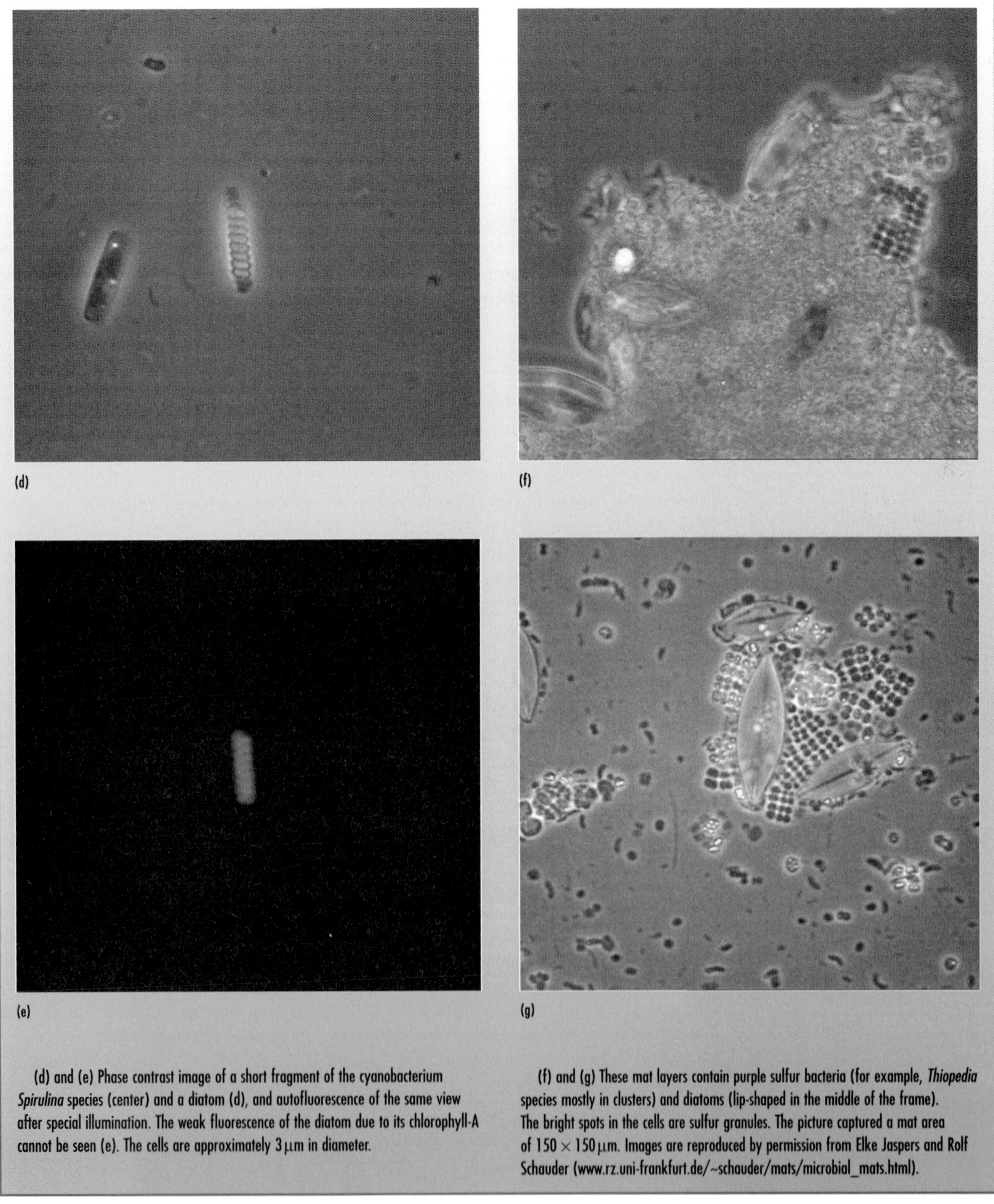

(d) and (e) Phase contrast image of a short fragment of the cyanobacterium *Spirulina* species (center) and a diatom (d), and autofluorescence of the same view after special illumination. The weak fluorescence of the diatom due to its chlorophyll-A cannot be seen (e). The cells are approximately 3 μm in diameter.

(f) and (g) These mat layers contain purple sulfur bacteria (for example, *Thiopedia* species mostly in clusters) and diatoms (lip-shaped in the middle of the frame). The bright spots in the cells are sulfur granules. The picture captured a mat area of 150 × 150 μm. Images are reproduced by permission from Elke Jaspers and Rolf Schauder (www.rz.uni-frankfurt.de/~schauder/mats/microbial_mats.html).

At the next depth level are the chemolithotrophic sulfur bacteria that use the redox energy generated by the gradient between oxygen produced by the cyanobacteria and the reduced sulfur compounds produced by sulfate reduction:

$$H_2S + 2O_2 \rightarrow H_2SO_4 + \text{Energy}$$

Chemolithotrophic bacteria, including the purple sulfur bacteria, form the main components of the pink layer of microbial mats. These bacteria are typically motile because they have to locate themselves at the redox gradient, which shifts in response to the diurnal cycle of photosynthetic activities.

Phototrophic sulfur bacteria conduct anoxygenic photosynthesis by using incident infrared radiation, which unlike visible light, penetrates deeper into the mat. Anoxygenic photosynthesis fixes carbon dioxide without producing oxygen:

$$4CO_2 + 2H_2S + 4H_2O \rightarrow 4\langle CH_2O\rangle + 2H_2SO_4$$

Rothermich and colleagues (2000) discovered the formation of polymeric hydroxyalkanoate (PHA) by purple sulfur bacteria in stratified photosynthetic microbial mats of the Great Sippewisette Salt Marsh in Cape Cod, Massachusetts. The production and accumulation of hydroxyvalerate and hydroxybutyrate in these mats followed a diel pattern, indicating a linkage to dark energy metabolism, which is not influenced by the availability of organic nutrients (Rothermich *et al.*, 2000).

Like aerobic heterotrophic microorganisms in the mat, fermentative microorganisms consume the carbohydrates produced by the cyanobacteria, except that they do it without oxygen, and therefore, the products are incompletely degraded substituted carbohydrates such as alcohols:

$$6<CH_2O> \rightarrow 2C_2H_5OH + 2CO_2$$

Anaerobic sulfate reducers exist close to the bottom of the mat where they engage in anaerobic respiration to consume organic matter using a variety of alternative electron acceptors such as nitrate, sulfate, iron, and manganese:

$$SO_4^{2-} + CH_3COOH \rightarrow H_2S + 2HCO_3$$

Approximately one-third of all organic carbon produced in microbial mats is consumed through anaerobic respiration by the sulfate reducers (Visscher *et al.*, 1999). Sulfate reducers and methanogenic microorganisms use the organic acids produced by fermenters in anaerobic respiration to produce carbon dioxide and methane, respectively. Methanogens are extremely important in the context of global climate change because methane is a potent greenhouse gas, and microbial emissions of methane from anaerobic environments such as mats are considered a significant contribution to the global methane budget:

$$CH_3COOH \rightarrow CH_4 + CO_2$$
$$CO_2 + 4H_2 \rightarrow CH_4 + 2H_2O$$

Beneath the viable portion of the microbial mat is the mineral phase with a size that depends on the "age" of the mat because the mineral deposits represent unmetabolized components of the mat. The development of Precambrian iron formation by microbial mats was investigated by Pierson and Parenteau (2000) at the Chocolate Pots Hot Springs in Yellowstone National Park. In mats existing at high temperatures (48–54°C), they found that the predominant species were *Synechococcus* and *Chloroflexus* or *Pseudanabaena* and *Mastigocladus*, whereas at lower temperatures (36–45°C), *Oscillatoria* species dominated. These organisms exhibited gliding motility, filamentous structures, and mineral encrustation that likely contribute to the accumulation of iron deposits.

The ubiquitous distribution of contemporary microbial life suggests that microbial mats should occur in a wide variety of environments and ecosystems. However, field observations contradict this hypothesis because microbial mats tend to be concentrated in ecosystems that are defined as "extreme environments", including hypersaline salt marshes, hyperthermal springs, desiccated temperate deserts, and the cold dry environments of the Polar Regions. Several reasons have been presented to explain the marginalization of contemporary microbial mats to extreme environments, but the most widely accepted view is that although microbial mats germinate everywhere, competition, predation, and disturbance by plants and grazing animals restrict the proliferation of mats to environments that are inhospitable to more complex organisms. Therefore, the relative abundance of mature mats should be inversely proportional to overall ecosystem biodiversity, but this theory has not yet been adequately explored. The recognition of the physiological diversity of microbial mats and their sustainability in extreme environments has increased interest in understanding the molecular mechanisms that support the survival of microbial life in these relatively inhospitable environments, and in using them as models for investigating the possibility that microbial mats can be found on planets where the environmental conditions may resemble these on Archaean Earth, when microbial mats dominated both land and seascapes.

MICROBIAL LIFE AND EVOLUTION IN EXTREME ENVIRONMENTS

The resurgence of interest in astrobiology, and the recognition that microbial mats thrive in extreme environments, have engendered vigorous research programs on the nature of "extremophiles" as a distinct category of microorganisms (Edwards, 1990; Gould and Corry, 1980; Heinrich, 1976; Horneck and Brack, 2000; Horneck and Baumstark-Khan, 2002; Horikoshi and Grant, 1998; Kushner, 1978; Seckbach, 1999; Wiegel and Adams, 1998). However, the category of extremophiles is populated by a rather diverse array of species because the mechanisms of microbial survival under "extreme" environmental conditions are no less subject to the driving forces of evolution by natural selection than are the mechanisms which support survival under mild conditions inhabited by complex multicellular organisms. The most commonly investigated extreme conditions are those associated with electromagnetic radiation, temperature, salt concentration, atmospheric pressure, pH, and drought. Occasionally, toleration of high concentrations of metal ions and organic solvents are included in the extremophile research paradigm (Horikoshi and Grant, 1998).

The typical presentation of "global average" environmental conditions on the Earth's surface considerably masks the range of conditions in which viable organisms are actually observed. Furthermore, the concept of "optimum" growth conditions for specific organisms is biased in favor of those conditions that best support observable phenotypes. For example, the global average temperature has been measured at 15 +/– 0.6°C, but the range of temperatures in which living organisms grow starts from below 0°C to above 100°C, which spans the freezing to boiling temperature range for water molecules that are absolutely essential for terrestrial life. The ability of microorganisms to flourish at various conditions of temperature and water activity (a_w) has been used to define several groups of microorganisms, but the distinctive characteristics of these groups are not as rigorous as they may appear at first because of the multifaceted interactions among several environmental factors that support microbial growth (Tables 6.1 and 6.2).

It seems reasonable then to suppose that microorganisms capable of surviving in extreme environments dominated ancient Earth, and that their genetic lineage should be "older" on an evolutionary scale than organisms which have adapted to the environmental conditions that are now commonly found on Earth. This supposition is upheld by the observation that populations of microorganisms belonging to the Archaea phylogenetic branch dominate

Table 6.1 Categories of microorganisms defined by temperature conditions for growth.

	Minimum growth temperature (°C)	Optimum growth temperature (°C)	Maximum growth temperature (°C)	Example of organisms
Psychrophiles	−10	10–15	20	*Methanogenium frigidum* (minimum growth temperature is −10°C; optimum growth temperature is 15°C).
Psychotrophs	−10	20–40	45	Food-borne pathogen *Listeria monocytogenes* (minimum growth temperature is −0.4°C; optimum growth temperature between 30 and 37°C; and maximum growth temperature is 45°C).
Mesophiles	5	28–43	52	*Escherichia coli* (minimum growth temperature is 8°C; optimum growth temperature is 37°C; and maximum growth temperature is 45°C).
Thermophiles	30	50–65	70	*Bacillus stearothermophilus* (minimum growth temperature is 20°C; optimum growth temperature is 50°C; and maximum growth temperature is 70°C).
Hyperthermophiles	65	80–105	110	*Pyrodictium* (inhabiting geothermally heated areas of the seabed has a minimum growth temperature of 82°C; optimum growth temperature of 105°C; and maximum growth temperature of 110°C).

Table 6.2 Categories of microorganisms defined by the availability of water required for growth. Water activity (a_w) represents the ratio of the water vapor pressure of the growth conditions to the vapor pressure of pure water under the same conditions. The data represent water activity (a_w) levels that can support the growth of particular groups of microorganisms.

Minimum a_w	Growth environment	Example of microorganisms
1.0	Freshwater	*Spirillum* species
0.9	Marine water	*Bacillus marinus* ATCC29841 isolated from marine sediment from a depth of 237 m, North East Atlantic
0.8	Estuarine water	*Vibrio* species
0.7	Hypersaline lakes	*Halobacterium* species
0.65	High carbohydrate substrates	Osmophilic organisms, e.g. *Zygosaccharomyces richteri*
0.61	Deserts, dry indoor dust	Xerophilic fungi, *Wallemia sebi*

extreme environments with respect to hyperthermophiles, halophiles, and methanogens (Fig. 6.7). However, there is no convincing independent evidence to suggest that the Archaea are "older" than the Bacteria, and some investigators have argued that the Archaea branch emerged from the Gram-positive Bacteria, although the contribution of relatively recent horizontal gene transfer has not been fully investigated. Therefore, conclusions based solely on genetic homologies are preliminary (Olendzenski *et al.*, 2001; see Table 6.3). In addition,

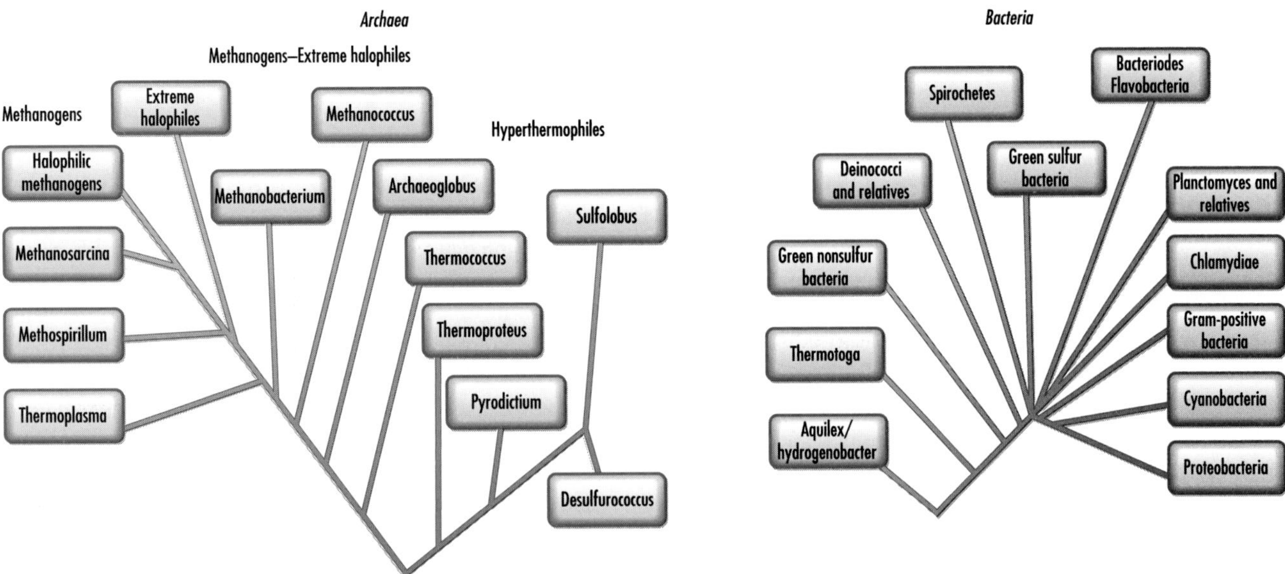

Fig. 6.7 Phylogenetic trees representing major groups in the Archaea and Bacteria domains with the position of organisms commonly found in extreme environments: methanogens, halophiles, and thermophiles.

Table 6.3 Comparative characteristic features of the phylogenetic groups.

Characteristic	Bacteria	Archaea	Eukarya
Methanogenesis	No	Yes	No
Chlorophyll-based photosynthesis	Yes	No	Yes
Reduction of S^0 to H_2S	Yes	Yes	No
Nitrogen fixation	Yes	Yes	No
Cell wall	Muramic acid present	Muramic acid present	Muramic acid present
Membrane lipids	Ester-linked	Ether-linked	Ester-linked
Membrane-bound nucleus	Absent	Absent	Present
Ribosomes	70S	70S	80S (70S organelles)
Initiator tRNA	Formylmethionine	Methionine	Methionine
Introns in tRNA genes	Rare	Yes	Yes
Operons	Yes	Yes	No
Capping and poly-A tailing of mRNA	No	No	Yes
Plasmids	Common	Yes	Rare
Protein synthesis sensitive to diphtheria toxin	No	Yes	Yes
RNA polymerases	One-type (4 subunits)	Several (8–12 subunits)	3 (12–14 subunits)
Rifampicin inhibition of RNA polymerase	Yes	No	No
Sensitivity to chloramphenicol, streptomycin, kanamycin	Yes	No	No

certain members of the Bacteria such as *Deinococcus radiodurans* are known to possess the ability to survive in extreme conditions of electromagnetic radiation that are not generally encountered under contemporary environmental conditions on Earth (Fig. 6.7; Zhang *et al.*, 2003).

THE EMERGENCE OF MULTICELLULARITY AND EUKARYOSIS, AND THEIR CONSEQUENCES FOR ENVIRONMENTAL EVOLUTION

The history of life on Earth is unquestionably dominated by unicellular microorganisms, and in terms of relative biomass and physiological diversity, the contributions of microbes to global environmental change far outweigh the contributions of complex multicellular macro-organisms. However, the distinction between unicellular and multicellular organisms has been questioned by some investigators who express the view that multicellular organisms are in fact complex communities of unicellular microorganisms, whereas organisms that are considered unicellular exist as complex colonies with emergent properties under natural conditions (Margulis and Sagan, 1991). It is important to distinguish the development of multicellularity from eukaryosis, or the origin of the eukaryotic phylogenetic lineage because certain prokaryotic organisms include in their life cycles, multicellular "fruiting" structures that are produced under stressful environmental conditions (Yan *et al.*, 2003; Fig. 6.8). Both multicellularity and eukaryosis have had important consequences for environmental evolution because of the consolidation of physiological process that, for example, led to the indispensable functions of organelles such as mitochondria, plastids, and the nucleus (Andersson and Kurland, 1999; Andersson *et al.*, 2003; Martin and Muller, 1998). Margulis (1996) has further proposed an internally consistent phylogenetic classification system for life in which reference to "Kingdoms" and "Domains" are replaced by the simple delineation of non-nucleated (Prokarya) and nucleated (Eukarya) organisms. Because it includes a plausible mechanism of evolutionary transition from one of these groups to the other, the proposal has the appeal of extricating discussions of co-evolution between life and environment from digressions associated with contemporary situations. Instead, a retracing of the natural history of environmental change is facilitated by investigating the genetic and ecological history of events which produced the emergence of multicellularity and eukaryosis.

Endosymbiosis

The impact of complex multicellular eukaryotes on global environmental processes should not be underestimated, and speculations about the origin of this kind of life form have engaged numerous investigators for many years (Bui *et al.*, 1996; Margulis, 1996; Taylor, 1974–2003). The rationale for these studies in the context of environmental evolution is that the processes which supported the emergence and overwhelming success of the eukaryotes have been accompanied by deep-rooted environmental change. The evidence is irrefutable that prokaryotes preceded multicellularity and eukaryosis in particular. Therefore, it is reasonable to accept the view that processes and interactions which were established by prokaryotes led directly to the emergence of the eukaryotes. The most thoroughly investigated approach in this respect is "the serial endosymbiosis theory" (SET) (Margulis, 2000; Margulis and Fester, 1991; Margulis and Sagan, 2002; Taylor, 1974). According to SET, cells with nuclei (eukaryotic cells including protoctists, fungi, animals, and plants) emerged indirectly from members of the Bacteria through a series of endosymbiotic steps. The first step involved the acquisition of motility through anaerobic symbiosis between spirochetes and other prokaryotic cells such as *Thermoplasma* species. The first nucleated cells emerged through the formation of membranous nucleocytoplasm from this original symbiotic complex as mastigote microbes or protists. In a subsequent step in SET, mitochondria were acquired by some of these eukaryotes. Mitochondria-rich mastigotes eventually emerged as fungi and animal cells, and further endosymbiotic events involving the acquisition of cyanobacterial colored photosynthetic plastids engendered the ancestors of contemporary plants and algae (see Chapter 1; Germot *et al.*, 1996; Margulis, 1996; Seckbach, 2002). These propositions have been amply supported by molecular investigations of the independent

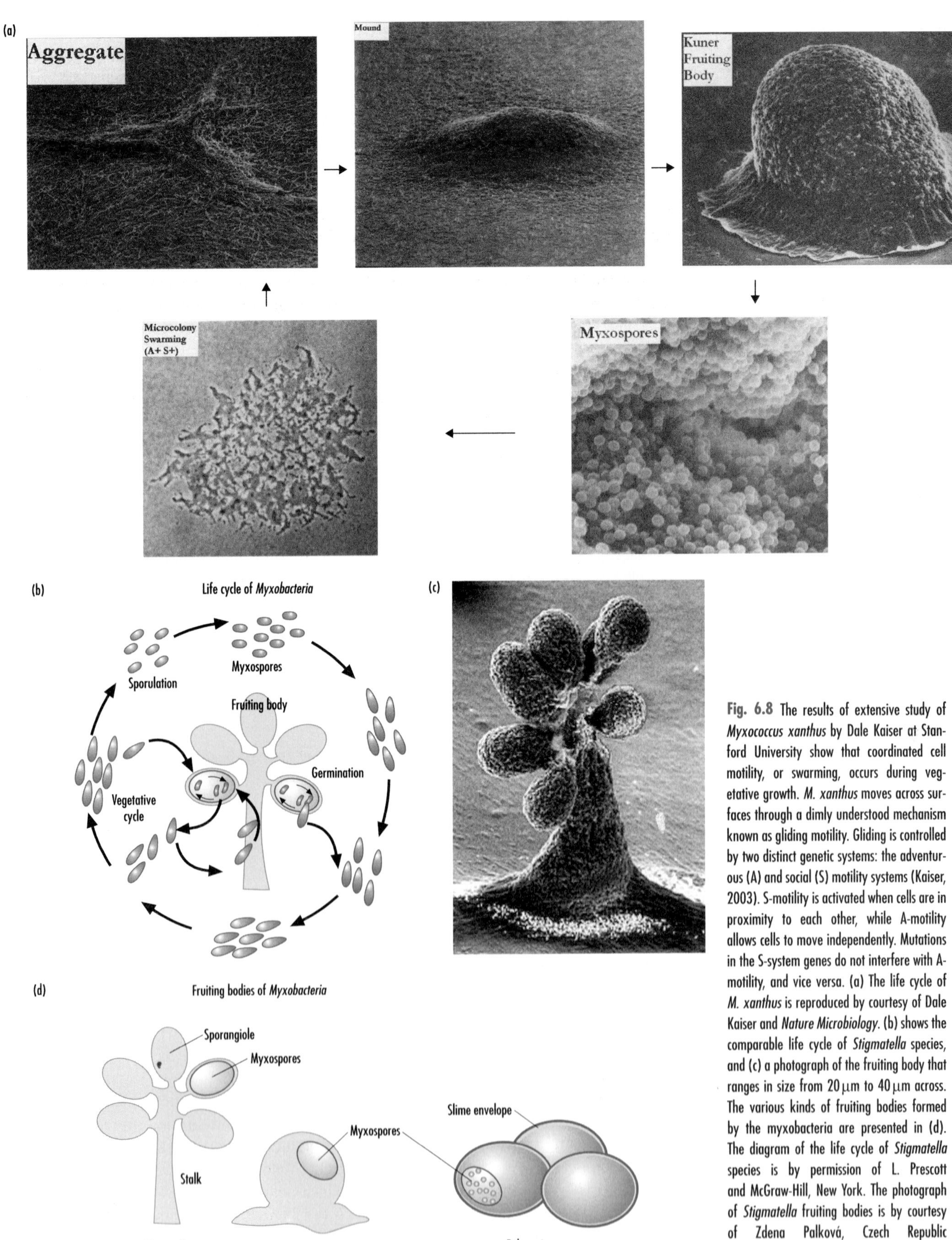

Fig. 6.8 The results of extensive study of *Myxococcus xanthus* by Dale Kaiser at Stanford University show that coordinated cell motility, or swarming, occurs during vegetative growth. *M. xanthus* moves across surfaces through a dimly understood mechanism known as gliding motility. Gliding is controlled by two distinct genetic systems: the adventurous (A) and social (S) motility systems (Kaiser, 2003). S-motility is activated when cells are in proximity to each other, while A-motility allows cells to move independently. Mutations in the S-system genes do not interfere with A-motility, and vice versa. (a) The life cycle of *M. xanthus* is reproduced by courtesy of Dale Kaiser and *Nature Microbiology*. (b) shows the comparable life cycle of *Stigmatella* species, and (c) a photograph of the fruiting body that ranges in size from 20 μm to 40 μm across. The various kinds of fruiting bodies formed by the myxobacteria are presented in (d). The diagram of the life cycle of *Stigmatella* species is by permission of L. Prescott and McGraw-Hill, New York. The photograph of *Stigmatella* fruiting bodies is by courtesy of Zdena Palková, Czech Republic (www.natur.cuni.cz/~zdenap/linksMS.html).

(a)

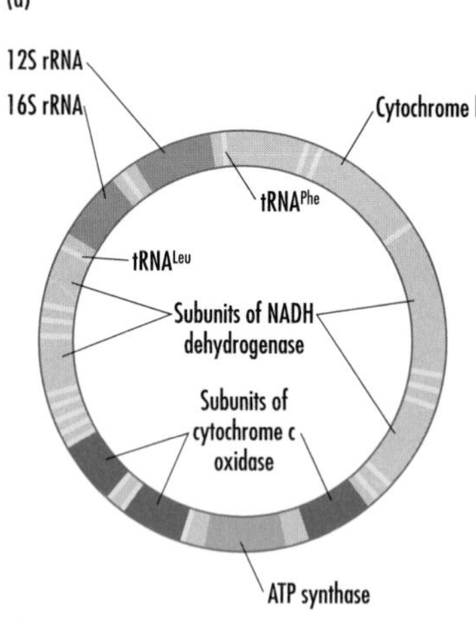

(b)

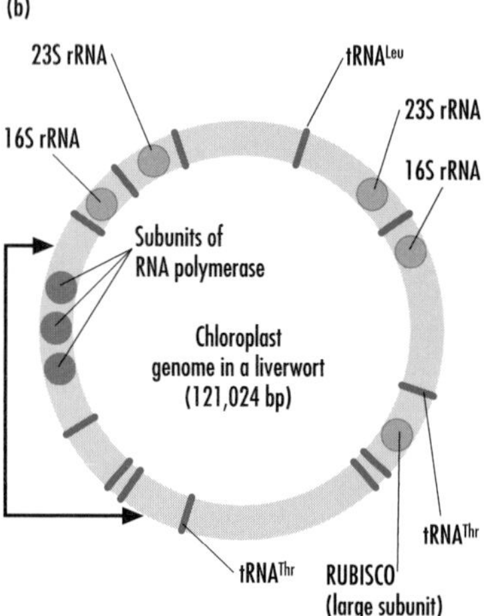

Fig. 6.9 Major genetic components of the covalently closed circular genomes of human mitochondria (16,569 base pairs; (a)) and liverwort *Marchantia polymorpha* chloroplast (b). Diagrams reproduced by courtesy of John Kimball (http://biology-pages.info).

genomes of mitochondria and chloroplasts, which invariably resemble prokaryotic genomes (Fig. 6.9).

To discover the microbial lineage of mitochondria, Andersson and colleagues (1998) sequenced the entire genome of *Rickettsia prowazekii*, consisting of 1,111,523 base pairs. The genome of this prokaryote includes the highest proportion (24%) of non-coding DNA sequence that has been observed in any organism. They found extensive similarities between the genome of this obligate parasite and the mitochondrial genome. Indeed, phylogenetic analyses indicated that the genome of *R. prowazekii* is more closely related to mitochondria than the genome of any other prokaryote sequenced so far (Andersson *et al.*, 1998). Chloroplasts appear to have evolved on three different occasions through the endosymbiosis of cyanobacteria, with the production of green algae and plants on one occasion, the red algae on a second occasion, and secondary endosymbiosis involving the engulfment of a photosynthetic eukaryote by another eukaryote producing glaucophytes (unicellular algae) on the third occasion (Lewin, 2002). In addition to the postulates regarding the origin of chloroplasts and mitochondria, SET addresses eukaryotic motility factors, thought to have arisen from endosymbiotic spirochete bacteria (see Chapter 1). The microtubular motility structure (undulipodia) which is ubiquitous in eukaryotes is absent in prokaryotes. The basal bodies from which undulipodia develop are presumed to have evolved into the mitotic spindle, making mitotic cell division in eukaryotes possible (Margulis, 2000).

In a different approach to SET that is of considerable importance for environmental evolution, Martin and Muller (1998) presented a new hypothesis on eukaryogenesis based on comparative biochemistry of energy metabolism. According to this view, eukaryotes emerged from a symbiotic association between autotrophic Archaea cells hosting Bacteria cells that are capable of anaerobic heterotrophic respiration, producing molecular hydrogen as a waste product. The perpetuation of symbiosis between these organisms must have depended on the Archaea cell's dependence on hydrogen. Bui and colleagues (1996) expanded the hydrogenesis approach by investigating heat-shock proteins (Hsp) in Trichomonads, one of the earliest eukaryotes to diverge from the main line of eukaryotic lineage. Trichomonads are facultative anaerobe protists that lack mitochondria and peroxisomes, instead, they possess hydrogenosomes which, like mitochondria, are double-membrane bounded organelles specializing in producing ATP via pyruvate as the primary substrate, but unlike mitochondria, hydrogenosomes do not contain DNA.

Bui and colleagues (1996) further demonstrated that hydrogenosomes contain heat-shock proteins (Hsp70, Hsp60, and Hsp10) with signature sequences that are conserved only in mitochondria and alpha-*Proteobacteria.* Phylogenetic analyses of heat-shock proteins in the hydrogenosomes showed that these proteins branch within a monophyletic group composed exclusively of mitochondrial homologs. Their data established that mitochondria and hydrogenosomes appear to have a common ancestor within the Bacteria lineage. However, Andersson and Kurland (1999) questioned the hypothetical linkage between hydrogenesis and the evolution of mitochondria by conducting phylogenetic reconstructions with the genes coding for proteins that are active in energy metabolism. Their analysis confirmed the simplest version of SET regarding the symbiotic origin of organelles, but their data did not support the hypothesis on hydrogenesis and syntropy as the origin of mitochondria.

Biotic effects on the evolution of Earth's atmosphere

Since the establishment of life on Earth, the reduction–oxidation (redox) potential of the atmosphere has shifted from a reducing condition (the logarithm of the reciprocal of the electron concentration (*pE*) was approximately −5 gram-molecules per liter) to an oxidizing condition (*pE* of +13 gram-molecules per liter) (Lovelock, 2001). This trend is generally attributed to physicochemical reactions and the consequences of biological activity (Catling *et al.*, 2001). Margulis and Sagan (1991) argued that the increasing demand for hydrogen initiated the "crisis of oxygen" pollution in the Archaean age. The emergence of living systems created a demand for carbon–hydrogen compounds thereby reducing the carbon dioxide content of the atmosphere. By comparison to the Earth's current 0.03% CO_2, the atmospheres of Mars and Venus contain at least 95% CO_2 (Fig. 6.1). The loss of hydrogen into outer space intensified the shortage of molecular hydrogen. However, non-biologically available hydrogen was abundant in water, and it took the emergence of photosynthetic cyanobacteria to develop a physiological mechanism for splitting water into its constituent atoms with the concomitant production of carbohydrates and elemental oxygen.

The emergence of plants that co-opted this microbial invention did much to entrench photosynthesis and its products as a major signature of biological life on Earth (Kasting and Seifert, 2002). The mineral record supports the view that oxygen build-up was sudden, but it is difficult to pinpoint the precise time when photosynthesis generated sufficient oxygen to permanently change the course of environmental evolution. Before the rise of oxygen, microorganisms also played a role in the evolution of the Earth's atmosphere. Anaerobic bacteria may have influenced global temperatures, thereby warming the Earth through their release of large amounts of methane into the atmosphere. However, one of the major repercussions of the oxygen build-up in the atmosphere is the evolution of non-photosynthetic sulfide-oxidizing bacteria, which occurred contemporaneously with a major shift in the isotopic composition of biogenic sedimentary sulfides between 0.64 and 1.05 billion years ago. These events were likely supported by a rise in atmospheric oxygen concentrations to levels higher than 5–18% of present levels (~21%), which could also have triggered the evolution of animals (Canfield and Teske, 1996; Lewis and Sheridan, 2001).

PRACTICAL ASPECTS OF MICROBIAL DIVERSITY AND ENVIRONMENTAL EVOLUTION

Hydrogenesis

There is considerable international interest in finding sources of energy to serve as alternatives to fossil fuels such as gasoline, natural gas, and coal because global supply of these fuels will ultimately be depleted, and they contribute substantially to declining air quality in urban centers. Furthermore, their combustion byproducts contribute to the phenomenon of global warming. Hydrogen has been proposed as a viable alternative to fossil fuels largely because it is perceived as "clean", and its ultimate source, water, is plentiful. Therefore research on the biological basis of hydrogenesis has intensified (Vanzin *et al.*, 2002). As discussed above, hydrogen production is probably one of the "oldest" metabolic capabilities in the prokaryotic phylogenetic branch, and the search for efficient hydrogen producers is focused on the hydrogenase-producing organisms in that category (Vignais *et al.*, 2001). Among the most promising hydrogen producers is *Carboxydothermus hydrogenoformans*, an anaerobic carbon monoxide bacterium that expresses a NiFeS–carbon monoxide dehydrogenase (Svetlitchnyi *et al.*, 2001). Hydrogen production is a genetically inducible process in many microorganisms, which allows a certain level of control for biotechnological development of this resource (Vanzin *et al.*, 2002). In addition, many hydrogen producers can grow on substrates that are essentially carbonaceous industrial waste materials, a fact that has been considered a double incentive for developing large-scale production of hydrogen. However, the heterogeneous nature of industrial biomass waste reduces the rate of biological hydrogen production. One solution to this problem is to convert heterogeneous biomass to a homogeneous synthesis gas containing carbon monoxide, which is subsequently converted to hydrogen via enzyme activity by organisms such as *Clostridium thermoaceticum* and *Rhodospirillum rubrum* (Maness and Weaver, 2001; Martin *et al.*, 1983):

$$CO + H_2O \leftrightarrow CO_2 + H_2$$

Rubrivivax gelatinosus, a phototrophic bacterium, contains a carbon monoxide shift pathway that is very active in darkness. This organism is capable of oxidizing hydrogen, but only under lighted conditions. Unlike the hydrogenases found in other microorganisms, the *Rubrivivax gelatinosus* enzyme is resistant to oxygen (Bonam *et al.*, 1989; Maness *et al.*, 2002). Research at the U.S. National Renewable Energy Laboratory is underway to employ genetic engineering for probing and overcoming the genetic steps that underpin the rate-limiting steps in hydrogen production. Most significant progress in this direction has been achieved for the carbon monoxide sensing factor in *Rhodospirillum rubrum* (He *et al.*, 1999). The rapid rate of progress in understanding biological hydrogenesis has engendered a call for caution because there is no information about the potential environmental implications of large-scale hydrogen production and use, if this gas were to replace fossil fuels (Alper, 2003). According to Tromp and colleagues (2003), widespread use of hydrogen could have detrimental environmental impacts through emissions of molecular hydrogen that are expected to increase the abundance of water vapor by up to one part per million by volume in the stratosphere. In this view, the consequence of hydrogen in that atmospheric compartment will be stratospheric cooling and the enhancement of heterogeneous chemical reactions that could contribute to the depletion of the ozone layer (Tromp *et al.*, 2003).

Methanogenesis

Interest in methanogenesis and its consequences for environmental evolution has gained considerable attention because of the recognition that methane is a potent greenhouse gas (Updegraff *et al.*, 2001; Welsh, 2000). The atmospheric concentration of methane exerts a strong influence on global climate due to its capacity to trap heat. Microbial processes in natural and cultivated wetlands, including rice paddies and ruminant digestive processes, are important sources of atmospheric methane (Agnihotri, 1999; Panikov, 1999). Cao and colleagues (1995 and 1998) demonstrated that methane emission rates varied significantly across space and time, making it difficult to estimate global methane emissions from measurements of several point sources. Nevertheless, they estimated global methane emissions at $145\,\mathrm{Tg\,yr^{-1}}$, with 63% coming from natural wetlands and the remainder from cultivated rice paddies. This estimate refers to net emissions because methane production, consumption, and oxidation are known to occur simultaneously in microbial consortia dominated by members of the Archaea and Bacteria.

Pancost and colleagues (2000) investigated methane oxidation in marine sediments collected from mud volcanoes on the Mediterranean ridge to discover that the methane-rich sediments contain high levels of methanogen-specific biomarkers as measured by depletion of ^{13}C. In the same sediments, the investigators showed the depletion of biomarkers specific for sulfate-reducing bacteria and other heterotrophic bacterial groups, with the conclusion that ^{13}C-depleted methane is consumed by methanogenic organisms in a reversal of the methane-producing metabolism supported by organisms using sulfate as the terminal electron acceptor (Pancost *et al.*, 2000).

Biotechnological control of methanogenesis and methane consumption might be a way to reduce the greenhouse gas warming potential from emission sources. Several lines of research have focused on the form, function, and evolution of methane monooxygenase (MMO) found in methanotrophic bacteria (Notomista *et al.*, 2003; Smith *et al.*, 2002). However, much remains to be learnt about the diversity of MMO in natural microbial populations and how this diversity influences the global balance of methane production and consumption. To produce a quantitative assessment of methanotrophy, Kolb and colleagues (2003) developed a real-time polymerase chain reaction (PCR) procedure based on the alpha subunit of the particulate methane monooxygenase gene (*pmo*A) and 16S ribosomal DNA of known methanotrophic organisms. Their analysis revealed five distinct subgroups of methanotrophs within the alpha and gamma subclasses of *Proteobacteria*: the *Methylococcus* group; the *Methylobacter* and *Methylosarcina* group; the *Methylosinus* group; the *Methylo-*

capsa group; and an unidentified forest cluster group (Kolb *et al.*, 2003). The population dynamics of these subgroups in different ecosystems linked to methane production must be thoroughly understood to provide meaningful information for large-scale numerical models being developed to predict the impact of methane on global climate change.

Carbon sequestration

Although the scientific community has long recognized the significance of different greenhouse gases such as methane, nitrogen oxides, and water vapor, the reduction of anthropogenic carbon dioxide emissions has dominated international policy debates on mitigating global climate change (Korner, 2003). The replacement of carbon dioxide-producing energy sources with cleaner fuels may be the most impacting strategy for reducing the global warming potential, but it is becoming increasingly clear that developing and improving the effectiveness of ecological sinks for atmospheric carbon is an important complementary strategy (Lackner, 2003). Microorganisms play crucial roles in carbon sequestration in both terrestrial and aquatic ecosystems; however, the most thoroughly discussed, yet controversial, proposal with respect to microbial diversity is "ocean fertilization". Phytoplankton in marine ecosystems play critical roles in regulating the global carbon cycle, but the ability of phytoplankton to fix atmospheric carbon is limited by the concentration of trace elements and iron. Therefore, ocean fertilization refers to the process of supplementing the natural concentrations of iron and other trace elements such as manganese in the ocean to increase phytoplankton productivity. It has been estimated that fertilizing the Antarctic Ocean with iron could allow the phytoplankton to convert all the available nutrients into new organic matter including up to 1.7–2.8 billion tons of carbon, with approximately 5% of that carbon sequestered in the deep waters (Lackner, 2003). Many investigators have questioned the feasibility of ocean fertilization as a means of sequestering atmospheric carbon (Buesseler and Boyd, 2003; Chisolm *et al.*, 2001; Johnson and Karl, 2002), and others have pointed to potential unintended biogeochemical repercussions of the process (Fuhrman and Capone, 1991).

The paucity of data on the linkage between microbial diversity of the oceans and global biogeochemical cycles makes it difficult to fully predict the potential impact of adding iron and trace elements to the marine environment. However, ocean fertilization with iron will likely lead to the development of low oxygen conditions and the mobilization of other non-carbon biogeochemical element cycles. Low oxygen conditions are associated with changes in phytoplankton diversity and increases in the emission of the greenhouse gases nitrous oxides and methane. In addition, ocean fertilization with iron can potentially affect the flux of dimethylsulfide, which has also been implicated in contributing to global climate change (Fuhrman and Capone, 1991).

CONCLUSION

Microorganisms and their activities have far-reaching consequences for the quality of the physical environment on Earth. We are only just beginning to understand the depth of microbial influences on the co-evolution of life and the environment. Despite the agreement that microbial life is ancient on Earth, there are still unresolved debates about its ultimate origin. The principal questions will eventually yield to a combination of continued exploration of remote "extreme" environments on Earth, and the exploration of other planets, which may be at the stage of early Earth in terms of environmental conditions. Even if certain aspects of the panspermia theory are proven correct, the question of the origin of unicellular life forms will remain unanswered. Understanding the history and physiological stages of microbial life on Earth can provide important lessons for the future trajectory of environmental evolution. The increasing impact of human activities on the environment creates an urgent need to investigate how processes invented by diverse categories of microorganisms several eons ago can assist in managing contemporary environmental problems.

QUESTIONS FOR FURTHER INVESTIGATION

1 Compare and contrast the key pieces of evidence for terrestrial versus extraterrestrial origins of microbial life on Earth.
2 Define stromatolites and describe the main features of the biological and physical-chemical processes that led to their formation.
3 What are microbial mats? Describe the diversity of metabolic process that sustains a self-contained microbial mat. If you reside at a location near a microbial mat (for example, salt marshes, hot springs, mines), take a field sampling trip and dissect the mat. Describe any differences or similarities with the mat depicted in this chapter.
4 Describe the four different kinds of "extreme environments" and the survival strategies used by the kinds of microorganisms that inhabit the environments. What can the discovery of microbial diversity in extreme environments tell us about the possibility of the origin of life on Earth and the possibility of the existence of life in space?
5 Describe the sequential endosymbiotic theory and discuss the main pieces of evidence that support this theory. In what way has the emergence of eukaryotes affected the course of environmental evolution?
6 Discuss three case studies where understanding of the co-evolution of microbial diversity and environmental parameters could benefit the implementation of biotechnological strategies for solving contemporary problems associated with global environmental change.

SUGGESTED READINGS

Bui, E.T., P.J. Bradley, and P.J. Johnson. 1996. A common evolutionary origin for mitochondria and hydrogenosomes. *Proceedings of the National Academy of Sciences (USA)*, **93**: 9651–6.

Canfield, D.E. and A. Teske. 1996. Late Proterozoic rise in atmospheric oxygen concentration inferred from phylogenetic and sulphur-isotope studies. *Nature*, **382**: 127–32.

Cao M., J.B. Dent, and O.W. Heal. 1995. Modeling methane emissions from rice paddies. *Global Biogeochemical Cycles*, **9**: 183–95.

Des Marais, D.J., M.O. Harwit, K.W. Jucks, J.E. Kasting, D.N.C. Lin, J.I. Lunine, J. Schneider, S. Seager, W.A. Traub, and N.J. Woolf. 2002. Remote sensing of planetary properties and biosignatures on extra-solar terrestrial planets. *Astrobiology*, **2**: 153–81.

Gold T. 1999. The Deep Hot Biosphere. New York: Springer-Verlag.

Guerrero R., M. Piqueras, and M. Berlanga. 2002. Microbial mats and the search for minimal ecosystems. *International Microbiology*, **5**: 177–88.

Hoffmann, A.A. and P.A. Parsons. 1997. *Extreme Environmental Change and Evolution.* Cambridge, England: Cambridge University Press.

Horikoshi, K. and W.D. Grant (eds.) 1998. *Extremophiles: Microbial Life in Extreme Environments.* New York: Wiley-Liss.

Horneck, G. and A.C. Baumstark-Khan (eds.) 2002. *Astrobiology: The Quest for Conditions of Life.* Berlin: Springer-Verlag.

Hoyle, F. and N.C. Wickramasinghe. 1984. *From Grains to Bacteria.* Cardiff, UK: University College Cardiff Press.

Joyce, G.F. 2002. The antiquity of RNA-based evolution. *Nature*, **418**: 214–21.

Kasting, J.F. 2001. The rise of atmospheric oxygen. *Science*, **293**: 819–20.

Kasting, J.F. and J.L. Siefert. 2002. Life and the evolution of earth's atmosphere. *Science*, **296**: 1066–8.

Lazcano, A. and S.L. Miller. 1994. How long did it take for life to begin and evolve to cyanobacteria? *Journal of Molecular Evolution*, **39**: 546–54.

Margulis, L. 1996. Archaeal–eubacterial mergers in the origin of *Eukarya*: Phylogenetic classification of life. *Proceedings of the National Academy of Sciences (USA)*, **93**: 1071–6.

Margulis, L., C. Matthews, and A. Haselton (eds.) 2000. *Environmental Evolution.* Cambridge, MA: MIT Press.

Margulis, L. and D. Sagan. 1991. *Microcosmos: Four Billion Years of Microbial Evolution.* New York: Simon and Schuster.

Margulis, L. and D. Sagan. 2002. *Acquiring Genomes: A Theory of the Origin of Species.* New York: Basic Books.

Nishioka, K. 1998. *Report on cosmic dust capture research and development for the exobiology program*, Report CR-97-207698. Washington, DC: National Aeronautics and Space Administration. Distributed by National Technical Information Service, Springfield, VA.

Taylor, F.J.R. 1974. Implications and extensions of the serial endosymbiosis theory of the origin of eukaryotes. *Taxon*, **23**: 229–58.

Taylor, F.J.R. 1979. Symbioticism revisited: A discussion of the evolutionary impact of intracellular symbiosis. *Proceedings of the Royal Society of London*, **204**: 267–86.

Taylor, F.J.R. "Max". 2003. The collapse of the two-kingdom system, the rise of protistology and the founding of the International Society for Evolutionary Protistology (ISEP). *International Journal of Systematic and Evolutionary Microbiology*, **53**: 1707–14.

Von Bloh, W., S. Franck, C. Bounama, and H.-J. Schellnhuber. 2003. Maximum number of habitable planets at the time of Earth's origin: New hints for panspermia? *Origins of Life and Evolution of the Biosphere*, **33**: 219–31.

Wiegel, J. and M.W.W. Adams (eds.) 1998. *Thermophiles: The Keys to Molecular Evolution and the Origin of Life?* Philadelphia, PA: Taylor and Francis.

CHAPTER 7

BIOGEOCHEMICAL CYCLING OF CARBON AND NITROGEN

Chapter contents

INTRODUCTION

The first section of this book focused on three questions regarding concepts and methods: "What are the units of microbial diversity?" "How do we recognize diversity in microbial communities?" And "How do we represent microbial diversity?" In this section we turn to questions about generalizable principles and applications: "Of what use is microbial diversity?" And "How can systematic knowledge of microbial diversity contribute to the effective management of global environmental problems?" There are many partial answers to these two questions, depending on the perspective taken and the geographic scale of reference. The situation is analogous to the anecdotal tale of several blind persons trying to describe the features of an elephant by feeling different parts of the animal's anatomy. It is arguable that microbial life ultimately influences all other life forms on Earth. One of the major ways through which microorganisms achieve this omnipotency is by influencing the biogeochemical cycling of elements. The diverse group of organisms and physiological reactions, which form the core of biogeochemical cycles, are of immense importance to how we understand and manage global environmental problems.

The abundance and activities of organisms on Earth modify the direction of flow and size of reservoirs for many chemical elements. Biogeochemists investigate the interactions between both living and dead biological systems and the chemical constituents of geological formations. The Earth is essentially a biogeochemically "closed" system, with the exception of incident electromagnetic radiation from the sun, the escape of a fraction of this radiation as heat, and the introduction of materials through meteoritic impacts. Consequently, we refer to biogeochemical "cycles" because most elements processed by living organisms are ultimately recycled through the inevitable cyclical processes of birth–growth–death–decay. Despite the complex nature of the major biogeochemical cycles, and the relatively small size of individual cells, it is increasingly recognized that microorganisms play disproportionate roles in stimulating and sustaining key biogeochemical processes.

Granting the predominance of microbial participation in biogeochemical cycling, human activities are also increasingly perturbing the cycles due to the rapid introduction of certain compounds from industrial activities into domains that are not able to rapidly process these pollutants (see Chapter 11 for more discussion of the connection between microbial diversity and environmental pollution). This chapter highlights the current state of knowledge on the diversity of microorganisms and their physiological processes which sustain the carbon and nitrogen cycles. The knowledge base is presented mainly to demonstrate that much needs to be learned about the linkages between microbial diversity and global biogeochemistry. There are persistent difficulties posed by scaling up information on microbial diversity obtained through the study of small-scale samples to global estimates of biochemical cycling. Researchers in numerical modeling are in constant need of revised parameters and quantitative information about chemical fluxes in the environment, but the disciplinary connection between modelers and microbiologists is still "under construction", and there is a need for more intensified cross-fertilization of ideas and sharing of technical language between these two research categories. In addition, the difficulties posed by unculturable groups of microorganisms (see Chapter 3) are yielding to molecular methods of taxonomy, however, practically nothing is known about the contribution of this largest category of microorganisms to ecosystem functions. For example, 16S rRNA gene clone libraries frequently contain approximately 50% of sequences belonging to groups with no known cultured representatives, and similar groups of uncultured sequence representatives occur within recognized cultured groups. Furthermore, many estimates of the kinetic features of biogeochemical cycling rates conducted by specific microbial groups are based on the physiological characteristics of a relatively small number of cultivated microorganisms. We presume, however, that the apparent diversity existing in the vast category of uncultured organisms is reflected in physiological differences upon which the forces of natural selection act to produce speciation and further expansion of specific processes within the biogeochemical cycles. Therefore, some of the observed diversity may be related to microbial characteristics that are only remotely related to biogeochemistry, for example those associated with survival in peculiar niches.

Finally, the paucity of information linking taxonomic diversity to physiological diversity in microbial communities limits our ability to adequately measure and interpret functional redundancy with respect to the long-term stability of biogeochemical cycles and their resistance to human-caused perturbation. To accelerate progress in this direction will require long-term commitment to the interdisciplinary research paradigm initiated to understand Earth as an integrated ecological and biogeochemical system.

THE EARTH AS AN INTEGRATED BIOGEOCHEMICAL SYSTEM

Biogeochemistry is a multidisciplinary science that aims to translate a tremendous array of natural history data from biology, geology, and chemistry into coherent and predictive models designed to explain the interrelated fates and consequences of life on Earth at the molecular level. Biogeochemists study the transformation and trafficking of molecules through living organisms. Whereas ecosystem research is typically compartmentalized, biogeochemistry must be pursued in a global environmental context because the flows of materials are not generally bound by physical parameters that separate, for example, aquatic from terrestrial systems (Levia and Frost, 2003). The abundance and distribution of chemical elements on Earth do not reflect the cosmic abundance of the elements, because biological entities preferentially sequester certain categories of compounds and they tend to avoid others that are presumed to be generally toxic (Fig. 7.1). Among the more than 100 naturally occurring elements in the periodic table, only 26 are found naturally in biological systems, and 15 (H, C, N, O, Na, P, S, Mg, K, Ca, Mn, Fe, Cu, Zn, and Mo) are required for the growth of all living organisms that have been thoroughly evaluated, although the evidence is equivocal for Fe (Schlesinger, 1991; Weinberg, 1997). For example, certain species of *Lactobacillus* can abstain from Fe during active growth, using Mn instead (Imbert and Blondeau, 1998). The inert gases are virtually irrelevant for living systems except when they are radioactive; and 21 elements including Be, Br, Hg, Pb, and Rn are toxic to all organisms even at low concentrations.

Selective use of chemical elements by organisms suggests that biogeochemical cycles are constrained by the occurrence of specific physiological processes among phylogenetic groups. The ability of prokaryotes to colonize different environments depends directly on a broad range of stable enzyme functions, and despite their small individidual physical size, the diversity of prokaryotic activities within microbial communities impacts the direction of biogeochemical cycling. Carbohydrates, proteins, lipids, and nucleic acids represent the

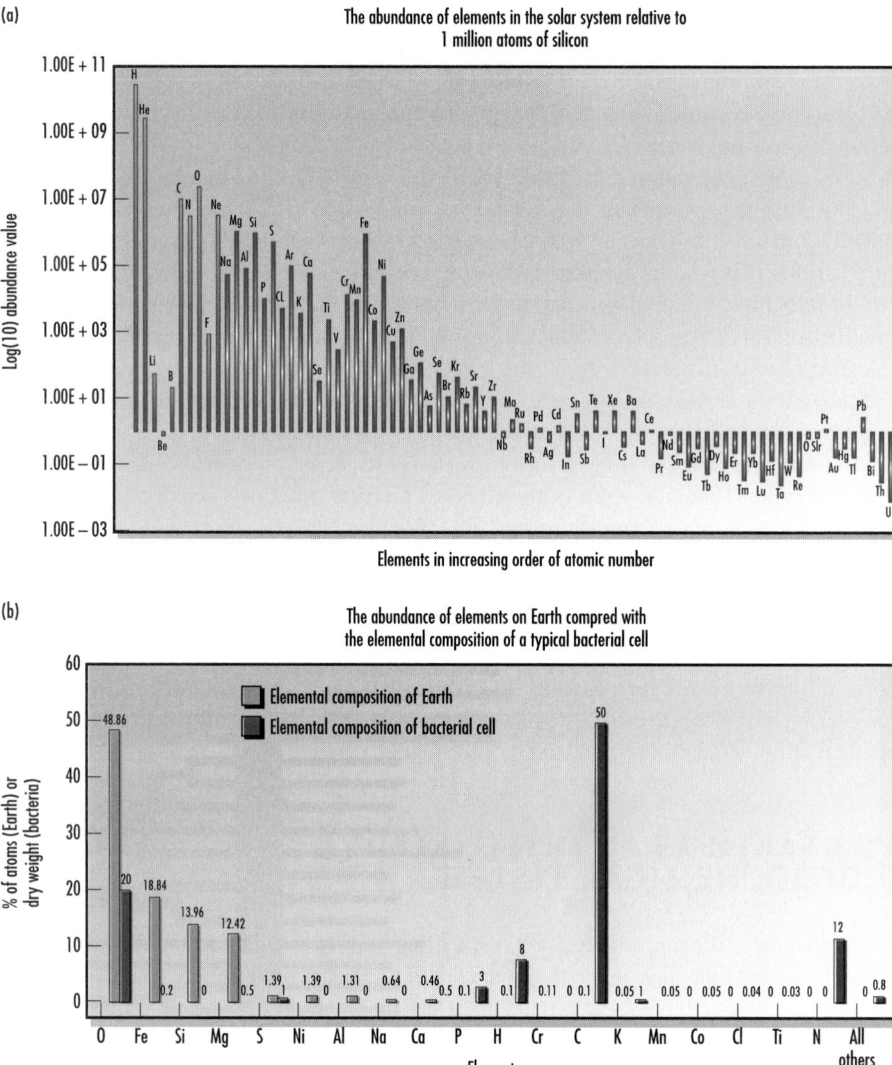

Fig. 7.1 The abundance of elements in the solar system (a) differs from the relative abundance of elements on planet Earth (b) as a result of geochemical events, which have placed constraints on the abundance of elements and molecules in the biosphere. Data in (a) from Butcher *et al.* (1992) and in (b) from Emiliani (1992) and Heldal *et al.* (1997).

core molecules that sustain cell structure and physiological processes. In addition to water, carbon, nitrogen, phosphorus, and sulfur are the major constituents of these life-sustaining molecules, and consequently, biogeochemical studies have focused on these four elements (Fig. 7.1). The cycling of trace elements, in particular metals, is also important because many enzymes require trace quantities of Mg, Mn, Fe, and Zn to function efficiently. Under certain circumstances, non-essential metals such as Hg and Pb exert their toxicity through interference with normal metal–protein interactions. Therefore, considerable attention has focused on the regulation of metal bioavailability through the contribution of microorganisms to the cycling of trace elements (Benjamin and Honeyman, 1992). This chapter focuses on carbon and nitrogen cycles because they are of primary importance to the growth of all organisms, and because of the large amount of information available about them at both local and global scales of understanding. In the next chapter, the cycling of phosphorus, sulfur, and trace elements is discussed.

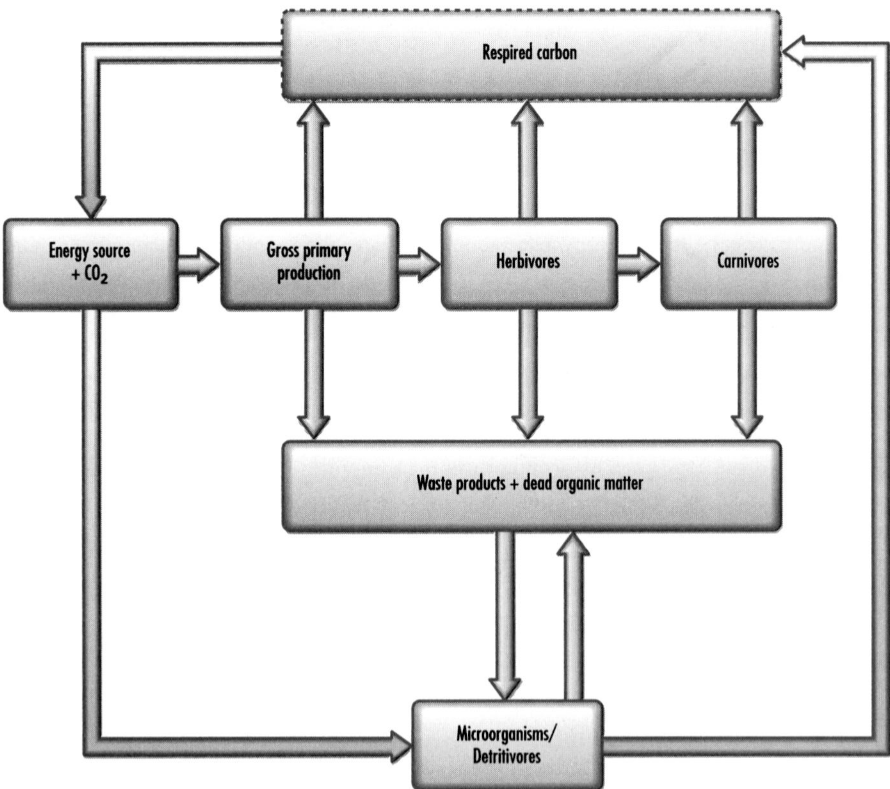

Fig. 7.2 Major trophic groups in stable biomes and the flow of energy among them. Trophic-level classifications are not stringent because many organisms belong to more than one group. Most estimates of net primary productivity are in the range of 45–65 petagrams (1 petagram = 10^{15} g) of carbon per year, representing gross primary production minus carbon released due to respiration (Schlessinger, 1991; Staley and Orians, 1992).

An average of 345 watts (with a range of 15.5–439 W) of energy radiated from the sun is intercepted by each square meter of the Earth's surface (the total surface area of the Earth is $510.3 \times 10^6\,km^2$, with approximately 71% covered by the oceans). This means that each year, the Earth intercepts 5×10^{24} joules of energy from the sun. Biological photosynthesis captures only a tiny fraction (3×10^{20} joules) of this amount (Staley and Orians, 1992). The hydrologic cycle is driven primarily by the heat derived from the conversion of a small fraction of incident solar radiation. Ecosystems, in which water is available and organisms capable of photosynthetic activity are present, play major roles in the fixation of atmospheric carbon. Carbon fixation underpins the cycling of other essential elements through intricate food webs at various trophic levels (Fig. 7.2). Most organisms generate energy through the biochemical transformation of C, H, and O, and these transformations influence the maturation and stability of large ecosystems known as biomes (Sellers *et al.*, 1997; Plate 7.1).

Several investigators have questioned the molecular basis of the homeostatic controls that regulate biogeochemical cycles, and how the understanding of individual cycles can contribute to the integrative knowledge of Earth as a coherent biogeochemical system (Falkowski, 1994; Falkowski *et al.*, 1998 and 2000; Lovelock, 1995; Newman and Banfield, 2002). Many questions remain to be answered. A variety of techniques, ranging from genomic analysis of natural microbial communities to remote sensing of large scale ecological systems, and numerical modeling are now available for shedding new light on the old questions. These techniques attempt to integrate data gathered from very different scales of analysis (Dewar, 1992; Goward and Williams, 1997; Janetos and Justice, 2000; Patil *et al.*, 2001).

The salient questions in this line of investigation are framed by three key concepts:

1 mass balance;
2 early warning; and
3 bioremediation.

Under these overarching concepts, we focus on how the understanding of microbial diversity can improve the state of knowledge about integrated large-scale biogeochemical cycling.

The concept of mass balance presumes that it is possible to develop a quantitative model where the sizes of the reservoirs for each element are known, and that the amount and direction of flow between the reservoirs are also known rather precisely. Unfortunately, even for well-studied elemental cycles such as carbon, there are still missing links and unreconciled data on the sizes and rates of change of various reservoirs. One of the reasons for the persistence of these problems is that relatively little information is available on the role of different microorganisms in transforming carbon compounds in different soils and different parts of the oceans. Improved knowledge of population and physiological diversity in microbial communities is expected to illuminate the current "black-box" situation of the numerical mass-balance models of biogeochemical cycling.

Early warning of imbalances in ecosystem processes is conceptually aimed at capturing the response of biological communities to environmental change before irreversible damage occurs. It has long been recognized that the microbial components of ecosystems respond fairly quickly to both natural and human-induced perturbations. Theoretically, major shifts in the size of reservoirs or the direction of flow of chemical elements between reservoirs should be reflected by changes in structural and physiological diversity of microbial communities. However, there are currently no generalizable approaches for assessing the physiological responses of microbial communities to environmental stress factors. Stress response in microorganisms is an extremely diverse process, and only a small subset of characterized responses is directly relevant to biogeochemistry. Furthermore, very few microbial stress-response systems have been investigated under natural field conditions (Ogunseitan, 2000; Storz and Hengge-Aronis, 2000). Therefore, reliance on microbial communities to indicate response to stress from environmental change is sometimes compromised by the diversity of individual responses without a succinct method of integrating these responses into a coherent index. To circumvent this stumbling block will require more comprehensive knowledge of microbial physiological diversity.

Bioremediation concepts focus on the deliberate use of living organisms to modify the biogeochemical cycling of elements. This biotechnological approach requires detailed and precise knowledge of the structural and physiological diversity of the microbiological communities that play crucial roles in natural biogeochemistry. For example, several proposals have been put forward on how to increase the efficiency of natural sinks for carbon dioxide and methane as a strategy to reduce the atmospheric concentration of greenhouse gases. Some of the proposals require the stimulation of photosynthetic activity in microbial communities. However, without proper understanding of the interconnectedness of biogeochemical cycling, deliberate stimulation of one aspect of the integrated system may have negative impacts on other aspects (Lackner, 2003). These gaps in the state of knowledge on how biogeochemical cycles are integrated have led to the establishment of multidisciplinary research programs described in the next section.

Integrative research on biogeochemical cycling

One of the more prominent research programs to address the multidisciplinary framework of the Earth's integrated biogeochemical system is NASA's Earth Science Enterprise, which emerged as a priority mission in 1991 (http: //earth.nasa.gov/). The understanding of global biogeochemical cycles is one of the major goals of the initiative entitled "Mission to Planet

Earth" (Box 7.1). NASA's program depends on several satellites, space shuttle flight missions, and ground observations to generate consistent datasets on biogeochemical processes, climate and hydrology, ecosystem functions, Earth system history, solid Earth processes, solar influences, and human interactions with the entire structure. Some satellites, particularly those focusing on the measurement of photosynthetic pigments such as chlorophyll, are needed to gather data which complement surface observation of photosynthetic activities in diverse microbial communities. The extensive spatial and temporal scales covered by satellite measurements facilitate the quantitative integration of microbial nutrient-cycling activities into comprehensive models of biogeochemical cycling. In addition, satellite-based information is indispensable for understanding the global distribution of a diverse group of pathogenic microorganisms whose survival in the environment depends on ecosystem productivity (see Chapter 10 for a discussion of remote sensing and *Vibrio cholerae*). For this and other reasons, it is increasingly important for investigators focusing on the broad context of microbial diversity to be familiar with the products of large-scale ecosystem assessment tools such as the satellites represented in Box 7.1.

The initiatives under Mission to Planet Earth include the Earth Observing System (EOS) and Earth Probes. Three major objectives under these two initiatives are particularly relevant to assessments of biogeochemical processes involving microorganisms, namely the carbon cycle, atmospheric chemistry, and the hydrologic cycle. EOS coordinates long-term measurements of interactions among the atmosphere, oceans, Earth crust, and hydrologic and biogeochemical cycles. The assessment of biogeochemical dynamics under NASA's EOS is focused on atmospheric chemistry and the carbon cycle, and the interactions between marine and terrestrial biospheres and the atmosphere are emphasized within these focus areas. The emphasis on the carbon cycle is driven by concerns about anthropogenic forcing of climate change due to the increasing concentration of carbon dioxide, methane, and other greenhouse gases in the atmosphere, and to the changes in the mobility of dissolved carbon sources in the environment (Pastor *et al.*, 2003). To fully understand the carbon cycle requires the elucidation of how the biogeochemical dynamics of elements are coupled to carbon utilization in the biosphere. For example, the elemental composition of photosynthetic phytoplankton is in the order of $C_{106}N_{16}P$, and nitrogen or phosphorus is frequently observed to limit phytoplankton growth and carbon fixation (Staley and Orians, 1992). Therefore, it is necessary to gain an appreciation of how the nitrogen and phosphorus cycles are coupled to the carbon cycle before reliable predictive models can be produced.

The dominance of remote sensing in NASA's program also means that the participation of microbial communities may not be adequately assessed, apart from those engaged in photosynthetic activities by means of pigments visible to the multispectral sensors used by Earth-imaging satellites (Justice *et al.*, 1998). Technological limitations currently hamper the analysis of microbial diversity and its relevance to biogeochemical cycling at the global scale. Many investigators consider the lag in quantitative research on microbiological contributions to the global cycling of carbon and other elements as a major impediment to the generation of numerical models. In response to these concerns, several research initiatives have been launched to better characterize the role of microorganisms in biogeochemical processes in different ecosystems. Among these initiatives are the "Microbial Observatories" (MO) and "Biocomplexity in the Environment" (BE) programs sponsored by the U.S. National Science Foundation.

The MO program began in 1998 to support teams of investigators focusing on "long-term ecological research" (LTER) sites that provide opportunities for monitoring changes in environmental parameters and how these changes affect structural and functional diversity in microbial communities. LTER emphasizes the discovery of new taxonomic groups, characterization of phylogenetic relationships, microbe–microbe interactions, and the description of novel physiological properties across environmental gradients (National Science Foundation, 2000). The BE program was launched in 1999 to promote integrative investigations of environmental systems using advanced scientific and engineering methods. As an emerging concept, biocomplexity focuses on the capacity of the biosphere for

BOX 7.1

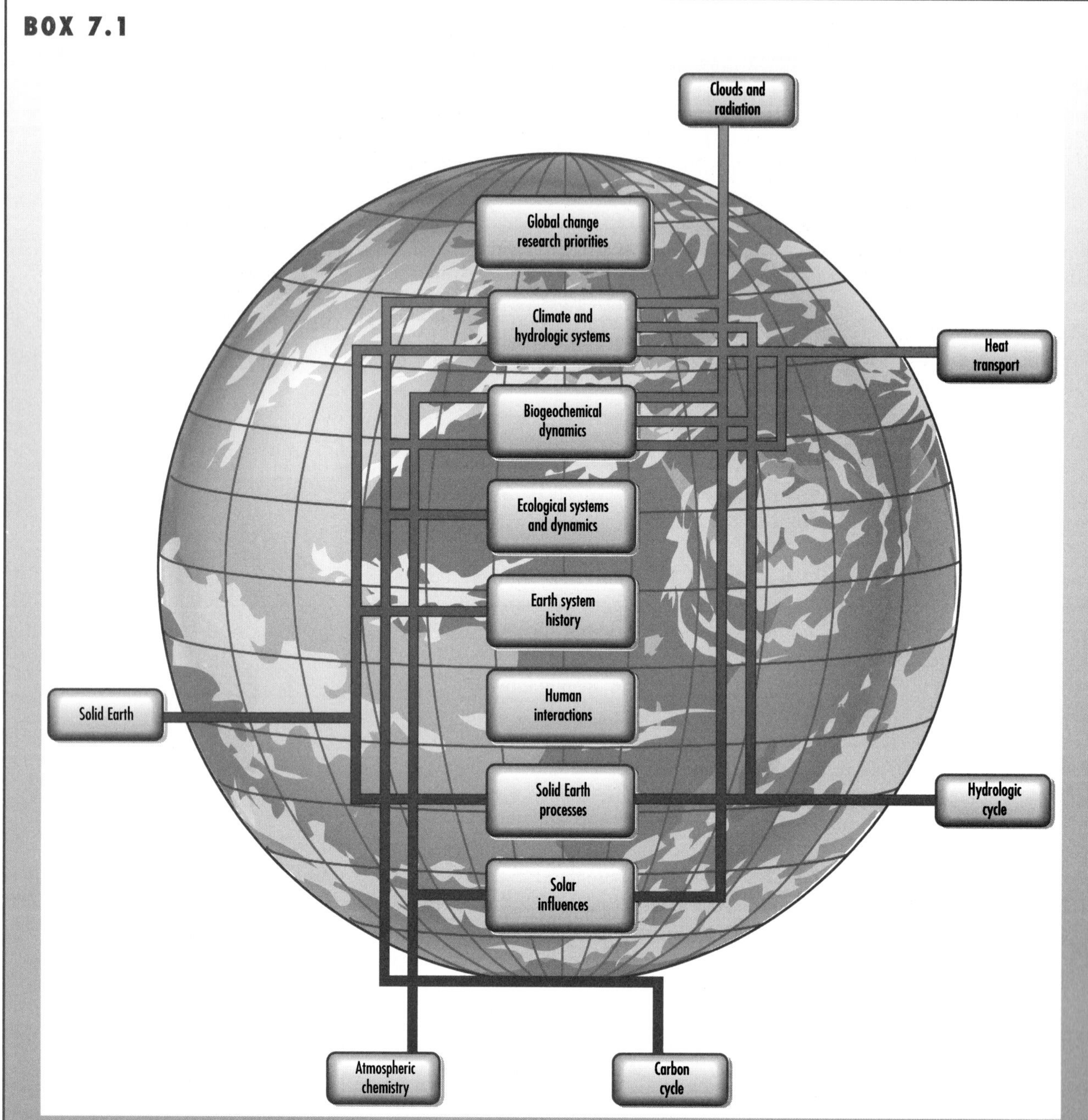

In an unprecedented effort to understand the Earth as an integrated system of interactions among biotic and physical-chemical factors, NASA launched the Mission to Planet Earth research program that identifies seven research priority areas including biogeochemical dynamics. In addition, the program covers six specific topics including the carbon cycle, hydrologic cycle, and atmospheric chemistry. The data being collected span a wide range of processes and scales of analysis. A substantial portion of the data requires satellite instruments for global coverage. The features of the most relevant satellites are presented in the adjoining table. The contribution of remote sensing to the assessment of microbial diversity is only recently becoming clear. For example, the data collected by the SeaWiFS satellite on photosynthetic pigments in the ocean are relevant to the observation of relative photosynthetic activities in different ecological zones. The global coverage of satellite data allows subsequent pinpointing of locations for surface-level analysis of microbial community structure and specific functions. Image is by courtesy of NASA. For more information on specific satellites, mission profiles, and information updates on the The Earth Observing System, visit the program website at http://eospso.gsfc.nasa.gov/.

Characteristics of selected satellites employed for remote sensing under NASA's EOS program.

Satellite name	Function	Launch date
SeaWiFS/OrbView-2	The Sea-viewing Wide Field-of-View Sensor (SeaWiFS) provides quantitative data on global ocean bio-optical properties. The data are represented as a scale of ocean colors signifying the quantity and diversity of marine phytoplankton, with inferences for ecosystem productivity.	August 1, 1997
TRMM	NASA and the National Space Development Agency of Japan jointly operate Tropical Rainfall Measuring Mission. The satellite monitors tropical rainfall and the associated release of energy that drives global atmospheric circulation that influences global climate and local weather systems.	November 27, 1997
Landsat-7	Landsat-7 continues the classic tradition of the Landsat series, first launched in 1972, making it the longest running program to acquire images of the Earth from space. The Enhanced Thematic Mapper provides repetitive, multispectral, high-resolution imagery of land surfaces. The data can be used to generate a normalized difference vegetation index (NDVI), which is useful for estimating the magnitude of carbon sinks.	April 15, 1999
Terra	The Terra satellite is considered the flagship of the EOS program. It monitors the interactions among solar radiation, atmospheric, land, and oceanic parameters. The mission involves five instruments including the Advanced Thermal Emission and Reflection Radiometer (ASTER), which provides high spatial resolution (15–90 m) multispectral images of the Earth's surface, and the Moderate-Resolution Imaging Spectroradiometer (MODIS), designed to measure global-scale biological (including chlorophyll fluorescence) and physical processes every 1–2 days.	December 18, 1999
ACRIMSAT	This program measures the total amount of the Sun's energy that falls on the Earth's land and ocean surface, and the proportion absorbed by the atmosphere by means of a series of active cavity radiometer irradiance monitors.	December 20, 1999
NMP/EO-1	The Earth-Observing-1 satellite is part of NASA's New Millennium Program aimed at validating technologies that contribute to cost reductions in follow-up Landsat missions. EO-1 depends on the "Hyperion" instrument, which employs 220 spectral bands ranging from 0.4 to 2.5 micrometers (instead of the 10 multispectral bands used by the traditional Landsat satellites) to monitor complex land ecosystem processes. The spatial resolution is 7.5 km by 100 km of land.	November 21, 2000
SORCE	The Solar Radiation and Climate Experiment (SORCE) uses a spectral irradiance monitor to accurately measure solar ultraviolet, far ultraviolet, visible, and near infrared solar irradiance up to 20 nanometers with implications for ecosystem changes on Earth.	January 25, 2003
AURA	Uses four major instruments, including the High Resolution Dynamics Limb Sounder (HIRDLS) to measure atmospheric trace gases such as ozone, water vapor, methane, nitrogen oxides, chlorofluorocarbon compounds and aerosols in the upper troposphere, stratosphere, and mesosphere.	January 2004
OCO	The Orbiting Carbon Observatory measures atmospheric carbon dioxide concentration from space, as part of NASA's high priority to model the carbon cycle as accurately as possible. OCO is revolutionizing the strategies for identifying the sinks for anthropogenic carbon dioxide.	Pending
VCL	The Vegetation Canopy Lidar will use a multibeam laser-ranging device to provide a global inventory of the structure of forests on Earth, including accurate estimates of global biomass.	Pending

Source: From http: //eospso.gsfc.nasa.gov/.

adaptation and homeostatic self-organization. The biocomplexity concept has emphasized interdisciplinary research on complex environmental systems including interactions of non-human biota or humans on systems that potentially exhibit nonlinear characteristics. Four specific areas identified under the BE program underscore the central role of biogeochemical dynamics in the understanding of the Earth system, namely dynamics of coupled natural and human systems; coupled biogeochemical cycles; genome-enabled environmental science and engineering; and instrumentation development for environmental activities (National Science Foundation, 2001). By emphasizing quantitative approaches and global perspectives, BE aims to improve science-based predictive capabilities to complement remotely sensed monitoring programs for modeling Earth as a coherent biogeochemical system (National Science Foundation, 2003). As a framework of research in assessments of microbial diversity, both MO and BE programs encourage the exploration of remote and peculiar ecosystems such as deep-sea hydrothermal vents and hyperthermal environments that harbor unique microorganisms and new metabolic processes. New instruments and methods for exploring microbial genomes and proteomes of uncultivated microorganisms are revealing the dominant role of microbial processes in major biogeochemical cycles, and how these processes are interconnected with other components of the biosphere.

THE CARBON CYCLE

Carbon is the cornerstone element supporting the perpetuation of life on Earth. Carbon atoms exist in nature at oxidation states ranging from +4 (CO_2) to −4 (CH_4). This range of valences contributes to the existence of more than one million known carbon compounds,

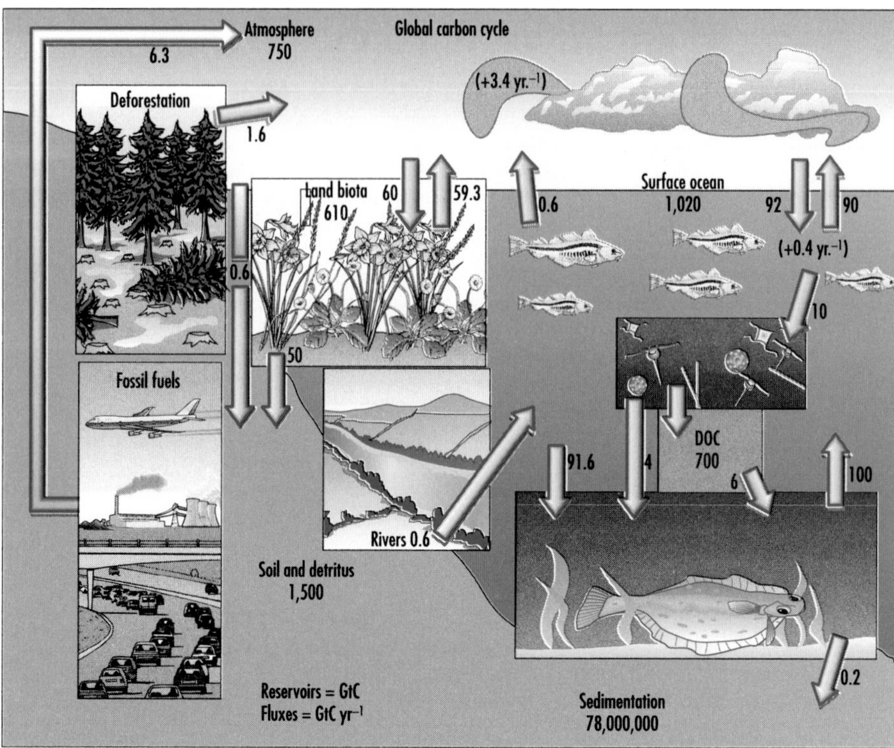

Fig. 7.3 Major components of the global carbon cycle. Reservoir (boxes) sizes are in petagrams of carbon, and fluxes (arrows) in petagrams per year. Composite estimates are from studies up to the year 2000. The diagram is reproduced with permission from Richard A. Feely, Pacific Marine Environmental Laboratory/NOAA, Seattle, WA 98115. National Oceanographic and Atmospheric Administration (2003) (http://www.pmel.noaa.gov/co2/co2-home.html).

of which thousands are essential for maintaining cellular physiological processes. The biogeochemical cycling of carbon is complicated by the molecular diversity of carbon biochemistry, and not much is known about the relationships between the diversity of carbon substrates and the diversity of the organisms interconverting them. Despite its complexity, the carbon cycle is among the best understood of the elemental cycles (Fig. 7.3). The carbon cycle is defined by the relative sizes of different reservoirs, and the efficiency of fluxes among carbon sinks and sources. Microorganisms are involved in every aspect of the carbon cycle either through specialized enzymatic processes as depicted in Fig. 7.4, or through fossilization processes that created the largest reservoir of carbon in sedimentary carbonate rock (Holmen, 1992; Maier, 1999; but see Gold, 1999, who favors the abiogenic theory which posits that petroleum originated from the initial materials that formed the Earth). All microorganisms participate in the global carbon cycle, but distinct categories of participation are recognized on the basis of how carbon flow is coupled to the cycling of other elements. The nomenclature on carbon utilization is also linked to the source of energy used by organisms to support growth. Therefore, the classification of microorganisms according to their metabolism considers whether they use organic versus inorganic sources of carbon, and whether they use light or the biochemical transformation of chemical compounds as energy sources.

With respect to carbon source, **autotrophic** microorganisms use inorganic carbon compounds (e.g. carbon dioxide, bicarbonate) for biosynthesis, whereas **heterotrophic** microorganisms use organic forms of carbon. With respect to energy source, **phototrophic** microorganisms use light to generate energy, whereas **chemotrophs** use the energy released from the breakage of either inorganic bonds (**lithotrophs**) or organic bonds (**organotrophs**). These terms are usually linked to capture both the source of carbon and the source

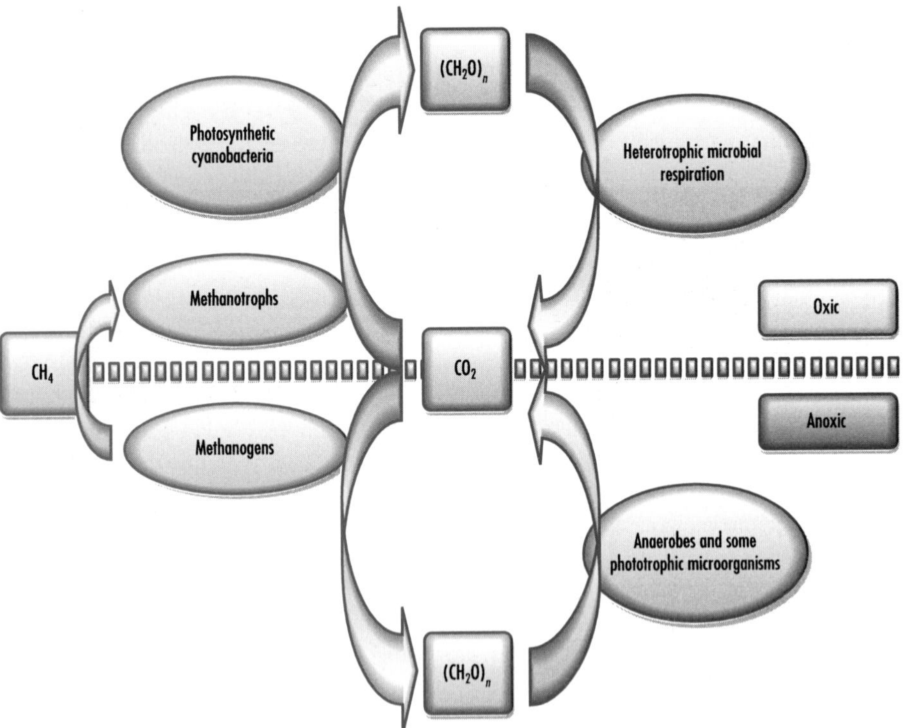

Fig. 7.4 All microorganisms participate in the carbon cycle, but in different ways depending on the availability of oxygen. Photosynthetic cyanobacteria, methanotrophs, and aerobic heterotrophs recycle carbon under aerobic conditions; and methanogens, and some phototrophic anaerobes are active under anoxic conditions. This is a simplified model of the carbon cycle because of the very limited understanding of the diversity of carbon biochemistry; it also interrelates with the diversity of microorganisms and their physiology.

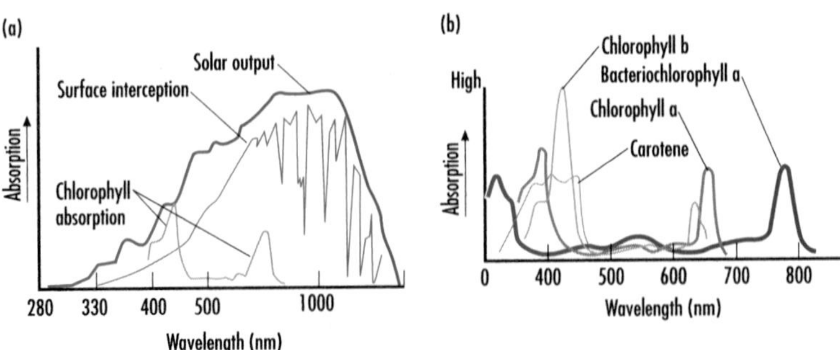

Fig. 7.5 The carbon cycle is initiated by the capacity of photosynthetic organisms to trap solar electromagnetic radiation that is not intercepted by the atmosphere (a). Different kinds of light-gathering arrays have adapted to different segments of the electromagnetic spectrum in order to maximize the utilization of different ecological niches. For example, bacteriochlorophylls absorb at wavelengths near ultraviolet and infrared, which are not covered by plant chlorophylls "a" and "b" (b). By courtesy of David Webb, University of Hawaii.

of energy in the same word. For example, **photoautotrophs** have the capacity to fix carbon dioxide through photosynthetic reaction centers that use light-gathering arrays, which are adapted to specific segments of the electromagnetic spectrum (Fig. 7.5). **Chemoautotrophic** organisms are non-photosynthetic autotrophs that obtain energy through the oxidation of inorganic chemicals. **Chemolithothrophic** organisms include the carboxydobacteria, a diverse group of bacteria that have the capacity to grow on carbon monoxide as the sole carbon and energy source under aerobic conditions. Most of these organisms are Gram-negative but several Gram-positive carboxydobacteria have been described, including species of *Arthrobacter*, *Bacillus*, *Streptomyces*, *Sarcina*, *Nocardia*, and *Corynebacterium*, and *Actinoplanes*, *Microbispora*, and *Mycobacterium* (O'Donnell *et al.*, 1993; Park *et al.*, 2003). Aerobic **heterotrophic** bacteria gain energy by respiration using oxygen and organic carbon compounds, thereby returning carbon dioxide to the atmosphere. Under anaerobic conditions, **methanogenic** microorganisms produce methane instead of carbon dioxide, and the methane can be oxidized by **methanotrophs** that gain energy in the process.

Photosynthesis

The diversity of photosynthetic prokaryotes is best understood by considering the variety of photosynthetic reaction centers (PRCs) and the molecular diversity of key enzymes involved in carbon fixation. PRCs are complexes of proteins and metallic cofactors that catalyze the first steps in the conversion of light energy to chemical energy during photosynthesis (Fig. 7.6). Photosynthetic pigments (chlorophylls, bacteriochlorophylls, and carotenoids) exhibit high absorption coefficients (10^4–$10^5\,M^{-1}\,cm^{-1}$) and characteristic absorption spectra with multiple peaks in the visible and far-red region (see Fig. 7.5). The specificity of bacteriochlorophyll absorption spectra has been used as a fingerprinting technique for assessing the diversity of phototrophic organisms inhabiting ecosystems where photosynthesis is partitioned into different ecological zones (Frigaard *et al.*, 1996). The photosynthetic mem-

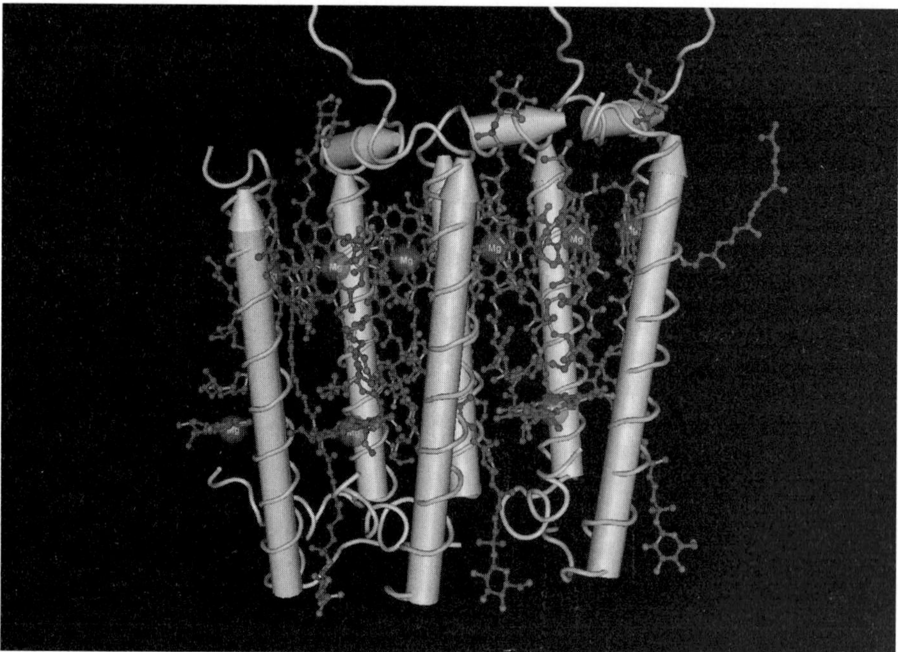

Fig. 7.6 The structure of the light-harvesting complex 2 from *Rhodopseudomonas acidophila* strain 10050. Apoproteins (isolated from non-polypeptide moieties with which they naturally form functional complexes) are arrowed large rods pointing upward (alpha-apoprotein and beta-apoproteins). The 800 nm and 850 nm absorbing bacteriochlorophyll molecules and the carotenoid (rhodopsin–glucoside) molecules are interspersed among the arrowed rods in ball and stick format. The magnesium at the center of each bacteriochlorophyll molecule is shown as a sphere. The periplasm (the region near or immediately within bacterial or other cell wall) is at the top of the image with protruding lines. The three-dimensional structure of the protein was created and manipulated from amino acid sequence data by means of the Cn3D software, version 4.1. For more information on the comparative structural analysis of prokaryotic light-harvesting complexes, see the website of Richard Cogdell at the University of Glasgow, Scotland (http://www.gla.ac.uk/ibls/BMB/rjc/lh2start.html).

branes of certain members of *Proteobacteria* (purple bacteria) contain two light-harvesting complexes, LH-I and LH-II. Light energy is absorbed directly by LH-II, which transfers the energy first to LH-I then to the reaction center. There is notable similarity between the PRC structure of two major taxonomic groups of photosynthetic bacteria, **purple sulfur bacteria** (*Chromatiaceae*) and **purple non-sulfur bacteria** (*Rhodospirillaceae*) (Pierson and Olson, 1987). The structures of the LH-II of two related purple bacteria, *Rhodospirillum* and *Rhodopseudomonas acidophila* have been deduced from x-ray crystallography measurements (Fig. 7.6). In both cases, a ring of symmetrically arranged bacteriochlorophyll-a was identified, with *Rhodospirillum* containing 16 molecules and *Rhodopseudomonas* containing 18 molecules which absorb light in the 850 nm wavelength band. The PRCs of these organisms consist of at least three protein subunits called L (light), M (medium), and H (heavy). In *Rhodopseudomonas viridis* and *Thiocapsa pfennigii*, the four-heme cytochrome *c* is the fourth (and the largest) protein subunit (Deisenhofer and Michel, 1992). The subunits L and M bind bacteriochlorophylls, bacteriopheophytins, quinones, a ferrous iron ion, and a carotenoid as prosthetic groups. The PRCs from different microorganisms also differ in the types of cofactors used, and the difference influences the absorption spectrum specific for different types of light-gathering pigment (Fig. 7.7). For example, the PRC of *Rhodobacter*

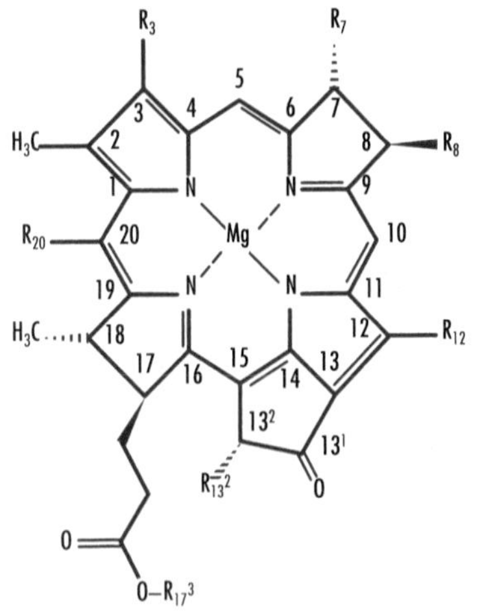

(a)

Bacteriochlorophyll class	R_3	R_7	R_8	7,8-bond	R_{12}	R_{13}^{2}	R_{20}	Main R_{17}^{3}
a	—CO—CH_3	M	E	Single	M	—CO—O—CH_3	H	P, (GG)
b	—CO—CH_3	M	=CH—CH_3	Single	M	—CO—O—CH_3	H	P
c[1]	—CHOH—CH_3	M	E, Pr, I	Double	M, E	H	M	F
c[2]	—CHOH—CH_3	M	E	Double	M	H	M	S, (other)
d	—CHOH—CH_3	M	E, Pr, I, N	Double	M, E	H	H	F
e	—CHOH—CH_3	—CHO	E, Pr, I, N	Double	E	H	M	F
g	—CH=CH_2	M	=CH—CH_3	Single	M	—CO—O—CH_3	H	F
Plant chlorophyll *a*	—CH=CH_2	M	E	Double	M	—CO—O—CH_3	H	P
Plant chlorophyll *b*	—CH=CH_2	—CHO	E	Double	M	—CO—O—CH_3	H	P

[1]Green sulfur bacteria; [2]Green filamentous bacteria. R groups: E = ethyl; F = farnesyl; GG = geranylgeranyl; H = hydrogen; I = isobutyl (2-methylpropyl); M = methyl; N = neopentyl (2,2-dimethylpropyl); P = phytyl; Pr = propyl; S = stearyl. Adapted with permission from Niels-Ulrik Frigaard, Pennsylvania State University, and from Scheer (1991).

(b)

Fig. 7.7 The structure of chlorophyll molecules (a), showing the diversity of reactive groups (b) that influence the light absorption spectra of light-gathering arrays (c). By courtesy of Niels-Ulrik Frigaard, Pennsylvania State University.

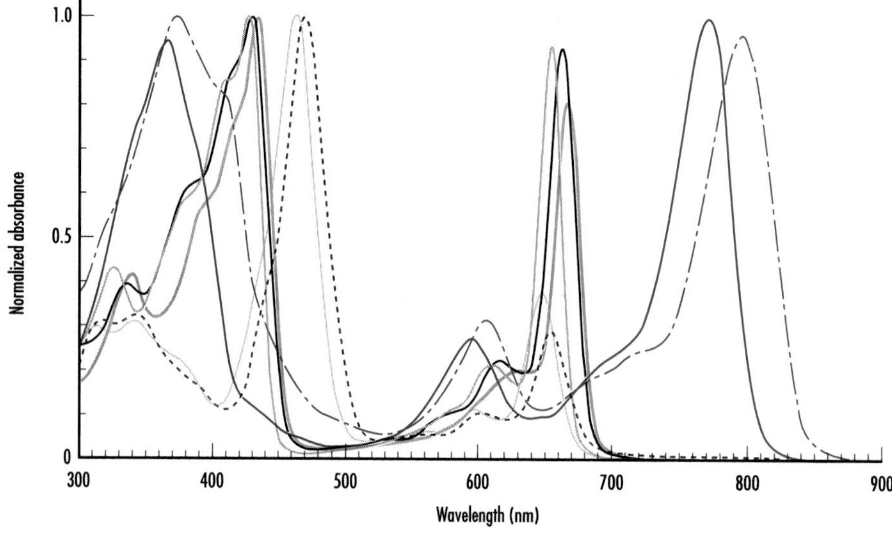

	Chlorophyll pigment class (BC = Bacteriochlorophyll)						
	A	B	BC-A	BC-B	BC-C	BC-D	BC-E
Line	———	———	———	—·—	-------	———	———
Maximum absorption wavelengths (nm)	430, 663	463, 648	364, 770	373, 795	434, 666	427, 655	469, 654
Maximum absorption wavelengths for corresponding pheophytin (nm)	408, 664	No data	357, 746	367, 776	412, 666	406, 657	435, 665

(c)

Fig. 7.7 *Continued*

sphaeroides and *Rhodobacter capsulatus* contain bacteriochlorophyll-a and bacteriopheophytin-a, whereas the PRC of *Rhodopseudomonas viridis* and *T. pfennigii* contain bacteriochlorophyll-b and bacteriopheophytin-b. The absorption spectrum of photosynthetic membranes of the green filamentous bacterium *Chloroflexus aurantiacus* also differs from the absorption spectrum of the purple non-sulfur bacterium *Rubrivivax gelatinosus*, although both contain a protein complex containing bacteriochlorophyll-a which absorbs around 800 and 860 nm (Fig. 7.8). The main pigments in *Chloroflexus* are large aggregates of bacteriochlorophyll-c, which absorbs around 740 nm and are organized without protein. The PRCs of both organisms also contain carotenoids, which absorb in the 400–500 nm region (Frank, 1993).

The diversity of PRCs is one of the remarkable features of stratified microbial communities such as microbial mats (see Chapter 6). In these systems, the variety of microbial physiological systems creates environmental heterogeneity which, in turn, feeds back to increase microbial diversity through the establishment of spatio-temporal variability of biochemical gradients. For example, the colorful layers of microbial mats correspond to the predominance of photosynthetic pigments, with green and pink layers dominated by cyanobacterial bacteriochlorophyll-a, and carotenoids, the major pigments of phototrophic purple sulfur bacteria inhabiting the photic zone from 2–5 mm beneath the surface.

The assessment of molecular diversity in autotrophic microbial communities is increasingly performed by analyzing the key enzymes involved in carbon fixation. Take for example, the enzyme responsible for carbon fixation in the Calvin–Benson–Basham reductive pentose

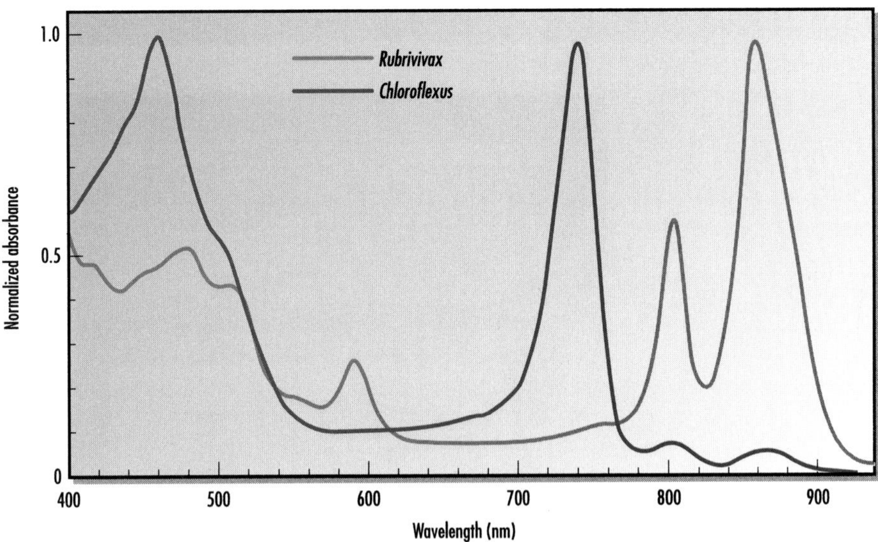

Fig. 7.8 The absorption spectra of photosynthetic membranes of the green filamentous bacterium *Chloroflexus aurantiacus* compared with that of non-sulfur bacterium *Rubrivivax gelatinosus*. By courtesy of Niels-Ulrik Frigaard, Pennsylvania State University and K.V.P. Nagashima for the spectrum of *Rubrivivax*.

phosphate pathway, namely, ribulose-1,5-bisphosphate carboxylase/oxygenase (E.C. 4.1.1.39; or RuBisCO). RuBisCO catalyzes the carboxylation of ribulose-1,5-bisphosphate to create two molecules of 3-phosphoglycerate (see Box 7.2). Several different forms of RuBisCO exist among the Bacteria, Archaea, and Eukarya phylogenetic domains. All versions of RuBisCO that have been investigated are clearly related, but there is considerable diversity in molecular structure as defined by amino acid sequence and three-dimensional protein conformation, as well as in catalytic properties and substrate specificity. Reports of two major types of RuBisCO, namely type I and type II, differentiated on the basis of amino acid sequence similarity (25–30%) have dominated the literature, although there are some reports of an even more distantly related RuBisCO in Archaea inhabiting anoxic environments (Tabita, 1995; Watson *et al.*, 1999). Type I RuBisCO has been considered the most abundant enzyme on Earth, being present in plants and algae where it accounts for 16% of chloroplast protein content (Ellis, 1979). Type I RuBisCO is also present in cyanobacteria, and in phototrophic and chemoautotrophic prokaryotes. Type II RuBisCO is found primarily in the *Proteobacteria*. The two types differ structurally, with type I RuBisCO containing eight large and eight small subunits; whereas type II RuBisCO has only large subunits which are also structurally different from the large subunits of type I RuBisCO (Watson and Tabita, 1997).

Extensive phylogenetic analysis has been performed on the basis of the molecular structure of the large subunits of type I RuBisCO. The analysis shows four subtypes, represented by IA, IB, IC, and ID. Subtypes IA and IB are associated with organisms producing green chlorophyll photosynthetic pigments, such as plants, green algae, and cyanobacteria, and several other groups of the Bacteria domain. Subtypes IC and ID are mostly associated with organisms producing carotenoid photosynthetic pigments, such as eukaryotic red algae, and "purple" bacteria. For an example of phylogenetic analysis of autotrophic carbon fixation using RuBisCO genes, see Box 7.2.

Methanogenesis

Methanogenic microorganisms play an important role in the carbon cycle because methane is a major product of energy-generating metabolic reactions derived from the utilization of various inorganic and organic substrates under anaerobic conditions. Table 7.1 presents

BOX 7.2

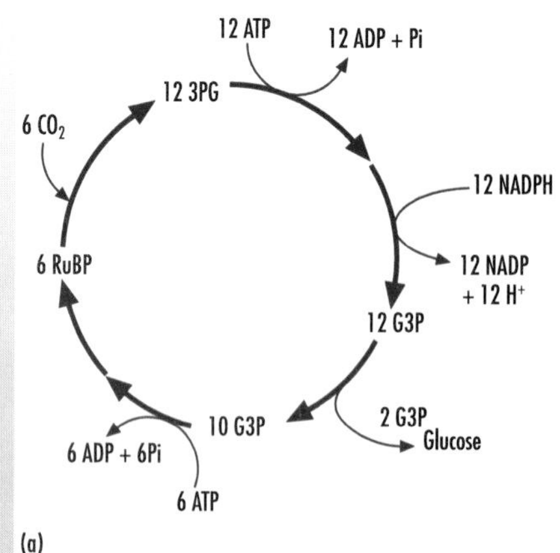

(a)

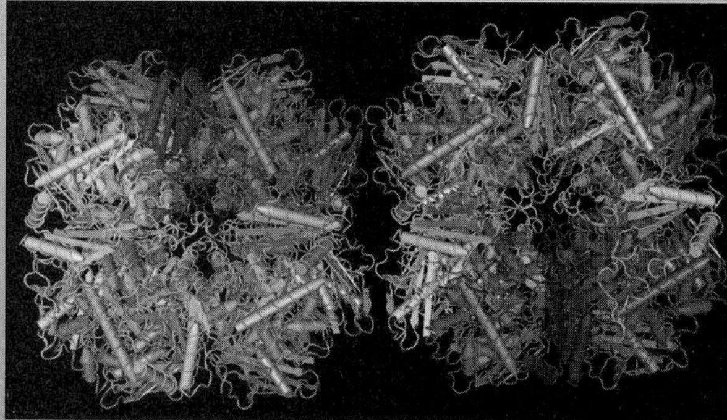

(b)

(c)

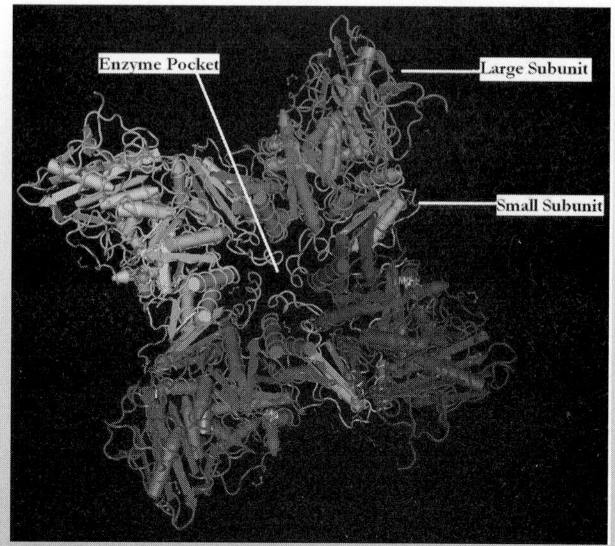

(d)

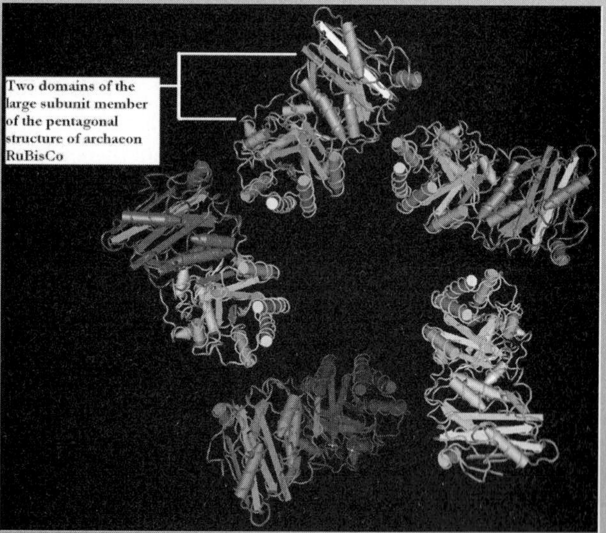

(e)

The assimilation of carbon dioxide through the Calvin–Benson–Basham (CBB) cycle (a) is one of the major pathways used by microorganisms to fix carbon. Other pathways include the reductive tricarboxylic acid cycle, the 3-hydroxypropionate cycle, and the non-cyclic reductive acetyl-CoA pathway (Alfreider *et al.*, 2003). The primary reaction in the CBB cycle is the addition of carbon dioxide to ribulose bisphosphate (RuBP) to produce 3-phosphoglycerate (3PG). This reaction is catalyzed by RuBisCO (b), an enzyme whose structural and functional diversity has been proposed for reconstructing phylogenetic relationships. The most common form of RuBisCO (type I) consists of eight large and eight small subunits in two tetrad arrangements (c) (also see Mizohata *et al.*, 2002). For most photosynthetic members of the Bacteria, the tetrad form exhibits an enzyme pocket enclosed by the subunits (d). Type I RuBisCO is further subdivided into "red or type IA and IB" (eukaryotic red algae and "purple" bacteria) and "green or type IC and ID" (plants, green algae, and bacteria) varieties, on the basis of phylogenetic analysis conducted on the amino acid sequence of the large subunit. Type II RuBisCO is found mostly in *Proteobacteria*. Jordan and Ogren (1981) suggested that the type II is the ancestral form of RuBisCO because of its relatively poor affinity for carbon dioxide. Type II RuBisCO functions best under environmental conditions with high concentrations of carbon dioxide, and low oxygen levels, a situation reminiscent of early Earth's atmosphere. However, the structure of RuBiscO in some members of the Archaea is remarkably different. The RuBisCO of *Pyrococcus kodakaraensis* strain KOD1 exists as a pentagonal structure consisting of the large subunit only (e) (also see Maeda *et al.*, 1999 and Kitano *et al.*, 2001). Protein structures were created and

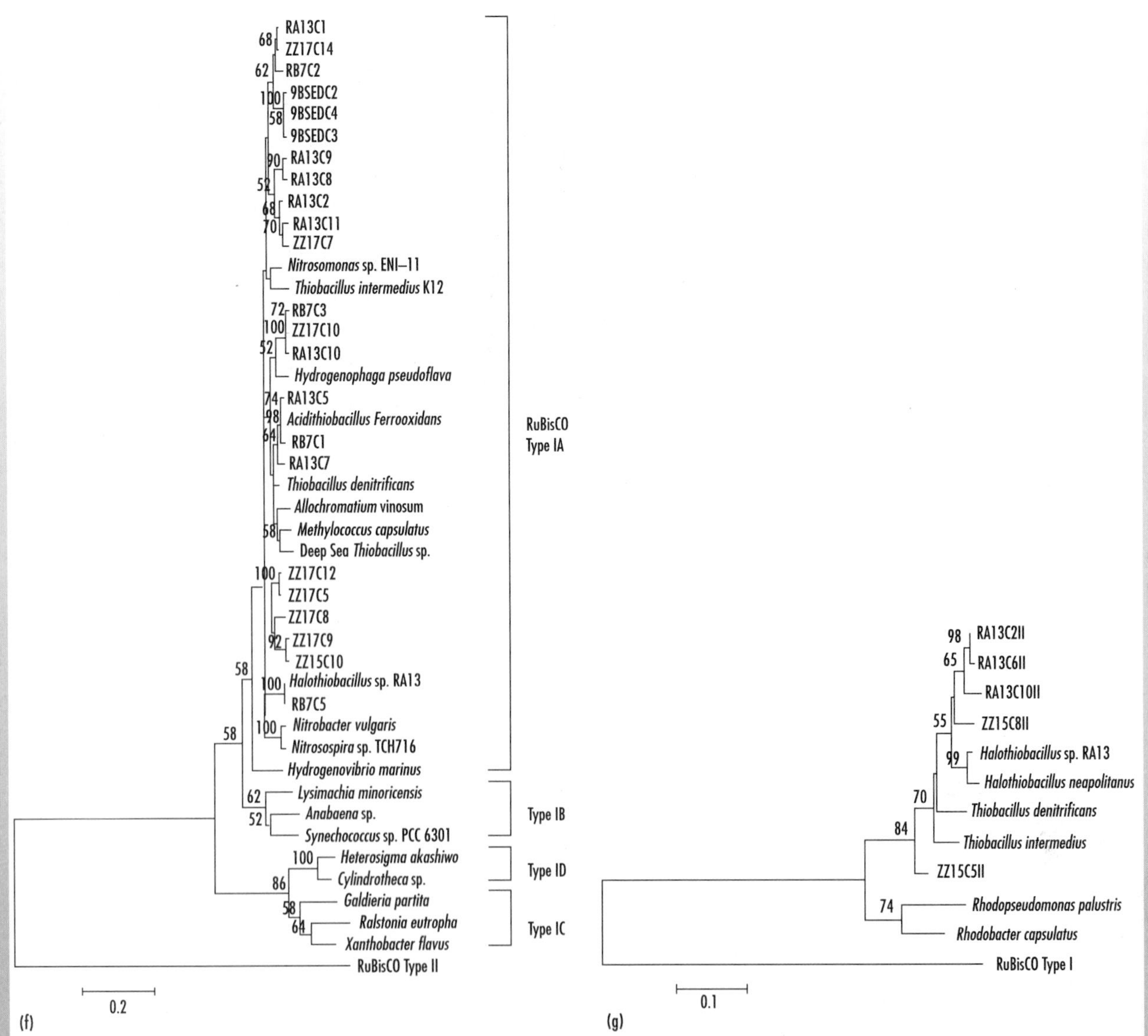

manipulated by means of the Cn3D software, version 4.1. For a discussion of the connection between RuBisCO activity and global warming, see Chapter 11.

Alfreider and colleagues (2003) used molecular techniques to investigate the phylogenetic diversity of the large subunit genes of types I and II RuBisCO in subterranean environments. They used the neighbor-joining method (see Chapter 5) to construct trees based on deduced amino acid sequences of genes encoding for the large subunit of type I RuBisCO (*cbbL* gene) derived from clone libraries, *Halothiobacillus* sp. RA13 isolated at the study site (shown in bold), and sequences from the database of the National Center for Biotechnology Information (NCBI). The gene encoding for the large subunit of type II RuBisCO sequence (*ccbM*) of *Thiobacillus denitrificans* was used as outgroup in the data presented in (f). One hundred bootstrap replicates were performed, and values over 50% are indicated on nodes. The scale bar in (f) represents 20% estimated change. The phylogenetic tree in (g) represents similar experiments but using deduced amino acid sequences of *cbbM* genes derived from the clone libraries, the isolate *Halothiobacillus* sp. RA13 obtained from the study site (shown in bold), and sequences from the NCBI database. The type I RuBisCO sequence (*ccbL*) of *Hydrogenophaga pseudoflava* was used as outgroup. One hundred bootstrap replicates were performed, and values over 50% are indicated on nodes. The scale bar in (g) represents 10% estimated change. At the subterranean site, diverse sequences related to the cluster of green-like type IA RuBisCO sequences were detected among obligate and facultative chemolithoautotrophic *Proteobacteria*, whereas type II RuBisCO sequences were related to *Thiobacillus* species. The newly discovered *Halothiobacillus* species is unusual because it possesses the two types of RuBisCO. The study was the first to demonstrate that microbial autotrophic fixation of carbon dioxide through the CBB cycle plays an important role in subsurface systems with little or no incident light and aeration. Phylogenetic trees are modified from Alfreider *et al.* (2003).

Table 7.1 The variety of reactions in microbial methanogenesis.

Substrate	Product	Energy production $\Delta G_0'$ (kJ/mol of methane)[1]
Hydrogen and carbon dioxide (carbonate)	Methane and water	−135.6
Formic acid	Methane, carbon dioxide, and water	−130.1
2-propanol and carbon dioxide	Methane, acetone, and water	−36.5
Acetic acid	Methane and carbon dioxide	−31.0
Methanol	Methane, carbon dioxide, and water	−104.9
Methylamine and water	Methane, carbon dioxide, and ammonium	−75.0
Dimethyl sulfide	Methane, carbon dioxide, and hydrogen sulfide	−73.8

[1]Reaction pathways and energetics data from Jones (1991) and Gottschalk (1989).

comparable data for seven different substrates used by microorganisms to produce methane. Most of the methane that occurs in nature is biogenically derived from acetate, even though this pathway generates the least amount of energy (Gootschalk, 1989). The reason is that methanogenesis from acetate is probably one of the most primitive metabolic pathways since acetate would have been present in the early oceans from abiotic synthesis (Schlesinger, 1991). The chemoheterotrophic organisms that invented this pathway probably also scavenged the products of obligate heterotrophs. In contrast to the early emergence of the acetate pathway, methanogenesis from carbon dioxide reduction is more physiologically complex, and emerged much later through the cooperation of heterotrophic members of the Bacteria that convert organic matter to substrate, and the Archaea cells that convert the acetate to methane (Wolin and Miller, 1987).

The current atmospheric concentration of methane is 1.7 parts per million by volume (ppmv), making it the most abundant organic gas in the atmosphere. The methane content of the atmosphere has increased steadily over the past 300 years from the pre-industrial level of 0.65 ppmv (Cicerone and Oremland, 1988; Tyler, 1991). The build up of methane in the atmosphere is strongly implicated in global climate change because although there is a lot more carbon dioxide than methane in the atmosphere, a molecule of methane is 22 times more effective than a molecule of carbon dioxide in trapping infrared radiation that would otherwise escape into space (Tyler, 1991). Natural methane production is dominated by microbial activity, and the process occurs in a wide variety of ecological systems where oxygen is limited, including wetlands, anaerobic sediments, animal and insect intestines, geothermal vents, and artificial municipal landfills. All methanogens are strict anaerobes, belonging to a phylogenetically diverse branch within the Archaea domain (Boone and Maestrojuán, 1993; Boone *et al.*, 1993) (Fig. 7.9).

More than 60 distinct species of methanogens have been isolated in pure culture or as members of microbial consortia. *Methanococcus janaschii*, *Methanobacterium thermoautotrophicum*, and *Methanosarcina acetivorus* are among the best-studied examples of methanogens, and their genomes have been completely sequenced (Bult *et al.*, 1996; Smith *et al.*, 1997; Galagan *et al.*, 2002). Members of the phylogenetic family *Methanosarcineae* are the most metabolically diverse methanogens, and they have been isolated from a wide range of ecological systems. The *Methanosarcineae* are also unique among the Archaea in their ability to form complex multicellular structures (Galagan *et al.*, 2002). Raskin and colleagues (1994) developed a technique based on group-specific 16S rRNA hybridization probes to identify methanogens. Eight probes were required to circumscribe methanogens available as pure cultures with the exclusion of the family *Methanothermaceae*. The hybridization probes were specific for members of the taxonomic orders *Methanobacteriales*, *Methanococcales*, and *Methanomicrobiales* (Table 7.2).

An alternative to the use of 16S rRNA for phylogenetic analysis of methanogens is the potentially more specific molecular signatures associated with methyl-coenzyme M

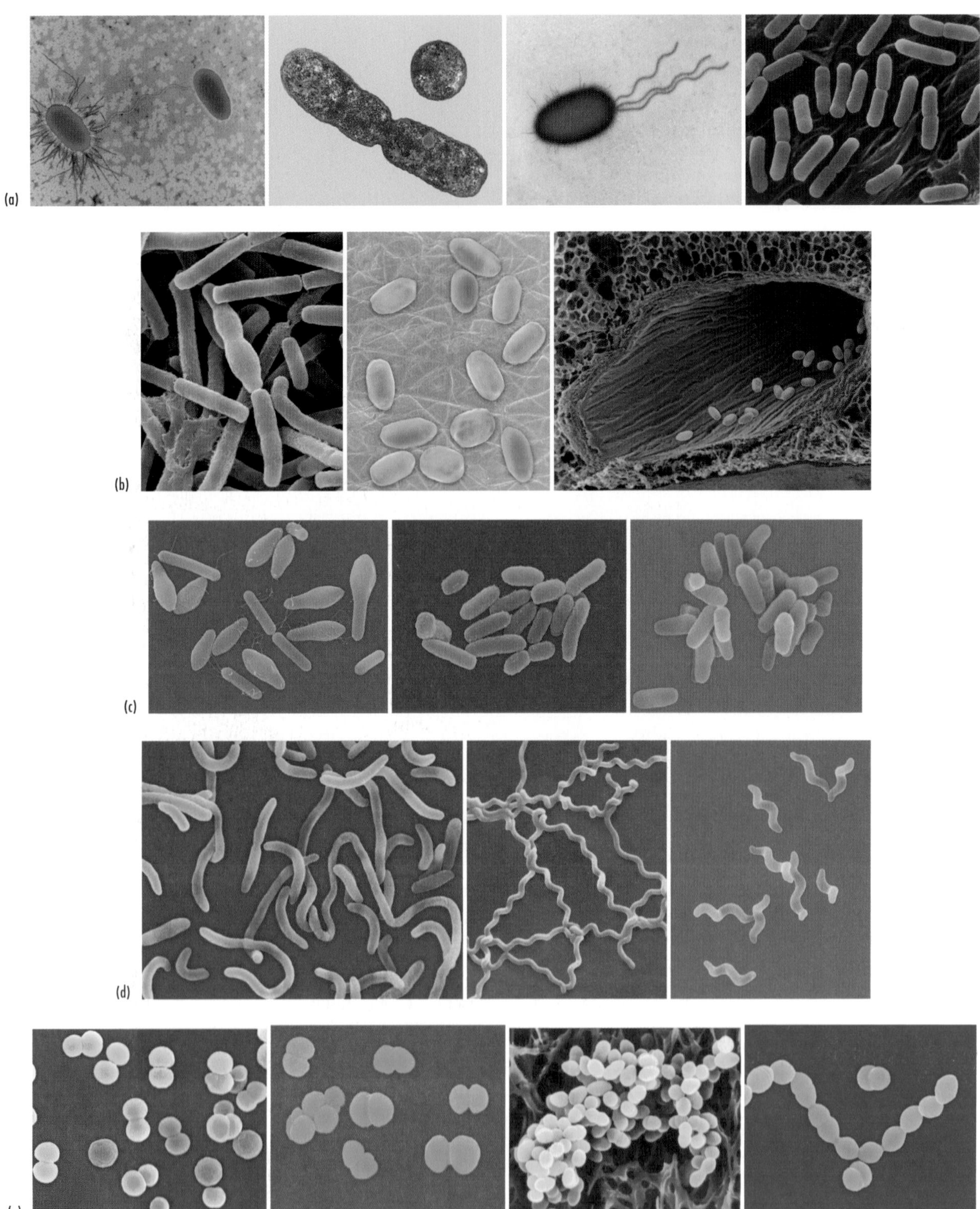

Plate 2.1 (a)–(e).

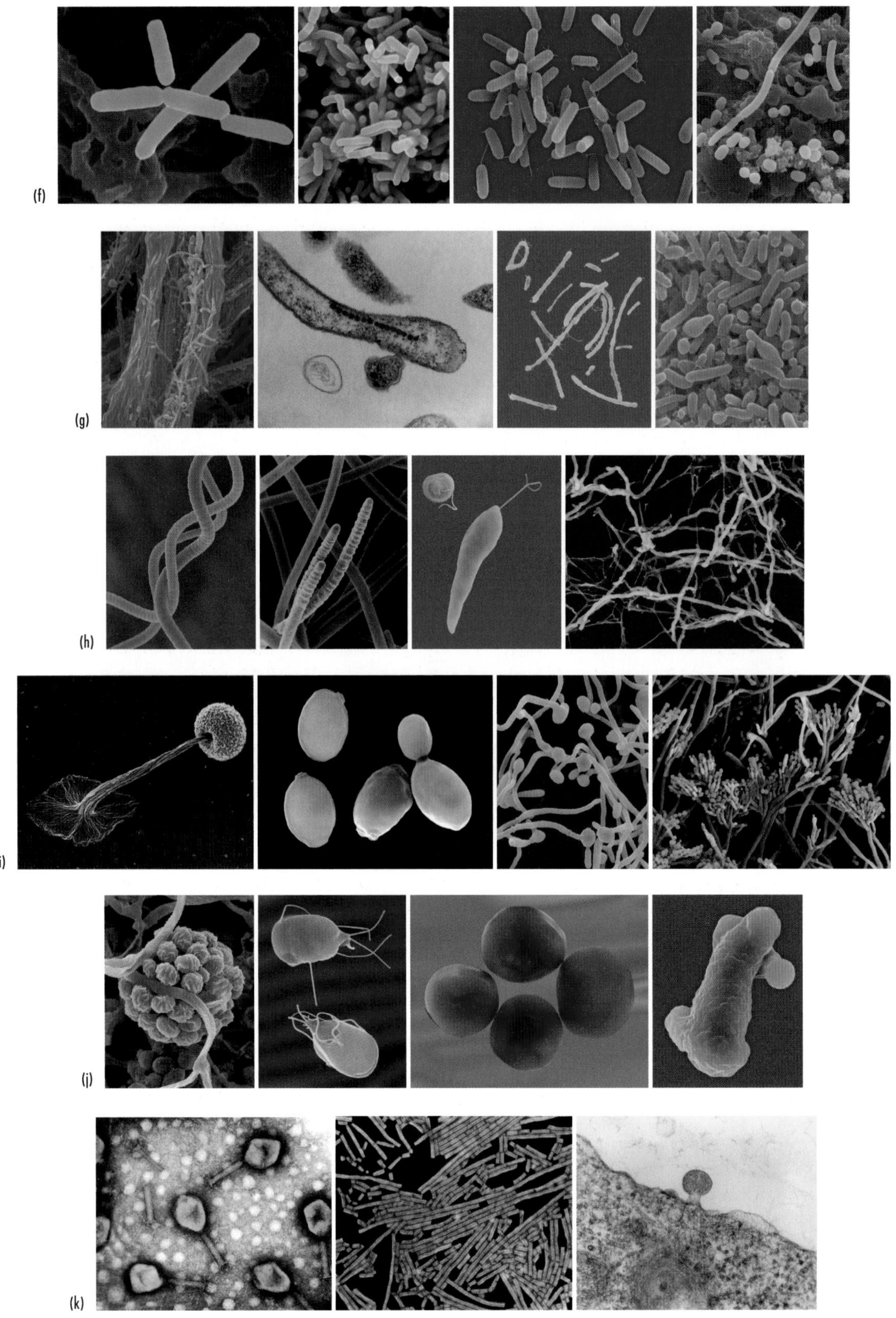

Plate 2.1 (f)–(k).

Plate 2.1 Morphological diversity recorded by scanning and transmission electron microscopy. Images are copyright Dennis Kunkel Microscopy, Inc. and reproduced by courtesy of Dennis Kunkel (for technique, see Kramer and Kunkel 2001; Tomb and Kunkel, 1993. For electronic access to additional images, see ASM, 2002). Descriptions are from left to right panels.

(a) First panel shows two cells of the well-studied bacterium, *Escherichia coli*, captured in a state of conjugative genetic exchange (magnification × 3,645). Genetic material moves from the donor to the recipient cell through the clearly visible conjugation tube. Second panel shows a transmission electron micrograph of thin sections of *E. coli* cells depicting cell wall and internal cell structures (magnification × 12,205). Third panel shows polar flagellation in *E. coli* (magnification × 3,515). Fourth panel depicts cells of pathogenic *E. coli* 0157:H7 found in contaminated food materials (magnification × 3,000). The acquisition of pathogenicity by innocuous *E. coli* strains that are typically in harmless commensal relationship with human hosts can occur through genetic exchange mechanisms that can lead ultimately to speciation (Souza *et al.*, 2002; Reid *et al.*, 2000).

(b) Vegetative and sporulating cells of *Bacillus anthracis* (magnification × 2,000). Purified spores of *B. anthracis* (magnification × 5,000) shown on human skin in the second panel have been used as extremely hazardous biological weapons. Thin sections of infected lung tissue shown in the third panel (magnification × 1,410) show the spores lodged in air vessels. The inhaled form of anthrax is usually fatal.

(c) Other supremely hazardous microbial pathogens include, in the first panel, vegetative cells and spores of *Clostridium botulinum*, the causative agent of botulism caused by the most potent naturally occurring toxin known (magnification × 1,750). In contrast to *B. anthracis* spores, which are produced in the middle of the cell, *C. botulinum* spores are formed at one end, producing "drumstick" morphology. *Yersinia pestis*, the causative agent of bubonic plague, is depicted in the second panel (magnification × 3,250). *Y. pestis* has been responsible for considerable mortality in the human population. Current concerns about global environmental change, encroachment of human settlements into forested regions, acquisition of multiple antibiotic resistance, and bioterrorism have increased risks of epidemics from *Y. pestis* and *Mycobacterium tuberculosis* (third panel; magnification × 6,250).

(d) Spiral and curved bacterial morphology include a variety of human pathogens such as *Vibrio cholerae* (first panel; magnification × 2,130). *Leptospira interrogans* (magnification × 4,000) and *Campylobacter jejuni* (magnification × 3,400) are depicted in the second and third panels, respectively.

(e) Light microscopy can aid bacterial diagnostics, but morphology alone cannot provide definitive species identification for related organisms that cause seemingly very different diseases. For example, *Neisseria meningitidis* (first panel; magnification × 3,250) looks very similar to *Neisseria gonorrhoeae* (second panel; magnification × 4,250), but they cause very different diseases due to physiological specialization that allows them to favor different modes of infection and to colonize different tissues. In contrast, morphological differentiation has been helpful in distinguishing clustered coccoid cells of *Staphylococcus aureus* (third panel; magnification × 3,025), the causative agent of skin pimples and boils, from the chain-link arrangement observed for *Streptococcus pneumonia*, which causes systemic infections (fourth panel; magnification × 3,750).

(f) Bacterial metabolic activities support the geochemical cycling of many elements. *Bradyrhizobium japonicum* shown in the first panel (magnification × 5,000) is one of the few bacterial species capable of fixing atmospheric nitrogen into compounds that can be assimilated by living organisms. *Nitrosomonas* species depicted in the second panel (magnification × 2,200) are important for the process of nitrification, the oxidation of ammonium salts to nitrite, and the further oxidation of nitrite to nitrate. Nitrification proceeds in the reverse direction of nitrogen fixation, and it is essential in the process of reducing the environmental burden of waste products discharged by human communities. Member species of the genus *Pseudomonas* (third panel; magnification × 3,000) complete the denitrification process by converting nitrate to nitrogen gas. *Acinetobacter* species in the fourth panel (magnification × 1,400) are capable of removing phosphorus from aquatic environments through the formation of intracellular polyphosphate granules that allow the organism to grow under nutrient-poor conditions (Pauli and Kaitala, 1997). Certain species are also capable of accumulating poly-beta-hydroxybutyrate, which has potential application for use in the manufacture of biodegradable plastics (Rees *et al.*, 1993).

(g) Additional bacterial species important in the biogeochemical cycling of elements include filamentous iron-oxidizing bacteria observed with the scanning electron microscope forming dense rope-like structures that grow in fresh water on the surface of submerged rocks, or epilithon (first panel; magnification × 800). *Aquaspirillum magnetotacticum* (second panel; magnification × 13,535) is shown with magnetosomes (iron oxide granules) aligned inside the cell to aid magnetotaxis according to the Earth's magnetic field. The third panel depicts extremely halophilic *Halobacterium* species (magnification × 1,600) that can grow in 2–4M salt concentrations. Other halophilic bacteria include *Haloarcula* species that have survived long-term exposure to extraterrestrial space conditions. Fourth panel depicts *Alicyclobacillus* species (magnification × 2,400), an acidophilic and thermophilic spore-forming bacteria important in nutrient cycling under extreme environmental conditions.

(h) Preliminary microscopic examination may suggest erroneously that some prokaryotes are large multicellular eukaryotic organisms. For example, the "blue-green algae", now known collectively as cyanobacteria include *Spirulina pacifica* in the first panel (magnification × 260) and *Scytonema* species in the second panel are aquatic photosynthetic prokaryotes (magnification × 255). Note the corrugation produced by cell replication at the tip of the filaments. A true eukaryotic, albeit unicellular, green alga, *Euglena gracilis* is depicted in the third panel (magnification × 440). Filamentous prokaryotes are not limited to the aquatic environment. Soil-inhabiting prokaryotic *Streptomyces* species depicted in the fourth panel (magnification × 725) forms filaments that are suggestive of eukaryotic fungi.

(i) The slime mold, *Didynium* species, depicted in the first panel (magnification × 28) is also not a true fungus, but it produces "fruiting" bodies that facilitate survival under stressful environmental conditions. True fungi exhibit extreme morphological diversity, ranging from the unicellular budding yeast, *Saccharomyces cerevisiae* in the second panel (magnification × 3,025), to the combined hyphal and yeast structures of *Candida albicans* (third panel), and the true filamentous fungus *Penicillium roqueforti* shown in the fourth panel (magnification × 205).

(j) Microscopy has been indispensable for identifying unique morphological characteristics of eukaryotic microorganisms that threaten air and water quality. The toxic mold *Stachybotris* species depicted in the first panel (magnification × 400) is a risk factor in "sick-building syndrome". Water quality health hazards are associated with parasitic protozoa *Giardia lamblia* (second panel; magnification × 1,000), *Cryptosporidium parvum* (third panel; magnification × 2,310), and *Entamoeba histolytica* (fourth panel; magnification × 800).

(k) Viruses have been discovered for most biological species that have been studied throughout the phylogenetic tree. For their small size and basic structure of protein coat enclosing nucleic acid genome, viruses are morphologically diverse, but as for bacteria, microscopic assessment of morphology alone cannot yield definitive identification. Bacterial viruses infecting many different species may look exactly like the *E. coli* phage T4 depicted in the first panel. The tobacco mosaic virus (second panel; magnification × 27,300) was among the first virus particles to be purified. The human immunodeficiency virus (HIV) shown infecting a lymph tissue cell in the third panel (magnification × 27,630) continues to wreak public health havoc in many parts of the world.

Global net primary productivity

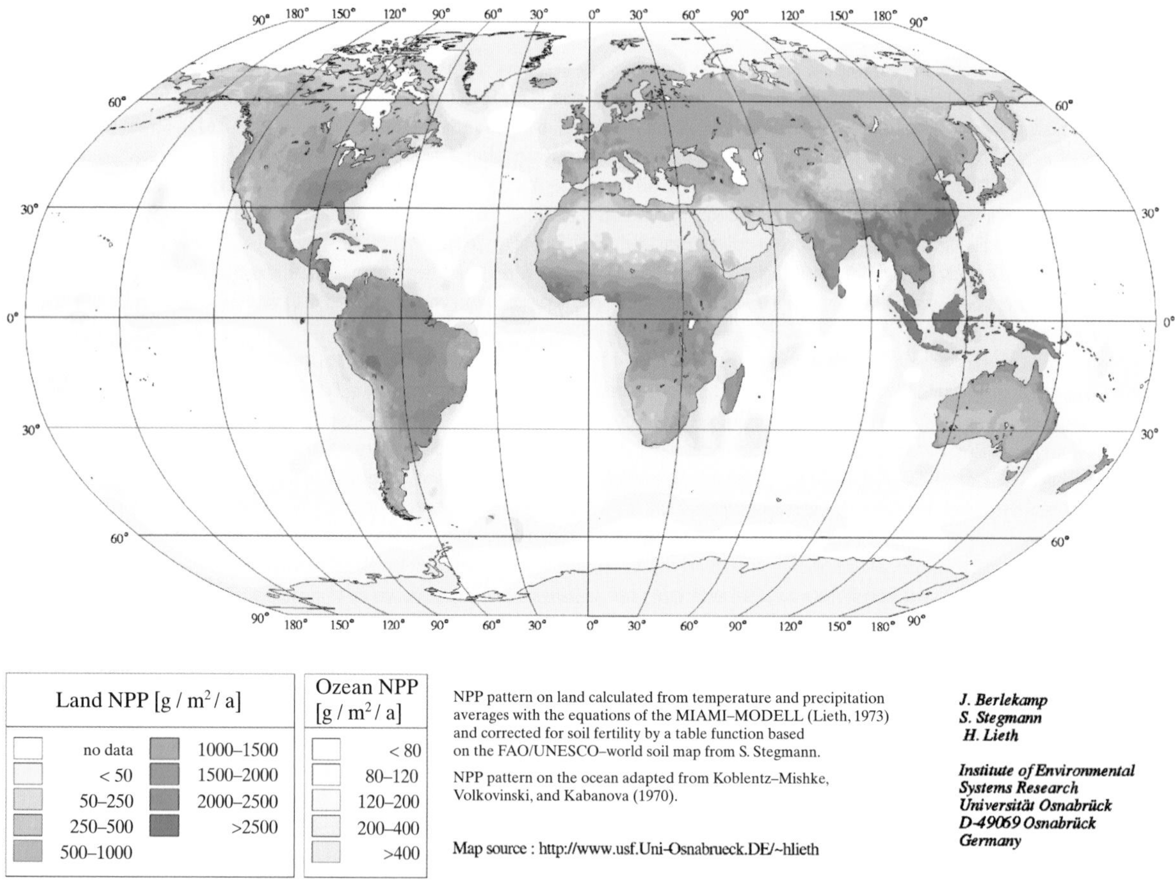

(a)

Plate 7.1 Global map showing geographical distribution of net primary productivity (a) in the major biomes (b). According to some estimates, coral reefs are the most productive ecosystems, yielding 4,900 g of dry organic matter per square meter per year. In comparison, the value for coastal seawater is 200 g per square meter per year, and for tropical rainforests up to 2800 g per square meter per year (Atlas and Bartha, 1998). The map in (a) is by courtesy of J. Berlekamp, S. Stegman, and H. Lieth at the Institute of Environmental Systems Research, Osnabrück University, Germany. Map source is http://www.usf.Uni-Osnabrueck.DE/~hlieth. Map in (b) is by courtesy of the United States Department of Agriculture, Natural Resources Conservation Service. http://www.nrcs.usda.gov/technical/worldsoils/mapindex/biomes.html.

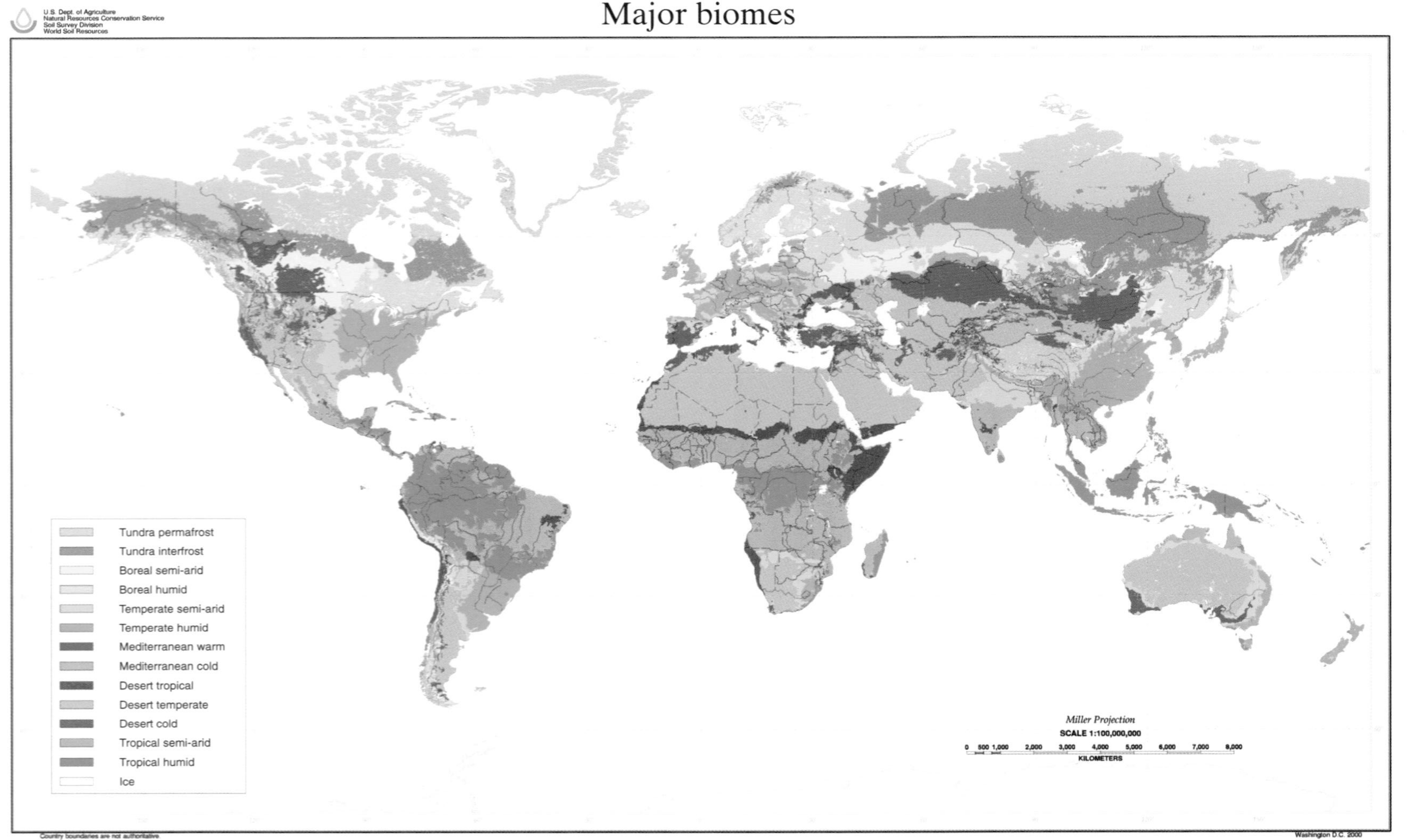
Major biomes
U.S. Dept. of Agriculture
Natural Resources Conservation Service
Soil Survey Division
World Soil Resources
Tundra permafrost
Tundra interfrost
Boreal semi-arid
Boreal humid
Temperate semi-arid
Temperate humid
Mediterranean warm
Mediterranean cold
Desert tropical
Desert temperate
Desert cold
Tropical semi-arid
Tropical humid
Ice
Miller Projection
SCALE 1:100,000,000
0 500 1,000 2,000 3,000 4,000 5,000 6,000 7,000 8,000
KILOMETERS
Country boundaries are not authoritative
Washington D.C. 2000

(b)

Plate 7.1 *Continued*

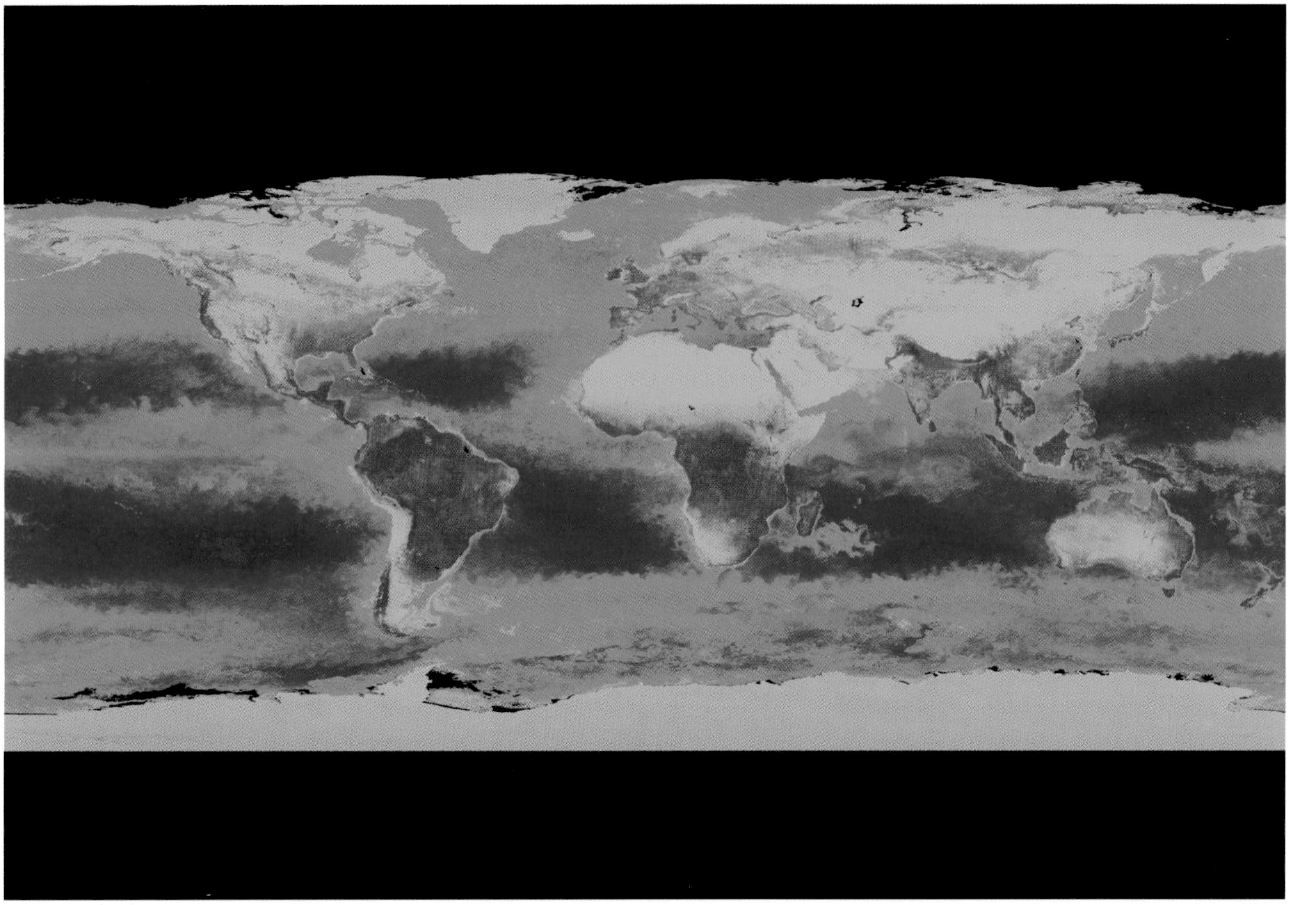

(a)

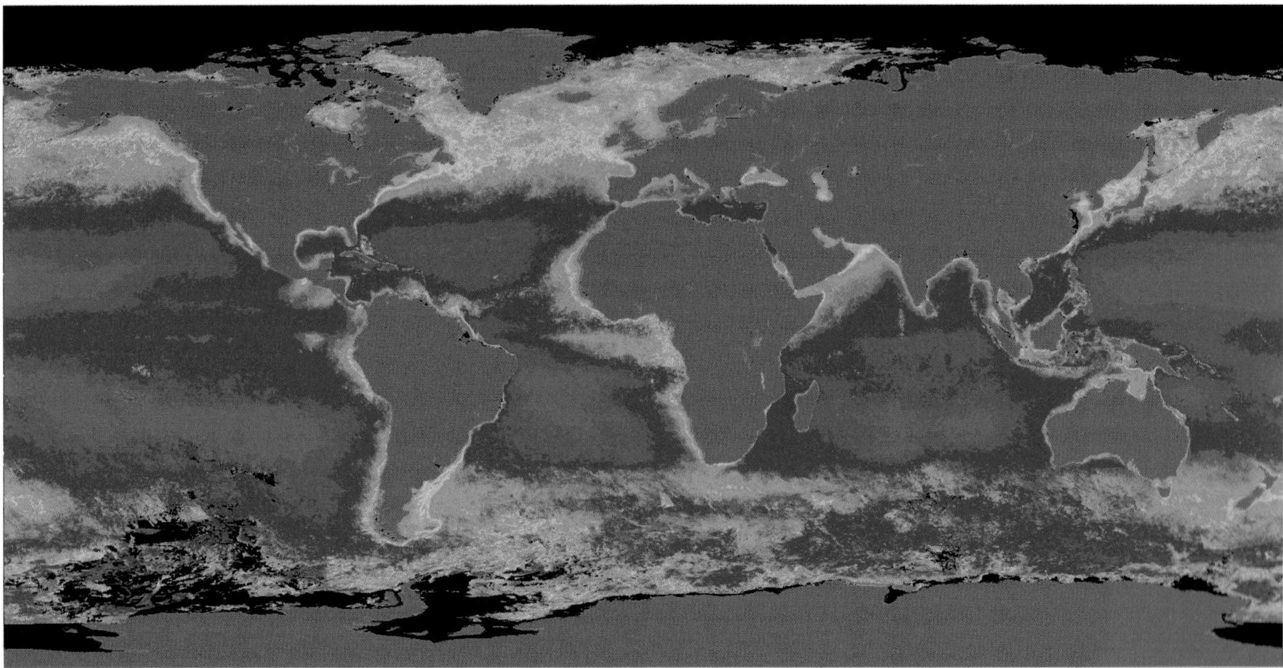

(b)

Plate 7.2 (a) SeaWiFS satellite composite image of phytoplankton productivity measured by chlorophyll-a concentration during July 2003. The image clearly shows the high level of microbial growth around coastal regions due to nutrient (primarily nitrogen and phosphorus) loading from urban environments. (b) The map displays the composite of all Nimbus-7 Coastal Zone Color Scanner data acquired between November 1978 and June 1986. Approximately 66,000 individual 2-minute scenes were processed to produce this image. Images provided by ORBIMAGE. © Orbital Imaging Corporation and processing by NASA Goddard Space Flight Center.

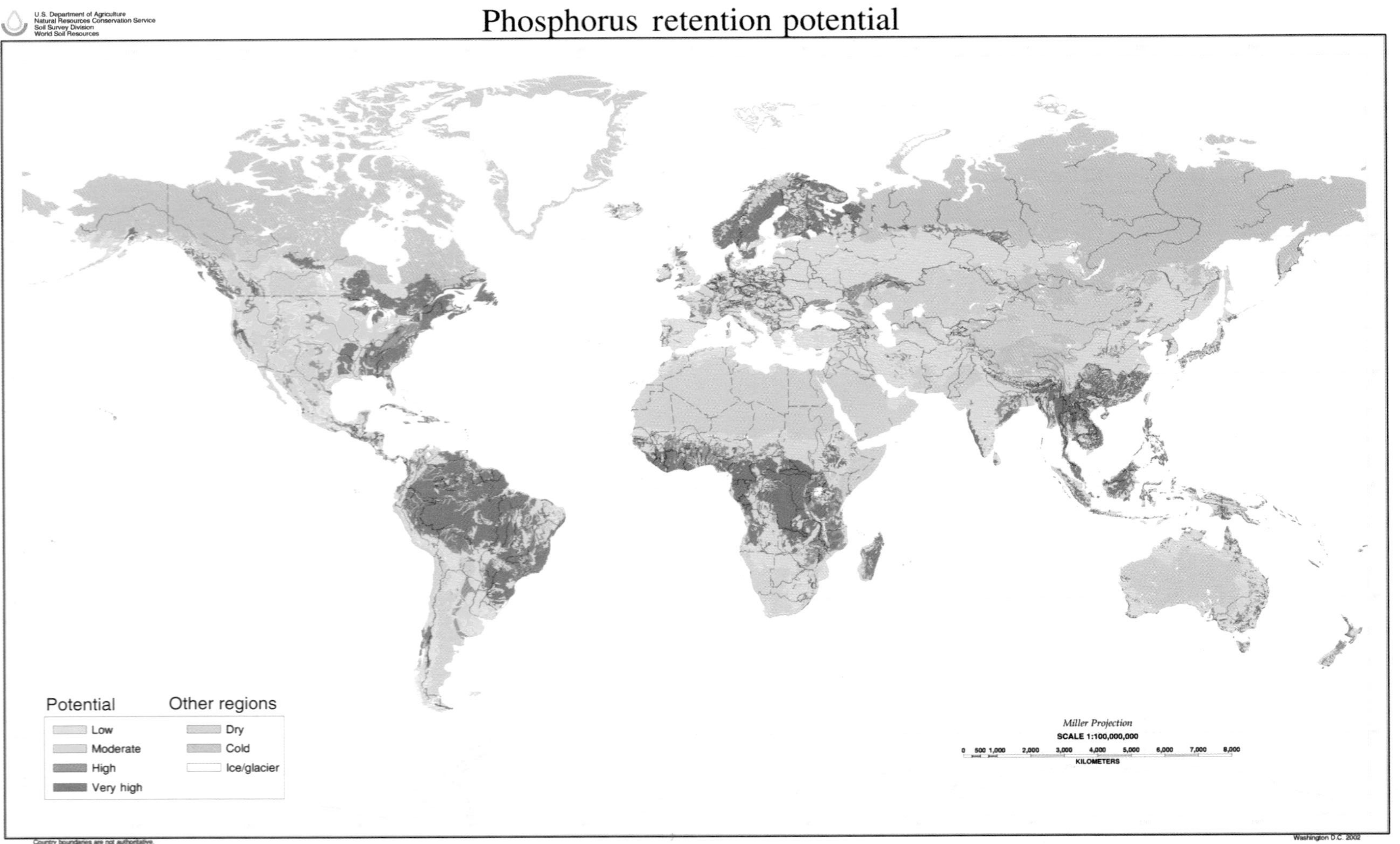

(a)

Plate 8.1 Global distribution of phosphorus resources represented as soil phosphorus retention potential (PRP; (a)), based on climate and soil quality data. PRP is generally higher in the tropical regions than the temperate regions because of higher soil temperatures (b). However, the quality and quantity of vegetation cover in the temperate regions also influences PRP. Map is reproduced by courtesy of United States Department of Agriculture, Natural Resources Conservation Service, Soil Survey Division, World Soil Resources, Washington, DC (http://www.nrcs.usda.gov/technical/worldsoils/).

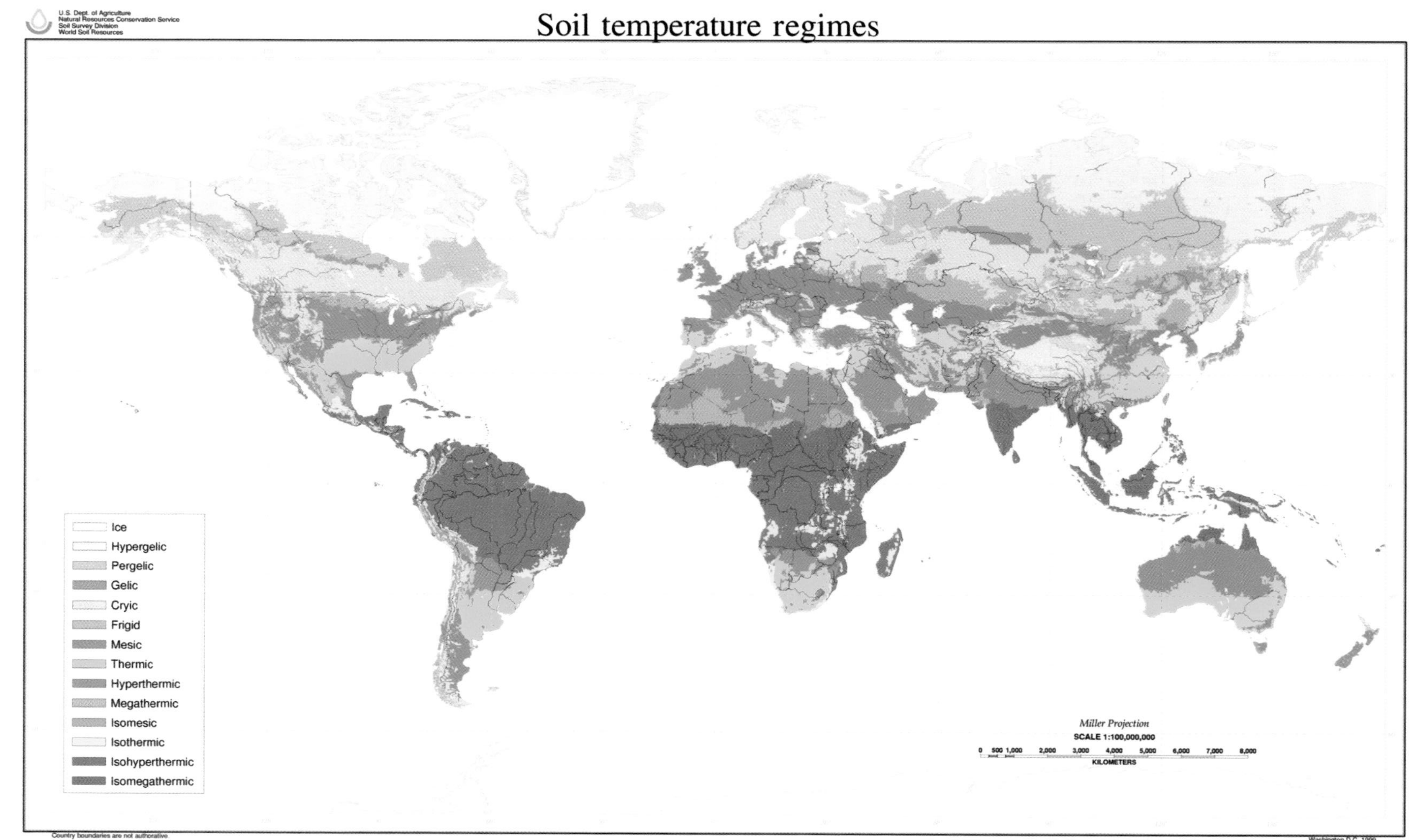
U.S. Dept. of Agriculture
Natural Resources Conservation Service
Soil Survey Division
World Soil Resources
Soil temperature regimes
Ice
Hypergelic
Pergelic
Gelic
Cryic
Frigid
Mesic
Thermic
Hyperthermic
Megathermic
Isomesic
Isothermic
Isohyperthermic
Isomegathermic
Miller Projection
SCALE 1:100,000,000
0 500 1,000 2,000 3,000 4,000 5,000 6,000 7,000 8,000
KILOMETERS
Country boundaries are not authoritative.
Washington D.C. 1999

(b)

Plate 8.1 *Continued*

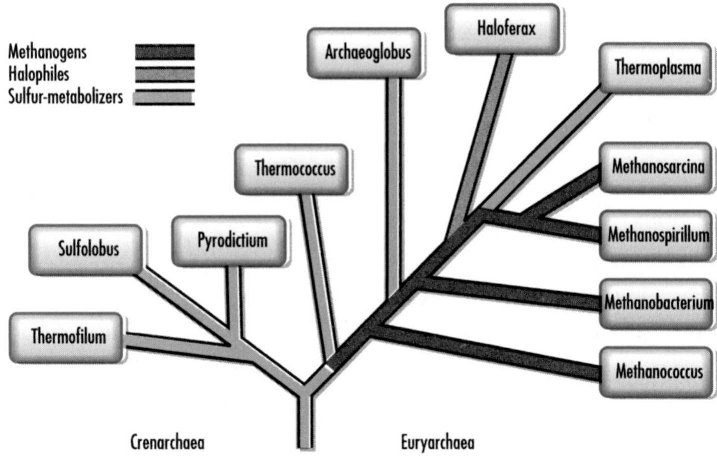

Fig. 7.9 Phylogenetic tree showing the position of methanogens within the Archaea domain.

Table 7.2 Major taxonomic groups describing the methanogens and salient ecological characteristics.

Phylogenetic order	Families (example species)	Optimum growth temperature (°C)	Range of mole % G + C	Natural habitat
Methanobacteriales	***Methanobacteriaceae*** (*Methanobacterium thermoautotrophicum*)	30–65	23–61	Anaerobic sediments; intestines
	Methanothermaceae (*Methanothermus fervidus*)	83	33–34	Thermal springs
Methanococcales	***Methanococcaceae*** (*Methanococcus jannaschi*)	32–85	29–34	Hydrothermal vents
Methanomicrobiales	***Methanomicrobiaceae*** (*Methanomicrobium mobile*)	20–55	39–61	Intestines; marine sediments
	Methanocorposculaceae (*Methanocorpusculum parvum*)	37	48–52	Anaerobic digesters
	Methanosarcinaceae (*Methanosarcina barkeri*)	25–40	36–52	Marine sediments

reductase (MCR), the enzyme catalyzing the terminal step in methanogenesis. This nickel-containing enzyme is one of 10 highly specific enzymes involved in the methanogenic pathway based on carbon dioxide and molecular hydrogen substrates (Ermler *et al.*, 1997; Shima *et al.*, 2002). MCR contains the coenzyme F_{430} that is related to heme, vitamin B_{12}, and chlorophyll, but differs from these other porphyrin structures because it contains a nickel ion at its core (Fig. 7.10). The enzyme catalyzes the final step of methane formation in the energy metabolism of all methanogenic Archaea. Methyl-coenzyme M and coenzyme B are converted to methane plus the heterodisulfide of coenzyme M and coenzyme B.

To investigate the putative linkage between the molecular diversity of MCR and the optimum temperature of methanogen growth, Grabarse and colleagues (2000 and 2001) compared the crystal structures of MCR from *Methanosarcina barkeri* (optimum growth temperature of 37°C) and *Methanopyrus kandleri* (optimum growth temperature of 98°C)

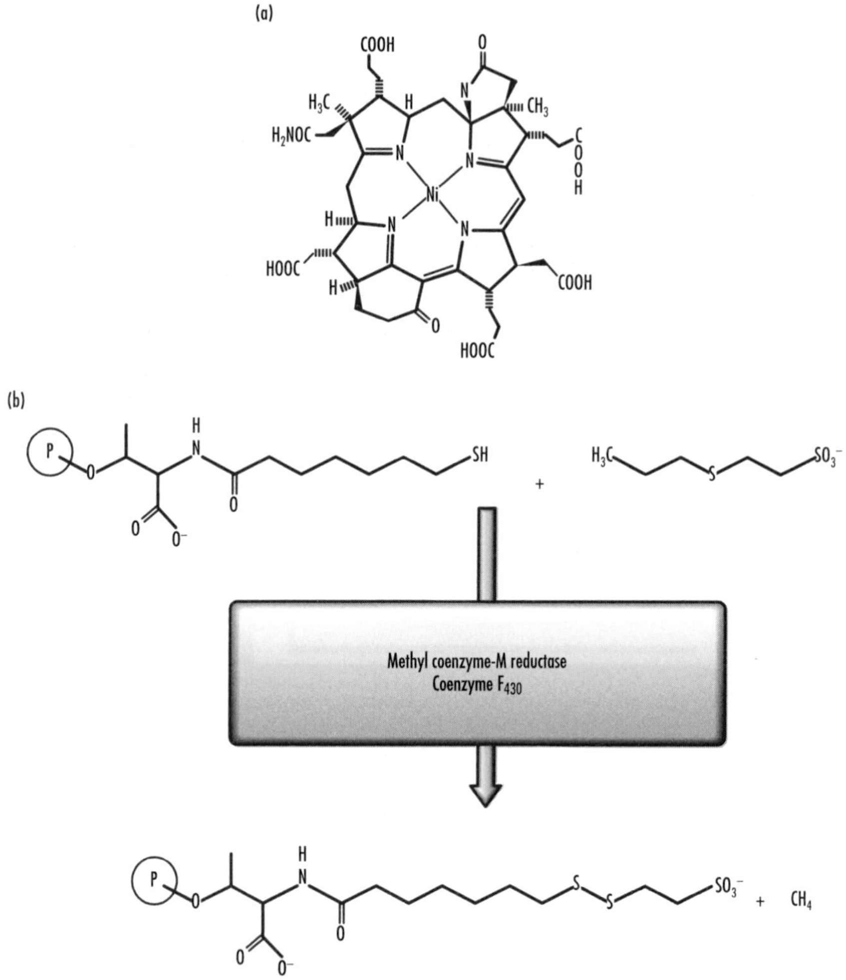

Fig. 7.10 The key enzyme in methanogenesis is methyl coenzyme M reductase with an active site that contains the hydroporphinoid nickel complex coenzyme F_{430} (a). Porphyrins and other closely related tetrapyrrolic pigments occur widely in nature, and they play very important roles in various biological processes. The basic structure of porphyrin consists of four pyrrole units linked by four methine bridges. The presence of a metal in the porphyrin cavity changes the light absorption spectrum of the molecule, a characteristic that is often used to determine certain features of enzymes containing porphyrins. The nickel center of coenzyme F_{430} catalyzes the unusual methane-producing reaction shown in (b). With permission from Bernhard Jaun, ETH Zurich, Switzerland.

with the known structure of MCR from *Methanobacterium thermoautotrophicum* (optimum growth temperature of 65°C). The active sites of MCR exhibit modified amino acid residues including a thiopeptide, an S-methylcysteine, a 1-N-methylhistidine, and a 5-methylarginine, which, in addition to high intracellular concentration of lyotropic salts, may facilitate adaptation to growth under high temperature conditions in *M. thermoautotrophicum* and *M. kandleri*.

Luton and colleagues (2002) developed a molecular signature based on the amino acid sequence inferred from the methyl coenzyme-M reductase (*mcr*A) gene. The polymerase chain reaction (PCR) primers developed by this group of investigators were tested against 23 species representing all five recognized orders of methanogens within the Archaea domain. The products of PCR analyses of these diverse taxonomic groups ranged in size from 464 to 491 base pairs. Comparisons between the *mcr*A and 16S small subunit rRNA gene sequences using PHYLIP (see Chapter 5) demonstrated that the topologies of phylogenetic trees created by both types of sequences were similar. The analysis revealed a far greater diversity in the methanogen population within landfill material than has been seen previously.

Hallam and colleagues (2003) developed this approach further in their assessment of the possibility that methane-oxidizing Archaea have co-opted key elements of the methanogenic pathway, thereby reversing many of the methanogenesis steps to oxidize methane anaerobically. To test this hypothesis, they investigated the occurrence and genomic conservation of *mcr*A in Archaea species isolated from marine environments. These investigators also demonstrated phylogenetic congruence between *mcr*A and small subunit rRNA tree topologies. Further analysis of the different *mcr*A sequences associated with two recognized methanogen-like archaeal groups (ANME-1 and ANME-2) indicated that the conservation of catalytic activity is maintained through the specificity of amino acids at the enzyme active site. These results also provide additional support for the proposal to identify methanotrophic Archaea with *mcr*A sequences in the context of establishing a functional genomic link between methanogenic and methanotrophic Archaea. The development of this and other molecular signatures for methanogens has facilitated the improved characterization of methanogenesis in diverse ecosystems, including blanket bog peat (Hales *et al.*, 1996), deep gas hydrate sediments (Marchesi *et al.*, 2001), Antarctic sediments (Purdy *et al.*, 2003), and wetland soils (Utsumi *et al.*, 2003).

Methanotrophy

The reverse of methanogenesis, or the capability of microorganisms to utilize methane as a substrate for energy generation, is an ecologically important process with respect to biogeochemical cycling and potential impacts on global environmental change. The activity of methane-oxidizing microorganisms (methanotrophs) is a considerable sink for atmospheric methane (Reeburgh, 1980 and 1982; Ward *et al.*, 1987; Whalen and Reeburgh, 1990). Methanotrophs are ubiquitous in the environment where they play important roles in the ecology of terrestrial, marine, and freshwater systems. In general, methanotrophs oxidize methane through methanol to formaldehyde, and ultimately to carbon dioxide (Morris *et al.*, 2002).

The common biochemical pathway for methanotrophy suggests that the process should be limited to aerobic conditions. Indeed, based on their main substrates, methane-oxidizing microorganisms can be divided into two groups comprising the autotrophic ammonium-oxidizing bacteria (AAOB) and the methane-assimilating bacteria (MAB), which are the widely recognized methanotrophs. Whereas the AAOB belong to the β subgroup of the *Proteobacteria* group, most methanotrophs that have been isolated are phylogenetically related to the γ and α subgroups, representing the *Methylococcaceae* (type I methanotrophs) and the *Methylocystaceae* (type II), respectively (Kolb *et al.*, 2003). These organisms are generally Gram-negative, obligately aerobic, catalase and oxidase producers with intracytoplasmic membranes (Topp and Hanson, 1991). However, methanotrophic microorganisms are not limited to aerobic environments and there is considerable geochemical evidence that methanotrophy occurs under anoxic conditions (Alperin and Reeburgh, 1985) (Table 7.3). Orphan and colleagues (2002) recognized that no microorganism capable of anaerobic growth on methane as the sole carbon source has yet been cultivated, but they used fluorescent *in situ* hybridization with rRNA gene probes and secondary ion mass spectrometry analyses to demonstrate the occurrence of methanotrophic archaeal groups and consortia in anoxic sediments. Furthermore, Teske and colleagues (2002) used 16S rRNA sequencing and carbon isotope analysis of archaeal and bacterial lipids to explore the microbial communities in hydrothermally active sediments in the Gulf of California, Mexico. Their analysis revealed the occurrence of two distinct lineages of uncultivated *Euryarchaeota* within the anaerobic methane oxidation group (ANME-1). The main evidence consisted of the detection of the archaeal lipids that are diagnostic for the non-thermophilic *Euryarchaeota*, and sn-2-hydroxyarchaeol, which has been associated with cultivated members of the phyloge-

Table 7.3 Major groups of methanotrophic microorganisms.

Category	Representative species	Range of mole % G + C	Habitat (reference)
Aerobic methanotrophs			
Type I	*Methylomonas* species	50–54	Hypersaline and alkali lakes (Trotsenko and Khmelenina, 2002)
Type II	*Methylosinus* species	62.5	Acidic peat bogs (Dedysh *et al.*, 2003); rice paddies (Kolb *et al.*, 2003)
Type X	*Methylococcus capsulatus*	62.5	Flooded rice fields (Kolb *et al.*, 2003)
Yeasts	*Rhodotorula* species	50.7–55.4	Deep sea floor (Nagahama *et al.*, 2001)
Anaerobic methanotrophs			
ANME archaeal groups	*Methanosarcina* species	36–52	Hydrothermal vents; marine sediments

netic order *Methanosarcinales*. The δ-^{13}C values for these lipids were consistent with those of anaerobic methanotrophic Archaea (see Chapter 4 for details on the use of lipids for assessing microbial diversity).

In addition to the anaerobic methane oxidizers, which belong to the Archaea domain, four other categories of aerobic methanotrophic microorganisms are recognized (Table 7.3). All known methanotrophs synthesize methane monooxygenase (MMO), a three-component enzyme that catalyzes the selective oxidation of methane to methanol (Fig. 7.11). Two types of MMO are recognized based on their fractionation within the soluble (sMMO) or membrane-associated, particulate (pMMO) components of cell extracts. Much more is known about the form and function of sMMO than pMMO because of the difficulties associated with recovery of pMMO while maintaining enzymatic activity. However, Lieberman and colleagues (2003) successfully solved this problem by purifying the enzyme and demonstrating that it exists as a dimer with both mononuclear copper and a copper-containing cluster.

Type I methanotrophs, including the genera *Methylomonas* and *Methylobacter* are notable because they appear not to have the capacity to synthesize sMMO, Type X utilize the ribulose monophosphate pathway for formaldehyde assimilation, and, like Type I organisms they have an incomplete tricarboxylic acid (TCA) cycle, but unlike Type I methanotrophs, they synthesize sMMO under copper-limited conditions. *Methylococcus capsulatus* is the best-studied member of the type X methanotrophs. The broad spectrum of sMMO enzyme activity in this organism has endeared it to investigators interested in the biodegradation of recalcitrant toxic compounds such as trichloroethylene, which is also a substrate for *M. capsulatus* sMMO (Brusseau *et al.*, 1990). Type II methanotrophs differ from type I and type X methanotrophs in that they possess a complete TCA cycle, and they are generally not thermophilic organisms. *Methylosinus* and *Methylocystis* are the major phylogenetic genera describing type II methanotrophs. In addition to the three prokaryotic groups of methanotrophs, unicellular fungi species belonging to the genera *Rhodotorula*, *Candida*, and *Sporobolomyces* are capable of growing slowly on methane as the sole energy source. Methanotrophic yeasts capable of growth at pH 3.5 were isolated from lake water and soil, as well as ruminant feces, although their environmental significance is not clear (King, 1992; Saha and Sen, 1989; Wolf and Hanson, 1979).

Several attempts have been made to quantify the contribution of microbial methanotrophy to the reduction of atmospheric methane. The initial stages of such studies focused on determining the population densities of methanotrophs using real-time polymerase chain reactions targeted at functional genes such as the phylogenetically conserved *pmo*A, which encodes for the alpha subunit of the particulate methane monooxygenase. Using such techniques, Kolb and colleagues (2003) demonstrated that up to five million molecules of *pmo*A occur per gram of flooded rice paddy soils, however, the technique was less useful for quan-

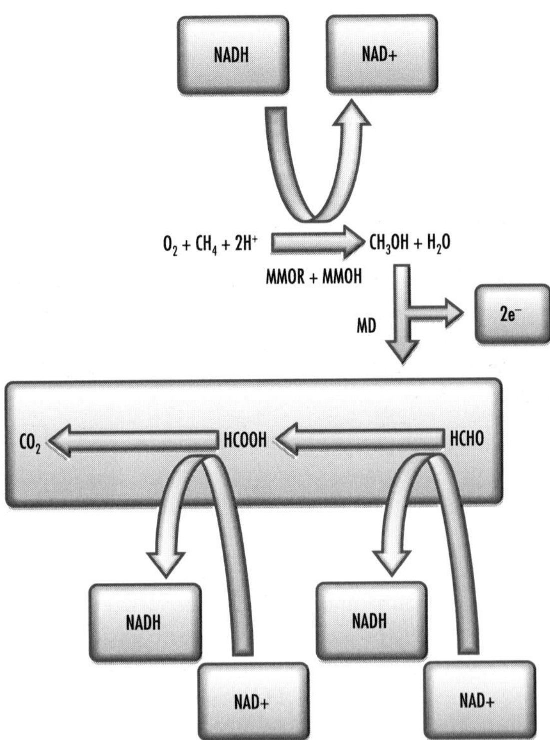

Fig. 7.11 The biochemical pathway for methanotrophy involves the oxidation of methane by methane monooxygenase (MMO), which exists as either particulate, soluble, or both in methanotrophic microorganisms. MMO contains three large components: a monomeric reductase named MMOR (MW = 38 kDa), a hydroxylase named MMOH (MW = 248 kDa), and a small structural protein named MMOB (MW = 15 kDa) which does not play a direct role in enzyme function but is known to have some key structural functions. A conformational change occurs in MMOH when MMOB is present, leading to a 1000-fold increase in the rate of oxygen reaction at the MMOH active site. Methanol is further oxidized to formaldehyde through the action of methanol dehydrogenase (MD).

tifying the potential for methanotrophy in forest soils. In addition to quantifying the genetic potential for methanotrophy in natural ecosystems, it is also important to reliably estimate the species composition of the methanotrophic community to account for differences in growth rate and methane consumption. Dedysh and colleagues (2003) used 16S rRNA-targeted fluorescent oligonucleotide probes to differentiate methanotrophic bacteria in acidic peatlands from the Sphagnum-dominated wetlands of Siberia and northern Germany. They discovered that the *Methylocystis* group (type II) of methanogens represented 60% and 95% of the total methanotrophic community in these ecosystems respectively, whereas the type I methanotrophs represented less than 1% of the community.

Progress is also being made in integrating data gathered at the cellular, microbiological level concerning the global consequences of methane balance. For example, Le Mer and Roger (2001) estimated that 60% to more than 90% of the methane produced in anaerobic zones of wetlands is re-oxidized by aerobic methanotrophs in the rhizosphere and in oxidized soil–water interfaces. Furthermore, the consumption of methane by aerobic soils is very low and the microorganisms involved are not well characterized, so it is difficult to provide quantitative estimates of methanotrophy in such environments, although the work of Phillips and colleagues (2001) suggests that the rate of methanotrophy in coniferous forest soils is influenced negatively by carbon dioxide enrichment and by the application of ammonium fertilizers. To facilitate the explicit representation of methanotrophy in models of methane transfer between soils and the atmosphere, Grant (1999) developed a mathematical model based on the kinetics of methane oxidation as controlled by the biomass and specific activity of aerobic methanotrophs, soil temperature, and oxygen transfer. The preliminary conclusion was that methanotrophy is generally limited by the diffusion of methane through soils. Finally, there is considerable interest in using methanotrophs for converting the vast untapped methane (natural gas) resources to methanol, which is easier to transport and less dangerous than methane as a fuel source (Lieberman *et al.*, 2003). However, such projects require more detailed understanding of the environmental controls over the expression of genes encoding for methanotrophy, and of genetic engineering toward alcohol tolerance in the organisms.

Heterotrophy

The degradation of organic carbon to carbon dioxide is the reverse of photosynthesis and consequently it is an important process in the biogeochemical cycling of carbon. Under aerobic conditions, carbon respiration is very efficient, and it is responsible for the recycling of plant litter in many forested ecosystems. It has been estimated that in the absence of carbon respiration, photosynthesis could consume the global supply of carbon dioxide and grind to a halt in fewer than 300 years (Maier, 1999). The term "heterotrophy" implies the ability of organisms to derive nourishment from a wide variety of heterogeneous or exogenous carbon substrates. Most of the fixed carbon in oxic environments is in the form of organic polymers, including cellulose (15–60% of dry plant mass), hemicellulose (10–30%), and lignin (5–30%). Additional carbon-based polymers that are produced as a result of carbon fixation are proteins and nucleic acids, which make up 2–15% of dry plant mass, starch, chitin, peptidoglycan, and humic and fulvic compounds (Wagner and Wolf, 1998). These carbon polymers exhibit different levels of resistance to degradation and ultimate recycling to carbon dioxide. The most important determinant of recalcitrance is whether they are carbohydrate-based polymers (such as cellulose), which are easier to degrade under a variety of environmental conditions, or phenylpropane-based polymers (such as lignin), which are much more difficult to degrade and can only be recycled under aerobic conditions (Fig. 7.12).

Lignin is the material that provides structural strength to woody plants; it consists of tyrosine and phenylalanine units converted to randomly polymerized phenylpropene aggregates (Fig. 7.12). The complex structure and heterogeneity of lignin make it difficult for microorganisms to degrade, and few extracellular enzyme systems are known to attack lignin. Among the best-studied non-substrate specific enzymes that catalyze the degradation of lignin is lignin peroxidase, a metalloenzyme that depends on hydrogen peroxide to work. The free oxygen radicals generated by this enzyme system react with lignin polymers to produce phenylpropene units such as coniferyl alcohol, coumaryl alcohol, and sinapyl alcohol (Mester and Field, 1997; Morgan *et al.* 1993; Srebotnik *et al.*, 1994). Lignin peroxidase is primarily a product of filamentous fungi, particularly those associated with "white rot" of woody tissue. *Phanerochaete chrysosporium* and *Bjerkandera* species have been investigated extensively for the broad spectrum of their lignin peroxidase (LiP) activities and the potential that they have demonstrated in the biodegradation of toxic environmental pollutants (Mester and Field, 1997; Blodig *et al.*, 2001). Apparently, not all fungi use LiP to degrade lignin. Srebotnik and colleagues (1994) demonstrated unequivocally that *Ceriporiopsis subvermispora*, a fungus lacking LiP, was capable of degrading various lignin compounds and that this organism has mechanisms for the degradation of nonphenolic lignin that are as efficient as those in *P. chrysosporium* but that do not depend on LiP.

The degradation of lignin by these organisms is linked to the production of one of the most stable carbon reservoirs in soils, namely humic and fulvic acids, and humin, collectively known as humus (Fig. 7.12). Humus polymers exist as three-dimensional sponge-like structures with molecular weights ranging from 700 to 300,000 (Maier, 2000). The turnover rate of humus under most climatic conditions is less than 5% per year (Haider, 1992; Wagner and Wolf, 1998). It is not clear how large a reservoir for global carbon is represented in humus but it is known that the proportionate humus content of most soils is stable over time, and the existence of soil humus in a state of dynamic equilibrium reduces the potential for the impact of humus degradation or formation on short-term global carbon fluxes. The primacy of humus as the most stable reservoir for soil carbon has been challenged by a relatively recent discovery. In 1996, Sara Wright and colleagues discovered an abundant soil protein, which is now implicated in several important soil properties and biogeochemical functions. The protein, named glomalin for its association with arbuscular mycorrhizal fungi belonging to the order *Glomales*, is extremely stable in soils, where it accumulates to several mg per g of soil, reaching up to 100 mg per g in some Hawaiian soils (Wright and Upadhyaya, 1996 and 1998; Wright *et al.*, 1996). Glomalin is known to be associated with carbohydrates (glycoproteinous) containing between 30% and 40% carbon by weight, and 1% to 9% of tightly bound iron, but beyond that, not much else is known about its structure. In addition, neither the environmental cues for the production of glomalin nor its specific cellular biological role and processing in the fungi that produce it are well understood. However, because it is detectable by means of immunological techniques, several studies have demonstrated glomalin's ubiquitous occurrence in soils across many geographical zones and land-use patterns (Knorr *et al.* 2003; Rillig *et al.* 2002a and 2002b; Steinberg and Rillig 2003).

Several ecological functions have been attributed to glomalin. These ecological functions include its role as "glue" between soil particles thereby explaining many features regarding soil texture, soil aggregation, water-holding capacity, autochthonous microbial diversity, and the stability or turnover rate of soil carbon sinks. Glomalin has been shown to account for up to 27% of soil carbon. Therefore, this protein is also a major component of soil organic matter, weighing two to 24 times more than humic acid, which was previously thought to be the main contributor (approximately 8%) to soil carbon. Furthermore, glomalin

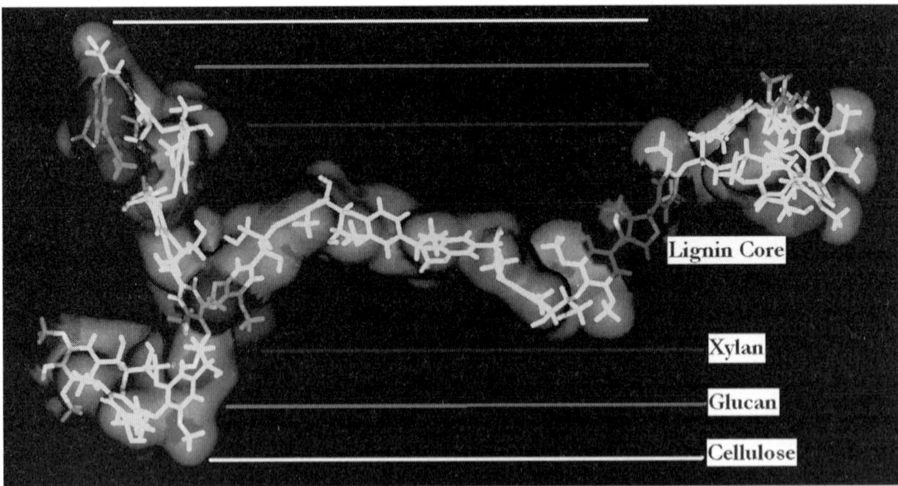

(a)

Fig. 7.12 The aerobic degradation of lignin is an important step in the biogeochemical cycling of carbon. (a) shows the arrangement of carbonaceous polymers in plant tissue, including the situation of lignin molecules among other polymers namely, cellulose, glucan, and xylan. These polymers are also major reservoirs of carbon in the biosphere. The lignin structure is derived from data generated by NMR spectra (Marita *et al.*, 2001 and Baucher *et al.*, 2003). The structure was visualized by means of the Accelry ViewerLite software. The model representing the layered sequence of other polymers surrounding lignin is by courtesy of Patrick Hatcher, Ohio State University. (b) Molecular structure of lignin with monomeric unit in parentheses. By courtesy of Jussi Sipilä, Helsinki, Finland. http://www.helsinki.fi/~orgkm_ww/lignin_structure.html. (c) A colony of white-rot fungus growing on wood surface. (d) Hyphae of lignin peroxidase producer *Phanerochaete chrysosporum* penetrating wood tissue. Images in (c) and (d) by the Fungi group, National Institute of Education, Nanyang Technological University, Singapore (members.tripod.com/~decomposers/woodfungi.htm). (e) Three-dimensional structure of lignin peroxidase from *P. chrysosporum* showing the location of the heme centers and the calcium ions (Blodig *et al.*, 2001). The enzymatic degradation of lignin by microorganisms contributes to the formation of one of the most stable carbon reservoirs in the biosphere, humic materials (f).

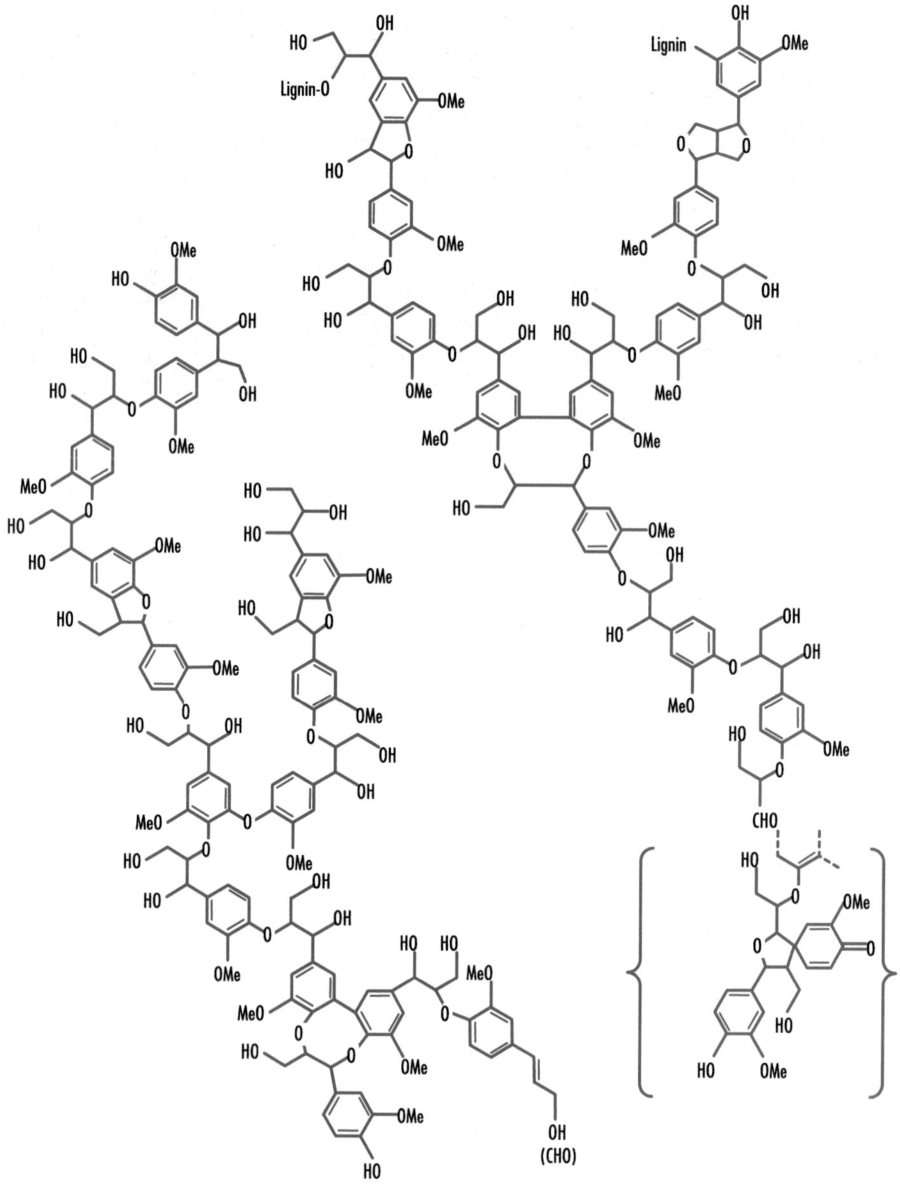

(b)

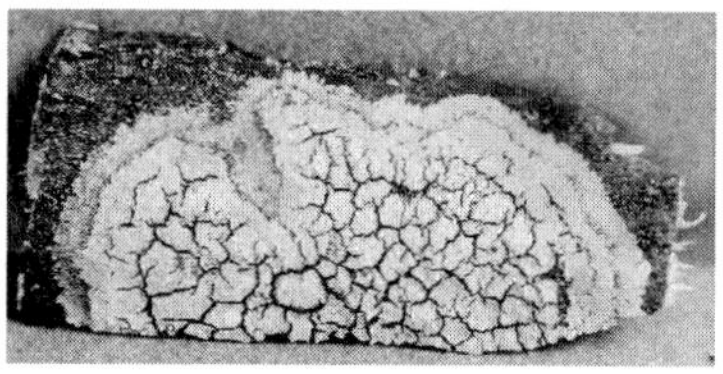

(c)

(d)

Fig. 7.12 *Continued*

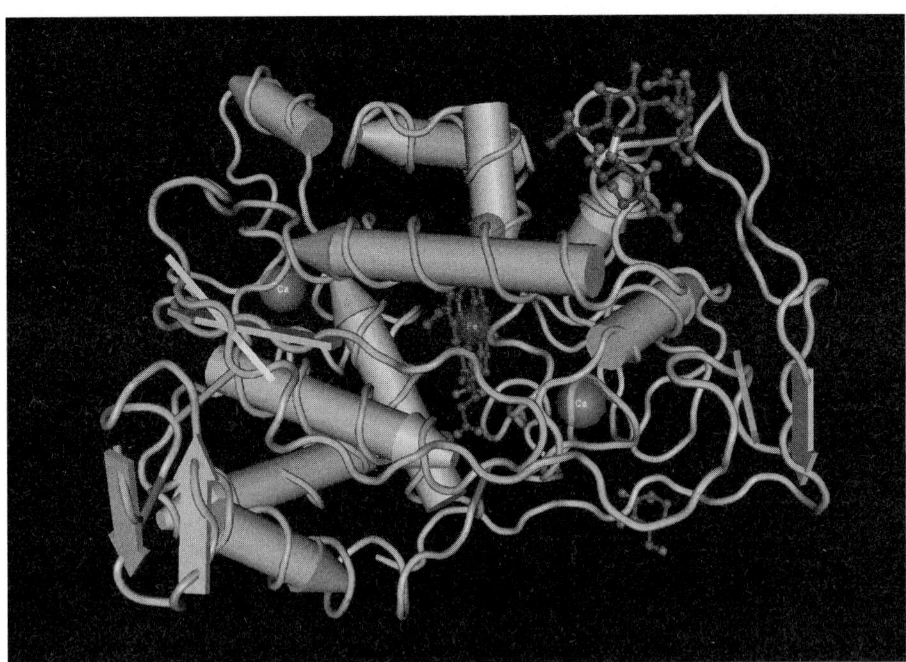

(e)

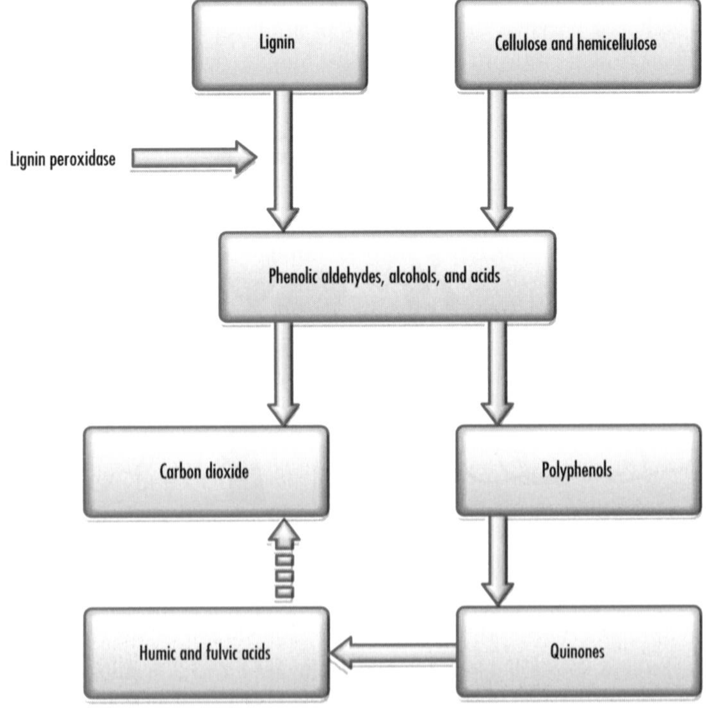

(f)

Fig. 7.12 *Continued*

molecules have been estimated to survive in soils for seven to 42 years, depending on soil temperature and humidity conditions. Apparently, higher levels of atmospheric carbon dioxide stimulate the growth of glomalin-producing fungi, and consequently the level of glomalin found in soils (Rillig *et al.*, 2003). This and other observations have raised the profile of glomalin management in soils as a potential indicator of the ecological effects of global climate change, but much more needs to be learned about the fate of this intriguing protein, and its influence on microbial diversity.

Unlike the situation for lignin and glomalin, many microorganisms produce extracellular enzymes that are capable of degrading cellulose, which is unquestionably the most abundant polymer on Earth. Cellulose molecules are not very soluble, they have molecular weights as high as 1.8 million consisting of up to 10,000 β-1,4 linked glucose monomers. Therefore, the microbial degradation of cellulose occurs outside the cell through the activity of extracellular enzymes including β-1,4-endoglucanase, which randomly attacks cellulose polymers; β-1,4-exoglucanase, which requires access to the reducing end of the molecule to release disaccharide cellobiose; and β-glucosidase, which hydrolyzes cellobiose to glucose. These enzymes are ubiquitous in microbial communities including both prokaryotes and eukaryotes such as microscopic and filamentous fungi. However, there is considerable polymorphism in their forms, functions, and degradative kinetics. The cellulase system of *Trichoderma reesei* is one of the most widely investigated in terms of technological applications because there is considerable interest within the energy industry for the conversion of cellulose to renewable energy resources such as ethanol.

The U.S. Department of Energy's Office of Energy Efficiency and Renewable Energy has an extensive research program to explore strategies for reducing the cost and increasing the effectiveness of cellulase enzymes in the bioethanol production process. The current cost estimate for cellulase ranges from 10 to 30 cents per liter of ethanol produced, but biotechnological approaches can presumably reduce cellulase cost to less than 1 cent per liter of ethanol. This requires a tenfold increase in specific activity or production efficiency, or both (Sheehan and Himmel, 1999). A 90:9:1 mixture of a cellobiohydrolase from *T. reesei* (CBH I), a heat-tolerant endoglucanase from *Acidothermus cellulolyticus* (EI), and a β-D-glucosidase was capable of complete saccharification of cellulose in acidified yellow poplar after five days. This discovery represented an important breakthrough because it demonstrated for the first time that polymorphic cellulase systems can be engineered for specific pretreated biomass materials (Baker *et al.*, 1998).

Biochemical and phylogenetic range of heterotrophy

Cellulose and lignin are, of course, only two of the extremely diverse group of chemical substrates suitable for heterotrophic metabolism by an equally diverse group of microorganisms. The lack of comprehensive understanding of how the complexity of substrates is matched by heterotroph diversity is one of the major remaining questions about carbon biogeochemistry. Plants, animals, microorganisms, and human industrial processes release a wide range of carbon substrates into the environment. Many of these substrates are ultimately recycled through heterotrophic respiration. In some instances, the substrates are not only recalcitrant but also toxic. The tremendous physiological diversity among heterotrophic microorganisms has aided efforts to reduce the potential ecotoxicological impacts of carbon-based compounds of industrial origin, but the problem seems to be growing at a rate that far exceeds the current level of understanding of the ecological context of heterotrophy (see Chapter 11 for further discussion of diversity and toxic chemical pollution).

Several techniques have been used to investigate the diversity of heterotrophic carbon metabolism under field and laboratory conditions. For example, the "heterotrophic plate count" is a standard culture technique developed for the quantitative determination of culturable aerobic and facultative anaerobic microorganisms that obtain carbon and energy from organic compounds in environmental samples. In natural waters the majority of prokaryotes recovered by such methods have not been identified, and most of the identified species are Gram-negative bacteria belonging to the genera *Acinetobacter*, *Aeromonas*, *Citrobacter*, *Enterobacter*, *Proteus*, and *Pseudomonas* (Jackson *et al.*, 2000; LeChevallier *et al.*, 1980). To circumvent this apparent taxonomic bias, some investigators have developed more direct assessments of heterotrophic activity based on specific enzyme assays, although it is not entirely certain that such methods do not introduce other kinds of bias. Among the most widely used enzyme activities is 4-methylumbelliferyl-heptanoate hydrolase (MUHase), but there is little information on the variation of MUHase occurrence in different heterotrophic organisms and how the activities of this enzyme are influenced by various environmental conditions (Tryland and Fiksdal, 1998).

Another direct strategy for linking the diversity of organic carbon substrates to the physiological and phylogenetic diversity of microbial communities is to introduce various radiolabeled (^{14}C) substrates independently into environmental samples, while monitoring the evolution of $^{14}CO_2$. This strategy has several advantages including its usefulness in determining the potential for metabolism of xenobiotic (non-natural) carbon sources that may be toxic and in estimating the heterotrophic potential of a microbial community through the determination of the turnover of naturally occurring substrates. Radioisotope tracer experiments are also very sensitive because meaningful data can be collected following short incubation periods. However, the information derived from such experiments can be limited because it is often difficult to link biodegradative activity to specific phylogenetic groups without needing to culture organisms (Ouverney and Fuhrman, 1999; Radajewski *et al.*, 2000). In addition, the concentrations of chemicals typically used in such experiments exceed the expected concentrations in nature, therefore it is difficult to extrapolate data on enzyme kinetics and the potential for substrate co-metabolism. Furthermore, the lack of detection of radioactive CO_2 does not necessarily mean that the substrate is not partially metabolized; it only means that mineralization did not occur.

Carbon isotopes are also used in a different context to link microbial diversity and heterotrophy. The slight preference of enzymatic reactions for different naturally occurring isotopes (^{12}C versus ^{13}C) provides an important opportunity for assessing heterotrophic metabolism and comparing it with other carbon-cycling processes in natural microbial communities. For example, the carbon-fixation enzyme, ribulose bisphosphate carboxylase, has a greater affinity for $^{12}CO_2$, resulting in organic carbon that is approximately 2.8% depleted in ^{13}C relative to its abundance in dissolved bicarbonate. Andrews and colleagues (2000) used the carbon isotope discrimination approach to explore the response of microbial communities to atmospheric carbon dioxide enrichments as part of an attempt to understand the implications of projected greenhouse gas impacts on the climate. The investigators measured the interrelatedness between microbial community composition and δ-^{13}C of respired carbon dioxide from soils under different temperature conditions. They found that the CO_2 produced by soil heterotrophs at 4°C was 2.2 to 3.5 permill (‰) enriched in ^{13}C relative to CO_2 respired at 22°C and 40°C and was similarly enriched relative to bulk soil carbon. There was no isotopic difference between CO_2 produced at 22°C and 40°C. Microbial community composition, as measured by the differences in populations of morphology types, was found to differ across the temperature range. Only 12% of microbial morphotypes were common to all three incubation temperatures, while unique morphotypes were found at each incubation temperature. Species richness, inferred from morphotypes, was significantly lower at 4°C than at higher temperatures. The change in microbial community structure caused a shift in the pool of mineralizable carbon, resulting in different isotopic compositions of CO_2 respired at different temperatures (Andrews *et al.*, 2000).

The extremely high diversity of heterotrophic microorganisms means that it is difficult to develop specific molecular genetic methods for assessing their occurrence, distribution, and activities in the environment. Due to the relatively more precise nature of autotrophism, attempts to quantify the distribution of heterotrophic microorganisms have taken the approach of comparing their population to the population of autotrophs in the same habitat. However, this approach is often confounded by the occurrence of facultative heterotrophs. For example, Kolber and colleagues (2001) interpreted the vertical distribution of bacteriochlorophyll-a, the numbers of infrared fluorescent cells, and the variable fluorescence signal at 880 nm wavelength, to indicate that photosynthetically competent anoxygenic phototrophic bacteria are abundant in the upper open ocean and comprise at least 11% of the total microbial community. However, these organisms are facultative photoheterotrophs, metabolizing organic carbon when available, and reverting to photosynthetic light utilization when organic carbon is scarce. They are globally distributed in the euphotic zone and represent a major component of the microbial community engaged in the cycling of both organic and inorganic carbon in the ocean.

The comparative assessment of autotrophism versus heterotrophism in marine picoplankton has been the subject of considerable debate, especially among marine microbiologists. Giovannoni and colleagues (1990) showed that the diversity of autotrophic prokaryotes in oceanic euphotic zones was surprisingly very low when compared to the diversity of heterotrophic prokaryotes, despite the observation that the contribution of these two groups to the overall microbial biomass are very similar. In fact, only four genera of marine cyanobacteria were considered to be globally significant, namely *Prochlorococcus*, *Synechococcus*, *Trichodesmium*, and the recently discovered N_2-fixing *Synechocystis*. In addition to these four groups, two newly recognized facultative phototrophs belonging respectively to the α- and γ-*Proteobacteria* are of potential significance (Vaulot *et al.*, 2002). In contrast there exists a remarkable diversity of marine heterotrophic prokaryotes observed within many groups such as the *Proteobacteria* and the *Cytophaga–Flavobacterium–Bacteroides* cluster or the *Actinobacteria* (Giovannoni *et al.*, 2000).

The situation for microbial eukaryotes is somewhat more complicated. Using 18S rDNA signature profiles, Moreira and López-García (2002) examined the level of diversity among marine protists of average size between 2–3 micrometers. In oceanic euphotic zones, the abundance and biomass of pico and nanoplanktonic heterotrophic eukaryotes appear to be commensurate with those of the autotrophic ones. However, recent work being conducted under the framework of the European biodiversity assessment program (PICODIV) revealed that the diversity of heterotrophic microbial eukaryotes could be much higher than the diversity of microbial eukaryotic autotrophs as observed for the prokaryotic community. In the equatorial Pacific Ocean at 75 m depth, Moon-van der Staay and colleagues (2001) used unique 18S rDNA sequences that could be attributed to heterotrophic organisms to demonstrate that the diversity of these organisms may be twice as high as the diversity of comparable autotrophic organisms.

Vaulot and colleagues (2002) presented substantiated arguments in response to the question of why the diversity of heterotrophs should be higher than that of autotrophs. For prokaryotes, the diversity of heterotrophs can be attributed to their involvement in the degradation of a diverse array of organic molecules produced by photosynthetic organisms. Each taxonomic category probably specializes in the utilization of specific substrates, although a one-on-one match of phylogenetic diversity to chemical diversity has not been substantiated. In contrast to the situation with heterotrophs, the niches available to autotrophs are likely to be much lower given the restrictions imposed by light availability and substrate diffusion. The impetus for high diversity is different for heterotrophic eukaryotes. According to the concept of the "microbial loop" (see Box 7.3) where models of nutrient cycling in marine ecosystems include a major role for the participation of microorganisms in intricate food webs, heterotrophic microbial eukaryotes play the role of non-specialized grazers feeding on the prokaryotic community. As such, nutrient availability should not be a major driving force behind the phylogenetic diversity of heterotrophic eukaryotes. However, the apparently high diversity suggests that these heterotrophic eukaryotes play more complex roles in the microbial loop than previously recognized. For example, discrimination among prokaryotic prey on the basis of physical size may play a larger role in fractionating the prokaryotic community with respect to the degradation of different kinds of organic molecules. Furthermore, microbial heterotrophic eukaryotes may also use such organic carbon molecules to supplement their nutrient supplies. This possibility has not been part of the traditional models of the microbial loop, but it certainly warrants further investigation.

BOX 7.3

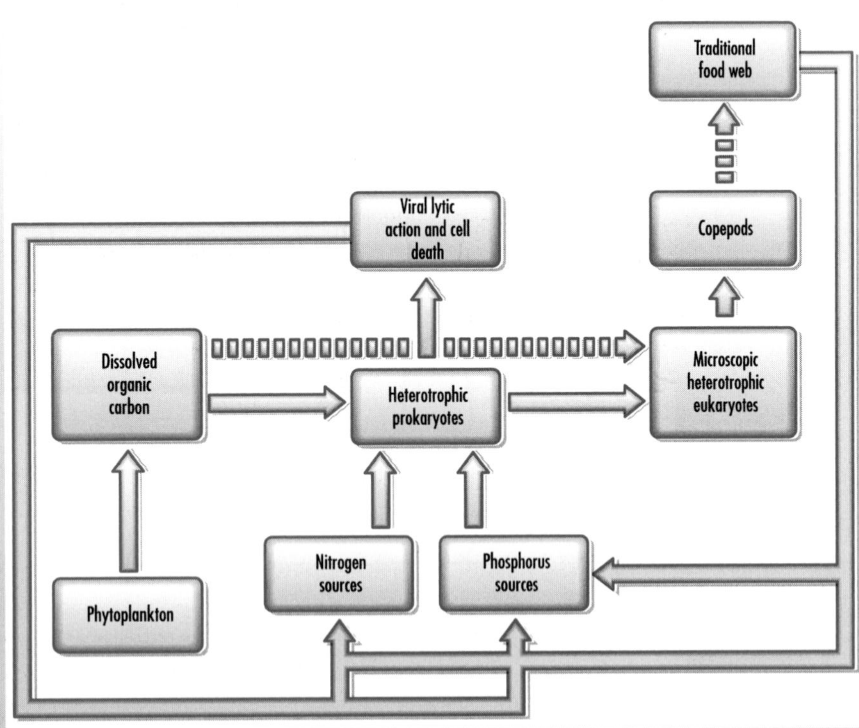

The concept of the microbial loop addresses carbon cycling by heterotrophic microorganisms and integrates microorganisms into the more traditional food webs in the marine ecosystem (Azam *et al.*, 1983). The microbial loop is essentially a food chain that works within, or parallel to, the classical food chain. In the microbial loop, heterotrophic bacteria obtain their carbon and energy sources directly from dissolved inorganic materials. These organisms, which are too small to be preyed on directly by copepods, are grazed by flagellates and ciliates. Ciliates are then consumed by copepods and the process continues up the classical food chain. The microbial loop concept is considered a major milestone in marine biology. In the idealized view of the microbial food web presented below, interactions between heterotrophic bacteria and the traditional food web are described by grazing loss of biomass, viral lyses, and the limitation of growth rate by the available concentrations of organic carbon, inorganic and organic nitrogen, and phosphorus nutrients. In a landmark study that promises to engender major advancement of knowledge on marine microbial diversity, Venter and colleagues (2004) applied whole genome shotgun sequencing to nucleic acid molecules extracted directly from microbial populations in sea water samples collected from the Sargasso Sea near Bermuda. The analyses resulted in a total of 1.045 billion base pairs of non-redundant sequences estimated to derive from at least 1,800 "genomic species" including 148 novel bacterial phylotypes. The authors claim to have identified more than 1.2 million new genes in the marine samples, including more than 782 new rhodopsin-like photoreceptors, which are important for photosynthesis. This represents a tripling of the number of photoreceptors which had been reported previously for all species. The implications of such tremendous microbial diversity for nutrient cycling and for the relationships among different trophic systems in the marine environment will doubtlessly require an equally tremendous level of effort to unravel. But the rewards, in terms of contribution to the understanding of major concepts in microbial ecology, are likely to reach far beyond the microbial loop.

THE NITROGEN CYCLE

One of the salient differences between Earth and the other inner planets of the solar system is the dominant abundance of nitrogen gas in the atmosphere. Nitrogen is four times more abundant than oxygen in the atmosphere, but oxygen is about 10,000 times more abundant than nitrogen in the entire Earth system including soils and the oceans (Fig. 7.1). These differences are due to the composition of the materials from which Earth is originally derived and the processes that have contributed to Earth's stabilization as a planet. Oxygen is a major component of the solid Earth, however, nitrogen is not stable as part of a crystal lattice, therefore it is not incorporated into the solid Earth, and instead it is enriched in the atmosphere relative to oxygen. Molecular nitrogen is relatively inert, and it is useless for most living

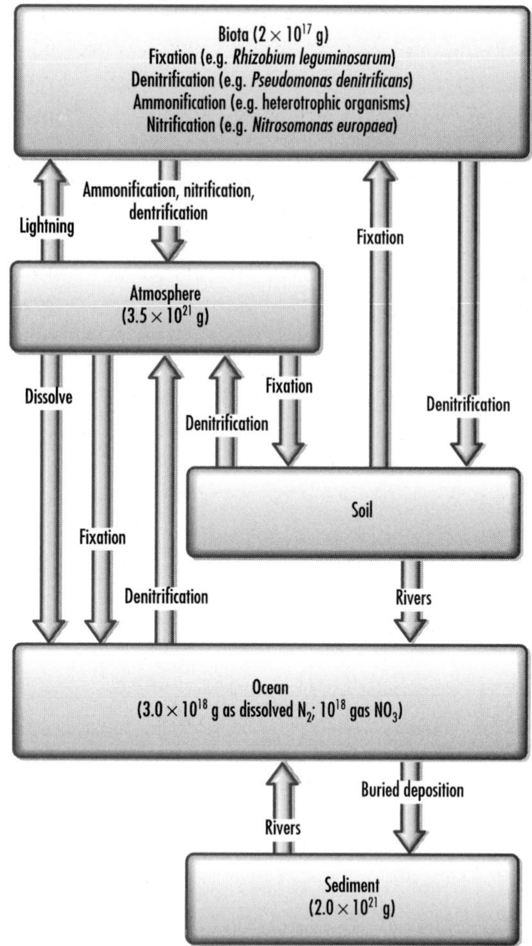

Fig. 7.13 Major reservoirs and fluxes in the biogeochemical cycling of nitrogen. Data on reservoir sizes from Jaffe (1992), Maier (2000), and Davidson (1991).

organisms, and unlike oxygen, nitrogen is stable in the atmosphere and its involvement in atmospheric chemical reactions is very limited. However, nitrogen is capable of existing in various reduced and oxidized forms that span oxidation states between +5 (NO_3^-) and −3 (NH_4^+). These nitrogen compounds are extremely valuable for living systems in part because all amino acids have a nitrogen end group, which is required for the formation of polypeptide chains. Nitrogen makes up approximately 12% of microbial cell dry weight (see Fig. 7.1). Therefore, the availability or scarcity of nitrogen minerals effectively controls biomass production despite the primary importance of carbon sources (Arp, 2000). Nitrogen limitation is responsible for the overwhelming dependence of the modern agricultural industry on soil fertilization and also for the eutrophication of coastal water systems due to urban and agricultural run-off into the ocean (Plate 7.2). The atmosphere contains an estimated 3.9×10^{15} metric tons (3.9×10^{6} petagrams) of nitrogen, however, this largest reservoir of nitrogen on Earth is not actively cycled whereas the mineral nitrogen content of the oceans (0.052 petagrams in the biomass, 300 petagrams in dissolved and particulate organics, and 690 petagrams in soluble salts) is actively cycled. Also the 25 petagrams of nitrogen in the biotic component of lands is actively recycled, while the 110 petagrams of nitrogen in dead organic matter is very slowly recycled (Jaffe, 1992). Four major microbial processes dominate the biogeochemical cycling of nitrogen, namely nitrogen fixation, nitrification, denitrification, and nitrogen mineralization (Fig. 7.13). The microorganisms involved in some transformations of nitrogen belong to very specific phylogenetic groups, but the physiological processes involved are very diverse representing some of the most fascinating microbial ecological interactions, including the symbiotic relationship between bacteria and leguminous plants that is essential for nitrogen fixation. One of the critical questions relating microbial diversity and Earth system science is the enigmatic—and controversial—claim of a rapid increase in the atmospheric concentration of nitrous oxide (Sowers *et al.*, 2003). Nitrogen oxides are important greenhouse gases, and considerable research effort has been dedicated to understanding the sources and sinks for these gases in the Earth system (Davidson, 1991; McIsaac and David, 2003; Whitman and Rogers, 1991).

Another emerging concern regarding global environmental change and the nitrogen cycle is the increasing concentration of nitrogen in soils from the enhanced deposition of atmospheric nitrogen and fertilization (Bohlke *et al.*, 1997; Adams, 2003). A lively debate has surfaced regarding the question of whether nitrogen can rapidly accumulate at rates exceeding 25 kg per hectare per year in ecosystems that lack nitrogen-fixing microorganism–plant symbiosis (Bormann *et al.*, 2002). Solutions to these and other conceptual and practical problems posed by anthropogenic forcing of the nitrogen cycle will require better understanding, and possibly biotechnological manipulation, of microorganisms involved in specific reactions within the cycle. Innovations in technical strategies for exploring microbial diversity associated with nitrogen cycling must be integrated with mass balance analysis, isotopic studies of element flows, and numerical modeling to resolve critical questions on nitrogen biogeochemistry (Lin *et al.*, 2000). Genomic research is revolutionizing the information databases for many organisms that play central roles in the cycling of nitrogen. For example, the genomes of at least two rhizobial species involved in nitrogen fixation, *Mesorhizobium loti* and *Sinorhizobium meliloti* have now been completely sequenced (see Chapter 4 and Appendix therein). Furthermore, the sequence tags representing expressed genes from major plant partners in symbiotic nitrogen fixation (soybean, *Medicago truncatula*, and *Lotus japonicus*) are now publicly available. These genomic and proteomic approaches are being used to develop rapidly accessible microarray systems that promise to contribute substantially to current understanding of key bottle-necks in the nitrogen cycle (Colebatch *et al.*, 2002; Montoya *et al.*, 2002; Taroncher-Oldenburg *et al.*, 2003).

Nitrogen fixation

Nitrogen fixation is the process by which atmospheric nitrogen is made available as ammonia to the biosphere. It is one of the best-understood processes in the nitrogen cycle because of its essential role in maintaining soil fertility and agricultural production:

$$N_2 + 6e^- \rightarrow 2NH_3 (\Delta G'_0 = +150\,\text{kcal}\,\text{mol}^{-1} = 630\,\text{kJ}\,\text{mol}^{-1})$$

or

$$N_2 + 16ATP + 8e^- + 8H^+ \rightarrow 2NH_3 + 16ADP + 16P_i + H_2$$

Biological nitrogen fixation on a global scale is estimated to account for about 0.24 petagrams of nitrogen fixed per year, however, there is considerable geographical variation on the relative importance of biological nitrogen fixation in sustaining ecosystem primary productivity (Fig. 7.14). The amount of nitrogen fixed in any given situation depends on the interaction between environmental conditions and the local diversity of biological system(s) capable of fixing nitrogen. The rate of nitrogen fixation may vary from barely detectable to hundreds of kilograms per hectare per year at specific locations depending on the qualitative and quantitative diversity of azotrophic (nitrogen-fixing) microorganisms (Hubbel and Kidder, 2003). For example, it has been estimated that the marine environment contributes between 40% and 80% of the total 0.24 petagrams of nitrogen fixed per year, but most of the marine nitrogen fixation is carried out by a single genus, namely *Trichodesmium* (Berman-Frank *et al.*, 2003).

Evolutionary history of biological nitrogen fixation

Biological nitrogen fixation is an enzymatic process that is unique to prokaryotes. Most nitrogen-fixing microorganisms (diazotrophs) belong to the Bacteria domain, but a few are Archaea (Chien *et al.*, 2000). The most important category of nitrogen fixation is carried out largely by symbiotic bacteria, which is limited to a few phylogenetic groups, but non-symbiotic nitrogen fixation, also known as "free-living" fixation is well documented and can be accomplished by a wider range of phylogenetic groups of microorganisms (Table 7.4). Nitrogen-fixing prokaryotes belong to a broad range of ecological niches, including obligate

Table 7.4 Diversity of nitrogen-fixing systems including symbiotic and free-living microorganisms.

Microorganism genus	Symbiotic host	Amount of nitrogen fixed (Kg N ha^{-1} yr^{-1})	Habitat
Rhizobium	Legumes	200–300	Soils
Bradyrhizobium			
Anabena	Azolla	100–120	Leaves, aquatic systems, heterocyst-forming microorganisms
Aphanizomenon	Cycads		
Calothrix			
Cylindrospermum			
Gloeotrichia			
Nostoc			
Scytonema			
Tolypothrix			
Streptomyces	Alder (*Alnus* species)		Sand-dune environments
Frankia	Casuarina Buckthorn (*Hippophae rhamnoides*)		
Oscillatoria	Marine sponges (e.g. *Dysidea* sp.)	30–40	Non-heterocystic microorganisms
Lyngbya			
Microcoleus			
Trichodesmus			
Azospirillum	Free living	1–2	Tropical grass soils, root surfaces, and anaerobic sediments
Azotobacter			
Bacillus			
Beijerinckia			
Clostridium			
Chlorobium			
Chromatia			
Desulfotomaculum			
Desulfovibrio			
Klebsiella			
Methanosarcina			
Pseudomonas			
Rhodomicrobium			
Rhodopseudomonas			
Rhodospirillum			
Thiobacillus			
Vibrio			

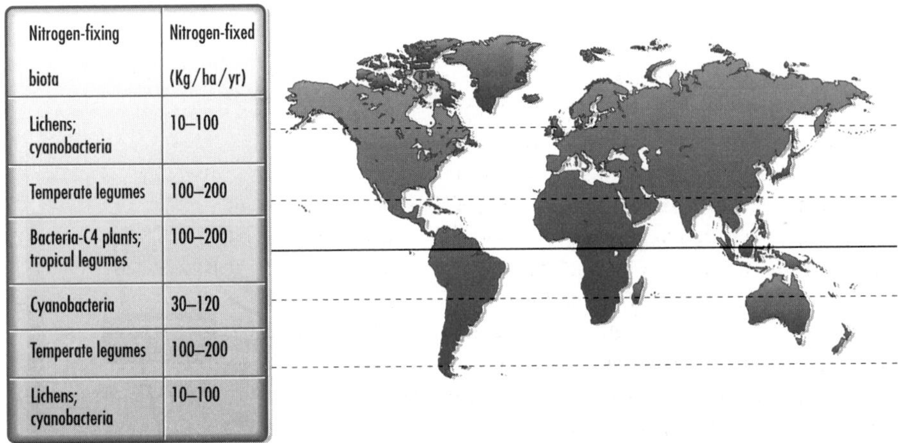

Nitrogen-fixing biota	Nitrogen-fixed (Kg/ha/yr)
Lichens; cyanobacteria	10–100
Temperate legumes	100–200
Bacteria-C4 plants; tropical legumes	100–200
Cyanobacteria	30–120
Temperate legumes	100–200
Lichens; cyanobacteria	10–100

Fig. 7.14 The geographical distribution of biological nitrogen fixation and its contribution to local nitrogen economy is a function of climatic factors, the diversity of biological systems capable of fixing nitrogen, and the demand of the biological community for nitrogen. The data are from Hubbell and Kidder (2003).

anaerobes (e.g. *Clostridium pasteurianum*), facultative anaerobes (e.g. *Klebsiella pneumoniae*, a close relative of *E. coli*), photosynthetic bacteria (e.g. *Rhodobacter capsulatus*), cyanobacteria (e.g. *Anabena*), obligate aerobes (e.g. *Azotobacter vinelandii*), and methanogens (e.g. *Methanosarcina barkeri*). All microorganisms which are capable of fixing nitrogen produce nitrogenase, an oxygen-sensitive metalloenzyme that requires ATP (as indicated by the high positive value of $\Delta G'_0$), reduced ferredoxin, and in some instances, other cytochromes and coenzymes (see Box 7.4). One of the most fascinating case studies of the co-evolution of life and the environment is to unravel the natural history of the emergence and persistence of nitrogen fixation among the highly diverse phylogenetic categories of prokaryotes inhabiting various ecological niches under different oxygen stress conditions.

Nitrogenase is a highly conserved enzyme complex. All known versions of the enzyme have been shown (through phylogenetic analyses) to derive from a single common ancestor. Furthermore, analysis of the catalytic subunits of nitrogenase suggests that they existed before oxygenation of the Earth's atmosphere (Berman-Frank *et al.*, 2003; Broda and Prescheck, 1983) (Fig. 7.15). The hypothesis that nitrogen fixation is an ancient process that existed before the Earth's atmosphere became aerobic is not entirely consistent with the random pattern of occurrence of nitrogen fixation among various phylogenetic lineages of prokaryotes. This is because an ancient origin would predict a pattern of universal occurrence of the capacity for nitrogen fixation among the various lineages, but in fact we know that nitrogen fixation is not as common as it should be, if it emerged early, and if it confers additional fitness to microbial survival. The apparent random distribution of the capacity for nitrogen fixation among microbial lineages can also be due to a multiple independent origin of nitrogen-fixation genes; the loss of a nitrogen-fixation phenotype in organisms that found a better way to obtain reduced nitrogen; or the acquisition of a nitrogen-fixation gene through horizontal genetic exchange instead of vertical inheritance through phylogenetic lineage. These possibilities have been framed as working hypotheses by various research groups striving to understand the evolution of nitrogen fixation and to use biotechnological strategies for reducing the negative impacts of human interactions with the nitrogen cycle.

Nitrogen-fixing organisms have evolved various mechanisms for protecting nitrogenase from oxygen toxicity. One of the most common strategies is the production of leghemoglobin by plants in root nodules that harbor nitrogen-fixing microorganisms. In a function similar to the action of mammalian hemoglobin, leghemoglobin scavenges free oxygen away

BOX 7.4

Gene	Role
nifA	Positive regulator
nifB	Required for the synthesis of the Mo–Fe cofactor
nifD	The alpha subunit of nitrogenase
nifE	May function as a scaffold upon which the Fe–Mo cofactor is built
nifF	Flavodoxin electron donor to *nifH*
nifH	Dinitrogen reductase
nifJ	Ferredoxin oxidoreductase for electron transport to nitrogenase
nifK	Beta subunit of dinitrogenase
nifL	Negative regulator
nifM	Maturation of *nifH*
nifN	Required for the synthesis of Mo–Fe cofactor
nifQ	Mobilization of molybdate for the synthesis of Mo–Fe cofactor
nifS	Mobilization of sulfur for the synthesis of Fe–S cluster
nifT	Unknown
nifU	Mobilization of sulfur for the synthesis and repair of Fe–S cluster
nifV	Homocitrate synthase involved in the synthesis of Mo–Fe cofactor
nifW	stability of dinitrogenase and protection of oxygen inactivation
nifX	Synthesis of Mo–Fe cofactor
nifY	Insertion of Mo–Fe cofactor into apodinitrogenase
nifZ	Synthesis of P-cluster and Fe–S cofactor

(a)

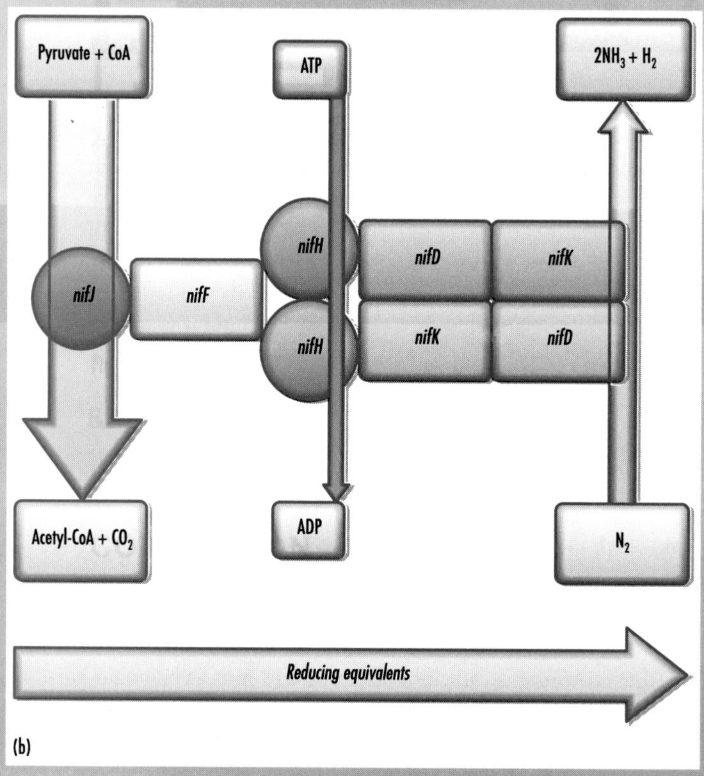

(b)

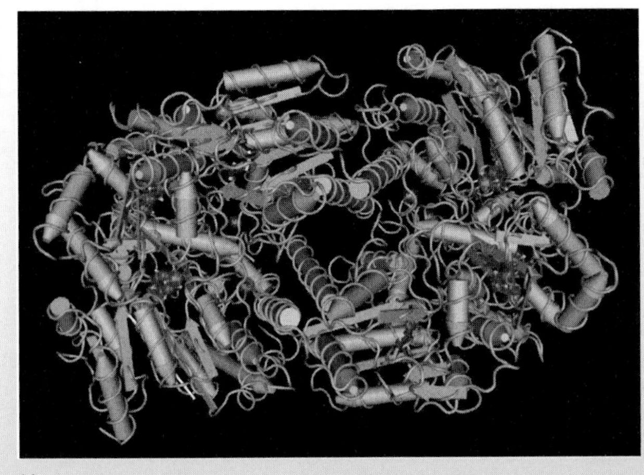
(c)

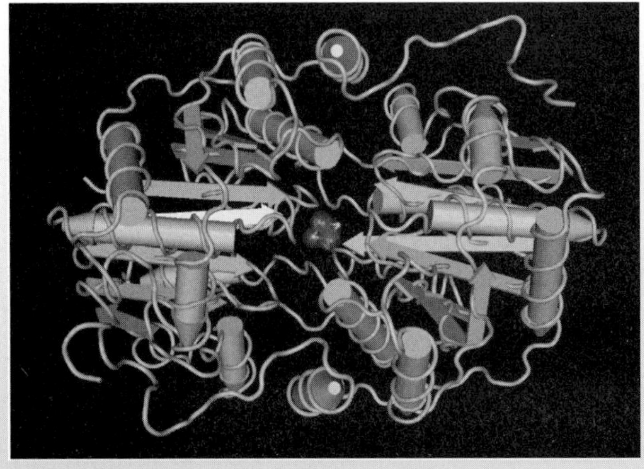
(e)

Protein
Homocitrate
CH_2COO^-
$CH_2CH_2COO^-$
Cys-S — Fe — S — Fe — S — Fe — S — Mo — O_2C
N-His
Protein

(d)

At least 20 genes are involved in biological nitrogen fixation (see table). Six of these genes are required for the synthesis and catalytic activity of nitrogenase, which comprises up to 10% of total cellular proteins in many diazotrophs (b).

The enzyme complex of *Klebsiella pneumonia* featuring 10 domains of the alpha and beta subunits, including the metal-binding domains is shown in (c). The complex consists of two metalloprotein components namely molybdoferredoxin, which requires molybdenum and iron ((d) shows the structure of the FeMoco prosthetic group in nitrogenase); and azoferredoxin, which requires iron only (e). The diversity of nitrogenase complexes among azotrophic microorganisms includes the substitution of molybdenum by vanadium. Vanadium-nitrogenases, with a 1.5 times lower specific activity, are less catalytically efficient than the molybdenum variety (Berman-Frank *et al.*, 2003). There is currently no unified theory to explain the distribution of vanadium versus molybdenum varieties according to ecological niche specialization. The divergence is likely to have occurred early in the evolution of the enzyme complex.

In general, the nitrogenase complex is not specific for nitrogen because it can catalyze the reduction of other triple- and double-bond molecules such as acetylene, hydrogen azide, hydrogen cyanide, and nitrous oxide. The flexibility in the use of different metallic cofactors and non-specificity of substrate indicate that the evolutionary history of nitrogenase has been influenced by various environmental selective pressures. The two domains and location of the metal center of the gamma subunit of Fe-nitrogenase (dinitrogenase reductase) of *Clostridium pasteurianum* are shown in (e).

The three-dimensional images were created with the C3nD software (version 4.1) using data from the National Center for Biotechnology Information (http://www.ncbi.nlm.nih.gov/) as described in Schlessman *et al.* (1998) and Schmid *et al.* (2002).

from the nitrogenase reaction site in order to avoid irreversible enzyme inactivation (Kundu and Hargrove, 2003). In another case, *Azotobacter* species protect nitrogenase by having the highest known rate of respiratory metabolism of any organism, leading to the maintenance of a very low level of oxygen in their cells. The nitrogenases of *Rhizobium* species are protected both by the plant leghemoglobin and the physical security provided by root nodules. In addition, *Rhizobium* produces large amounts of extracellular polysaccharide slime layers, which limit the diffusion of oxygen to the cells. Some cyanobacteria protect nitrogenase from oxygen by cellular differentiation and fixing nitrogen in heterocysts which possess only photosystem

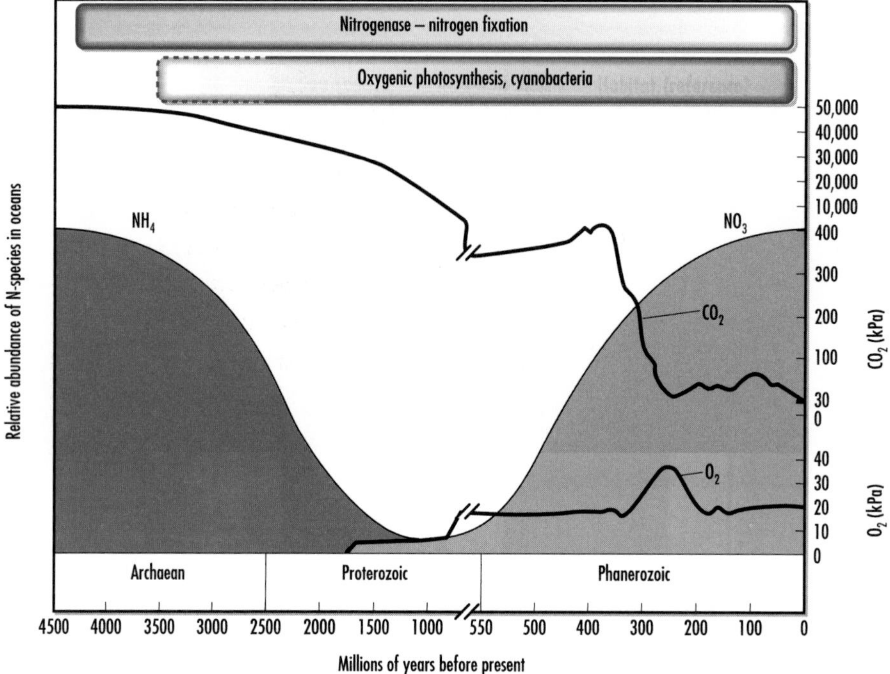

Fig. 7.15 Based on phylogenetic analyses of nitrogenase and geochemical evidence for the early environmental conditions on Earth, the emergence of nitrogen fixation is estimated to predate the emergence of photosynthesis. Contemporary distribution of diazotrophs across various ecological niches and phylogenetic groups reflects the co-evolution of the biogeochemical cycles of oxygen, CO_2, and nitrogen (ammonia and nitrate). The discontinuous segment for oxygenic photosynthesis indicates the debated origins of this process in cyanobacteria. The diagram is modified with permission from Berman-Frank *et al.* (2003).

I (used to generate ATP by light-mediated reactions) whereas the other cells have both photosystem I and photosystem II (which generates oxygen when light energy is used to split water to supply H_2 for the synthesis of organic compounds). Lastly, it is important to mention the as yet unclear ecological ramifications of a report by Ribbe and colleagues (1997) of aerobic nitrogen fixation coupled to carbon monoxide oxidation by *Streptomyces thermoautotrophicus*. This organism occurs naturally in enriched soils beneath burning charcoal piles. The pertinent enzyme is molybdenum-dinitrogenase which acts in concert with a manganese-dependent oxidoreductase by coupling nitrogen reduction to the oxidation of superoxide generated from oxygen by a molybdenum-containing carbon monoxide dehydrogenase.

Nitrogenase is clearly one of the most valuable enzymes in the natural history of the biosphere because of its relative rarity (compared, for example, with the photosynthesis enzyme ribulose bisphosphate carboxylase) and the importance of nitrogen fixation as a crucial linkage point in global biogeochemical cycles. The nitrogenase system of Bacteria is encoded and regulated by a complex of at least 20 genes (Box 7.4). The nitrogenase protein contains two components. Component 1 is defined by an iron–sulfur center and a molybdenum–iron cofactor encoded by *nifD* and *nifK* genes. In some organisms, molybdenum is replaced by vanadium in component 1, and the genes for these vanadium enzymes are referred to as *vnf* genes. Other organisms have component 1 systems that rely only on iron, and *anf* genes encode these. Component 2 contains an iron–sulfur center encoded by *nifH* (Belay *et al.*, 1984; Chien *et al.*, 2000; Dilworth *et al.*, 1987; Kessler *et al.*, 1997; Scherer, 1989). The *Methanoarchaea* contain *nifH* genes that are only distantly related to bacterial *nifH* clusters and are presumed to play ecological roles different from nitrogen fixation; there is also some evidence that genes related to *nifH* in the Bacteria domain participate in photosynthesis (Burke *et al.*, 1993).

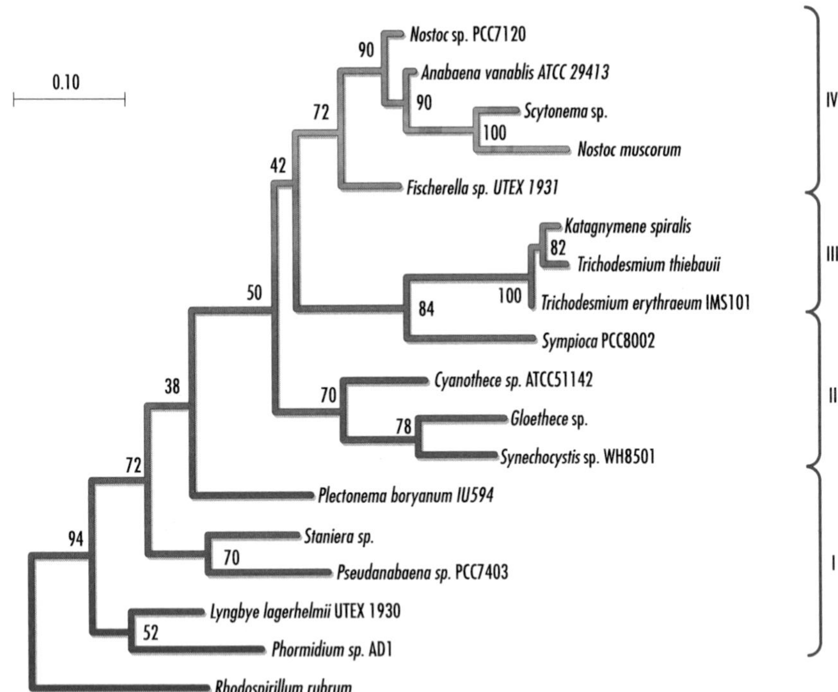

Fig. 7.16 Phylogenetic analysis of dinitrogen reductase gene (*nifH*) sequences among cyanobacterial species. The tree is based on a maximum-likelihood method (see Chapter 5). The scale represents the expected number of changes per sequence position. The numbers depict bootstrap values obtained for a bootstrap sampling of 100. Numbers at the side correspond to different N_2-fixation strategies: I = microaerobic (e.g. *Plectonema* species); II = aerobic (e.g. *Lyngbya* species); III = Partial cellular differentiation (e.g. *Trichodesmium* species); and IV = full cellular differentiation (e.g. *Anabaena* and *Nostoc* species). 324 homologous nucleotides were used to infer the phylogenetic position of the *nifH* genes. The tree is reproduced by courtesy of Paul Falkowski (Berman-Frank *et al.*, 2003).

Nitrogen fixation and environmental change

Although much is known about the molecular underpinnings of nitrogen fixation, there are unresolved questions regarding the ecological ramifications of the process and its sensitivity to global environmental change. The energy-intensive nature of microbiological nitrogen fixation requires that tight controls are placed on the pertinent gene expression and enzyme activity. In addition to the function of regulatory genes, feedback inhibition is one of the most effective control factors because ammonia (the product of nitrogen fixation) represses the reaction. In aquatic systems where cyanobacteria such as *Oscillatoria* and *Lyngbya* fix nitrogen without the aid of heterocysts, process economics has been achieved through temporal and spatial stratification. For example, in certain cases, nitrogen fixation is limited to nocturnal cycles when energy resources that have accumulated through photosynthesis during daytime are used to support the activity of nitrogenase, and oxygen production has been shut down. In other systems, nitrogen fixation occurs simultaneously with photosynthesis, but at locations within a microbial mat that are remotely beneath the surface layers and where oxygen tension is much reduced (Atlas and Bartha, 1998).

Nitrogen fixation by cyanobacterial species deserves special consideration for additional reasons. This group of prokaryotes are the only known nitrogen fixers that also produce oxygen as a byproduct of photosynthesis. Therefore, protection against oxygen stress is particularly important in the context of nitrogenase activity, and it is reasonable to suppose that strategies for avoiding oxygen stress are major driving forces behind evolutionary divergence and niche specialization among these organisms (Fig. 7.16). Cyanobacteria are an extremely

Fig. 7.17 Lichen-covered rock at Badger Island, Tasmania, Australia. This environment is a prime habitat for the establishment of lichens, a symbiotic system featuring fungi and photosynthetic organisms such as nitrogen-fixing cyanobacteria. Lichens are famous for their colonization of nutrient-depleted environments and marginal habitats. They occur in several pioneer terrestrial habitats from Arctic and Antarctic to tropical areas and in many deserts where they are able to form long-lived and stable communities. Cyanobacterial lichens are responsible for the fixation of up to 100 kg of nitrogen/ha/yr in the polar regions of the planet. Photograph is by courtesy of Laurie Ford.

diverse group of organisms with an evolutionary history dating back to approximately 3.5 billion years ago. Their physiological diversity facilitates the colonization of diverse ecosystems, including deserts, freshwater, salty marine systems, hot springs, and cold Antarctic and sub-Arctic environments. The ability of some members of the cyanobacteria to engage in both photosynthesis and nitrogen fixation allows them to colonize extremely nutrient-poor environments. To achieve this level of hardiness, cyanobacteria sometimes engage in symbiotic relationships with filamentous fungi to produce lichen, which is among the sturdiest biological systems known (Fig. 7.17). The wide distribution of cyanolichen (fungi plus cyanobacteria) in many ecosystems, coupled with their participation in the major biogeochemical cycles (carbon through photosynthesis; nitrogen through nitrogen fixation, and phosphorus through phosphate solubilization) has endeared these organisms to investigators seeking reliable ecological indicators of global environmental change, including acid precipitation (Goward and Arsenault, 2000), changes in vegetation cover and plant diversity (Cornelissen *et al.*, 2001), and climate change (van Herk *et al.*, 2002).

The integrity of eco-physiological processes that characterize nitrogen fixation is sensitive to the disruption of the interconnectedness of the biogeochemical cycling of different elements. For example, Billings and colleagues (2002) used measurements of the natural balance of ^{15}N to explore the alterations of nitrogen dynamics under conditions of elevated carbon dioxide in an intact Mojave Desert ecosystem. They showed that ^{15}N was enriched in vegetation but not in soil under elevated CO_2. The results led to the hypothesis that microbial activity increases with elevated CO_2, thereby enriching pools of plant-available nitrogen but not necessarily through newly fixed atmospheric nitrogen. This indicates that long-term exposure to elevated CO_2 may result in significant perturbations to the nitrogen cycle in soils.

In another example, Montoya and colleagues (2002) explored questions regarding the connection between nitrogen availability and ecosystem productivity in oligotrophic marine environments such as the Sargasso Sea. They used nitrogen isotope ratio analyses to demonstrate that the observed low δ-^{15}N in suspended particles and zooplankton cannot arise through isotopic fractionation associated with nutrient uptake and food web processes; instead it is consistent with a significant input of new nitrogen to the upper water column through nitrogen fixation. The results provided direct evidence that nitrogen fixation contributes significantly to the nitrogen budget of oligotrophic ocean systems.

Acea and colleagues (2003) explored questions about the recovery of nitrogen cycling in heat-damaged soil systems through a biotechnological approach involving inoculation of heated soils with cyanobacterial species *Oscillatoria*, *Nostoc*, and *Scytonema*, separately and in combination. Their results showed the importance of a diverse population of nitrogen-fixing organisms for the recovery of damaged ecosystems. Further research into the impact of forest fires on nitrogen fixation is warranted, not just in soils directly affected by episodic heat but also in ecosystems that may serve as distant receptors of nitrogen deposition (Adams, 2003; Neff *et al.*, 2002).

Finally, Ohkuma and colleagues (1999) did pioneering work to characterize the diversity of nitrogen-fixation genes in the symbiotic microbial community inhabiting the guts of diverse termites that specialize in nitrogen-poor diets. Methane generation from termite colonies is of some concern regarding greenhouse gas emissions and climate change. Therefore, termitariums present a unique system in which to investigate the linkages between the nitrogen and carbon cycles. These investigators conducted phylogenetic analysis using a segment of the dinitrogenase reductase gene (*nifH*) on nucleic acids extracted directly from termites' microbial community. Their analysis revealed the presence of diverse *nifH* sequences within and between individual termite species. The *nifH* sequences found in termites, which showed significant levels of nitrogen fixation activity, belonged to the anaerobic *nif* group (consisting of clostridia and sulfur reducers) or the alternative *nif* methanogen group among the *nifH* phylogenetic groups. In the case of termites which exhibited low levels of nitrogen fixation activity, the majority of sequences were assigned to the most divergent *nif* group, probably functioning in processes other than nitrogen fixation and being derived from methanogenic Archaea. They also found several sequence clusters that are unique to termites, possibly belonging to novel microbial species.

Ammonification and nitrification

Atmospheric nitrogen is replenished in part by the capability of biological systems to decompose fixed nitrogen. This decomposition occurs in several steps, the first of which is nitrogen mineralization, the production of ammonia from organic nitrogen compounds (this step is also simply called ammonification). The second step is the production of nitrogen oxides from ammonium ions (nitrification), and the final step is the reduction of nitrous oxides to elemental nitrogen (denitrification).

Ammonification

Microorganisms that carry out nitrogen mineralization or ammonification use a variety of enzymes, including urease and glutamine dehydrogenase (Fig. 7.18). Urease is a metalloenzyme that requires either nickel or manganese, depending on the species (Yamaguchi *et al.*, 1999). Urease production is common among environmental and pathogenic microorganisms, and soil urease activity limits the effectiveness of nitrogen-based fertilizers (Benini *et al.*, 1999). Ammonification results in an estimated annual precipitation of 0.038 to 0.085 petagrams of ammonium at the global scale (Sorderlund and Svensson, 1976). Many organisms including eukaryotes and prokaryotes can perform ammonification and many more can perform the reverse process, which is to assimilate ammonium into amino acids and other organic nitrogen compounds (Sepers, 1981; Jana, 1994). However, few organisms, primarily soil-inhabiting autotrophic prokaryotes, can oxidize ammonium further to nitrogen oxides, primarily nitrite and nitrate in the process called nitrification.

Nitrification

The central role of nitrification in the nitrogen cycle makes the process of critical importance to problematic issues in global environmental change. Nitrification is linked to the eutrophication of aquatic systems and to the production of trace concentrations of greenhouse gases in the atmosphere (Robertson and Kuenen, 1991). Nitrification is a two-step reaction beginning with the oxidation of ammonium to nitrite:

$$NH_4^+ + 1.5O_2 \rightarrow NO_2^- + 2H^+ + H_2O(\Delta G'_o = -66\,\mathrm{kcal\,mol^{-1}} = -277.2\,\mathrm{kJ\,mol^{-1}})$$

In the next reaction, nitrite is oxidized to nitrate:

$$NO_2^- + 0.5O_2 \rightarrow NO_3^-(\Delta G'_o = -17\,\mathrm{kcal\,mol^{-1}} = -71.4\,\mathrm{kJ\,mol^{-1}})$$

At the global scale, most ammonia oxidation is carried out by autotrophic microorganisms, which cannot oxidize nitrite. Ammonia- and nitrite-oxidizing bacteria are typically spatially isolated. Nitrite does not accumulate in nature because it is extremely reactive (Atlas and Bartha, 1998). Nitrification is catalyzed in part by ammonia monooxygenase (AMO), a membrane-bound metalloenzyme that is evolutionarily related to methane monooxygenase (MMO) (Holmes *et al.*, 1995). AMO is responsible for the conversion of ammonium to hydroxylamine (NH_2OH), which is then used as substrate by hydroxylamine oxidoreductase to produce nitric acid, water, and hydrogen. A terminal oxidase produces nitrous acid that lowers the pH of the nitrification microenvironment. Nitrifying organisms are chemolithotrophs, meaning that they support the assimilation of carbon dioxide with the energy derived from nitrification. About 100 moles of nitrite or 35 moles of ammonia need to be oxidized in order to fix one mole of carbon dioxide.

$$HN_2 - \underset{\underset{O}{\|}}{C} - NH_2 + H_2O \rightarrow 2NH_3 + CO_2$$

(a)

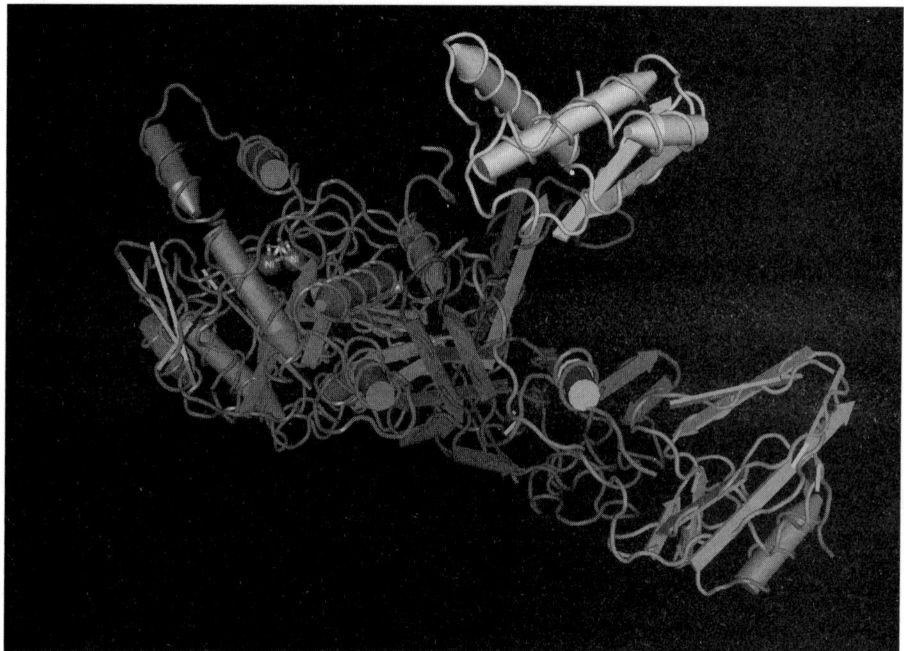

(b)

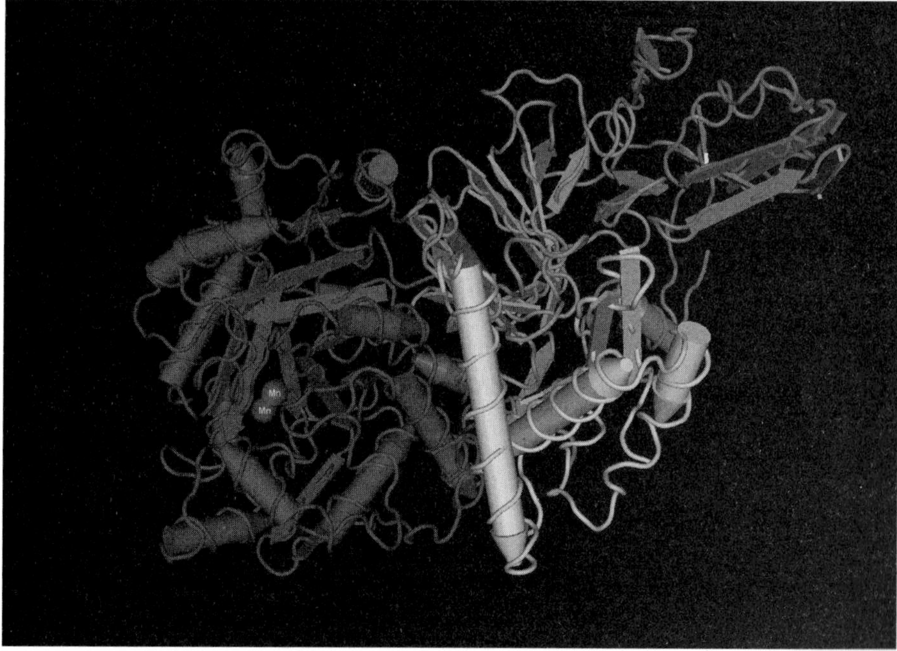

(c)

Fig. 7.18 Ammonification partially recycles fixed nitrogen by decomposing organic nitrogen compounds to ammonia (a). In the simple case of urea, the process is catalyzed by the metalloenzyme urease, which exists in different molecular forms in diverse organisms. For example, urease requires nickel in *Bacillus pateurii* ((b); Benini *et al.*, 1999) but the enzyme is active with manganese instead of nickel in *Klebsiella aerogenes* ((c); Yamaguchi *et al.*, 1999). Metal-binding domains are very similar, however, the selective pressures that lead to preferences for different metallic cofactors have not been established. Three-dimensional protein structures were created with Cn3D version 4 from the National Center for Biotechnology Information (http://www.ncbi.nlm.nih.gov/) based on data from Benini *et al.* (1999) and Yamaguchi *et al.* (1999).

Ammonia monooxygenase is encoded by the *amo* operon consisting of the three genes *amoA* (enzyme active site), *amoB* (enzyme subunit), and *amoC* (membrane protein). The *amo* operon is found in culturable members of the β and γ subdivisions of the *Proteobacteria*, where it occurs in multiple (2–3) and single copies, respectively (Norton *et al.*, 1996; Klotz and Norton, 1998; Utåker *et al.*, 1995). It is generally recognized that the population density of culturable microorganisms capable of oxidizing ammonia cannot account for the total rate of ammonia oxidation measured in ecosystems. Therefore, 16S rRNA sequences have been used extensively to directly explore the distribution and abundance of ammonia oxidizers in natural environments (Head *et al.*, 1993; Kowalchuk *et al.*, 2000; Purkhold *et al.*, 2000; Stephens *et al.*, 1996 and 1998). The studies based on 16S rRNA sequences have also been complemented with the direct assessment of functional characteristics based on polymerase chain reactions using sequences specific for the *amo* genes (Norton *et al.*, 2002). Among the culturable soil microorganisms, the most commonly found ammonia oxidizers belong to the genera *Nitrosomonas*, although some fungi (e.g. *Aspergillus* species) can convert ammonium to nitrate, but at rates thousands of times slower than bacterial ammonification (Focht and Verstraete, 1977). Other bacterial genera that have been associated with ammonia oxidation include *Nitrosococcus*, *Nitrospira*, *Nitrosolobus*, and *Nitrosovibrio* (Maier, 2000). The special case of ammonium oxidation under anaerobic conditions deserves particular mention. Schmidt and colleagues (1997) demonstrated that under the presence of gaseous NO_2 *Nitrosomonas eutropha*, an obligately lithoautotrophic bacterium, was able to oxidize ammonia with the production of nitrite and nitric oxide, and hydroxylamine occurred as an intermediate. Between 40% and 60% of the nitrite produced was further denitrified to nitrogen gas via nitrous oxide. In a nitrogen atmosphere supplemented only with 25 ppm NO_2 and 300 ppm CO_2, the cells performed physiological processes required for growth and proliferation while oxidizing ammonia. In this case, anaerobic ammonia oxidation was inhibited by NO. However, anaerobic ammonia oxidation does not appear to be inhibited by NO in other species. A lithothrophic anaerobic ammonia oxidizer named *Brocadia anammoxidans* resisted up to 600 ppm NO presumably by detoxification through further oxidation (Schmidt *et al.*, 2002). The role of anaerobic ammonia oxidizers in natural ecosystems is not yet clear; however, their occurrence adds an exciting dimension to the physiological diversity of microbial involvement with the nitrogen cycle.

Not all nitrifying bacteria are capable of ammonia oxidation as is the case for *Nitrobacter* species, which are the dominant organisms capable of oxidizing nitrite in soils, as determined by the use of polyclonal antibodies targeted at nitrite oxido-reductase (Maron *et al.*, 2003), fluorescent microscopy (Bartosch *et al.*, 2002), and fatty acid analysis (Lipski *et al.*, 2001). The genus *Nitrobacter*, including *N. winogradskyi*, *N. hamburgensis*, and *N. vulgaris* are distinguishable by their possession of vaccenic acid as the main fatty acid, and the lack of hydroxyl fatty acids (Lipski *et al.*, 2001). Other commonly found nitrifying prokaryotes include members of the genus *Nitrococcus*, *Nitrospina*, and *Nitrospira*. Not surprisingly, there is considerable genetic relationship between ammonia- and nitrite-oxidizing prokaryotes, and it is quite possible that the observed niche specialization is a reflection of adaptation to allochthonous sources of nutrients in particular habitats (Feray and Montuelle, 2002; Regan *et al.*, 2003; Teske *et al.*, 1994).

One of the understudied aspects of nitrification is the occurrence of heterotrophic nitrification by a group of aerobic organisms that can use oxygen and nitrite in the presence of an organic substrate, even under conditions where the dissolved oxygen concentration approaches saturation. The most notable example of a heterotrophic nitrifying organism is *Thiosphera pantotropha*, a wastewater inhabitant that uses enzymes involved in nitrogen oxidation reactions to convert ammonium to nitrate in the presence of acetate or other organic substrates. There is speculation, but insufficient molecular evidence, that these enzymes are identical in form or function to ammonium monooxygenase and hydroxylamine oxidoreductase (Robertson and Kuenen, 1991). In cases where heterotrophic nitrification has been documented, the process is coupled with the capacity for aerobic denitrification, the process through which the product of nitrification (nitrate) is reduced to nitrous oxide and ultimately to molecular nitrogen. Heterotrophic nitrification is slower than autotrophic nitrification by a factor of 100–1000, and it appears to have evolved as a mechanism through which rate-limiting steps in the flow of electrons to oxygen can be overcome, although the ecological significance of this process is not clear. However, if the capacity for heterotrophic nitrification is nearly as widespread as the occurrence of heterotrophic microorganisms, then the potential contribution of this process to global nitrogen cycling needs more serious investigation.

Denitrification

Denitrification is an extremely important process for the global impact of nitrogen cycling because it controls the return of nitrogen to the gas phase through the dissimilatory reduction of nitrate to nitric oxide, nitrous oxide, and nitrogen:

$$2NO_3^- + 5H_2 + 2H^+ \rightarrow NO_2^- \rightarrow NO \rightarrow N_2O \rightarrow N_2 + 6H_2O\ (\Delta G'_o = -212\,\text{kcal per } 8e^- \text{ transfers})$$

Four major enzymes are responsible for the four steps involved in complete denitrification. Nitrate reductase, a membrane-bound Mo–Fe protein catalyzes the conversion of nitrate to nitrite and it is very sensitive to oxygen (Fig. 7.19). Therefore, denitrification occurs under conditions of reduced oxygen tension, although aerobic denitrification, a topic that was controversial for more than a century, has been documented in a peculiar group of denitrifiers including *Thiosphaera pantotropha*, *Pseudomonas denitrificans*, *Pseudomonas aeruginosa*, *Alcaligenes fecalis*, and *Zooglea ramigera*. Examples of organisms that obligately require anoxic conditions for denitrification include the lithotrophs *Thiobacillus denitrificans* and *Paracoccus denitrificans* (Robertson and Kuenen, 1991).

The reduction of nitrite to nitric oxide is catalyzed by the periplasmic copper or heme enzyme nitrite reductase. There is no definitive explanation for the existence of these two forms of nitrite reductase, as they are both commonly found in the environment and are unique to denitrifying prokaryotes (Brown *et al.*, 2000a and 2000b). The synthesis of nitrite reductase is subject to substrate induction, and like nitrate reductase, it functions in most organisms under conditions of reduced oxygen tension. The third enzyme in the sequence of denitrification is nitric oxide reductase, a membrane-bound enzyme that generates nitrous oxide from nitric oxide. Atmospheric nitrous oxide is the third most significant contributor to global warming, after carbon dioxide and methane. Therefore, it is important to understand the function of nitrous oxide reductases and their role in eliminating nitrous oxide from the biosphere. These enzymes consist of 65 kDa dimers containing a copper atom, which functions similarly to the metal's role in cytochrome-c oxidase (Ellis *et al.*, 2003). Nitrous oxide reductase is extremely sensitive to oxygen, however, protection mechanisms yet to be discovered must exist to ensure the functioning of this enzyme in microorganisms capable of aerobic denitrification as described above. Dissimilatory denitrification is not the only pathway for nitrate reduction. Assimilatory denitrification, known to occur in a diverse cohort of organisms, results in the production of ammonia from nitrate:

$$NO_3^- + 4H_2 + 2H^+ \rightarrow NH_4^+ + 3H_2O(\Delta G'_o = -144\,\text{kcal per } 8e^- \text{ transfers})$$

The ammonia produced is relatively easily incorporated into biomass and the process is not inhibited by oxygen, however, it is sensitive to ammonia concentrations through feedback inhibition. A wide range of organisms including microscopic prokaryotes and eukaryotes (fungi and algae) can perform assimilatory denitrification, and it predominates in environments containing a high concentration of organic matter.

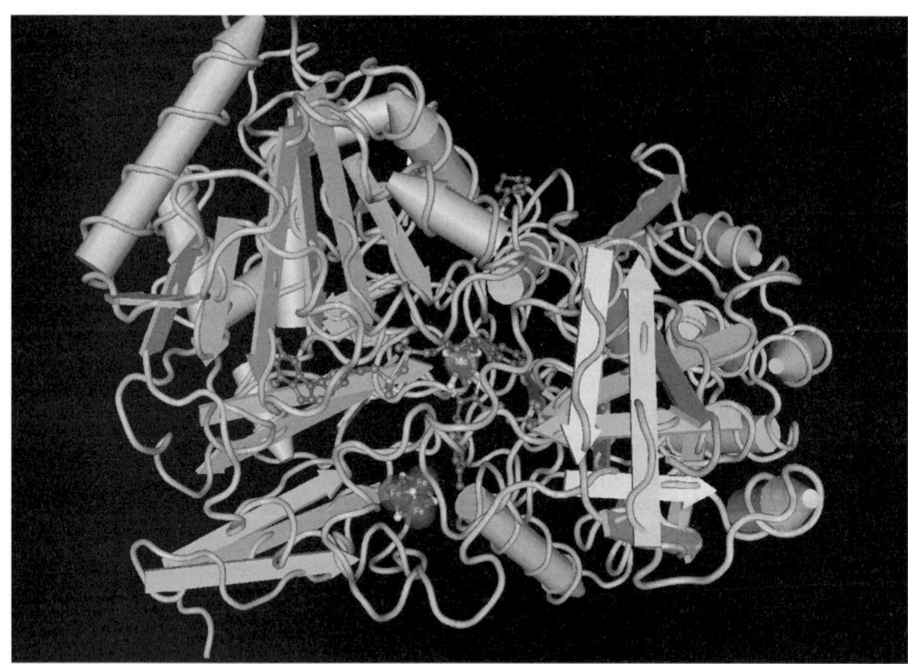

(a)

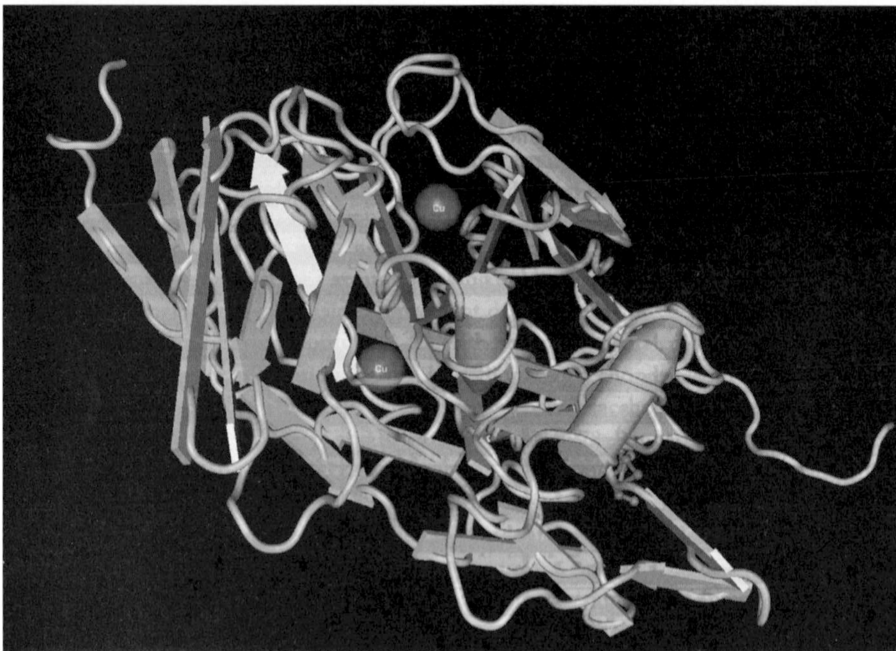

(b)

Fig. 7.19 The enzymes involved in denitrification are all metalloenzymes. The membrane-bound nitrate reductase of *Desulfovibiro desulfuricans* (a) requires iron and molybdenum (Dias *et al.*, 1999); whereas the periplasmic nitrite reductase of *Alcaligenes xylosoxidans* (b) requires copper (Ellis *et al.*, 2003); and the periplasmic copper–calcium clustered nitrous oxide reductase of *Paracoccus denitrificans* (Brown *et al.*, 2000a). The nitrous oxide reductase of *Pseudomonas nautica* (*Marinobacter hydrocarbonoclasticus*) exhibits a CuZ center, which belongs to a new type of metal cluster, in which four copper ions are bound by seven histidine residues. N_2O binds to this center via a single copper ion. The remaining copper ions likely act as an electron reservoir to ensure rapid electron transfer and the prevention of dead-end products (Brown *et al.*, 2000b). The structure of membrane-bound nitric oxide reductase has not been fully resolved. The requirement for different varieties of metal cofactors is one of the most promising ways to understand the molecular diversity of enzymes involved in denitrification. Three-dimensional protein structures were generated with the Cn3D-4.1 software using data from the National Center for Biotechnology Information (http://www.ncbi.nlm.nih.gov/).

The global dimension of the nitrogen cycle: Prospects and challenges

Several attempts have been made to produce quantitative estimates of global rates of denitrification in the context of other nitrogen transformation processes for the purpose of refining numerical models of the nitrogen cycle. The use of remote-sensing tools promised to be a rewarding approach for regional scale analysis of denitrification, but so far it has been difficult to bridge the scale of analysis between global monitoring programs and local assessment of denitrification. For example, Robertson *et al.* (1993) considered remote sensing to be an effective way of narrowing the range of uncertainty in extrapolating nitrogen emissions from small-scale to large-scale terrestrial ecosystems. Correlations between denitrification activity, soil moisture, and soil thermal infrared emissions were monitored in two different agricultural soils dominated by loam and silty clay, respectively. Thermal infrared emissions were found to be useful for estimating the denitrification rate in soil within a limited range of soil moisture levels, but the authors concluded that estimates of denitrification activity based on soil texture and moisture are not likely to be a fruitful approach to generating large-scale nitrogen fluxes.

In the marine environment, Capone and colleagues (1998) demonstrated the application of remote-sensing satellite (AVHRR) data to map a 2 million-km^2 bloom of *Trichodesmium erythraeum*, a nitrogen-fixing cyanobacterium, in the central Arabian Sea. The analysis suggested that nitrogen fixation could account for an input of about $1\,Tg\,N\,yr^{-1}$ into the ecosystem, an amount that is relatively minor when compared with estimates of the removal of nitrogen through denitrification. Finally, Groffman and Turner (1995) and Turner and Millward (2002), working with the International Satellite Land Surface Climatology Program (ISLSCP), used a remote-sensing based index of plant productivity to explore the relationships between plant productivity and annual fluxes of nitrogen and nitrous oxide in a prairie landscape dominated by tall-grass. In the watershed study area, mean landscape gas fluxes were $0.62\,g\,N\,m^{-2}\,yr^{-1}$ for nitrogen, and $0.66\,g\,N\,m^{-2}\,yr^{-1}$ for nitrous oxide. Matson and Vitousek (1990a and b) used a similar approach to investigate the influence of soil moisture on plant community composition and nitrous oxide fluxes.

Anthropogenic influences on the global nitrogen cycle probably exceed those on other major biogeochemical cycles, but the scientific foundation for predicting how ecosystems will ultimately respond to these influences is not firmly developed. Although the northern temperate forest ecosystems have experienced the greatest changes in nitrogen inputs from the atmosphere, the impact is rapidly spreading to other biomes, particularly the semi-arid and tropical regions. Asner and colleagues (2001) developed an integrated terrestrial biophysics–biogeochemical process model named *TerraFlux* to test the relative importance of factors that may strongly influence the productivity response of both humid tropical and semi-arid systems to anthropogenic nitrogen deposition. These kinds of integrated assessments of nitrogen fluxes have been very useful for characterizing regional scale changes in environmental conditions that affect the nitrogen cycle. However, there is still a major disconnection between the macro-scale changes and the diversity of microorganisms that control nitrogen fluxes, and some level of consistency of terminology and methodologies will be required of future studies (Krug and Winstanley, 2002; Silva *et al.*, 2000). Bridging this gap across the different scales of analysis will require the development of innovative methods for generating quantitative estimates of diversity at the molecular level, including enzyme polymorphisms, feedback controls, and relative rates of catalyses. Furthermore, the molecular level data need to be verified by quantitative measurements of microbial gas fluxes on a scale that is compatible with models being developed to quantify the relationships between large-scale nitrogen fluxes and physical-chemical parameters that control atmospheric chemical reactions.

CONCLUSION

Improved understanding of the diverse roles played by microorganisms in the biogeochemical cycling of elements has implications for elucidating and reconstructing the history of the co-evolution of life and the environment on Earth (Beerling, 1999; Farquhar *et al.*, 2000; Hoehler *et al.*, 2001), and for projecting the future evolution of life on Earth, with the potential for the survival of life on other planets (Nealson, 1999; Sarmiento and Bender, 1994). The extensive corpus of scientific information available on the carbon and nitrogen cycles reflects the consensus that these elements are central to the interactions between living systems and the physical-chemical environment. However, the genetic, physiological, and ecological determinants of carbon and nitrogen biogeochemistry have not kept pace with the urgent need for predictive numerical modeling of the reservoirs and fluxes of these element cycles. Molecular level assessment of microbial diversity is a necessary component of the emerging research agenda to construct a bridge between the historical biogeochemical events and predictive capabilities of integrated models that should translate such detailed information into global understanding.

QUESTIONS FOR FURTHER INVESTIGATION

1 The question of scale is central to the understanding of biogeochemical cycles and the interconnectedness of elemental reservoirs and fluxes. In this regard, the smallest scale deals with metabolic processes at the level of individual cells, and the largest scale concerns the global level dimension of major cycles. Select one microorganism each for the carbon and nitrogen cycles and trace how the organisms' contributions to these cycles proceed from the cellular level to the various interactions that mobilize the global cycles.

2 How will the carbon cycle change if microbial life on Earth were to become suddenly extinct? Reconsider this question if the following categories of microorganisms became extinct, one at a time: (a) phototrophic microorganisms; (b) methanogens; (c) aerobic heterotrophic microorganisms. What would be the expected time frame for each of these changes?

3 How will the nitrogen cycle change if microbial life on Earth were to become suddenly extinct? Reconsider this question if the following categories of microorganisms became extinct, one at a time: (a) nitrogen-fixing microorganisms; (b) denitrifying microorganisms; (c) nitrifying microorganisms. What would be the expected time frame for each of these changes?

SUGGESTED READINGS

Adams, M.B. 2003. Ecological issues related to N deposition to natural ecosystems: Research needs. *Environment International*, **29**: 189–99.

Alfreider, A., C. Vogt, D. Hoffmann, and W. Babel. 2003. Diversity of ribulose-1,5-bisphosphate carboxylase/oxygenase large-subunit genes from groundwater and aquifer microorganisms. *Microbial Ecology*, **45**: 317–28.

Arp, D.J. 2000. The nitrogen cycle. In E.W. Triplett (ed.) *Prokaryotic Nitrogen Fixation: A Model System for the Analysis of a Biological Process*, chapter 1. Norfolk, England: Horizon Scientific Press.

Beerling, D.J. 1999. Quantitative estimates of changes in marine and terrestrial primary productivity over the past 300 million years. *Proceedings of the Royal Society Biological Sciences (Series B)*, **266**: 1821–7.

Berman-Frank, I., P. Lundgren, and P. Falkowski. 2003. Nitrogen fixation and photosynthetic oxygen evolution in cyanobacteria. *Research in Microbiology*, **154**: 157–64.

Colebatch, G., B. Trevaskis, and M. Udvardi. 2002. Symbiotic nitrogen fixation research in the postgenomics era. *New Phytologist*, **153**: 37–42.

Dedysh, S.N., P.F. Dunfield, M. Derakshani, S. Stubner, J. Heyer, and W. Liesack. 2003. Differential detection of type II methanotrophic bacteria in acidic peatlands using newly developed 16S rRNA-targeted fluorescent oligonucleotide probes. *FEMS Microbiology Ecology*, **43**: 299–308.

Giovannoni, S.J., T.B. Britschgi, C.L. Moyer, and K.G. Field. 1990. Genetic diversity in Sargasso Sea bacterioplankton. *Nature*, **345**: 60–3.

Grant, R.F. 1999. Simulation of methanotrophy in the mathematical model ecosystems. *Soil Biology and Biochemistry*, **31**: 287–97.

Hoehler, T.M., B.M. Bebout, and D.J. Des Marais. 2001. The role of microbial mats in the production of reduced gases on the early Earth. *Nature*, **412**: 324–7.

Lackner, K. 2003. A guide to CO_2 sequestration. *Science*, **300**: 1677–8.

Maron P.-A., C. Coeur, C. Pink, A. Clays-Josserand, R. Lensi, A. Richaume, and P. Potier. 2003. Use of polyclonal antibodies to detect and quantify the NOR protein of nitrite oxidizers in complex environments. *Journal of Microbiological Methods*, **53**: 87–95.

Moon-van der Staay, S.Y., R. De Wachter, and D. Vaulot. 2001. Oceanic 18S rDNA sequences from picoplankton reveal unsuspected eukaryotic diversity. *Nature*, **409**: 607–10.

Nealson, K.H. 1999. Post-viking microbiology: New approaches, new data, new insights. *Origins of Life and Evolution of the Biosphere*, **29**: 73–93.

Newman D.K. and J.F. Banfield. 2002. Geomicrobiology: How molecular-scale interactions underpin biogeochemical systems. *Science*, **296**: 1071–7.

Orphan, V.J., C.H. House, K.-U. Hinrichs, K.D. McKeegan, and E.F. DeLong. 2002. Multiple archaeal groups mediate methane oxidation in anoxic cold seep sediments. *Proceedings of the National Academy of Sciences (USA)*, **99**: 7663–8.

Sowers, T., R.B. Alley, and J. Jubenville. 2003. Ice core records of atmospheric N_2O covering the last 106,000 years. *Science*, **301**: 945–8.

Srebotnik, E., K.A. Jensen, Jr, and K.E. Hammel. 1994. Fungal degradation of recalcitrant nonphenolic lignin structures without lignin peroxidase. *Proceedings of the National Academy of Sciences (USA)*, **20**: 12794–7.

Taroncher-Oldenburg, G., E.M. Griner, C.A. Francis, and B.B. Ward. 2003. Oligonucleotide microarray for the study of functional gene diversity in the nitrogen cycle in the environment. *Applied and Environmental Microbiology*, **69**: 1159–71.

Venter, J.C., K. Remington, J.F. Heidelberg, A.L. Halpern, D. Rusch, J.A. Eisen, D. Wu, I. Paulsen, K.E. Nelson, W. Nelson, D.E. Fouts, S. Levy, A.H. Knap, M.W. Lomas, K. Nealson, O.White, J. Peterson, J. Hoffman, R. Parsons, H. Baden-Tillson, C. Pfannkoch, Y.-H. Rogers, and H.O. Smith. 2004. Environmental genome shotgun sequencing of the Sargasso Sea. *Science*, **0**: 10938571–0 (published online 4 March 2004).

Watson, G.M. and F.R. Tabita. 1997. Microbial ribulose 1,5-bisphosphate carboxylase/oxygenase: A molecule for phylogenetic and enzymological investigation. *FEMS Microbiology Letters*, **146**: 13–22.

BIOGEOCHEMICAL CYCLING OF PHOSPHORUS, SULFUR, METALS, AND TRACE ELEMENTS

INTRODUCTION

Acidic rainfall and eutrophication are two of the major ecological problems associated with the sulfur and phosphorus cycles, respectively. Imbalances in the cycling of metals and trace chemical compounds produce ecotoxicological impacts that cut across phylogenetic domains in the case of metal toxicity and geographic domains in the case of stratospheric ozone-depleting compounds. The global cycling of phosphorus, sulfur, and metals is intimately connected to the carbon and nitrogen cycles. For example, most of the acidity in wet and dry precipitation is linked to the combustion of carbon-based fuels that have considerable amounts of sulfur. Likewise, eutrophication of freshwater systems is linked to high input loads of fertilizers containing both phosphorus and nitrogen (these two elements are usually limiting for the biological assimilation of carbon). Microbial physiological processes underpin the interconnectedness of biogeochemical cycles and in many cases the physiological bridge between the cycles of the major elements is formed by metals. The reason is that most metabolic processes are catalyzed by metalloenzymes, which require small amounts of trace metals for optimum functioning. Many enzymes are also inhibited by toxic metal ions. Therefore, the concentrations of metals inside and around microbial cells are tightly regulated. Imbalances in trace metal dynamics in ecosystems have considerable implications for the integrity of other biogeochemical cycles. This chapter continues the presentation of the linkages between microbial diversity and biogeochemical cycles with particular emphasis on:

1 Microbiological forms and functions that sustain the phosphorus, sulfur, metals, and trace element cycles.
2 Research activities aimed at modeling and projecting the biogeochemical cycles with the goal of correcting shifts that have occurred due to human industrial activities.

Chapter contents

THE PHOSPHORUS CYCLE

Anthropogenic soil fertilization usually takes the form of adding nitrogen and phosphorus to cultivated soils—the supply of these elements often limits biological production. Phosphorus is essential in the biosphere because it is an indispensable component of nucleic acids where it is involved in phosphate diester bonds that stabilize the "backbone" of nucleotide polymers. Phosphorus is also a major component of the "energy currency" of cells, as adenosine triphosphate (ATP) and the diphosphate (ADP) and monophosphate (AMP) varieties, and of the redox-generating molecules nicotinamide adenine dinucleotide phosphate and its reduced form (NADP and NADPH, respectively). Finally, phosphorus is a major component of cell membranes as phospholipid layers containing hydrophilic phosphate groups. Given these crucial roles, the pool of biologically available phosphorus controls ecosystem productivity to a considerable extent, and it has been demonstrated that phosphorus limitation restricts microbial growth in natural aquatic and terrestrial ecosystems, and in managed environments (Carlsson and Caron, 2001; Cleveland *et al.*, 2002; Yu *et al.*, 2003).

In natural environments, the biological availability of phosphorus is determined by the abundance of metal ions such as Ca^{2+}, Fe^{3+}, and Mg^{2+}, which cause phosphorus to precipitate at pH values above 7.0. Plate 8.1 shows the global distribution of soil "phosphorus retention potential" based on an integrated model of variables (e.g. soil temperature and vegetation cover) that influence phosphate solubility and precipitation. Microbial activities affect the phosphorus retention potential through enzymatic reactions and the capacity to alter the pH of their growth environment. For example, phosphatase and phytase activities mineralize organic phosphate, and chemolithotrophic microorganisms including *Thiobacillus* and *Nitrosomonas* are capable of mobilizing inorganic phosphate through the production of sulfuric acid and nitrous acid, respectively (Atlas and Bartha, 1998). Major fluxes of the phosphorus cycle are shown in Fig. 8.1. Most transfers within the phosphorus cycle do

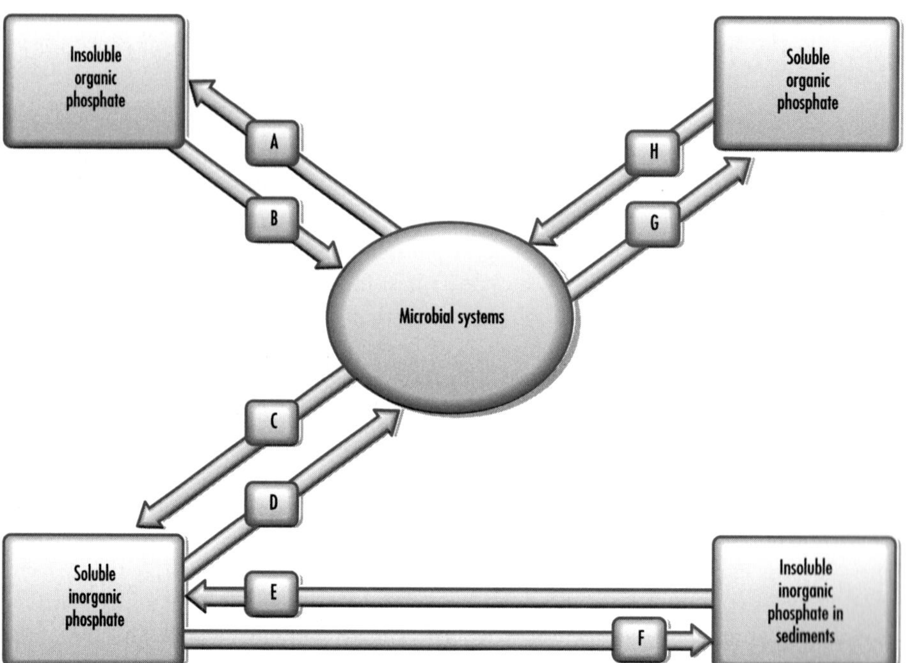

Fig. 8.1 Microbiological participation in the global phosphorus cycle. Most natural sources of biologically available phosphate are in the soluble inorganic forms, which are assimilated to produce biomass (D, A, G). Microbial extracellular phosphatase activities convert organic phosphate sources to assimilable phosphate, including the decay of animal and plant biomass (H, B, C). Industrial processing converts insoluble rock phosphate to soluble biologically available forms (E), and precipitation reactions reverse this trend (F).

not alter the oxidation state of phosphate, and the only gaseous form of phosphate that enters the atmosphere is phosphine (PH_3).

The global phosphorus cycle is not balanced due to the active mining of phosphate from sedimentary rock at a level that is greater than the rate of formation of phosphate-rich sediments. The estimated global reserve of phosphate rock is 50 billion metric tons, and annual mining production of phosphate for use primarily in agriculture and detergents is about 130 million metric tons (Fig. 8.2). These uses have contributed to the eutrophication of water systems that receive effluents from farm run-off and domestic wastewater, whereas phosphorus continues to be limiting in soils that require periodical fertilization to keep the level of food production from declining (Sundareshwar *et al.*, 2001).

Among the best-studied microbial processes in the phosphorus cycle is the function of acid and alkaline phosphatases, two enzymes that are distinguished only by the optimum pH of activity with acid phosphatase predominating among the fungi. Alkaline phosphatase is widely distributed among the prokaryotes, where it functions as a non-specific phosphomonoesterase through the formation of a covalent phosphoseryl intermediate. The enzyme also catalyzes the phosphoryl transfer reaction to various alcohols (Kim and Wyckoff, 1991). In *Escherichia coli*, alkaline phosphatase is a homodimer with 449 residues per monomer. It is a metalloenzyme with two Zn^{2+} ions and one Mg^{2+} ion at each active site (Fig. 8.3). Alkaline phosphatase is subject to feedback inhibition because the product of phosphatase activity, inorganic phosphate, is a strong competitive inhibitor at the active site (Le Du *et al.*, 2002).

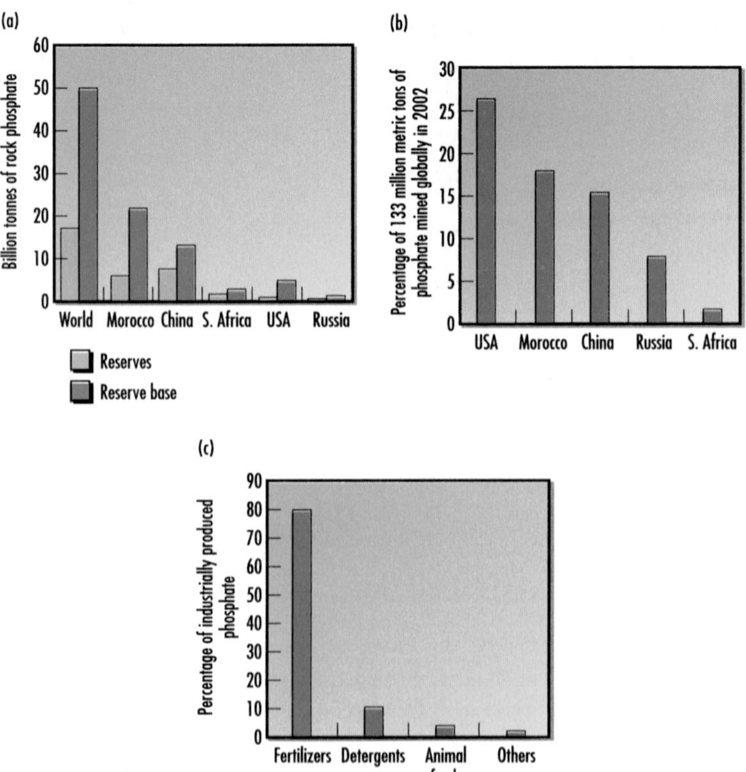

Fig. 8.2 (a) Guano deposits in Morocco, Western Sahara, and China account for more than 50% of global reserve and much more of the reserve base for rock phosphate, however, the United States dominates mining production (b) and consumption (c) of phosphate used in various sectors. If current mining trends continue and phosphorus cycling is hampered, known global phosphate reserves could be depleted within 100 to 400 years. There are no substitutes for phosphate in agriculture. Data from the United States Geological Survey (2003).

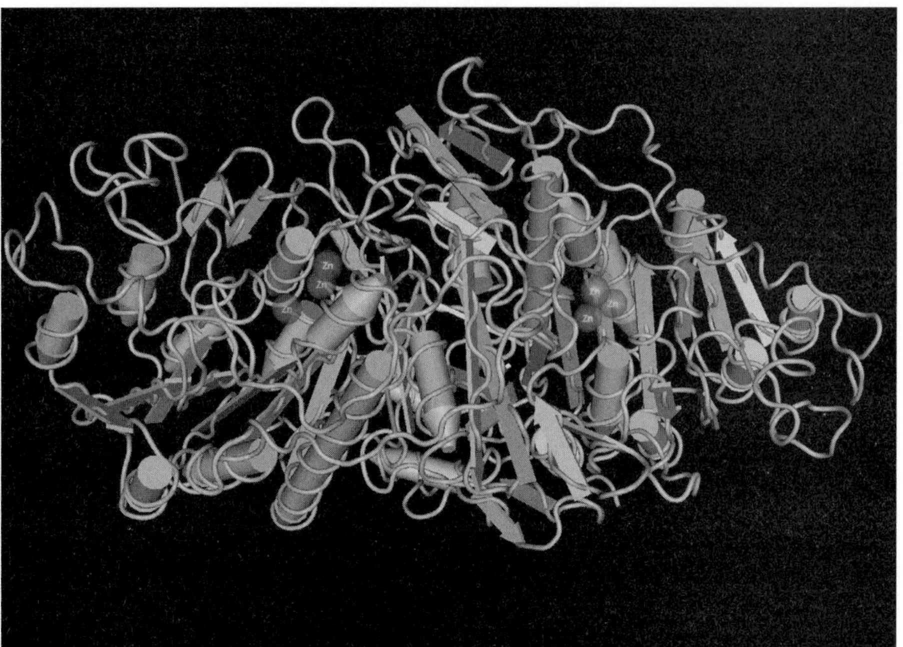

(a)

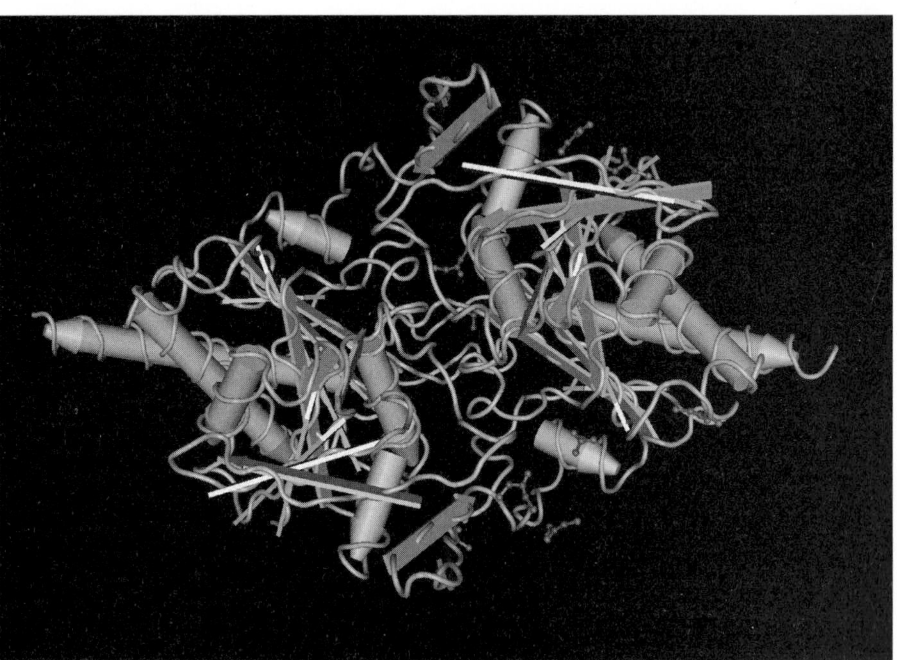

(b)

Fig. 8.3 Three-dimensional structure of metalloenzyme, alkaline phosphatase from *Escherichia coli* (a) compared with the Survival Protein E (SurE) of *E. coli*, also with phosphatase activity (b), but lacking a metallic co-factor. (c) Some SurE proteins from hyperthermophilic Archaea express the phosphatase activity at temperatures in excess of 90 degrees. The ecological functions of these enzymes have not been fully characterized, but their roles in phosphorus cycling are practically certain. Structures were generated using the Cn3D-4.1 software using data from the National Center for Biotechnology Information (http://www.nlm.ncbi.nig.gov/).

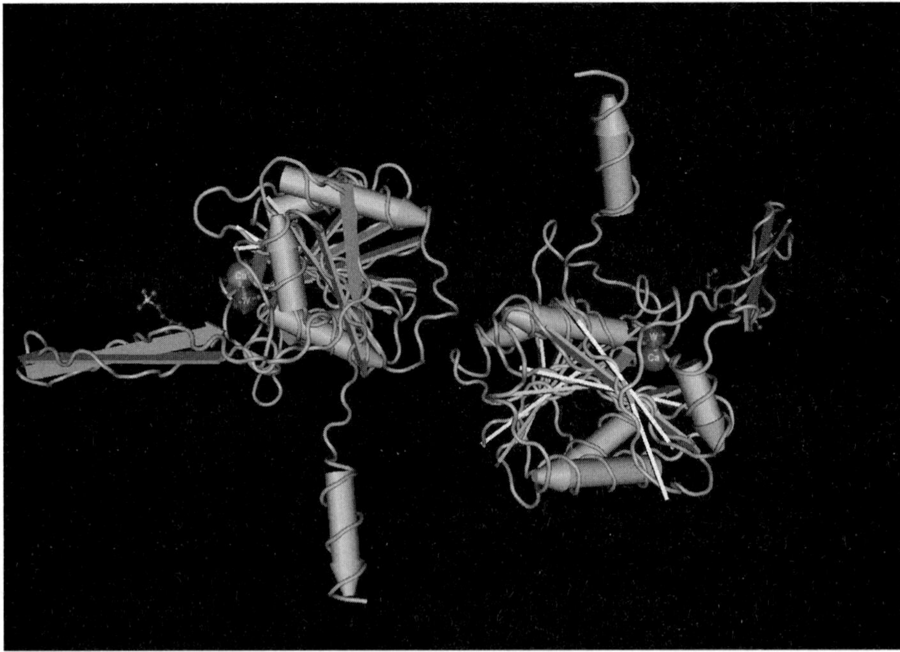

(c)

Fig. 8.3 *Continued*

A family of stress-response proteins known as "Survival Protein E" (*SurE*) is known to express acid phosphatase activity. *SurE* was originally discovered in the stationary phase survival of *E. coli*, and although the spectrum of substrates is still unclear, some lines of evidence indicate that guanosine-5′-monophosphate acts as a substrate (Mura *et al.*, 2003; Zhang *et al.*, 2001). These proteins are ancient and highly conserved, and they have been confirmed to be present in the archaeal organisms *Thermotoga maritima* and *Pyrobaculum aerophilum*, where they express phosphatase activity at temperatures up to 90°C and at the optimum pH range of 5.5–6.2 (Mura *et al.*, 2003). Although the ecological roles of these proteins are not fully understood, the characteristics that they display strongly suggest that they play active roles in recycling phosphorus in microbial communities.

In addition to phosphatase activity, many prokaryotes are known to express polyphosphate kinase, which catalyzes the production of phosphate polymers (polyphosphate granules). These polymers can be observed through the use of toluidine blue staining as inclusion bodies in several bacterial species (Onda and Takii, 2002). Polyphosphate kinase activity has been researched extensively in the context of wastewater purification because it is the key enzyme that reduces the phosphorus load of wastewater effluents to levels that do not lead to the eutrophication of receiving aquatic ecosystems. A version of polyphosphate kinase (PPK-1), which has been investigated extensively in *E. coli*, is a magnesium-requiring enzyme that catalyses the reversible synthesis of inorganic polyphosphate from the terminal phosphate of ATP (Zhu *et al.*, 2003).

A second variety denoted as PPK-2 was discovered in *Pseudomonas aeruginosa*; it differs from PPK-1 in that it uses either guanosine triphosphate or adenosine triphosphate to synthesize polyphosphate and it has a preference for manganese instead of magnesium. The putative molecular mass of PPK-2 is 40.8 kDa and homologous versions have been associated with 34 other prokaryotes, including members of the Archaea, species of *Rhizobium* and *Streptomyces*, and some cyanobacteria (Zhang *et al.*, 2002). It has been reported that there is a high degree of polyphosphate kinase activity in activated sludge adapted to the achievement of enhanced phosphorus removal. McMahon and colleagues (2002) used fluorescent *in situ* hybridization and 16S rRNA screens to detect four different versions of PPK, with two showing high degree of homology to the PPK gene of *Rhodocyclus tenuis*, a member of the β-*Proteobacteria* commonly found in wastewater environments.

Most studies concerned with microbial enzymes involved in phosphorus cycling aim at biotechnological applications in local environments. Therefore, it is currently not easy to extrapolate the results of these studies to conclusions about the reciprocal influences of microbial activity on global-scale changes in the phosphorus cycle. For example, Le Du and colleagues (2002) were able to produce a genetically engineered version of *E. coli* alkaline phosphatase with a 40-fold increase in the rate of enzyme activity. The difference between the hyperactive mutant and the wild type was in the metal-binding domain of the enzyme, and further work is needed to determine whether such mutations have been selected in natural ecosystems. The characterization of enzyme diversity and different rates of functioning in various ecological contexts should allow a better integration of information on microbial activities into global models of the phosphorus cycle. One area in which progress has been rapid in this respect is the cycling of phosphine, the only recognized atmosphere-mediated component of the phosphorus cycle (Glindemann *et al.*, 2003).

Phosphine cycling

Early speculations about the location of sinks for phosphorus in the environment focused on the idea that aerosolized phosphate, bound to dust particles, plant pollens, and ocean spray, is the only significant atmospheric component of the phosphorus cycle (Graham and Duce, 1979). However, Lewis Jr. and colleagues (1985) showed that there is a more substantial atmospheric link in the phosphorus cycle. They hypothesized that a terrestrial source and flux of a volatile phosphorus compound into the atmosphere could account for the observation of a flux at least 10 times larger than can be accounted for by phosphate carried on plant pollen and dust. Devai and colleagues (1988) subsequently detected levels of phosphine in the atmosphere that could not be explained by previous conceptions of the phosphorus cycle or by the industrial use of phosphine in food preservation. The phosphine hypothesis has largely been verified by the discovery of microbial contributions to the production of phosphine under anaerobic conditions (Glindemann *et al.*, 1996; Han *et al.*, 2000; Roels and Verstraete, 2001).

Phosphine is toxic to most living organisms, however, with a gas–water partitioning coefficient of approximately 5, it is not very soluble in water at ambient temperatures (15–20°C). Phosphine has been detected in the nanogram per m^3 range in remote air samples worldwide, particularly in the lower troposphere at night when its removal by oxidation with atomic oxygen (rate constant of 5×10^{-11} cm^3/s) is halted. The average half-life of phosphine in the lower troposphere is 28 hours, but can be as low as 5 hours in the presence of high concentrations of hydroxyl radicals produced under conditions of intense sunlight (Glindemann *et al.*, 2003). Oxidized phosphine leads to the production of phosphate ion, which is returned to the Earth's surface for continuation in the phosphorus cycle. Several hypotheses have been proposed to explain the microbiological origins of phosphine. Presumably, phosphate can serve as a terminal electron acceptor in the absence of sulfate, nitrate, and oxygen. Under such circumstances, the final product of phosphate reduction is phosphine (Atlas and Bartha, 1998; Glindemann *et al.*, 1999). However, phosphine is likely to ignite in the presence of oxygen or methane, and as such, it is currently not clear how phosphine microbiologically produced under anaerobic conditions in soils and sediments can accumulate in the atmosphere (Eismann *et al*, 1997; Glindemann *et al.*, 1996).

Claims about the capability of microorganisms to be able to reduce phosphate to phosphine were controversial for most of the twentieth century, and theorists claiming thermodynamic inconsistencies refuted early claims by experimentalists who presumably observed this process (Burford and Bremner, 1972). However, Jenkins and colleagues (2000) gathered convincing data on the microbial basis for reductive generation of phosphine, which could account for the presence of the toxic gas in natural anaerobic environments. They detected phosphine in the headspace gases of mixed cultures under conditions promoting fermentative growth of bacteria on mixed acid and butyric acid. Phosphine generation was not necessarily linked to methanogenesis. Among the pure cultures shown to produce phosphine are the mixed acid fermenters *Escherichia coli*, *Salmonella gallinarum*, and *Salmonella arizonae*, and the solvent fermenters *Clostridium sporogenes*, *Clostridium acetobutyricum* and *Clostridium cochliarium*. It is likely that the coexistence of these Bacteria species with *Methanoarchaea* is responsible for the apparent correlation between methanogenesis and phosphine emissions in nature. Finally, phosphine accumulation in the upper troposphere is of particular interest to modelers of global climate change because phosphine could be a vector for phosphorus oxy-acids, which may serve as condensation nuclei for cloud formation in the high troposphere and stratosphere (Glindemann *et al.*, 2003). Clearly, more research is needed to link the generation of phosphine by diverse organisms in the biosphere to these sorts of global environmental impacts.

THE SULFUR CYCLE

Sulfur is an essential element for all biological systems primarily because of its role as part of the sulfhydryl group in the amino acids cysteine and methionine, and in coenzymes. The strong affinity of sulfur for metal ions also makes this element very important for maintaining metal homeostasis regarding toxicity issues and for supporting the activities of several metalloenzymes. In addition to these intracellular roles, sulfur exists over a wide range of oxidation states in the environment, from the reduced sulfides and mercaptans with the oxidation state of −2 to the oxidized sulfites and sulfates with oxidation states of +4 and +6, respectively. This versatility contributes to the role of sulfur in influencing the ambient pH of microbial growth environments, a parameter that is capable of exerting a strong selective pressure on microbial diversity. Sulfur is relatively abundant in the Earth's crust, commonly found in both soluble and precipitated forms. Consequently, the biological availability of sulfur does not generally limit biological growth, even though sulfur accounts for about 1% of the average bacterial cell (Jorgensen, 1983).

The sulfur contained in various reservoirs is either actively or slowly recycled, except for the sulfur contained in the Earth's crust, which is also the largest reservoir, containing inert sulfur deposits and metallic sulfide ores (Fig. 8.4). Oceanic sulfate is only slowly recycled, however, terrestrial and atmospheric sulfur are actively recycled with substantial anthropogenic influence from the mining of metal sulfide ores and the combustion of fossil fuels leading to the concomitant emissions of sulfur dioxide. Microorganisms participate in the cycling of sulfur through oxidation and reduction reactions, and through the enzymatic mineralization of organic sulfur compounds. These reactions are generally characterized as desulfuration, oxidative sulfur transformations, and assimilative sulfate reduction (Fig. 8.5).

Desulfuration

The decomposition or mineralization of organic sulfur, also known as desulfuration, is one of the most consequential aspects of the sulfur cycle with respect to global environmental change. In soils and sediments, microorganisms produce specific enzymes to catalyze the degradation of organic sulfur compounds such as cysteine to hydrogen sulfide and mercaptans. For example, the degradation of cysteine is catalyzed by cysteine desulfhydrase:

$$SH\text{-}CH_2\text{-}CH(NH_2)\text{-}COOH + H_2O \rightarrow OH\text{-}CH_2\text{-}CH(NH_2)\text{-}COOH + H_2S$$

This is a commonly found process in microbial communities and it occurs under both aerobic and anaerobic conditions (Simo, 2001). The decomposition of organic sulfur com-

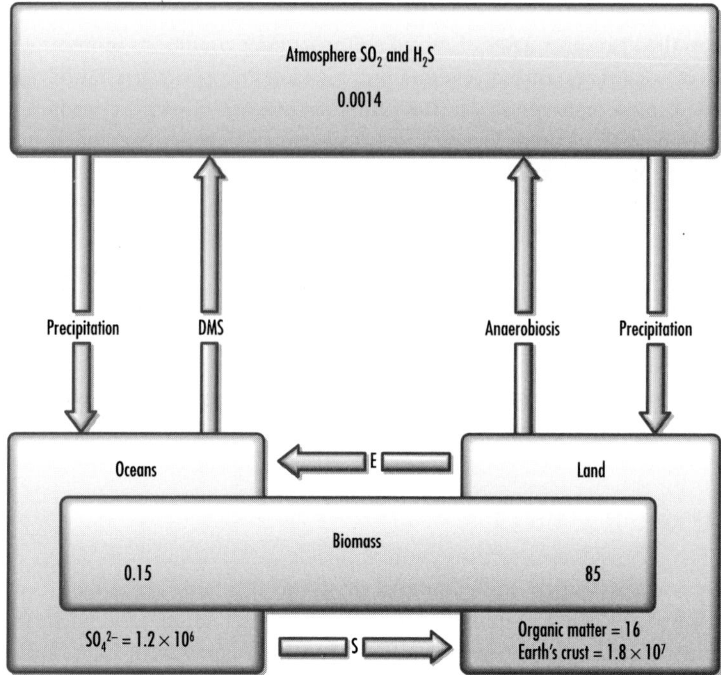

Fig. 8.4 The global sulfur cycle. The largest reservoir of sulfur is the Earth's crust, but it is a stable stock, whereas biomass, dissolved, volatile, and organic matter sulfur are actively cycled. Emissions of dimethylsulfide (DMS) from the ocean have global climate implications. The anaerobic generation of hydrogen sulfide is dominated by microbial activity. Erosion (E) moves considerable amounts of sulfur from land to the ocean, and sedimentation (S) processes including interactions of sulfur and metals return sulfur to the terrestrial domain. Human industrial activity and volcanoes modify the flux of sulfur between the major reservoirs and have the largest impact on the atmosphere. Unit for the values of sulfur in reservoirs is petagrams.

pounds in the ocean leads to the production of dimethylsulfide (DMS), a major natural source of sulfur to the atmosphere:

$$(CH_3)_2S^+CH_2CH_2COO^- \rightarrow CH_3SCH_3 + CH_2 = CHCOO^- + H^+$$

Dimethylsulfoniopropionic acid (DMSP), the substrate for DMS production, is produced by marine algae and functions both as a regulator of internal osmotic environment and as an antioxidant (Yoch, 2002; Sunda *et al.*, 2002). The reduced sulfur moiety of DMSP makes it very prone to microbial degradation, and it has been postulated to represent a major carrier of sulfur transfer through microbial food webs and organic sulfur cycling in the pelagic ocean (Simo, 2001). Dimethylsulfide contributes to the tropospheric sulfur load, and atmospheric sulfate aerosol particles that result from DMS emissions directly influence the global radiation balance through the upward scatter of solar radiation and indirectly through serving as cloud condensation nuclei (Fig. 8.6). The mean concentration of DMS in seawater is 2–4 nanomoles per liter, and emissions of this compound from the ocean account for approximately 15% of the total global sulfur emissions of 3.2 Tg/yr (Bates *et al.*, 1992). In what is generally known as the "CLAW hypotheses", Charlson and colleagues (1987) originally proposed a two-pronged feedback system connecting DMS and global climate. There is a considerable body of observational and experimental evidence to support the original hypothesis linking DMS emissions to enhanced cloud albedo, but there is no comparable level of evidence to support the idea that DMS synthesis is induced in phytoplankton communities in response to climate change (Bates and Quinn, 1997; Charlson, 1993).

Marine prokaryotes are thought to exercise control over DMS production efficiency by switching between two separate pathways for degrading DMSP, only one of which leads to

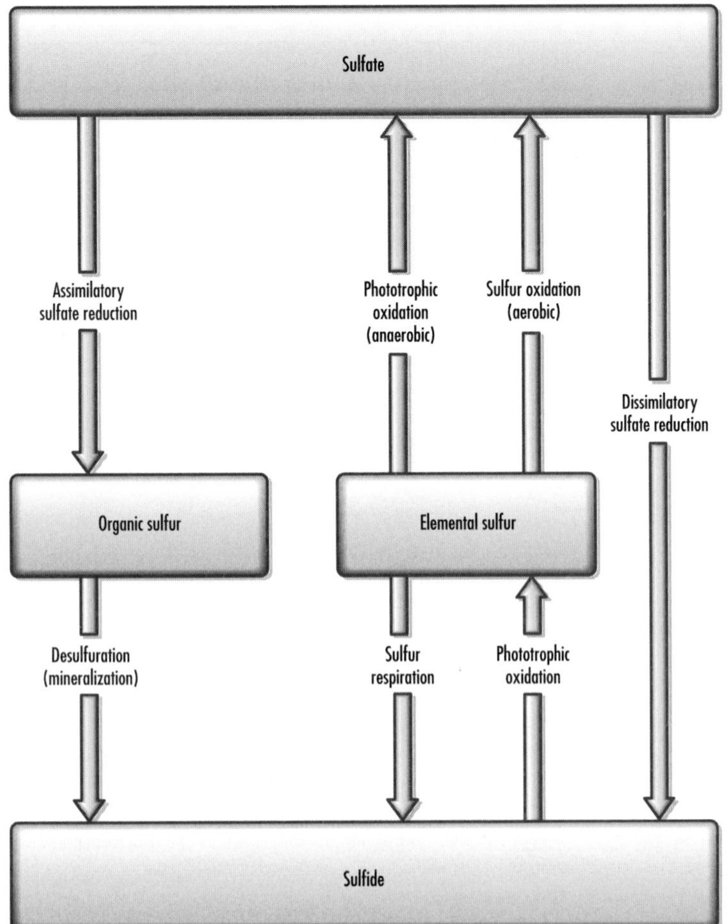

Fig. 8.5 A diverse set of microbial transformations in the global sulfur cycle are responsible for the mobilization of sulfur from the dissolved phase (sulfate) to the volatile phase (sulfide) and the solid phases including biomass (organic sulfur) and precipitates (elemental sulfur and metal sulfides).

the production of DMS (Simo, 2001). Exploration of this switching strategy has led to the questioning of DMSP's role as a universal substrate for many heterotrophic organisms whose population densities correlate with the turnover of dissolved DMSP. Alternatively, DMSP decomposition could be associated with a narrow range of bacteria such as the ubiquitous *Roseobacter* species belonging to the α-*Proteobacteria* group (González *et al.*, 1999 and 2000). Apparently, DMSP degradation is not limited to marine bacteria, however, there has been no thorough investigation of the ecological role of these inland freshwater organisms in environments that are not known to generate the substrate (Yoch, 2002). Horinouchi and colleagues (1999) demonstrated that certain aromatic hydrocarbon mono- and di-oxygenases commonly found in soil pseudomonads, including *Pseudomonas putida* and *P. fluorescens* can catalyze the oxidation of DMS under aerobic conditions. In addition, Ginzburg and colleagues (1998) demonstrated in the Sea of Galilee, that *Peridinium gatunense*, a freshwater dinoflagellate that dominates the phytoplankton population in the lake during the winter–spring season, stores up to 5.5 pg of DMSP per cell during stationary and declining growth phases. The DMSP undergoes bacterial and chemical degradation to release up to 0.1 mmol of DMS/m^2 per month in the Sea of Galilee late in the *Peridinium* bloom season. Lomans and colleagues (2002) proposed that in contrast to marine microorganisms, freshwater microorganisms may produce volatile organic sulfur compounds through the methylation of sulfide, which has been generated in anaerobic sediments, rather than through the

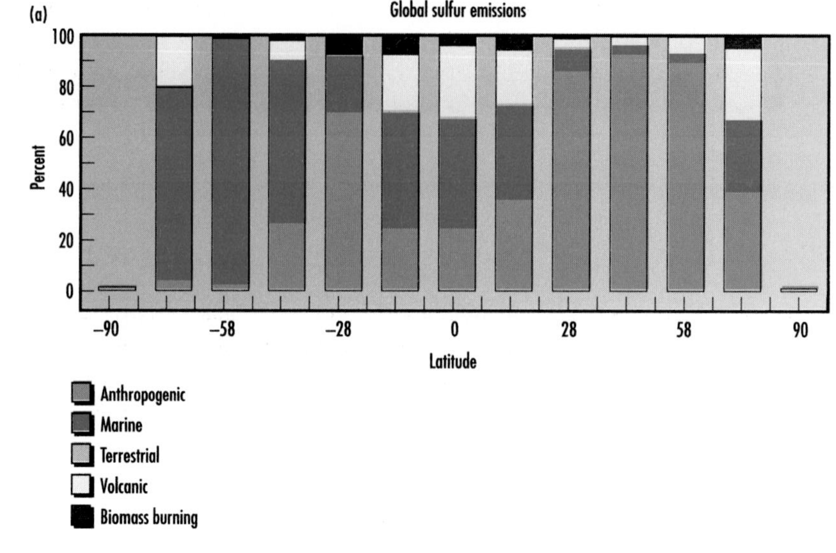

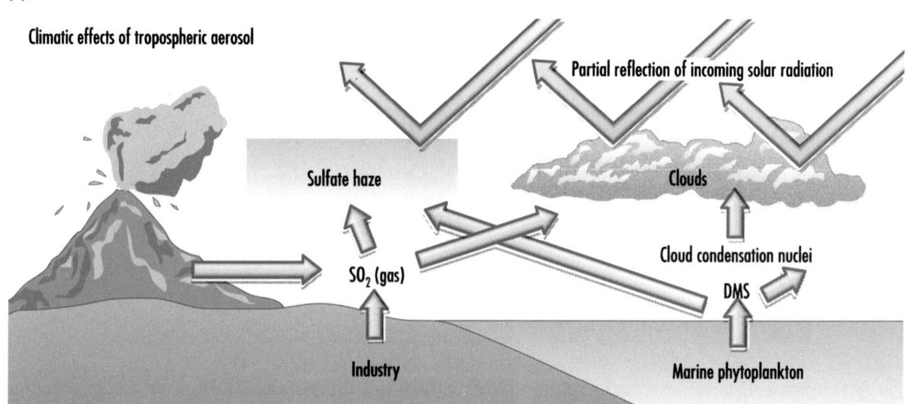

Fig. 8.6 (a) The sources of sulfur emissions in the global cycle show large variations in the proportion attributable to marine environments according to latitude. The total global sulfur emission has been estimated at 3.2 Tg/year (Bates *et al.*, 1992). (b) The principal impact from marine sulfur emissions is due to dimethylsulfide, which contributes to sulfate haze and cloud nucleation. By courtesy of Pacific Marine Environmental Laboratory, NOAA, Seattle, WA (http://www.saga.pmel.noaa.gov/review/dms_climate.html).

decomposition of organic sulfur compounds. Finally, Jonkers and colleagues (1996) showed that dimethylsulfoxide (DMSO) reduction to support growth with the release of DMS occurred in sulfate-reducing bacteria, including *Desulfovibrio desulfuricans* isolated from saline environments. These investigators were unable to demonstrate DMS production from DMSO by freshwater bacteria, but they speculated that sulfate-reducing bacteria could add substantially to the global emission of DMS.

Several attempts have been made to integrate information on microbial diversity, nutrient loading, and incident electromagnetic radiation to predict large-scale fluxes of DMS in marine environments (Belviso, 2000; Roelofs *et al.*, 2001). However, such models are necessarily incomplete without information on the thermodynamic aspects of DMS production and volatilization, and on the physiological diversity of microorganisms engaged with this aspect of the sulfur cycle (see Scholten *et al.*, 2003). Also, research needs to be intensified on the ultimate fate of DMS and the biogeochemical effects of stable intermediates in the reaction pathways for DMS recycling. One of the most notable compounds in this category is methanesulfonic acid (MSA), a very stable strong acid that is formed in large quantities from the chemical oxidation of DMS and deposited on the Earth's surface in wet and dry deposition. MSA can be degraded aerobically by a mono-oxygenase as a source of sulfur to support the growth of methylotrophic bacteria, and it is used by other microorganisms belonging to a wide range of phylogenetic groups, but so far, no anaerobic organisms have been shown to do the same (Kelly and Murrell, 1999).

Sulfur oxidation

The oxidation of reduced sulfur compounds is an important physiological process that supports the metabolism of chemolithotrophic microorganisms. These organisms are commonly found in extreme environments characterized, for example, by high acidity. Hydrogen sulfide is the most reduced form of sulfur and its oxidation proceeds in two steps:

$$H_2S + 0.5O_2 \rightarrow S^0 + H_2O\ (\Delta G'_o = -50.1\,\text{kcal mol}^{-1} = -210.4\,\text{kJ mol}^{-1})$$
$$S^0 + 1.5O_2 + H_2O \rightarrow H_2SO_4\ (\Delta G'_o = -149.8\,\text{kcal mol}^{-1} = -629.2\,\text{kJ mol}^{-1})$$

These reactions have been studied extensively in filamentous microaerophilic bacteria such as *Thermothrix*, *Beggiatoa*, *Thioploca*, *Thiothrix*, and *Thiobacillus* species. The organisms generally exist in the interface between oxic and anoxic environments because they require access to both oxygen and reduced sulfur compounds made by anaerobes. Some of the sulfur-oxidizing organisms deposit sulfur globules intracellularly and eventually oxidize these globules for energy generation and acid production (e.g. *Beggiatoa*). Less acid-tolerant species (e.g. *Thiobacillus thioparus*) deposit elemental sulfur, which is not further oxidized (Atlas and Bartha, 1998). In addition to the activities of microaerophilic chemolithotrophic bacteria, sulfur oxidation can also occur under anaerobic conditions by phototrophic bacteria, including species of *Chlorobium*, *Chromatium*, *Ectothiorhodospiria*, *Thiopedia*, and *Rhodopseudomonas*. Some aerobic heterotrophic bacteria and certain fungi are also known to oxidize sulfur, although they do not generate energy from this process and the ecological significance is not yet clear (Maier, 2000). *Thiobacillus denitrificans* is a notable exception to the typically obligately aerobic chemolithotrophic organisms that oxidize sulfur because it is a facultative anaerobe. *T. denitrificans* can use nitrate as a terminal electron acceptor instead of oxygen. The process is sometimes linked to the formation of metallic salts such as gypsum ($CaSO_4$):

$$5S^0 + 6NO_3^- + 2H_2O \rightarrow 5SO_4^{2-} + 3N_2 + 4H^+$$
$$5S^0 + 6NO_3^- + 5CaCO_3 + 6H^+ \rightarrow 5CaSO_4 + 3N_2 + 5CO_2 + 3H_2O$$

The optimum growth of *T. denitrificans* occurs around pH 7, therefore sulfur oxidation reactions that generate acidic conditions are efficient. Among the Archaea, *Sulfolobus* species can oxidize sulfur under hyperthermic, low pH conditions. The generation of acid from sulfur oxidation reactions leads to the leaching of metallic sulfide ores or to the solubilization of phosphorus deposits, thereby influencing the biological availability of metal ions and phosphorus, respectively (Green *et al.*, 2003; Adler and Sibrell, 2003). Excessive leaching of metal ores is a major environmental hazard associated with mining conditions, but under carefully controlled conditions, biomineralogists have explored the use of sulfur-oxidizing microorganisms for relatively environmentally benign mineral recovery (Curutchet *et al.*, 2001; Janssen *et al.*, 2001).

Phototrophic oxidation of sulfur is carried out by photosynthetic sulfur bacteria in anaerobic environments where the oxidation of hydrogen sulfide is linked to the fixation of carbon and the release of elemental sulfur:

$$CO_2 + H_2S \rightarrow (CH_2O) + S$$

This reaction contributes substantially to the cycling of carbon in hydrothermal vent systems deprived of the biochemically important driving forces provided by electromagnetic radiation (Kashefi *et al.*, 2003; Takai *et al.*, 2003).

Sulfur reduction

The reduction of elemental sulfur to sulfide (sulfur respiration) and dissimilatory sulfate reduction provide crucial pathways for the recycling of reduced sulfur compounds in the environment (Wagner *et al.*, 1998). Sulfur respiration is characteristic of anaerobic organisms growing on fermentation products. For example, *Desulfuromonas acetoxidans* uses sulfur reduction to grow, with acetic acid as the carbon source, although the reaction does not yield a high amount of energy (Correia *et al.*, 2002; Purdy *et al.*, 2003):

$$CH_3COOH + 2H_2O + 4S \rightarrow 2CO_2 + 4H_2S \; (\Delta G'_o = -5.7\,\text{kcal}\,\text{mol}^{-1} = -23.9\,\text{kJ}\,\text{mol}^{-1})$$

Inorganic materials can also be used to drive sulfur respiration. For example, in hydrothermal vents, certain hyperthermophilic Archaea, including *Pyrodictium occultum* and *P. abyssi* can synthesize hydrogen sulfide from hydrogen gas and sulfur using a hydrogen–sulfur oxidoreductase (Keller and Dirmeier, 2001).

Both organic carbohydrate compounds and molecular hydrogen can also be used to support sulfate reduction in a dissimilatory process:

$$4CH_3OH + 3SO_4^{2-} \rightarrow 4CO_2 + 3S^{2-} + 8H_2O$$
$$4H_2 + SO_4^{2-} \rightarrow S^{2-} + 4H_2O$$

In contrast to sulfur respiration, dissimilatory sulfate reduction is more common in the environment, and is likely to contribute more to the global sulfur cycle, especially in anaerobic sediments and waterlogged soils. Sulfate-reducing bacteria, including members of the genera *Desulfococcus*, *Desulfovibrio*, and *Desulfobacter* are strict anaerobes that have been implicated in economically important corrosion processes (Kjellerup *et al.*, 2003). Several microbial enzyme systems specialize in sulfur reduction, but one that is unique to this pathway is dissimilatory sulfite reductase (DSR). Wagner and colleagues (1998) used comparative assessment of DNA sequences specific to DSR to demonstrate that the inferred evolutionary relationship among organisms capable of sulfur respiration was nearly identical to relationships inferred on the basis of 16S rRNA comparisons. There is a high similarity between bacterial and archaeal DSRs, suggesting that the ancestral DSR was either present before the split between the domains Bacteria, Archaea, and Eukarya or laterally transferred between the prokaryotic domains soon after divergence.

Prospects and challenges of the sulfur cycle

Modeling and interpreting the historical global sulfur cycle provide a potentially powerful approach toward understanding the impacts of anthropogenic changes in the flux of sulfur at the global scale. However, modeling is limited by lack of a complete understanding of the organization of the physiological diversity of microorganisms interacting with the sulfur cycle. Some aspects of the historical record of the sulfur cycle are preserved in the isotopic composition of sedimentary deposits (Strauss, 1998). There are nine known isotopes of sulfur, but only four are stable, with ^{32}S constituting 95% of sulfur abundance in the Earth's crust. The other isotopes and their relative abundances are ^{33}S (0.76%), ^{34}S (4.22%), and ^{36}S (0.014%) (Charlson *et al.*, 1991). ^{35}S is most frequently used in radioisotope tracking

BOX 8.1

(a)

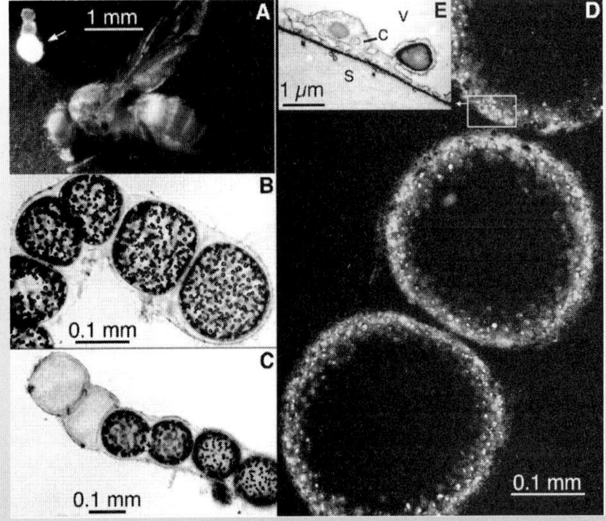

(b)

(a) In 2001, the SeaWiFS satellite observed highly visible microcrystalline particles of sulfur on the Namibian coast of Africa. These sulfur blooms occur when hydrogen sulfide is released from sediments containing large deposits of frustules or diatom shells. Microorganisms in the sediment layer consume dead phytoplankton, a process that depletes oxygen and leaves the environment essentially anaerobic. The anaerobic microorganisms use sulfate to oxidize organic matter with the concomitant release of hydrogen sulfide. Over long periods of time, large quantities of hydrogen sulfide bubble to the surface. The atmospheric oxidation of hydrogen sulfide at the surface produces elemental sulfur. The image clearly captures the two forms of sulfur, with the sulfide form appearing whitish (not to be confused with cloud cover) and the yellow sulfur crystals appearing green when seen through the bluish ocean water layer. This aspect of the sulfur cycle is particularly intriguing for public health because of the toxicity of hydrogen sulfide to marine life forms, and the odor problem that it creates for the coastal communities. For microbiologists, the situation presents a unique opportunity to investigate the microbial causes of this large-scale phenomenon. Apparently, the metabolic activity of the largest known prokaryote, *Thiomargarita namibiensis* (b) isolated from the sediment off the coast of Namibia is responsible in part for the sulfur bloom (Weeks *et al.*, 2002). Each cell of this organism is approximately 0.75 mm in diameter. Satellite image by courtesy of NASA Earth Science Enterprise (ESE), GSFC Earth Sciences (GES) Data, and Information Services Center (DISC)/Distributed Active Archive Center (DAAC). The photocollage of *T. namibiensis* is by courtesy of Heide Schulz and Manfred Schlosser at the Max Planck Institute for Marine Microbiology, Bremen, Germany.

experiments and has a half-life of 87.4 days. Marine evaporitic sulfate deposits and sedimentary pyrite are formed primarily by bacterial sulfate reduction. The isotopic compositions of these biominerals have been utilized to a limited extent to model the global sulfur cycle on Earth. However, the sulfur isotope records require further research to fill the gaps before the records can be useful for predictive modeling beyond local depositional and diagenetic effects (Geesey *et al.*, 2002).

One of the emerging and fascinating areas of research interest about the sulfur cycle is the upwelling of particulate sulfur blooms from the marine environment (see Box 8.1). Sulfur occupies a major position in microbial metabolism and its transformation is linked to redox reactions that influence the cycling of other elements through diverse metabolic networks. Understanding these metabolic networks is an essential component of modeling microbial consortia with the goal of quantifying the role of microorganisms in global elemental cycles (Vallino, 2003).

METALS AND TRACE ELEMENT CYCLES

Most of the enzymes used by microorganisms to catalyze the key reactions of the major biogeochemical cycles are metalloenzymes, meaning that they require metal ions positioned at or near the active site to function efficiently. Table 8.1 presents a few examples of microbial metalloenzymes. It has been estimated that at least a third of all proteins in the biosphere are metalloproteins, and several metals are commonly found associated with physiological systems in small concentrations (e.g. Ca, K, Na, Fe, Mo, Mn, Mg, Cu, Zn, Co, Ni, W, Al). For each of these "essential" metals, there are at least two other metals that can compete for their interaction with proteins. These competing "non-essential" metals are likely to be very toxic because the proteins to which they bind do not function properly. Consequently, organisms have evolved several different strategies for keeping toxic metals away from cell interiors and strategies to detoxify metals that cannot be kept out. In addition, the internal concentrations of essential metals are carefully regulated to avoid concentration-dependent toxicity. And here again, living systems have evolved strategies for maintaining metal homeostasis in situations where metals are required only in trace concentrations. Given that 99% of the total elemental mass on Earth is made up of eight elements (six of which are Al, Fe, Ca, Na, K, and Mg (the other two are O and Si)), there is considerable geochemical cycling of metals, and in many cases, the biosphere dramatically influences the size of the reservoirs and the direction of flow among the reservoirs (Morel and Price, 2003) (Fig. 8.7).

The biogeochemical cycling of trace metals is defined by the competitive solubilization of metallic minerals and facilitated uptake of metal ions through microbial release of acids and/or complexing factors that sequester metal ions for rapid uptake by cells (Table 8.2).

Table 8.1 Microorganisms engage with the biogeochemical cycling of metals partly because several crucial enzymes require trace quantities of metals to function properly. The biological availability of essential metal ions influences enzyme functions. Many enzymes are also inhibited by potentially toxic non-essential metals that can replace metals at the enzyme active site.

Metal	Enzyme examples	Relevant organisms and/or ecological function
Calcium	Collagenase; calpain protease	*Clostridium histolyticum* Pathogenesis
Cobalt	Halomethane methyltransferase	Facultative methylotrophs Chloromethane degradation
Copper	Copper amine oxidases	*Arthrobacter globoformis* Deamination of primary amines to corresponding aldehydes
Iron	Cytochrome P450; soluble methane monooxygenase	*Pseudomonas putida*. P450cam Polyaromatic hydrocarbon biodegradation
Magnesium	Magnesium chelatase	*Rhodobacter sphaeroides* Chlorophyll synthesis
Manganese	Manganese-dependent peroxidases	*Phanerochaete chrysosporium* Carbon cycling–lignin degradation
Molybdenum	Xanthine dehydrogenase; nitrate reductase; nitrogenase	*Pseudomonas putida* *Rhizobium* Purine (caffeine) degradation Nitrogen fixation
Selenium	Selenocysteine (UGA codon) enzymes; glutathione peroxidase; hydrogenases	*Escherichia coli* *Methanococcus voltae* Nitrogen respiration
Tungsten	"True W-enzymes"; aldehyde ferredoxin oxidoreductase	*Thermophilic* Archaea; *Pyrococcus furiosus* Molybdenum antagonist
Zinc	Aminolevulinate dehydratase	Several Bacteria and nearly all Archaea Porphyrin synthesis

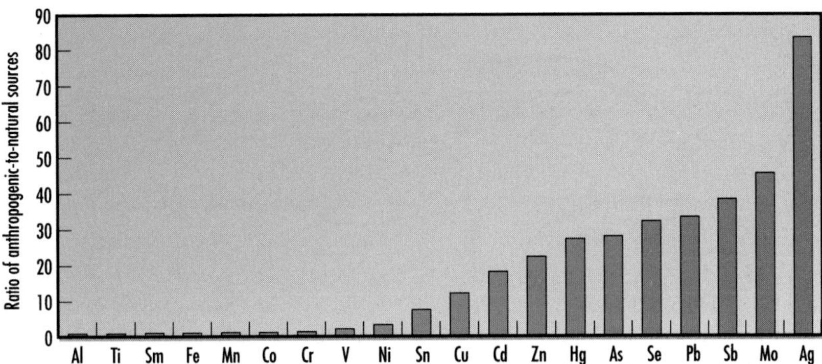

Fig. 8.7 The ratios of anthropogenic-to-natural sources of atmospheric metals reflect the extent of human perturbation of metal biogeochemistry. Organisms typically encounter physiologically detrimental concentrations of toxic metals, which are disproportionately represented in ecosystems affected by industrial activities. These metals can exert strong selective pressures on the molecular diversity of proteins that require specific metallic cofactors. The data are from Benjamin and Honeyman (1992).

Table 8.2 Production of metal-complexing factors by microorganisms affects the bioavailability of metals and their geochemical cycling (Morel and Price, 2003).

Metal-complexing factor	Targeted metals	Organisms
Siderophores	Fe	Heterotrophic bacteria; cyanobacteria
Phytochelatin	Cd	Diatoms
CuY and CuZ peptides	Cu	Coccolithopods and *Synechococcus*
Co-complexing agent	Co	*Prochlorococcus*
Pyoverdin	Fe	*Pseudomonas aeruginosa*

Iron is the fourth most abundant element in the Earth's crust. Consequently, it is not strictly considered a trace element, but its biological availability can be severely limited by oxidation–reduction reactions that mobilize iron from the ferrous (Fe^{2+}) to the ferric (Fe^{3+}) states with very different solubility properties. Therefore, microorganisms produce several different kinds of iron-acquisition molecules. For example, the siderophores of fluorescent pseudomonads are potent Fe^{3+} chelators (compounds that bind metals tightly) with complexing constants between 10^{24} and 10^{26} ($l \cdot mol^{-1}$) at pH 7.0 (Visca *et al.*, 1992; Weber *et al.*, 2000). Pyoverdins are among the best-studied siderophores, of which there are four major categories depending on the structure of the associated peptide chain. Pyoverdins containing N-hydoxy*(cyclo)*Orn (cOHOrn) at the C-terminus are the most common. Other varieties are characterized by an amide bond between the carboxyl group of the C-terminal amino acid and the ε-amino group of an in-chain lysine; a C-terminal *(cyclo)*depsipeptidic substructure formed by an ester bond between the carboxyl group of the C-terminal amino acid and an in-chain serine or threonine; or a C-terminal free carboxyl group (Fig. 8.8).

When iron is scarce, pyoverdins and iron-regulated outer membrane proteins (IROMPs), which serve as membrane receptors for the ferri-pyoverdins, are synthesized. Bound iron is released by reduction in the periplasmic space. In general, the variability of the pyoverdin peptide chain facilitates specificity of uptake such that a given ferri-pyoverdin can only be utilized by the producing strain (Weber *et al.*, 2000). The specificity of pyoverdins has been used for preliminary identification of members of the fluorescent *Pseudomonas* group (Meyer, 2000; Visca *et al.*, 1992).

In addition to the capacity to mobilize iron through the use of complexing agents, microbial interactions with iron are also linked directly to the cycling of other elements, includ-

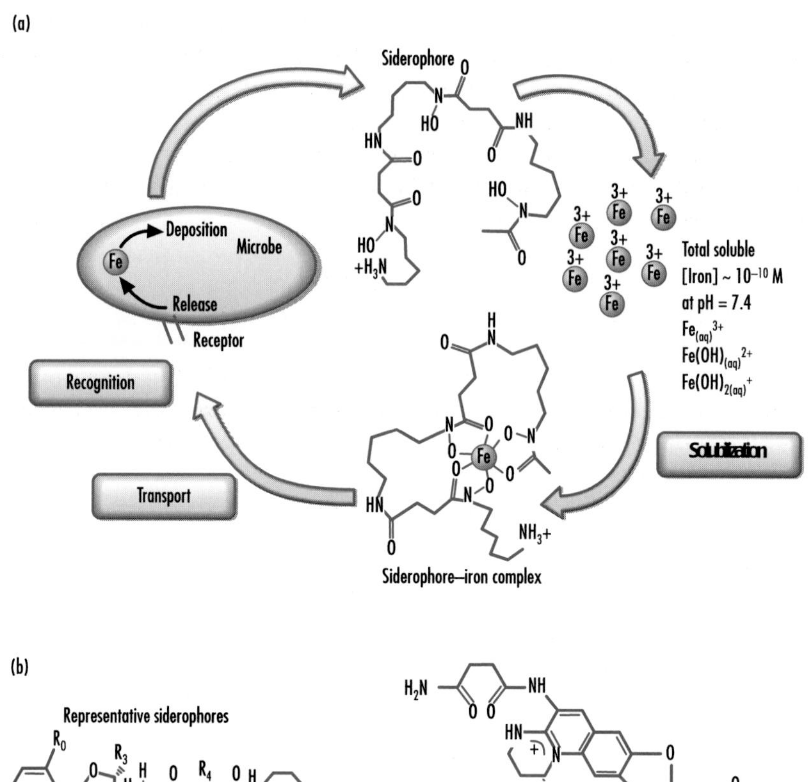

(b)

Representative siderophores

Mycobactins
(shown without iron)

Pseudobactin

Ferrichrome

Agrobactin, X=OH
Parabactin, X=H
(shown without iron)

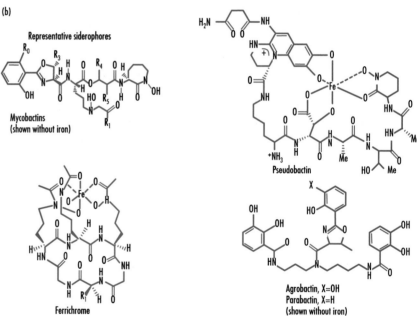

Fig. 8.8 (a) Mode of siderophore action to sequester iron. (b) The structure of microbial siderophores varies substantially, but their functions are very similar (Boukhalfa and Crumbliss, 2002). By courtesy of Al Crumbliss, Duke University and Marvin Miller, University of Notre Dame.

ing carbon and sulfur. In aerated environments at neutral pH, ferrous ion is not stable, and the predominant form of iron is ferric, usually precipitated as ferric phosphate. In aquatic systems, precipitated ferric compounds are deposited in relatively anoxic zones, where iron reduction can be used to drive anaerobic respiration with the release of more soluble ferrous compounds, which may then be re-oxidized by chemolithotrophic "iron bacteria" (Nealson and Myers, 1992; Nealson and Saffarini, 1994):

$$Fe^{3+} + (CH_2O) \rightarrow Fe^{2+} + CO_2$$

Shewanella putrefaciens, *Geobacter sulfurreducens*, and *Desulfuromonas acetoxidans* are among the best-studied microbial iron reducers (Atlas and Bartha, 1998). Sulfate-reducing bacteria in anaerobic environments are important participants in the reduction of ferric to ferrous ions in anaerobic sediments because the production of hydrogen sulfide can reduce iron oxyhydroxides to iron sulfides. The chemolithotrophic oxidation of ferrous ion is carried out by a well-known group of acidophilic bacteria that are associated with acid mine drainage, including *Thiobacillus ferroxidans*, *Leptospirillum ferroxidans*, and *Sulfolobus acidocaldarus*:

$$2Fe^{2+} + 0.5O_2 + 2H^+ \rightarrow 2Fe^{3+} + H_2O\,(\Delta G'_o = -6.5\,\text{kcal}\,\text{mol}^{-1} = -27.3\,\text{kJ}\,\text{mol}^{-1})$$

Ferrous ion oxidation under neutral conditions has been described for several bacterial genera within the "iron bacteria" group, including *Gallionella*, *Metallogenium*, *Siderococcus*, *Sphaerotilus*, *Leptothrix*, and *Pedomicrobium*, although the ecological significance of these processes is not clearly understood. Similarly, some heterotrophic bacteria, including *Alcaligenes*, *Bacillus*, and *Pseudomonas*, are reputed to conduct iron reduction but the reactions in these organisms are likely linked to the fortuitous activity of nitrate reductase (Lovley, 1993).

The biogeochemical cycling of iron has been used to exemplify the complexity of microbial interactions with trace metals both in terrestrial environments and in the deep ocean (Edwards *et al.*, 2003). It is increasingly clear that such complex interactions can have considerable implications for global environmental change. For example, the depletion of iron in oligotrophic marine environments has been used to explore hypotheses about "iron fertilization" of the ocean in order to increase carbon sequestration from the atmosphere. Although the predicted impact of such iron loading on global climate change is controversial, the large-scale experiments have demonstrated unequivocally that marine microorganisms dramatically influence—and respond to—the biogeochemical cycling of iron (Berger and Wefer, 1991; Chisholm, 2000). It is possible that similar effects can be demonstrated to a smaller extent with other metals (Coale, 1991; Morel *et al.*, 1991). For example, many of the same organisms that are involved in the transformation of iron also participate in reactions with manganese (Nealson and Myers, 1992). Like iron, manganese is microbiologically cycled between reduced manganous (Mn^{2+}) and oxidized manganic (Mn^{4+}) ions:

$$Mn^{2+} + 0.5O_2 + H_2O \rightarrow MnO_2 + 2H^+\,(\Delta G'_o = -7.0\,\text{kcal}\,\text{mol}^{-1} = -29.4\,\text{kJ}\,\text{mol}^{-1})$$

Manganic reduction has also been described in a few microbial species, including *Shewanella putrefaciens*:

$$Mn^{4+} + (CH_2O) \rightarrow Mn^{2+} + CO_2$$

The microbial reactions involving manganese have been associated with the formation of potentially commercial deposits of manganese and ferromanganese nodules in deep-sea environments (Nealson and Saffarini, 1994).

The literature is accumulating rapidly on less well-understood, but admittedly important, microbial involvement in the biogeochemical cycling of other metals including calcium and silicon. The precipitation of calcium carbonate is one of the most consequential biogeochemical cycling processes involving microbial (cyanobacterial) symbiotic activities in coral reefs. Microorganisms mediate the conversion of relatively soluble calcium bicarbonate ($Ca[HCO_3]_2$) to insoluble calcium carbonate ($CaCO_3$) through their participation in physiological reactions that alter the pH of surrounding environments. The respiratory release of carbon dioxide into aqueous environments produces carbonic acid, which in turn mediates the equilibrium between the carbonate and bicarbonate compounds of calcium. The sequestration of carbon dioxide for microbial photosynthesis shifts the equilibrium in favor of calcium carbonate accumulation, which finds use in the building of reefs by coral organisms (Shreeve, 1996). Microbial production of acids also influences the interactions between calcium and phosphorus to an extent that contributes more to the biological availability of phosphorus, perhaps more than any other process (Moutin, 2000). Microbial acids are also influential in the biogeochemical cycling of silicon through the weathering of siliceous rock. Silicon is used primarily in the biosphere for the construction of exoskeleton structures in diatoms. Treguer and colleagues (1995) estimated that up to 6.7 billion metric tons of dissolved silica is precipitated to form part of diatom and radiolaria exoskeletons annually.

In addition to knowledge of the cycling of iron, calcium, and silicon, the knowledge of microbial diversity and physiological processes involved in the biogeochemistry of trace elements is increasing rapidly for toxic metals such as mercury and lead. These metals are significantly enriched in the atmosphere through anthropogenic activities, and their toxicity to human and other organisms has led to the prioritization of research on their distribution and transformations on the global scale, and on biotechnological strategies for the remediation of locally polluted environments (Nriagu, 1996; Ogunseitan, 1998 and 2002; Ogunseitan *et al.*, 2000). One of the rapidly emerging areas of research involving microbial diversity and the biogeochemical cycling of trace elements involves the little known cycling of chlorine (Oberg, 2002). Contrary to widely held beliefs, chlorine participates extensively in a complex biogeochemical cycle involving more than 1,000 naturally produced chlorinated compounds and an equally extensive list of anthropogenic varieties. The depletion of the global ozone layer by chlorofluorocarbon compounds rekindled interest in the physicochemical fate of chlorine compounds, both in the stratosphere and in the troposphere (Platt and Honninger, 2003). Although the ratio of anthropogenic-to-natural sources of chlorine in the environment is high, many microorganisms have adapted to involve chlorinated compounds in physiological processes that affect the biogeochemical cycling of carbon (Kappler and Haderlein, 2003). The repercussion of these adaptations for the global cycling of chlorine has not been thoroughly investigated. The biogeochemical cycles associated with the microbial interactions with these trace elements will be discussed more fully in the final chapter of this book as part of the discussion on the role of microbial diversity in understanding—and potentially solving—contemporary global environmental problems.

CONCLUSION

The diversity of microbial physiological action and reaction is fundamentally important to the integrity of biogeochemical processes, including the cycling of phosphorus, sulfur, and several metals. It is only a matter of convenience that these cycles are discussed as if they are independent of one another. In reality, microorganisms must simultaneously regulate the intracellular concentrations of several elements in order to remain viable. In so doing, organisms are continuously engaged with mobilizing biogeochemical cycles through a vast network of redox reactions and other forms of enzyme-catalyzed chemical biotransformation. The ultimate fate of these cycles is likely influenced by a sensitive dependence on initial biochemical conditions that were established exclusively by microbial populations and communities. However, it is becoming increasingly clear that the course of major biogeochemical cycles is being altered substantially in human-dominated ecosystems (Beerling, 1999; Farquhar *et al.*, 2000; Hoehler *et al.*, 2001). These alterations are projected to have major implications for global environmental change on Earth, and new findings about the physicochemical conditions on other planets in the solar system are reinforcing this view (Nealson, 1999). Exploring the diversity of microbial forms and functions in ecosystems is essential not only for improved understanding of the interconnectedness of major biogeochemical cycles, but also for providing a biotechnological resource that could contribute to the solution of problems associated with human interference in the cycling of elements.

QUESTIONS FOR FURTHER INVESTIGATION

1 How will the sulfur cycle change if microbial life on Earth were to become suddenly extinct? Reconsider this question if the following categories of microorganisms became extinct, one at a time: (a) sulfur-oxidizing microorganisms; (b) sulfate-reducing microorganisms; (c) aerobic heterotrophic microorganisms. What would be the expected time frame for each of these changes?

2 How will the phosphorus cycle change if microbial life on Earth were to become suddenly extinct? Reconsider this question if the following categories of microorganisms became extinct, one at a time: (a) acid phosphatase-producing microorganisms; (b) alkaline phosphatase-producing microorganisms; (c) polyphosphate kinase-producing microorganisms. What would be the expected time frame for each of these changes?

3 Search the Internet and electronic journal databases for 10 different "metalloenzymes" not included in Table 8.1. What are the metal ions involved with these enzymes. How many of them are directly involved in biogeochemical cycles? How do the organisms harboring these enzymes participate in the biogeochemical cycling of the required metals to maintain access to the metals?

SUGGESTED READINGS

Beerling, D.J. 1999. Quantitative estimates of changes in marine and terrestrial primary productivity over the past 300 million years. *Proceedings of the Royal Society Biological Sciences (Series B)*, **266**: 1821–7.

Boukhalfa, H. and A.L. Crumbliss. 2002. Chemical aspects of siderophore mediated iron transport. *BioMetals*, **15**: 325–39.

Carlsson, P. and D.A. Caron. 2001. Seasonal variation of phosphorus limitation of bacterial growth in a small lake. *Limnology and Oceanography*, **46**: 108–20.

Cleveland, C.C., A.R. Townsend, and S.K. Schmidt. 2002. Phosphorus limitation of microbial processes in moist tropical forests: Evidence from short-term laboratory incubations and field studies. *Ecosystems*, **5**: 680–91.

Geesey, G.G., A.L. Neal, P.A. Suci, and B.M. Peyton. 2002. A review of spectroscopic methods for characterizing microbial transformations of minerals. *Journal of Microbiological Methods*, **51**: 125–39.

Glindemann, D., M. Edwards, and P. Kuschk. 2003. Phosphine gas in the upper troposphere. *Atmospheric Environment*, **37**: 2429–33.

González, J.M., R. Simo, R. Massana, J.S. Covert, O. Casamayor, C. pedros-Alio, and M.A. Moran. 2000. Bacterial community structure associated to a DMSP producing North Atlantic algal bloom. *Applied and Environmental Microbiology*, **66**: 4237–46.

Hoehler, T.M., B.M. Bebout, and D.J. Des Marais. 2001. The role of microbial mats in the production of reduced gases on the early Earth. *Nature*, **412**: 324–7.

Jenkins, R.O., T-A. Morris, P.J. Craig, A.W. Ritchie, and N. Ostah. 2000. Phosphine generation by mixed- and monoseptic-cultures of anaerobic bacteria. *Science of the Total Environment*, **250**: 73–81.

Kappler, A. and S.B. Haderlein. 2003. Natural organic matter as reductant for chlorinated aliphatic pollutants. *Environmental Science and Technology*, **37**: 2714–19.

Morel F.M.M. and N.M. Price. 2003. The biogeochemical cycles of trace metals in the oceans. *Science*, **300**: 944–7.

Nealson, K.H. 1999. Post-viking microbiology: New approaches, new data, new insights. *Origins of Life and Evolution of the Biosphere*, **29**: 73–93.

Nriagu, J.O. 1996. The history of global metal pollution. *Science*, **272**: 223–4.

Oberg, G. 2002. The natural chlorine cycle: Fitting the scattered pieces. *Applied Microbiology and Biotechnology*, **58**: 565–81.

Scholten, J.C.M., J.C. Murrell, and D.P. Kelly. 2003. Growth of sulfate-reducing bacteria and methanogenic archaea with methylated sulfur compounds: A commentary on the thermodynamic aspects. *Archives of Microbiology*, **179**: 135–44.

Takai, K., H. Kobayashi, K.H. Nealson, and K. Horikoshi. 2003. *Deferribacter desulfuricans* sp. nov., a novel sulfur-, nitrate- and arsenate-reducing thermophile isolated from a deep-sea hydrothermal vent. *International Journal of Systematic and Evolutionary Microbiology*, **53**: 839–46.

Weeks, S.J., B. Currie, and A. Bakun. 2002. Massive emissions of toxic gas in the Atlantic. *Nature*, **415**: 493–4.

CHAPTER 9

CROSS-SPECIES INTERACTIONS AMONG PROKARYOTES

Chapter contents

INTRODUCTION

Microbial species are generally defined by their participation in biotic and abiotic interactions, although these interactions are not always equally apparent. For example, the criteria for recognizing biotic interactions involving pathogens and parasites are, in certain circumstances, more easily observed than the criteria for recognizing abiotic interactions such as phototrophy or chemolithotrophy; whereas in other circumstances, it is easier to establish criteria for defining hyperthermophiles than it is to establish criteria for commensalism. There is a continuum of interactions, both biotic and abiotic, in which microorganisms constantly engage as they pursue the genetically encoded mandate to grow and reproduce.

Consider a single bacterial cell inoculated into a sterile environment. The immediate chance of survival of that cell and its potential to replicate depend on its capacity for interacting with the abiotic environment—the right combination of nutrients, the right pH, temperature, pressure, and so forth. If the abiotic conditions are optimal, the single cell begins to divide and replication proceeds exponentially. Consequently, the survival of this nascent "population" is determined by both abiotic factors and new biotic factors defined by intercellular action and reaction. Several categories of species interactions already described by ecologists can be observed in this hypothetical population, including competition, cooperation, and genetic exchange. After several generations, mutations will occur and these may persist or become extinct depending on their relative fitness in the particular environment. The biotic environment may be patchy or homogeneous with respect to the distribution of cells in the original habitat. Under a uniform pattern of cell distribution, the divergence of mutants could lead to the emergence of variants, strains, or ultimately, new species. Consider further that physical barriers enclosing this population are breached, leading to multifaceted biotic interactions with various species, and abiotic interactions with the metabolic products of different organisms. These are the interactions that define microbial communities, and every consequential microbial activity on Earth occurs within a framework of complex interactions. Therefore, it is important to understand the spatial, temporal, and molecular dimensions of biotic interactions within microbial communities, with the goal of identifying possible repercussions at the local level of ecosystem organization.

Eventually, this understanding should lead to the elucidation of multiple physicochemical and biotic networks that support the Earth system. This chapter aims to facilitate an appreciation of the reciprocal contributions of microbial diversity to:

1 Microbe–microbe interactions as understood through the phenomenon of quorum sensing.
2 Microbe–virus interactions.
3 The impact of genetic exchange mechanisms on biotic interactions in the microbial community.

4 Physiological manifestations of microbe–microbe interactions in consortia organizations.
5 The role of antibiotics production and resistance in microbe–microbe interactions.

These five topics are certainly a subset of a rather long list of possible interactions within microbial communities, however, they are selected here because of the relative wealth of information available, and because each case is associated with generalizable concepts in microbiology.

QUORUM SENSING

The phrase "quorum sensing" was coined by Peter Greenberg's research group in 1994 to describe population-density-responsive gene regulation by the regulatory systems governing bacterial luminescence (Fuqua *et al.*, 1994). Before the phrase was coined, the phenomenon described by the accumulation of molecular signals in environments that support densely packed microbial cells was described first by Tomasz (1965), who investigated the regulatory control of genetic exchange "competence" by hormone-like products in *Pneumococcus.* Soon afterwards, Nealson and colleagues (1970) investigated the cellular control of the synthesis and activity of bacterial luminescent systems. Since these early beginnings, quorum sensing has been generally understood as population interactive functions or "microbial communication" mediated by specific molecules, which are synthesized and excreted into the growth medium for the sole purpose of eliciting group responses to specific environmental stimuli (Fuqua and Greenberg, 2002; Manefield and Turner, 2002; Park *et al.*, 2003b). Quorum sensing has been investigated in several phylogenetic groups belonging predominantly to the *Proteobacteria* and Archaea. Both Gram-positive and Gram-negative bacteria are represented in the literature on quorum sensing (Korem *et al.*, 2003; Nakayama *et al.*, 2003; Bosgelmez-Tinaz, 2003; Paggi *et al.*, 2003). Quorum sensing is implicated in many microbial activities including biofilm formation and maturation, bioluminescence, biogeochemical cycling, swarming, genetic exchange, and disease causation in plants and animals (He *et al.*, 2003; Kim *et al.*, 2003; Guan and Kamino, 2002; Lesprit *et al.*, 2003; Molina *et al.*, 2003).

The general pattern of microbial population growth *in vitro* is well established for many groups of prokaryotes. For example, the inoculation of a few bacterial cells into supportive growth media results in a triphasic growth pattern consisting of lag, exponential, and stationary phase periods. Much information has been gathered from this growth pattern, but it is not entirely clear that such patterns exist under natural environmental conditions. There are many interpretations of events that occur during the lag phase of microbial growth, and some of these are particularly relevant to the discussion of intra-population cell-to-cell interactions (Carbonell *et al.*, 2002). According to one version, the establishment of cooperation is essential for microorganisms to successfully colonize a new environment. In particular, bacterial cell walls are semi-permeable, and the loss through diffusion of critical metabolites from a single cell may preclude or considerably retard active growth and the establishment of processes leading to colony formation. However, if an inoculum is sufficiently large, the growth inertia attributed to the diffusion of critical metabolites away from the cell can be overcome. Hence, the duration of the lag phase depends in part on the population density of the inoculum introduced into a new growth environment. This and the fact that many microorganisms exhibit complex growth requirements (i.e. are fastidious) are responsible in part for the shortcomings of employing cultivation methods to capture the diversity of microorganisms isolated from natural environments.

Cooperative growth results in the formation of microbial colonies, and there is substantial evidence that colony formation is not simply an artifact of cultivation under laboratory conditions. On solid surfaces in nature, cooperative growth produces microcolonies and biofilms that have important repercussions for the interactions of microorganisms with inanimate materials and with multicellular eukaryotes (Costerton *et al.*, 1995). In some aquatic environments, large-scale biofilms develop into microbial mats with stratified populations of diverse microorganisms engaged in intricate processes of nutrient cycling and energy transfer reactions. In extreme cases, cooperative growth produces complex secondary structures in swarming microorganisms such as the slime mold *Dictyostelium* (Dworkin, 1996). It is generally accepted that as population density increases, the occurrence of antagonistic events due primarily to competition for resources increases, even in a homogeneous population (Fredrickson and Stephanopoulos, 1981). The details of such antagonistic events have not been worked out entirely, and some of the early conjectures regarding population density-dependent antagonism in microbial populations may have been based on speculative extrapolations from the ecology of large multicellular organisms (Joint *et al.*, 2002; Rice *et al.*, 1999; Velicer, 2003). However, it is well documented that microbial growth rates are subject to feedback inhibition where the product of a metabolic process is toxic to cells, even in the presence of growth substrates and energy resources. For example, fermentation of sugars by yeasts is inhibited by high concentrations of ethanol, which slows growth. Similarly, hydrogen sulfide can inhibit sulfate reduction by interfering with the growth rate of sulfate reducers (Hines *et al.*, 2002).

Beyond the metabolite regulation of population growth, research efforts have been dedicated to elucidating the genetic basis of microbial competition and cooperation. Among the best-understood genetic programs regulating adverse interactions within microbial populations are the *hok–sok* system in *Escherichia coli*, and the DNA restriction/methylation system observed in diverse groups of microorganisms (Yarmolinsky, 1995). The *hok* and *sok* systems refer to "host killing" and "suppression of killing", respectively, where certain strains of *E. coli* have the repressible capacity to synthesize protein molecules that can damage the cell membrane. This genetic system is encoded on a satellite genome (plasmid), which must be retained in the cell to keep the host-killing function repressed. Loss of the plasmid leads to cell suicide. Therefore, the strategy appears to have been optimized for plasmid fitness, but the host cell derives benefits from the additional genes including those for antibiotic resistance that are typically encoded on plasmids (Gerdes *et al.*, 1986). A more widespread genetically encoded system for regulating interactions within microbial populations is attributed to quorum sensing.

To understand quorum sensing, attention has focused on the role of N-acyl derivatives of homoserine lactone (acyl-HSLs; e.g. N-3-oxo-hexanoyl-L-homoserine lactone) as the quintessential bacterial and archaeal quorum-sensing molecules (Ledgham, 2003; Paggi *et al.*, 2003). These extracellular signaling molecules are referred to as "autoinducers" or "pheromones" (Fig. 9.1). Although the function of acyl-HSLs discovered in different microorganisms is similar, there is considerable structural diversity among these signal molecules. In some cases, the same microorganism may synthesize structurally different types of acyl-HSLs. Acyl chain lengths occur in increments of two carbon units, ranging between the totals of four and 16, although exceptions to these generalizations have been reported in the literature (Fuqua and Greenberg, 2002). A family of enzymes known as acyl-HSL synthases synthesizes acyl-HSLs. These enzymes catalyze the reaction between the acyl segment, which is derived from fatty acid precursors conjugated to the acyl carrier protein, and the HSL segment, which is derived from S-adenosylmethionine. In addition to acyl-HSL synthesizing enzymes, many microorganisms also produce "quorum-quenching" enzymes, which inactivate acyl-HSLs (Zhang, 2003). For example, some *Bacillus* species produce lactonase enzymes that are capable of degrading the acyl-HSL molecules produced by *Erwinia carotovora* (Dong *et al.*, 2000). Efforts to exploit lactonase-supported quorum quenching through the biotechnological control of pathogenic microorganisms have so far proven inconclusive, however, the strategy is very promising (Molina *et al.*, 2003). Other organisms such as *Variovorax paradoxus* are able to metabolize acyl-HSLs as sources of energy, carbon, and nitrogen by cleaving the acyl group from the signal molecules (Leadbetter and Greenberg, 2000).

In addition to microbial degradation of signal quenchers, some eukaryotes can also produce quorum-quenching reagents that block microbial cell–cell communications, and thereby disintegrate infection processes that depend on the achievement of population

N-acyl homoserine lactone

Fimbrolide

(a)

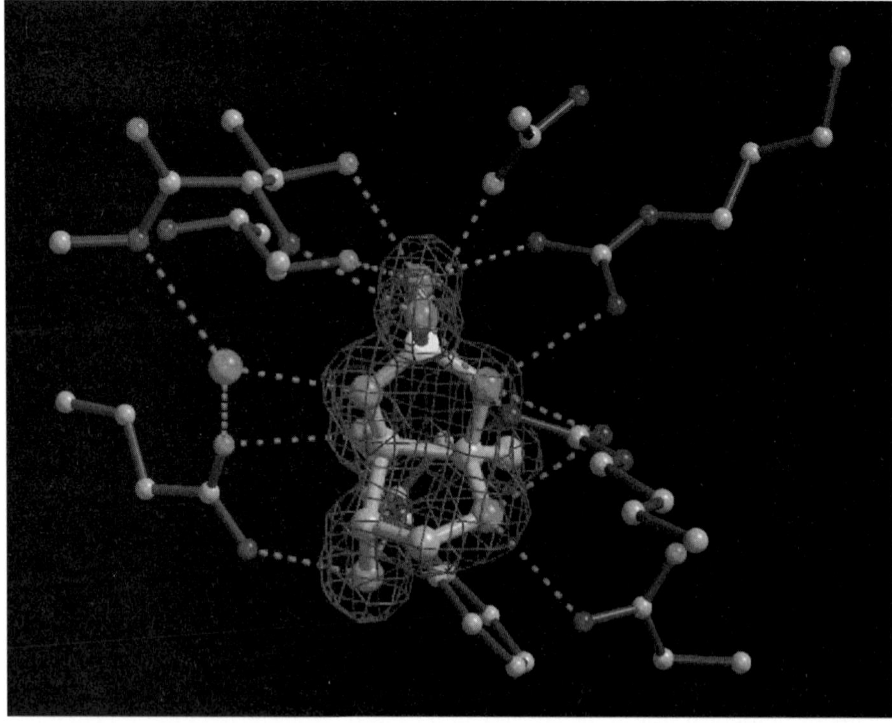

(b)

Fig. 9.1 Extracellular molecules known as autoinducers or pheromones mediate the phenomenon of quorum sensing in microorganisms. These molecules facilitate the coordination of gene expression in a population of cells during complex events, including migration, biofilm formation, and interactions with multicellular eukaryotes. (a) shows the general structure of two common signaling molecules, fimbrolide lactone and N-acyl homoserine lactone. (b) shows that the autoinducer AI-2, a universal signal for interspecies communication, has an unusual three-dimensional structure. The bound ligand is a furanosyl borate diester, which bears no resemblance to previously characterized autoinducers. The boron atom is in the center of the structure and oxygen atoms are linked to the central part with bonds represented by dashed lines. This is one of the rare occasions where the function of boron is known in microorganisms (Chen *et al.*, 2002). Image reproduced by courtesy of Fred Hughson and Bonnie Bassler, Princeton University.

density levels typically described as the "infectious dose" (Zhang, 2003). For example, *Delisea pulchra*, a red alga, is capable of inhibiting the function of acyl-HSL through the production of halogenated furanones (Manefield and Turner, 2002). The synthesis of acyl-HSL mimics by plants has been shown to interfere with population density associated activities in bacteria inhabiting plant surfaces (Teplitski *et al.*, 2000), and mammalian production of acylase causes the degradation of quorum-sensing molecules produced by biofilm bacteria (Xu *et al.*, 2003). Finally, soil microorganisms have been shown to degrade quorum-sensing molecules produced by plant pathogens, although the specific mechanisms of the interactions are not clear (Molina *et al.*, 2003).

Due to their small individual sizes and the apparently limited consequence of metabolic activities in each microbial cell, it seems reasonable to suppose that the capacity for quorum sensing underlies many of the microbial activities that result in large-scale changes, which can have considerable impact on the structure and integrity of complex ecological systems. Among the best-studied quorum-sensing systems is the LuxR–LuxI variety that controls light production in bacteria associated with marine eukaryotes (Fuqua and Greenberg, 2002). In this system, the benefit of light production appears to be solely for the eukaryote, whereas the prokaryote derives other benefits from the symbiotic relationship. Light production is induced in *Vibrio fischeri* when cell densities are up to 10^{10}–10^{11} cells per ml, a situation that occurs primarily on the surface of various marine fishes and squids. *V. fischeri* cells existing under other environmental conditions where high cell densities are not possible (e.g. a water column) cannot induce the quorum-sensing dependent bioluminescence system (Boettcher and Ruby, 1995; Kaplan and Greenberg, 1985). Presumably, lighted squid have an evolutionary fitness advantage over unlighted organisms, and are therefore favored for survival in the ocean. By implication, microbial communication systems are part of the overall ecological processes that sustain biological diversity.

The investigation of quorum sensing in *Vibrio cholerae* and *Pseudomonas aeruginosa* has revealed important insights into the series of events leading to biofilm formation (Vance *et al.*, 2003; Wagner *et al.*, 2003). Microbial biofilms mature following the initial attachment of free-living cells to either biotic or abiotic surfaces to form microcolonies. The attainment of quorum engenders additional reactions including the production of slime polysaccharides and the building of "mushroom" and "tower-like" structures up to 100 mm in height (Lawrence *et al.*, 2002). These events take up to eight days in *P. aeruginosa*, and the expression of genes associated with quorum sensing is required for the steps involving the conversion of microcolonies into mature biofilms (Chopp *et al.*, 2002; Nouwens *et al.*, 2003; Schuster *et al.*, 2003).

Microbial mats (see Chapter 6) are perhaps the most obvious ecosystems in nature where multiple quorum-sensing systems affecting biofilm formation and maturation can be expected to play major roles. However, beyond the exchange of biochemical substrates and products, very little is known about the cell-to-cell communication processes within microbial mats. Even less is known about the environmental and biotic factors contributing to the stability or demise of microbial mats over prolonged time periods. This is likely to be a very rewarding area of research in the future, and progress is already forthcoming in the analysis of quorum-sensing activities in even more challenging ecosystems. Guan and Kamino (2002) demonstrated that in the presence of 0.1 nM of synthetic N-(3-oxo)-hexanoylhomoserine lactone or Noctanoylhomoserine lactone, the total number of bacteria in seawater increased 3–8 fold as determined by microscope-aided enumeration and at least threefold by viable counting after seven days of incubation. In addition, there was a noticeable increase in the diversity of recoverable bacteria identified by unique 16S rDNA sequences. These observations clearly show the significance of quorum-sensing activities even in sparsely populated ecosystems.

The extent to which acyl-HSL-mediated quorum sensing defines microbial community structure by influencing interactions between different microbial populations remains to be fully understood. Six different outcomes of interactions between microbial populations are recognized on the basis of whether or not the interactions are detrimental to one or both populations in terms of growth rate (Fig. 9.2). The outcomes are extensions of the three basic categories of interaction, namely neutralism, antagonism, and cooperation. An individual population may experience any combination or all of these kinds of interactions within a densely populated natural microbial community such as those found in biofilms, mats, and soils. Given that most microbial communities contain more than two different microbial populations, describing the type of interaction between any two specific populations only serves the purpose of investigative convenience because the natural interactions between two different populations are likely to be influenced by the mere presence of a third microbial population.

Cross-population quorum sensing has been studied mostly for closely related microbial populations. For example, Kim and colleagues (2003) showed that the halophilic estuarine

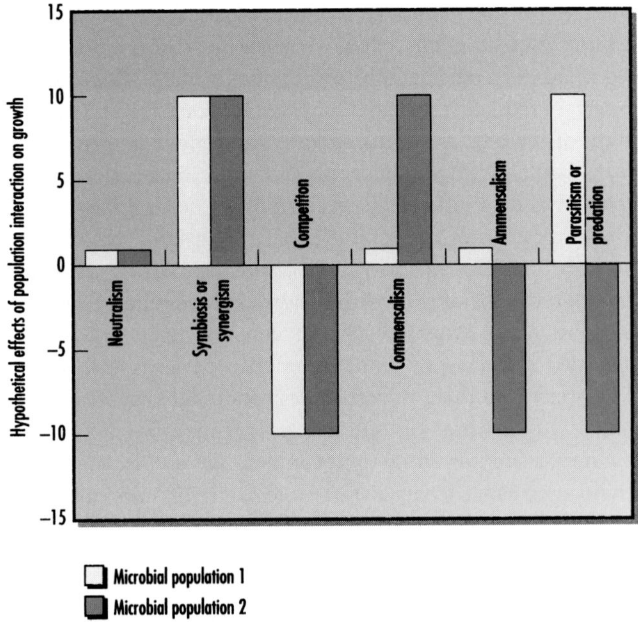

Fig. 9.2 Positive and negative outcomes of microbial population interactions occur primarily through the diffusion of metabolites, exoenzymes, or genetically encoded cross-population acting quorum-sensing molecules.

bacterium *Vibrio vulnificus* produces signaling molecules that stimulate quorum-sensing systems for the expression of luminescence in its close relative *V. harveyi*. In the photosynthetic bacterium *Rhodobacter sphaeroides*, a "community-escape response" is achieved by *CerR* and *CerI* genes, which encode for LuxR-type regulatory protein and the quorum-sensing linked enzyme acyl homoserine lactone synthase (Kho *et al.*, 2003). Repression of the community-escape response genes results in increased accumulation of the nutrient storage polymer poly-beta-hydroxybutyrate, which is typically synthesized under carbon and nitrogen starvation conditions. In their analysis of biological control agent *Pseudomonas aureofaciens*, Whistler and Pierson (2003) demonstrated that a cascade of quorum-sensing systems that respond to nutrient level and population density regulates the production of phenazine antibiotics in this organism. Finally, Paggi and colleagues (2003) investigated the induction of a potent extracellular protease by an acyl-HSL compound in the haloalkaliphilic archaeon *Natronococcus occultus*. The activities of the quorum-sensing factors correspond to the transition of the culture from exponential to stationary growth phase.

INTERACTIONS WITH VIRUSES

Microbial population densities are regulated by many different kinds of biotic factors, however, virus infection is one of the most prevalent forms of control over the fate of individual cells and the dynamics of microbial population densities in the environment. Comprehensive investigation of virus–microbe interactions in nature has been limited by the obligate parasitism of viruses and the difficulty encountered in cultivating the majority of microorganisms as pure cultures under laboratory conditions. There are four possible outcomes of virus infection among the prokaryotes: (a) **productive infection**, where a single virus infection leads to the lysis of the infected cells and production of progeny phage (the term phage is used to describe viruses infecting prokaryotes) according to burst size; (b) **lysis-from-without**, in which large numbers of virus particles attach to each host cell, resulting in the collapse of the host without production of progeny phage; (c) **lysogeny**, in which a single virus infects a cell, but instead of productive lysis, the virus genome integrates into the bacterial genome or exists indefinitely as a stable extra-chromosomal genetic entity with no yield of phage particles from the infected populations; and (d) **pseudolysogeny**, which is similar to lysogeny except that the host population grows at near optimum rates, and a certain proportion of the cells are routinely lysed to produce virus particles. So there is a nearly uniform yield of both progeny host cells and viruses in an infected population. These four potential outcomes have different repercussions for host population densities and the genetic structure of microbial communities (Noack, 1986; van Hannen *et al.*, 1999).

Sensitivity to host population density and physiological status are some of the factors postulated to influence the direction of virus life cycles (Ogunseitan *et al.*, 1990; Wiggins and Alexander, 1985). Productive virus infection by lytic and temperate (not obligately lytic) viruses is responsible for the large abundance of virus particles observed in aquatic environments, and is expected to predominate under conditions that favor a high growth rate of host cells, whereas lysis-from-without and lysogeny are favored by conditions under which host growth rates are low and population densities are declining (Weinbauer *et al.*, 2003). However, host resistance to specific virus infection is common and is attributable to various phenomena. Examples of host-resistance mechanisms include "resistance to superinfection" or "homoimmunity" following prior infection of a microbial population by a related, but genetically distinct virus; genomic "restriction/modification systems" in which host cells produce enzymes that are capable of selectively inactivating viral genomes, which are not protected by host-specific modification such as nucleic acid methylation; and host synthesis of extracellular polysaccharide layers that allow access to nutrients while preventing exposure of cell surfaces necessary for contact infection (Fig. 9.3). The role of viruses in causing widespread mortality of microbial populations has been investigated as a major factor in biogeochemical cycles, and in particular carbon-nutrient cycling through the "microbial loop" in the oceans (Azam *et al.*, 1983; Ducklow, 1983; Middledoe *et al.*, 1996; Murray, 1995; also see Chapter 7). Figure 9.4 summarizes current understanding of some interactions between viruses, their bacterial hosts, and carbon fluxes in environments populated by diverse virus–host systems (Bratbak *et al.*, 1990; Fuhrman, 1992; Fuhrman and Suttle, 1993; Suttle, 1994). Differences in the microspatial organization of aquatic and terrestrial systems have important ramifications for virus–host interactions, and these warrant separate discussion.

Aquatic viruses

Virus particles outnumber all other distinct genetic entities in aquatic environments. Although their individual contributions to total biomass are limited to sub-micrometer sizes consisting of nucleoprotein complexes, viral activities have major impacts on the flux of biomass attributable to all aquatic organisms (Mann, 2003; Paul *et al.*, 2002; Wilhelm and Suttle, 1999). The rate of decay and replenishment of virus particles in the aquatic environment depends on their physical stability between periods of episodic contact with susceptible hosts. Conversely, the abundance of host cells influences the character of viral growth cycles. Through these interactions with a wide variety of host cells, viruses exert major influences on aquatic nutrient cycling by affecting the population dynamics of organisms involved in biogeochemical and ecological processes. Viruses also affect the distribution of particle sizes and particle sinking rates in the oceans through acting upon bacterial and algal biodiversity and periodical blooms (Fuhrman, 1999; Wommack and Colwell, 2000).

The microbial loop in aquatic ecosystems consists of several compartments, including viruses, heterotrophic bacteria, heterotrophic nanoflagellates, and pigmented nanoflagellates; and has been modeled according to the quantitative estimates of productivity associated with each compartment, and the flows among them. In a study of an oligomesotrophic freshwater ecosystem, Bettarel and colleagues (2002 and 2003) estimated losses of the bacterial community to viral lysis as ranging from 30 million to 48 million cells per liter

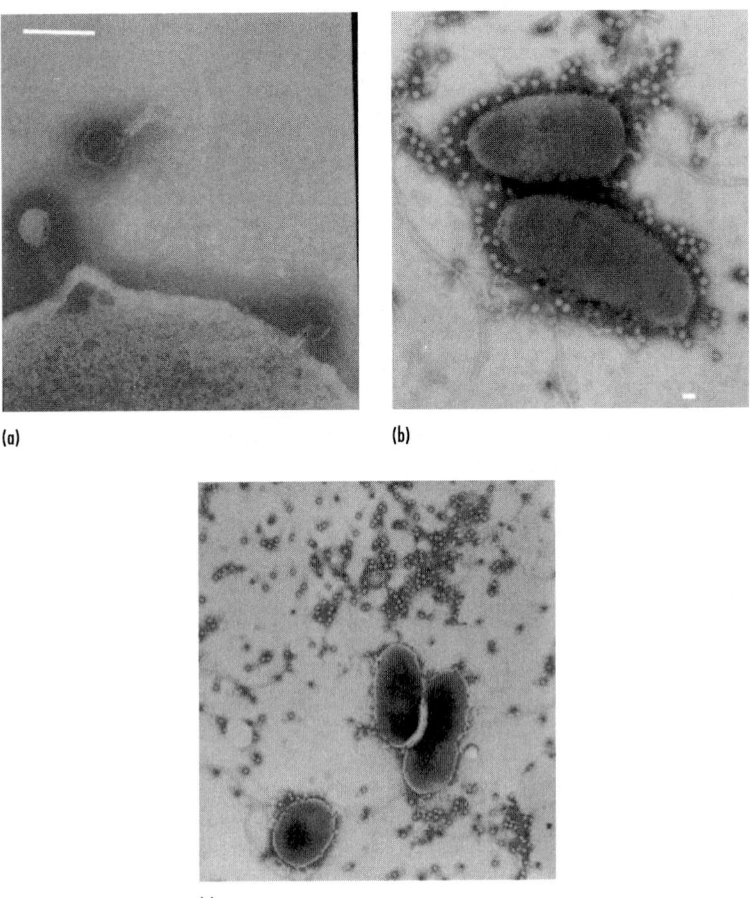

Fig. 9.3 Virus contact with host cells is important for productive infectious cycles, as well as for host-resistance mechanisms, as shown here for *Pseudomonas aeruginosa* and viruses isolated from a freshwater environment. (a) shows two virus particles on a bacterium cell, one of which (at lower right) has injected its nucleic acid into the infected cell, and the second one (at left) has attached itself, but presumably has been prevented from depositing its genome. (b) shows the rapid attraction of viruses to sensitive strains of host cells under conditions simulating a high multiplicity of infection (MOI; 200 viruses per cell). (c) depicts the same virus shown in (b), under the same MOI condition, but with a resistant strain of the same host species (*P. aeruginosa*). The resistant host is likely protected by the production of a polysaccharide layer. Photographs are from the research collections of the author. White bars are 100 nm long.

per day, whereas flagellates consumed four to six times more bacterial cells through grazing in the same time frame. These investigators estimated further that viral lysis was responsible for recycling at least 11% of bacterial carbon productivity, which amounted to 17.8% of total carbon productivity in the ecosystem. Similar findings have been reported for eutrophic and mesotrophic aquatic ecosystems (Fischer and Velimirov, 2002; Hennes and Simon, 1995).

In addition to exerting considerable influence on carbon dynamics, virus activities in the oceans are linked to other processes underlying global environmental change events. Bratbak and colleagues (1995) observed a positive correlation between the abundance of virus-like particles and the collapse of blooms of the coccolithophorid *Emiliania huxleyi*, and an inverse correlation between virus abundance and cell-specific calcification rates in coastal waters. *E. huxleyi* plays an important role in the release of dimethylsulfide (DMS). Therefore it is reasonable to expect that the collapse of a bloom could lead to large releases of DMS, but no significant relationships were observed between virus abundance and the con-

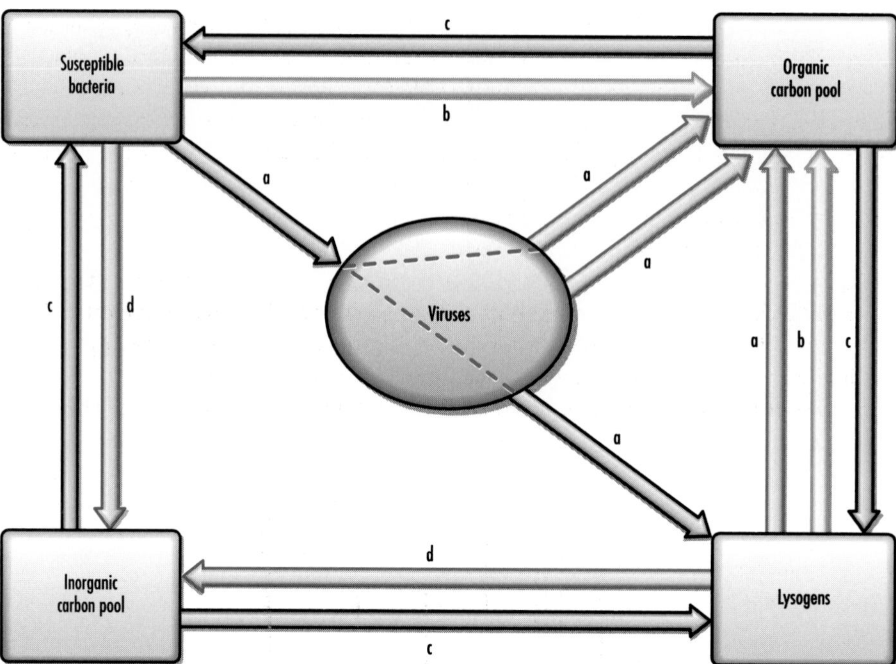

Fig. 9.4 Viruses play a major role in controlling bacterial populations in nature, and thereby contribute to the fluxes of carbon nutrients in the food web linking prokaryotic and eukaryotic phylogenetic branches. This is a simplified version of the microbial loop, focusing on viral activity. A fuller version of the loop is presented in Chapter 7. a represents lytic outcomes through virus activities; b represents bacterial disintegration from other factors; c represents uptake of carbon nutrients; and d represents the release of inorganic carbon through respiration. The dashed line within the viruses sphere indicates the direct connection from susceptible bacteria to lysogens and to organic carbon through lytic infections. Virus particles also contribute independently to the organic carbon pool after natural disintegration.

centrations of dissolved DMS or its precursor dimethylsulfoniopropionate. However, this result was explained by rapid bacterial degradation of DMS released from lysed cells.

In a different approach, Juneau and colleagues (2003) used pulse-amplitude-modulated fluorescence, a technique that allows rapid assessment of phytoplankton photosynthetic activity, to investigate the effects of viral infection on photosynthesis in the toxic bloom-forming raphidophyte alga, *Heterosigma akashiwo*. Virus infection impaired photosynthetic electron transport in infected algae with the concomitant generation of non-photochemical energy dissipated as heat. In this system, cell lysis by viruses was not dependent on light, demonstrating that the impact of viral activity in the oceans extends beyond the photic zone. In yet another twist on the repercussions of virus activities within the microbial loop, Murray (1995) explored a model describing the exudation of dissolved organic matter (DOM) by phytoplankton as an adaptive "cost-effective" strategy for reducing virus infection. The premise of the model is based on diffusion theory, which suggests that small particles adsorb solutes at a higher rate than larger particles. Therefore, colloidal viruses are more likely to encounter smaller bacteria than phytoplankton. So phytoplankton exudes DOM to increase the growth of bacteria, which in turn serve as non-specific adsorbents for viruses that would otherwise infect phytoplankton. The exudation of DOM increases with progression of phytoplankton blooming, and this strategy could reduce viral impact on photosynthetic activity.

The sensitivity of viral genomes to mutations induced by solar electromagnetic radiation has been used as an index for monitoring the biological consequences of incident ultraviolet light (Wilhelm *et al.*, 1998a and 2002; Weinbauer *et al.*, 1999). In the Gulf of Mexico, Wilhelm and colleagues (1998b) observed that during daylight, light-mediated decay rates of viral infectivity ranged from 0.7–0.85 per hour in surface waters, decreasing with depth in proportion to the attenuation of the natural peak of ultraviolet B (UVB) radiation at 305 nm. Virus inactivation by UVB is expected to be dependent on water depth, but the maintenance of virus concentrations at the observed ambient levels would require that 6–12% of the daily bacterial production is lysed even at the surface layers where the rate of virus inactivation is highest. This rapid turnover in infectious virus populations is due in part to host-mediated repair systems (photoreactivation), which restores infectivity to levels approaching 78% of damaged viral genomes.

Exposure of viruses to ultraviolet radiation can also produce a change in the virus infectious cycle from lysogeny to cell lysis, with consequences for biogeochemical cycling. Danovaro and Corinaldesi (2003) demonstrated that the addition of sunscreen products, which absorb UV radiation, to aquatic microbial communities induced the lytic cycle in up to 24% of bacteria. In addition to stimulating host–virus interactions, these chemicals either stimulated or repressed the activities of aminopeptidase, glucosidase, and phosphatase activities in the bacterial population, leading to changes in the rates of reaction in the biogeochemical cycling of carbon, nitrogen, and phosphorus. The proportion of naturally occurring prokaryotes that stably harbor temperate virus genomes has been under extensive investigation because of the implications for genetic transduction and sensitivity to environmental stress factors (Ogunseitan *et al.*, 1992; Saye *et al.*, 1987 and 1990; Ripp *et al.*, 1994).

To grasp the importance of lysogeny in nature, it is important to question the frequency of occurrence of lysogens (bacteria containing at least one virus genome) in natural microbial populations, and to estimate the number of virus particles that are produced by a lysogen after induction of lysis (burst size). The frequency of lysogenization varies according to the diversity of phage populations infecting a particular host species, and the prevalent environmental conditions (Ogunseitan *et al.*, 1992). Weinbauer and Suttle (1999) observed that in oligotrophic waters, lytic events produce 15 to 27 virus particles per cell (burst size), whereas in more productive systems, the burst size ranged from 33 to 64% of the values in oligotrophic waters. These investigators reported further that in the Gulf of Mexico, fewer than 5% of bacteria produced virus particles following induction by exposure to solar radiation. According to a steady-state model based on these observations, the induction of lysogenic bacteria by solar radiation could result in a maximum of 3.4% of the total bacterial mortality. These results imply that the induction of lysogenic phage production by solar radiation is not likely to be an important source of bacterial mortality in offshore or coastal waters.

Soil viruses

The interactions between microbes and viruses in soils have proven more difficult to investigate than similar couplets in aquatic environments because soil particulate matter adsorbs both host cells and viruses, thereby limiting opportunities for random contact. However, the extent of this limitation depends largely on soil properties, including moisture level, particle size (e.g. clay content), quantity of organic matter, and pH and temperature gradients (Dubiose *et al.*, 1979; Hurst and Reynolds, 2002). Although the adsorption of viruses to soil particles is reversible, direct observations of virus particles in soils are rare. Therefore, inferences on virus diversity in soil systems are tightly coupled to current knowledge of host cell diversity, which is based largely on very limited viable count assays. The ramifications of virus infection in the soil microbial community are likely to be as important for soils as they are for aquatic systems, including participation in host population dynamics and in genetic exchange events (Herron, 1995; Pantastica-Caldas *et al.*, 1992).

The technical difficulties encountered in the study of soil viruses have not entirely prevented the development of theoretical models useful for describing host–virus interactions in this complex environment. The models consider the patchy distribution of defining parameters such as nutrients and moisture in soils as an important point that differentiates the understanding of virus–host interactions in soils and in aquatic systems. Ashelford and colleagues (2000) demonstrated that the different temporal and spatial distributions of two physiologically distinct phage particles infecting *Serratia liquefaciens* in sugar beet rhizospheres could be explained by the application of a theoretical model based on optimal foraging theory. Similarly, Williams and colleagues (1987) developed a difference equation model to describe the interdependence characteristics of the population densities of hosts and virus particles in the soil environment. The roots of the model are defined by the following equations:

$$\mathrm{H}_{t+1} = sH_t \exp(r(1 - H_t \div K) - aP_t)$$
$$P_{t+1} = zcH_t(1 - \exp(-aP_t))$$

Where:

H_t and H_{t+1} represent host population densities measured as viable infectable units per gram of soil over one generation at times t and t+1, respectively.
P_t and P_{t+1} represent the number of virus particles present in soil at times t and t+1, respectively.
s is the fraction of host population that is not infectable, due either to dormancy or to the phenomenon known as "resistance to super-infection".
r is the intrinsic rate of natural increase of the microbial host cells, given the environment-carrying capacity of K.
a is the probability of an individual phage successfully replicating in an infected host. This is also known as the "infection efficiency".
z is the fraction of the free virus particles that are not inactivated by abiotic conditions of the soil, surviving until the next infectious cycle or phage generation.
c is the average burst size of the virus on specific hosts, or the number of progeny virus particles produced from one lytic infection cycle.

The exponent term in the first equation represents the fraction of the host population that escapes phage infection. The model equations are typically used to project the trends of virus and host cell populations under a set of environmental parameters. Williams and colleagues (1987) reported an application of the difference equation model to analyze data on the interactions between soil *Streptomyces* species and their specific viruses. The lowering of host population density by viral infection occurs within a background of dynamic fluctuations, ranging from equilibrium to chaos, which define microbial survival in natural soil environments. According to this model, infection efficiency (a) is the most influential parameter, essentially guaranteeing the co-existence of host and virus populations at values between 10^{-12} and 10^{-8}. Beyond this range, especially at high values of a, the populations of both host and virus are unstable. The difference equation model presented here does not address the incidence of lysogeny, which occurs at high frequencies in some bacterial populations, and its establishment can mediate the dynamics of host–virus interactions (Ogunseitan *et al.*, 1990 and 1992). Contrary to early speculations, lysogens are not necessarily less fit in nature than their virus-free counterparts because in certain environments, lysogenic conversion may endow host cells with advantageous genes, including antibiotic resistance and biodegradative enzymes (Ashelford *et al.*, 2000). The model also does not account for the exhibition of a broad host range by some viruses, which are presumed to be more influential in the structuring of microbial communities than host-specific viruses (Jensone *et al.*, 1998).

Attempts to quantify the influence of viruses on microbial community functions in soils are not as fully developed as comparable attempts in aquatic environments. This is due in part to the expectedly large impact of macro-invertebrate grazing as a dominant mechanism controlling microbial population densities in soils (Timms-Wilson *et al.*, 2002; Sherr and

Sherr, 2002). However, genetic exchange through transduction by viruses has been demonstrated in soils, although the evidence consists mostly of results from microcosm experiments where soils are inoculated with lysogenic bacteria and genetically distinct recipients (Zeph *et al.*, 1988). The complexity of virus–host interactions in soils makes it difficult to extrapolate data obtained through microcosm experiments to field conditions, and it is not clear how much genetic exchange is attributable to transduction when compared to the relatively high frequencies demonstrated for other genetic exchange mechanisms in soils (Timms-Wilson *et al.*, 2002).

PROKARYOTIC INTERACTIONS AND GENETIC EXCHANGE

It has been suggested that the major driving force behind microbial adaptation to a wide range of environmental conditions is not random mutation, but rather, the intercellular acquisition of genetic modules and recombination events that confer survival advantages (Guttman and Dykuizen, 1994; Sonea, 1988). Early microbiologists considered most microorganisms, particularly bacteria, to be "asexual" because they reproduce through cell division or budding. However, evidence for active intercellular genetic exchange mechanisms in bacteria emerged in 1928 with the observation by Griffith that virulence factors were exchanged between different strains of *Pneumococcus*. The mechanism underlying the exchange was not understood until 1944 when Avery and colleagues demonstrated the uptake of extracellular genetic material by cells through a process now known as transformation (Zinder and Lederberg, 1952). Natural sources of DNA for transformation include organism death or cell lysis, and the active exudation of DNA by certain microorganisms (Kloos *et al.*, 1994; Lorenz *et al.*, 1994).

Four additional mechanisms of horizontal genetic exchange have been demonstrated in natural environments, namely conjugation, transfection, lysogenization, and transduction (Ogunseitan, 1995; Trevors, 1999). The last three of these involve interaction with viruses, whereas conjugation, the genetically encoded process of organelle-mediated transfer of plasmid and chromosomal DNA, involves extensive direct interaction between bacterial cells (see Plate 2.1, panel A1). Transfection resembles transformation in the requirement for host competence, but it refers specifically to the uptake of viral genetic material, and it can be triggered in a microbial population by the death or lysis of infected cells. Transfection does not appear to be a major mechanism by which phage particles migrate among bacteria in nature because extracellular and extraviral nucleic acids are not expected to survive for long periods of time in the environment, although there are several influential factors on the extracellular stability of nucleic acids (Lee and Stotzky, 1999; Paget and Simonet, 1994; Ogram, 2000). In contrast, lysogenization and transduction can facilitate the exchange of genes between geographically distant hosts. Lysogenization differs from transduction because it refers to the mobilization of genes, which are specifically associated with the phage genome and are usually responsible for phenotypic conversion of the host, whereas transduction is the mobilization of host genes due to errors that occur during viral replication. Specialized transduction results in the mobilization and transfer of chromosomal genes adjacent to an integrated prophage (temperate phage within a host cell) genome at a higher frequency than distant genes. Conversely, generalized transduction results from the packaging of segments of a disintegrated host genetic material during the replication of prophage associated with pseudolysogeny. In the latter case, all genes are transduced at approximately the same frequency, depending on the molecular size of the gene, and the virus genome size.

Genetic exchange mechanisms in prokaryotes are credited with global impacts through the redistribution and dispersal of genes encoding for metabolic networks that determine, for example, the support of major biogeochemical cycles in microbial communities (see Box 9.1; Dunn and Handelsman, 2002; Klingmuller, 1991). If horizontal genetic exchange is

BOX 9.1

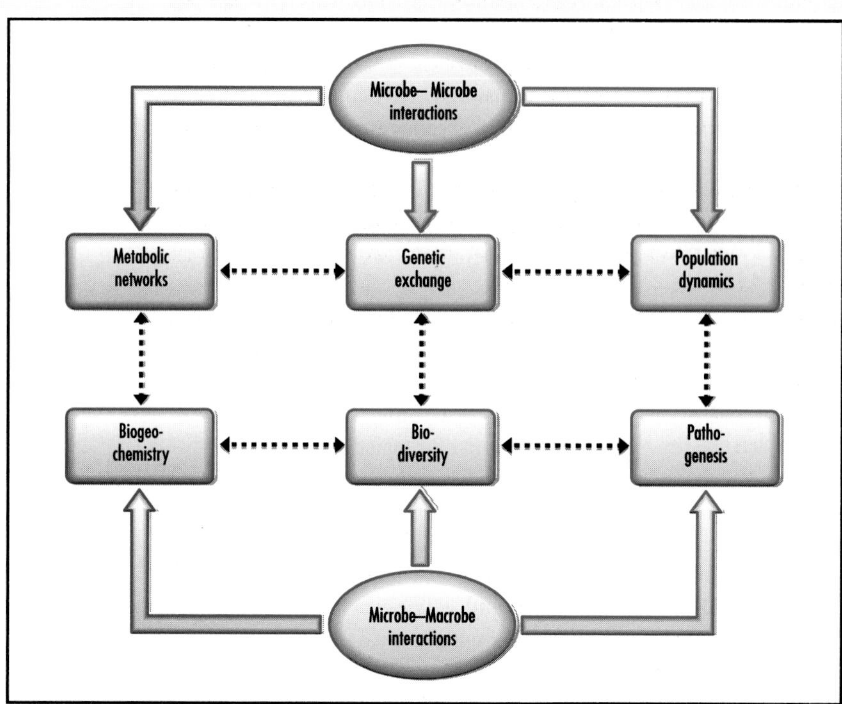

Genetic exchange is one of the most significant consequences of intercellular interactions among prokaryotes (microbe–microbe interactions). Genetic exchange influences microbial population dynamics, for example, through the dissemination of antibiotic resistance genes carried on plasmids (extra-chromosomal genetic elements). Intercellular gene transfer also influences the evolution of metabolic networks through the dissemination of genes encoding for metabolic enzymes. The evolutionary history of such enzymes can often be inferred from phylogenetic analysis of amino acid sequence information (see Chapter 5).

The effect of genetic exchange mechanisms in microbial communities ultimately influences the diversity of prokaryote–eukaryote (microbe–macrobe) interactions, for example, through the dissemination of genes associated with diseases and through controlling the rates of biogeochemical cycling (e.g. nitrogen fixation). These processes have profound influences on the qualitative aspects of biological diversity.

frequently reshuffling genetic potential, it follows that ultimately, microbial metabolic versatility will be evenly distributed across various ecological niches. This would contradict the observation of microbial specialization along ecological niches, and the findings of distinct clusters of nucleic acid sequence similarities among ecologically defined populations (Cohan, 1994; Palys *et al.*, 1997). In the definition of prokaryotic ecological populations, each adaptive mutation that emerges benefits only the genetic background of its original population, and adaptive mutants out-compete only members of their own population. In the absence of high frequency recombination events, the existing population-level genetic diversity at all loci will be replaced (purged) by the genotype of the adapted "winner". This phenomenon has been called the "selective sweep", and recurrent selective sweeps are referred to as "periodic selection" in prokaryotic populations (Guttman and Dykuizen, 1994; Levin 1981; Majewski and Cohan, 1999).

Several investigators have explored qualitative and quantitative aspects of the repercussions of various genetic exchange mechanisms for prokaryotic diversity. If the frequency of genetic exchange is negligible, then each prokaryotic population should diverge into a single DNA sequence "similarity cluster" that is associated with specific ecological roles. Conversely, it is also possible that globally adaptive mutations exist that could confer selective advantages to all ecological populations. Such mutations may arise simultaneously at distant locations, and their overall effect will be to purge divergence within and between populations. For example, Lawrence and Ochman (1998) employed a "molecular archaeology" approach for tracking the heterogeneous origins of specific traits that are represented in the mosaic structure of the *Escherichia coli* genome. By comparing the sequence of polymorphic genes in different strains and species, it is possible to speculate about historical genetic exchange events that have contributed to the configuration of metabolic pathways and the various connections among these pathways.

Majewski and Cohan (1999) used a modeling approach to investigate whether genetic exchange in prokaryotes is too rare to prevent the divergence of ecologically distinct populations according to the cumulative effects of random mutations that are neutral in their influence on adaptive fitness. The results of the model indicated that the effect of recurrent global adaptive mutations (global selective sweeps) on neutral sequence divergence depends largely on the mechanism of genetic exchange with respect to the size of the genome fragments recombined. Global selective sweeps can prevent populations from reaching high levels of neutral sequence divergence, but are not sufficient to produce genetically indistinguishable identities from two previously distinct populations. These results are supported by several consistent estimates of generally low rates of recombination in bacterial populations (Cohan, 1995; Maynard-Smith *et al.*, 1993).

The impact of genetic exchange mechanisms on microbial diversity depends both on the frequency of gene transfer events and on the size of genome fragment transferred with each event (Olson *et al.*, 1991). Transformation usually results in the uptake and incorporation of less than 1% of the genome, whereas transduction can mobilize DNA fragments equivalent to bacteriophage genome or approximately 10% of the host genome size; and conjugation can result in the transfer of more than 90% of the host genome in the case of events mediated by high frequency (*Hfr*) conjugative plasmids (Arber, 1994; O'Morchoea *et al.*, 1988; Zawadzki and Cohan, 1995). These size limitations on gene transfer are expected to correlate positively with the magnitude of the effect attributable to global periodic selection on sequence divergence in prokaryotic communities. In populations where transduction and conjugation are frequent, global periodic selection can reduce population divergence, but in populations where transformation, which requires natural competence, is the predominant mechanism of genetic exchange, global periodic selection should have little or no effect on sequence divergence (Majewski and Cohan, 1999). It is likely that other mechanisms influence population diversity in these organisms, and the observed level of diversity results from the balance of homogenizing effects of global periodic selection versus diversifying effects of mutations and recombination. Much research is needed to provide quantitative data at the level of ecological populations, which should ultimately facilitate predictions of how environmental conditions influence this balance to favor or restrict the dissemination of microbial genes in nature.

MICROBIAL CONSORTIA AND THE CRISIS OF ISOLATION

Microbial consortia represent a functional subset of microbial communities. Consortia are defined by their diversity and a seemingly coordinated effort to perform physiological functions, such as chemical biotransformation and the conduction of degradative pathways (Paerl and Pinkney, 1996). Environmental microbiologists accept, perhaps with a certain degree of frustration, that only a small fraction of naturally occurring microorganisms is culturable under laboratory conditions of isolation. Although the use of molecular techniques has predicted a high diversity of distinct microorganisms at the nucleotide sequence level, it has proven difficult to correlate the sequence-level clusters with ecological diversity in terms of the relative importance of microbial activities associated with specific ecological niches (Knight *et al.*, 1992; Pace, 1997).

Environmental biotechnologists have long considered the untapped natural microbial diversity as an almost infinite, self-replenishing resource of genetic potential that could be manipulated to support the aims of industrial ecology to serve human needs. But for microbial ecologists focusing on unraveling interactions and ecological functions, the inability to cultivate most naturally occurring microorganisms represents a major impediment to the understanding of how genetic potential is coordinated among different organisms and translated into the genetic expression that underpins the observed characteristics of microbial communities. One of the most widely acknowledged reasons for the difficulty of microbial cultivation under artificial conditions of isolation is that microbial cells exist in nature as part of consortia made up of diverse organisms engaged in intricate metabolic networks. Growth factors synthesized in very small concentrations by different organisms are made available for community use, and other organisms may consume waste products that would ordinarily limit the growth of their producers under artificial conditions. Hence, the dominant status of techniques based on microbial isolation in pure cultures represents a methodological crisis because such techniques are severely limited in providing insights into how microbial consortia operate. Although innovative methods are appearing in the literature to remove some of the hurdles that have limited the cultivation of naturally occurring organisms, these new techniques are more likely to encourage the formation of microcolonies from single species than to reveal the complexity of microbial growth in consortia (Zengler *et al.*, 2002; also see Chapter 3). The understanding of structural, genetic, and physiological characteristics of microbial consortia is necessary for the development of models that aim to account for the contribution of microbial activities as a major parameter in global environmental change and Earth system dynamics.

Many attempts have been made to elucidate the composition of microbial consortia involved in complex physiological processes, such as the biodegradation of recalcitrant chemicals (Al-Awadhi *et al.*, 2003; Rhee *et al.*, 2003). In cases where consortia are defined by complete metabolic pathways, the species composition may vary according to spatial and temporal variables associated with the enrichment process used for selecting the pathway. For example, Snellinx and colleagues (2003) isolated two different consortia capable of completely degrading 2,4-dinitrotoluene (2,4-DNT) in polluted soils. Although the consortia consisted of four and six species, respectively, both had two members each that were able to collectively degrade 2,4-DNT. The first consortium contained *Variovorax paradoxus* VM685, whereas the second consortium contained *Pseudomonas* sp. VM908, both of which can initiate the catabolism of 2,4-DNT by an oxidation step with the release of nitrite and 4-methyl-5-nitrocatechol (4M5NC). The growth of these two organisms on 2,4-DNT and subsequent metabolism of the intermediate chemicals were not efficient in the absence of other species, namely *Pseudomonas marginalis* VM683 for the first consortium; and *P. aeruginosa* VM903, *Sphingomonas* sp. VM904, *Stenotrophomonas maltophilia* VM905 or *P. viridiflava* VM907 for the second consortium. The metabolic relationships among the primary and secondary members of the consortia are not understood, but the lack of understanding did not prevent the technological application of the consortia for environmental bioremediation.

The stability of microbial consortia frequently depends on metabolic redundancy, where two or more members of a consortium have the capability to conduct a crucial process, using slightly different but phylogenetically related genes. For example, Song and Ward (2003) demonstrated the occurrence of five different nitrite reductase genes (*nirS*) during their investigation of the functional diversity of dissimilatory nitrite reductase in a microbial consortium selected for the ability to degrade 4-chlorobenzoate, an intermediate in the pathway of polychlorinated biphenyl (PCB) biodegradation. The phylogeny of the *nir* genes was not congruent with the 16S rRNA phylogeny of the genera associated with the consortium, namely *Acidovorax*, *Azoarcus*, and *Thauera*. In addition, it is difficult to predict the occurrence of specific varieties of nitrite reductase genes on the basis of the ecological and geographical origins of members of the consortium or the kinds of chlorinated substrates used for their enrichment. Microbial consortia can also maintain metabolic functions despite the sensitivity of some members to the toxicity of substrates and byproducts. Presumably, the sensitivity of some members to toxicity is ameliorated by the detoxification activities of other members of the consortium. For example, in comparing the abilities of sulfidogenic and nonsulfidogenic consortia to reduce chromium, Arias and Tebo (2003) demonstrated that the specific sensitivity of sulfate-reducing bacteria within the consortia to Cr (VI) toxicity did not prevent the complete reduction of chromium.

Syntrophy, or the co-metabolism of substrates by two or more species, is a commonly found feature of microbial consortia. For example, sulfate-reducing bacteria are commonly found as members of consortia functioning to oxidize methane in anoxic environments (Valentine, 2002). The syntrophic relationship is based on electron transfer between species, where the archaeal members of the consortium (anaerobic methane oxidizers groups 1 and 2; ANME-1 and ANME-2) apparently oxidize methane and shuttle reduced compounds to the sulfate-reducing bacteria. Although much is known about the identities of organisms involved in anaerobic methane oxidation, there remains considerable uncertainty about the nature and necessity of syntrophic associations in the consortia. Unraveling these syntrophic associations between methane oxidation and sulfate reduction has implications for the understanding of carbon cycling and the build up of carbonate structures during the early periods of Earth's history (Michaelis *et al.*, 2002).

Discoveries about the species membership and metabolic interactions within naturally occurring microbial consortia have proven useful for attempts to reconstruct consortia targeted at specific ecological functions (e.g. Alp *et al.*, 2002; Gilbert *et al.*, 2003). Innovative methods such as DNA microarray analysis and fluorescent *in situ* hybridization are facilitating comprehensive monitoring of the population dynamics and genetic expression in these artificial microbial consortia (Domingues *et al.*, 2002; Koizumi *et al.*, 2002). Specialized microbial consortia can include a large number of members, and as such, it can be difficult to reproduce the optimum physicochemical conditions for maintaining their activities under laboratory conditions. For example, Dennis and colleagues (2003) investigated a 34-member consortium specialized in the degradation of saturation concentrations of tetrachloroethylene. In this case, the artificial construction of microbial consortia is further complicated by the observation that spatial organization of member species can be as important as the required physiological diversity for optimum activity. Spatial variations in community structure of the consortium were observed relative to the source of tetrachloroethylene. Under natural conditions, a *Pseudomonas* species was predominant in a zone 30 cm from the chemical source, and a *Methanothrix* species was predominant at points beyond 85 cm from the source. Furthermore, a *Trichlorobacter* species was detected where chemical concentrations were highest, whereas *Dehalococcoides ethenogenes* was ubiquitous in zones up to 128 cm from the chemical source (Dennis *et al.*, 2003).

Alp and colleagues (2002) invented a strategy based on di-electrophoresis to solve some difficulties associated with the reconstruction of the spatial configuration required for optimizing microbial consortia functions. With this method, cells are sequentially attracted to planar microelectrodes to form layered aggregates with defined internal structures. The positions of the different microorganisms within the layers can be controlled by varying the electrode geometry, by introducing different organisms at different times, and by energizing different parts of the electrode structure at different times. The cell arrays can then be stabilized with the application of polymeric substances. These artificially structured microbial consortia do not exactly reproduce the configuration of naturally occurring varieties, but they have proven indispensable for realistic modeling and, potentially, for simulating novel metabolic interactions between microorganisms (Lorenz *et al.*, 2002).

NATURAL ANTIBIOSIS AND MICROBIAL DIVERSITY

Antibiosis, allelopathy, and amensalism are terms used to describe the inhibition of certain members of a microbial community by chemical products of other members of the community. Amensalism refers to the specific situation where chemical producers gain ecologi-

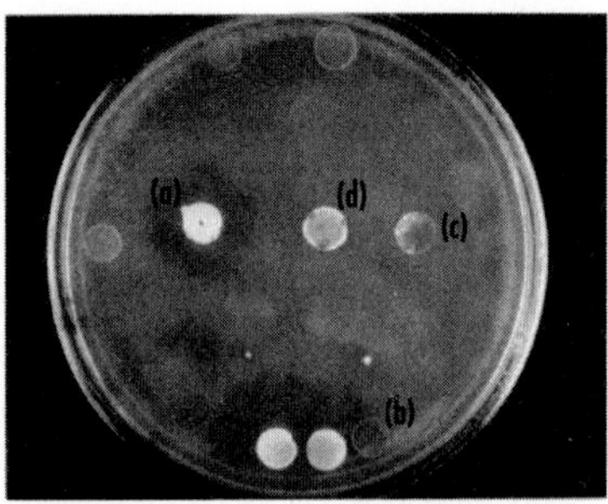

Fig. 9.5 Inter-strain antagonistic activities among pseudomonads isolated from a freshwater environment. Mixing a dense culture with cool agar before pouring into the Petri dish made a microbial lawn consisting of a pure culture of *Pseudomonas aeruginosa* isolated from a lake. Five microliter aliquots of independent cultures of 26 other bacterial strains isolated from the same lake were then spotted around the periphery and middle of the plate. After 48 hours of incubation, inhibition zones were observed in the *P. aeruginosa* lawn (a), whereas some natural environmental isolates are inhibited by the lawn bacterium (b). Other antagonistic activities are observed around microcolonies developing within other colonies (c), and some lytic activities are suggestive of the induction of latent virus infection (d). Photograph is from the research collection of the author.

cal advantage at the expense of affected populations, however, this is not always easy to observe in nature. The discovery of antibiotic production by filamentous fungi against bacteria revolutionized human ability to limit the impact of infectious pathogens on public health during the early part of the twentieth century. Societal benefits continue to accrue from this revolution, but it is becoming increasingly clear that the capacity of microorganisms to produce antagonistic substances for ecological advantage is equally matched by the ability of sensitive microbial populations to evolve resistance mechanisms to such antibiotics. It appears, therefore, that at least under artificial circumstances, and presumably also under natural conditions, a tortuous infinite cycle of antagonism, sensitivity, and resistance is fueled by a "gene-for-gene" evolutionary struggle among microorganisms (Levy, 1998 and 2002). It is certain that the observations of microbial responses to antibiotics during healthcare are not representative of responses in natural microbial communities. Very little is known about the quantitative aspects of the inferred ecological advantage conferred on antibiotic producers in nature (Fredrickson and Stephanopoulos, 1981).

Antibiotic production and its influences on microbial community structure and function have been investigated more extensively in soils than in other ecosystems (Thomashow *et al.*, 2002). The reason is that antibiotics are produced at very low concentrations, and the dilution factor in aquatic systems is probably too great to allow straightforward analysis of cause and effect, whereas soil particles may retain sufficient concentrations that make antibiotic detection relatively easy *in situ*. However, microbial culture methods have been used for identifying antibiotic producers and sensitive strains in environments where direct detection of antibiotic molecules is difficult (Fig. 9.5). The recent development of molecular tools, including nucleic acid probes based on genes encoding for specific antibiotic synthesis, has facilitated the quantitative assessment of antibiotic production in microbial communities (Raaijmakers *et al.*, 1997; Huddleston *et al.*, 1997; Weller and Thomashow, 1993).

The chemical structure of natural antibiotics varies widely (Table 9.1). Furthermore, antibiotics vary in their modes of action and in their influences on different populations. Some antibiotics induce a state of biostasis in sensitive populations by affecting the process of genetic expression or energy metabolism, whereas other antibiotics produce lyses by targeting the integrity of cell walls (Walsh, 2003). It has been shown that some microorganisms produce multiple antibiotic metabolites to achieve broad-spectrum impact on the biotic components of their environment. In certain cases, broad-spectrum activity is achieved by synergistic interactions among products, which individually either have a narrow spectrum of target organisms, or have low potency. For example, the non-obligate predator bacterium

Table 9.1 Examples of antibiotics, producers, and sensitive organisms in natural environments. Many of the producers in the list inhabit the rhizosphere environment, and their recognized role in producing antibiotics against many plant pathogens has accelerated biotechnological investments in optimizing these interactions for biological control (Thomashow *et al.*, 2002).

Antibiotic	Producer	Sensitive organism
Chaetomin	*Chaetomium globosum*	*Pythium ultimum*
2,4-diacetylphloroglucinol	*Pseudomonas fluorescens*	*Pythium ultimum*
Gliotoxin	*Gliocladium virens* *Aspergillus fumigatus*	*Mycobacterium smegmatis*
Herbicolin	*Erwinia herbicola*	*Fusarium culmorum*
Phenazine-1-carboxamide	*Pseudomonas chlororaphis*	*Fusarium oxysporum*
Phenazine-1-carboxylate	*Pseudomonas aureofaciens*	*Gaeumannomyces graminis*
Pyochelin	*Pseudomonas aeruginosa*	*Botrytis cinerea*
Pyoluteorin	*Pseudomonas fluorescens*	*Fusarium oxysporum*
Pyrrolnitrin	*Pseudomonas cepacia* *Serratia plymuthica*	*Sclerotinia sclerotiorum*
Streptomycin	*Streptomyces* species	Cyanobacteria and many Gram-negative bacteria
Surfactin	*Bacillus subtilis*	*Rhizoctonia solani*
Xanthobaccin	*Stenotrophomonas* species	*Rhizoctonia solani*

Aristabacter necator produces maculosin and banegasine, which display no antimicrobial activities alone, but together can potentiate the antimicrobial activity of pyrrolnitrin through synergism (Cain *et al.*, 2003). There are no data on the *in situ* characterization of synergistic interactions among antibiotics produced by microorganisms, and it is arguable that if synergism occurs, there is also a strong likelihood that antagonism also occurs whereby multiple antibiotics from different sources may interact to have reduced impact on sensitive populations.

It can be assumed that the evolution of genes encoding for antibiotic synthesis must always be accompanied by genes encoding for resistance in order to avoid adverse impacts on the producing organisms. The widespread occurrence of natural antibiotic resistance in microbial communities is reinforced by the selective pressure provided by the manufacturing and extensive use of antibiotics in public health and agriculture. This exogenous pressure makes it difficult to assess the extent of natural evolutionary constraints on antibiosis. For example, up to 53% of 1,400 *Escherichia coli* strains isolated from an aquatic environment were resistant to sulfmethoxazole, aminoglycosides, and beta-lactam (e.g. penicillins and cephalosporins) antibiotics (Park *et al.*, 2003a). The spread of antibiotic resistance in microbial communities is associated with genetic exchange interactions, and is facilitated by the apparent agglomeration of multiple antibiotic resistance genes on mobile genetic elements such as transposons (Park *et al.*, 2003a; Miller *et al.*, 1992 and 1993). Among the pathogenic organisms, the most common mechanism of resistance to aminoglycoside, beta-lactam and chloramphenicol antibiotics involves enzymatic inactivation through hydrolysis or through the formation of inactive derivatives. It is likely that such resistance determinants were acquired from the community "metagenome", which includes contributions from antibiotic-producing organisms (Davies, 1994; Davison, 1999).

The relationship between antibiotic production, antibiotic resistance, and microbial diversity has been investigated primarily for its commercial and technological value (Woodruff, 1996). The derivation of most antibiotics from natural sources, and the widespread distribution of resistance are taken as evidence for the importance of antibiosis in influencing the structure and function of microbial communities. However, the significance of natural antibiosis has to be evaluated within the overarching context of mutually beneficial and antagonistic relationships that permeate the microbial world and are expressed in many different forms, only a few of which are covered in this chapter. It is impossible to extricate the analysis of antibiotic resistance in natural environments from the selective pressure provided by societal production and use of antibiotics. The immediate threat of antibiotic resistance is to compromise public health strategies for controlling infectious diseases, but it is equally important to ask in what ways, and by what magnitude, this relatively new phenomenon is affecting the structure, function, and evolution of natural prokaryotic communities and their interactions with the other biotic components.

CONCLUSION

One of the most incisive strategies for "observing" interactions among microorganisms is to assay the fluxes of biochemical molecules in the vicinity of single cells, homogeneous populations, and heterogeneous communities of cells. This strategy has been employed with great success in the past century to establish major research themes and concepts in microbial ecology. The depths of knowledge within these themes vary, corresponding in part to the diversity of interactions mediated by a wide array of biochemical molecules. Three landmark discoveries in the twentieth century represent crowning achievements in this direction, and the contents of this chapter were presented to highlight the major discoveries.

First, the exchange of genetic material through transformation, transduction, transfection, and conjugation is a major pathway through which interactions among microorganisms occur. Most genetic exchange mechanisms have been observed within species boundaries, which, presumably, prevent promiscuity and preserve conservation of species identities. However, interspecies genetic exchange through one or more of the recognized mechanisms has been demonstrated convincingly by several investigators. The influence of physical and biochemical barriers on genetic exchange among prokaryotes cannot be overemphasized because it speaks directly to current concepts of species and the characteristics assigned to them. Without doubt, many of these barriers are yet to be identified. Furthermore, the relative flexibility of known boundaries is poorly characterized, and more research is needed to fully comprehend the consequences of natural genetic exchange among microorganisms. It is noteworthy that the biotechnological strategy of genetic engineering emerged from the rudimentary understanding of natural genetic exchange mechanisms. The full potential of this technology for effective deployment in natural environments will likely remain unrealized until we gain a more comprehensive understanding of how genes are mobilized between related and unrelated species.

Second, the discovery of antibiotics in the mid-twentieth century hinged upon the observation of interactions between fungi and bacteria. With respect to practical applications, this discovery revolutionized public healthcare, and it can be argued that, at least in this sector, the revolution is complete. However, with respect to the influence of the discovery of antibiotics on the theoretical concepts of microbial interactions, the revolution is far from complete. The reason is that there is still much to learn about the actual role of antibiosis in nature and about the development of resistance in sensitive populations. In a way, human manufacture of antibiotics in large quantities and the indiscriminate spread of various natural and artificial antibiotics into the environment may have compromised our ability to evaluate the real ecological role of antibiotics. Innovative experimental approaches are required to circumvent the limitations imposed by the pharmaceutical industry, and some of these innovations are already reaping benefits, including the discovery of another landmark in microbe–microbe interactions as described briefly in the following final paragraph.

The third discovery, coming later in the twentieth century, is generally discussed under the rubric of quorum sensing. Some biochemical molecules involved in quorum sensing have been characterized extensively, but it is likely that the discoveries so far represent the "tip of the iceberg" because many of the critical observations have been made through the investigation of laboratory cultures. The consequences of quorum formation in natural microbial populations are likely to further revolutionize current understanding of prokaryote–virus interactions, the establishment of metabolic consortia, and genetic exchange—all of which define the qualitative and quantitative features of microbial diversity.

QUESTIONS FOR FURTHER INVESTIGATION

1 The discovery of "quorum sensing" among prokaryotes resolved some, but not all, long-running controversies regarding intercellular communication among microorganisms. One of the remaining questions is whether or not quorum sensing is universal among the prokaryotes. The evolution of universal traits requires strong selection and unquestionable benefits. Find examples of prokaryotic life cycles where quorum sensing would not confer specific benefits, and may in fact be detrimental. Use the scientific literature databases to search for potential candidates that fit this conjecture.

2 Describe three major roles played by viruses in influencing the structure and function of bacterial communities in nature. What mechanisms are evolved by host systems to limit these viral influences?

3 Differentiate between a microbial community and a microbial consortium. Outline the major difficulties encountered in the study of microbial consortia under laboratory conditions. What kinds of methods are useful for observing microbial consortia *in situ*?

4 On September 21, 2003, NASA deliberately destroyed a $1.5 billion spacecraft that had been used for investigating the planet Jupiter for the previous 15 years. The reason was to avoid contaminating Jupiter with organisms from Earth, because environmental conditions on Jupiter are presumed to be able to support the growth and proliferation of organisms that evolved on Earth. Consider the opposite scenario where a hypothetical microorganism from another planet in our solar system was introduced to Earth. Describe the major types of possible interactions between the "alien" microbe and the prokaryotic organisms on Earth. What are the potential implications for the structural and functional integrity of different ecosystems on Earth?

SUGGESTED READINGS

Carbonell, X., J.L. Corchero, R. Cubarsi, P. Vila, and A. Villaverde. 2002. Control of *Escherichia coli* growth rate through cell density. *Microbiological Research*, **157**: 257–65.

Chopp, D.L., M.J. Kirisits, B. Moran, and M.R. Parsek. 2002. A mathematical model of quorum sensing in a growing bacterial biofilm. *Journal of Industrial Microbiology and Biotechnology*, **29**: 339–46.

Cohan, F.M. 1994. The effects of rare but promiscuous genetic exchange on evolutionary divergencies in prokaryotes. *American Naturalist*, **143**: 965–86.

Davies, J. 1994. Inactivation of antibiotics and the dissemination of resistance genes. *Science*, **264**: 375–82.

Davison, J. 1999. Genetic exchange between bacteria in the environment. *Plasmid*, **42**: 73–91.

Fuhrman, J.A. 1999. Marine viruses and their biogeochemical and ecological effects. *Nature*, **399**: 541–8.

Fuqua, W.C. and E.P. Greenberg. 2002. Listening on bacteria: Acyl-homoserine lactone signaling. *Nature Reviews Molecular Cell Biology*, **3**: 685–96.

Guttman, D.S. and D.E. Dykuizen. 1994. Clonal divergence in *Escherichia coli* as a result of recombination, not mutation. *Science*, **266**: 1380–3.

Joint, I., K. Tait, M.E. Callow, J.A. Callow, D. Milton, P. Williams, and M. Camara. 2002. Cell-to-cell communication across the prokaryote–eukaryote boundary. *Science*, **298**: 1207.

Levy, S.B. 1998. The challenge of antibiotic resistance. *Scientific American*, **218**: 46–53.

Levy, S.B. 2002. *The Antibiotic Paradox*. Boston: Perseus.

Majewski, J. and F.M. Cohan. 1999. Adapt globally, act locally: The effect of selective sweeps on bacterial sequence diversity. *Genetics*, **152**: 1459–74.

Manefield, M. and S.L. Turner. 2002. Quorum sensing in context: Out of molecular biology and into microbial ecology. *Microbiology*, **148**: 3762–4.

Ogunseitan, O.A. 1995. Bacterial genetic exchange in nature. *Science Progress*, **78**: 183–204.

Paerl, H.W. and J.L. Pinkney. 1996. A mini-review of microbial consortia: Their roles in aquatic production and biogeochemical cycling. *Microbial Ecology*, **31**: 225–47.

Suttle, C.A. 1994. The significance of viruses to mortality in aquatic microbial communities. *Microbial Ecology*, **28**: 237–43.

Trevors, J.T. 1999. Evolution of gene transfer in bacteria. *World Journal of Microbiology and Biotechnology*, **15**: 1–6.

Velicer G.J. 2003. Social strife in the microbial world. *Trends in Microbiology*, **11**: 330–7.

Woodruff, H.B. 1996. Impact of microbial diversity on antibiotic discovery, a personal history. *Journal of Industrial Microbiology and Biotechnology*, **17**: 323–7.

CHAPTER 10

INTERACTIONS BETWEEN MICROORGANISMS AND LARGE EUKARYOTES

Chapter contents

INTRODUCTION

No member of the multicellular eukaryotic phylogenetic lineage lives free from interactions with microscopic prokaryotes and/or microscopic eukaryotes. These interactions are more likely to be apparent when they result in detrimental impacts, for example, through the manifestation of disease conditions. In reality, the interactions between microorganisms and large multicellular eukaryotes cover the entire spectrum of all possible outcomes of biotic interactions, including symbiosis, parasitism, neutralism, amensalism, and commensalism. From the Earth system perspective, the most consequential interactions involve:

1 Impacts on total biodiversity through mortality events.
2 Impacts on ecological productivity through morbidity events.
3 Impacts on ecosystem integrity through mediation of biogeochemical pathways including, for example, symbiotic nitrogen fixation.

This chapter explores the diversity of microbial interactions primarily with large multicellular eukaryotic organisms and the impact of such interactions on ecosystems at various levels of organization. A perspective is presented on how changes in physical Earth system phenomena (such as climate) are influencing the emergence of new forms of biotic interactions.

MICROBIAL DIVERSITY AND GEOGRAPHY

Biogeography remains one of the few disciplines in biology where the dominant theorems have not been influenced dramatically through the study of microorganisms. The geographical dimension of biological diversity is not as apparent for microorganisms as it is for large multicellular eukaryotes such as plants and animals (Crisci *et al.*, 2003). This is due in part to the lack of adequate techniques for investigating the full spectrum of microbial geography. Furthermore, the small physical size and large numbers of microorganisms facilitate their rapid transportation and distribution at the global level (Zavarzin, 1994). Many interactions between prokaryotes and eukaryotes are exclusive, as in obligately parasitic or symbiotic relationships. Therefore, one may infer from this that factors which affect the geographical distribution of the eukaryotic hosts will be reflected in a recognizable geographical distribution of the associated microorganisms. This reasoning may hold for some microorganisms, however, it is not likely to represent a universal rule. For example, some pathogenic bacteria are widely distributed across the globe, but their interactions with hosts may be limited to narrower geographic ranges because of the prevalence of specific physicochemical and environmental conditions that influence host susceptibility (Graham, 2003; Gubler, 1998; Patz *et al.*, 1998).

New research tools for geographic information systems (GIS) have encouraged investigators to recognize patterns in the distribution of microbial pathogens and in the emergence of new infectious diseases. The use of GIS in the assessments of microbial diversity, for purposes other than mapping pathogen distribution, is steadily gaining acceptance. One of the most advanced efforts in this direction is the "biocartography" collaborative project between Idaho National Engineering and Environmental Laboratory, Yellowstone National Park, and Montana State University. The project aims to develop a geographically referenced database containing microbiological and geochemical data that can be used to support predictive models of biogeochemical interactions, and to improve access to biotechnologically important organisms. The initial coverage of the project is limited to the well-studied microbiologically rich habitat of the Yellowstone National Park, USA. An interactive online prototype system has been developed to facilitate the mapping of microbial distribution patterns according to user-specified query parameters. The database includes field data such as sampling information, sample type, weather conditions, and applicable methods that support the different schemes used to detect, identify, and classify microorganisms. Spatial data are represented on United States Geological Survey topographical maps, aerial photographs, and "ground truthed" GIS polygons including sample positions (Varley and Scott, 1998). For example, in 2003 the query database contained 47 named genus and species combinations and a wide range of temperature and pH combinations. A query for the location of ecosystems containing the genus *Chloroflexus* is linked to Bath Lake, Mushroom Pool, Painted Pool, Serendipity Springs, and Octopus Spring (Fig. 10.1).

The microbial cartography project described above is one of the few coordinated efforts that can produce a reliable database for non-pathogenic microorganisms. In contrast, several highly sophisticated surveillance programs exist that specialize in detecting, identifying, and monitoring the occurrence of pathogenic microorganisms which attack plants and animals (e.g. WHO, 2001–3). Societal investment in these surveillance programs is substantial because of the heavy toll that microbial infection of plants and animals exerts on agriculture and public health. These surveys are important precursors to disease eradication programs, although their success rests on overwhelming application of chemicals, such as pesticides and antibiotics, which are very effective in modifying the structural diversity and function of microbial communities. The ecological dangers posed by the widespread application of these toxic chemicals have led to an increasing tendency to find alternative strategies for controlling the distribution of economically important microbial species. For these biological control measures to succeed, it is important to develop a comprehensive knowledge base integrating microbial genetics, ecology, and geography (Zhou *et al.*, 2002). Furthermore, investigations connecting microbial diversity to the global pattern of plant and animal diseases have general implications for the stability of biodiversity in many ecosystems where food webs depend on the distribution and abundance of primary producers in balance with an equally complex distribution of consumers. The following sections deal with the conceptual and practical issues pertaining to the adverse and beneficial interactions of microorganisms with multicellular eukaryotes.

PLANT DISEASES

Plants, with their extensive root systems embedded in soils, are destined to interact intimately and ceaselessly with one of the most densely populated and diverse microbial communities on Earth. In many cases, these interactions are mutually beneficial; however, soils

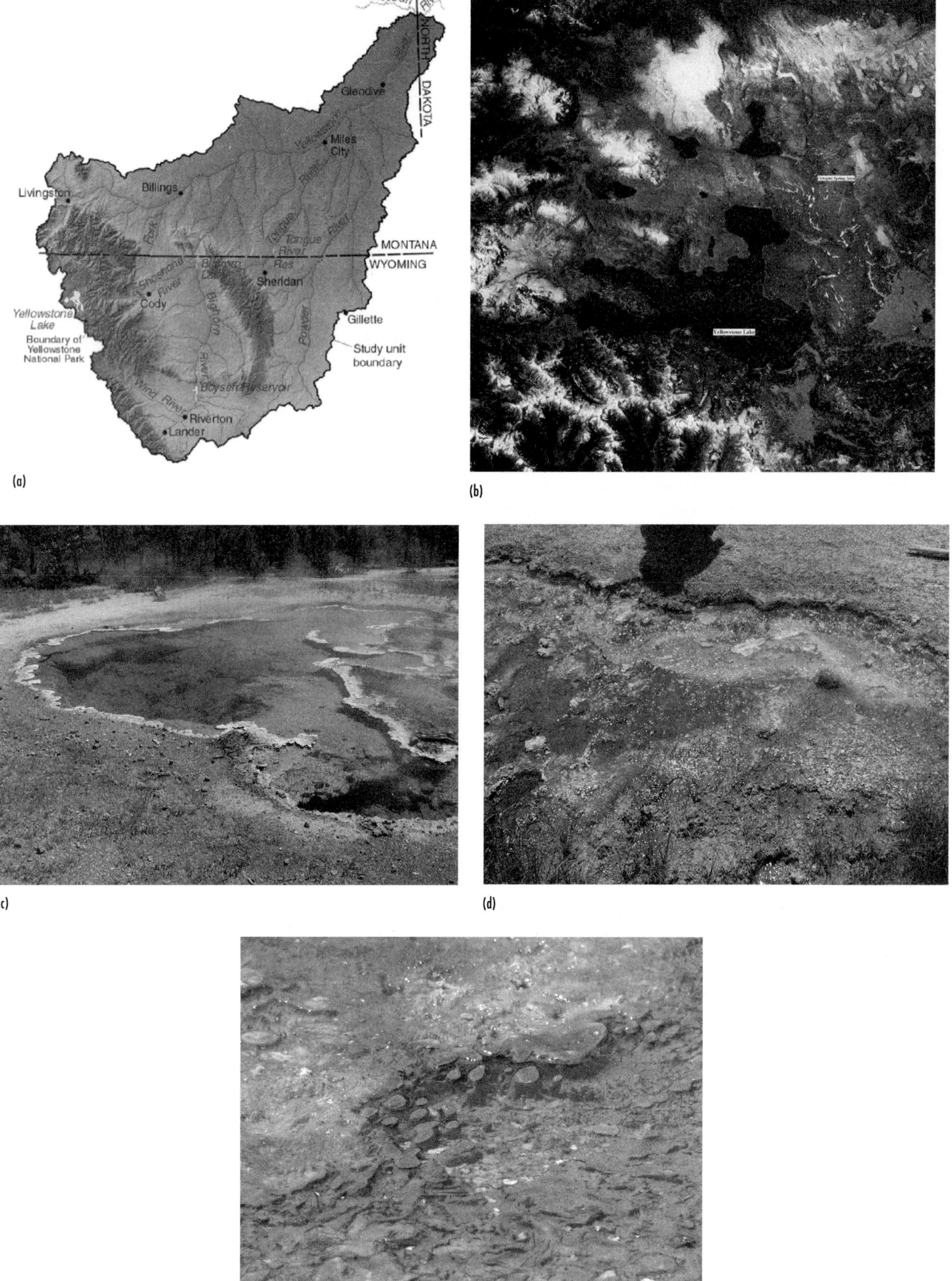
Missouri River
NORTH DAKOTA
Glendive
Miles City
Billings
Livingston
Fork
Tongue River
Tongue River Res
MONTANA
WYOMING
Shoshone River
Sheridan
Cody
Bighorn
Yellowstone Lake
Gillette
Powder
Boundary of Yellowstone National Park
Study unit boundary
River
Wind River
Riverton
Lander
Yellowstone Lake
(a)
(b)
(c)
(d)
(e)

are also the main reservoir for many microbial plant pathogens. Several commercial plants are at risk of infection by pathogenic fungi. The tendency of fungi to form extensive networks of vegetative mycelia, to survive adverse environmental conditions, and to disperse over wide surface areas through sporulation, extends the geographical range of their influence on plants. In contrast, bacterial infections of plants tend to be more localized epidemics, except where commercial activities lead to the dissemination of infected propagules by insect vectors across boundaries that would otherwise be inaccessible. Similarly, plant pathogenic viruses have limited autonomous transmission potential, although they frequently take advantage of plants' innate dissemination processes by infecting seeds and fruits and by transmission through herbivorous vectors (Desbiez *et al.*, 2002; Jiang and Zhou, 2002; Narayanasamy, 2002). Examples of plant diseases caused by viruses, fungi, and bacteria are presented in Tables 10.1, 10.2, and 10.3, respectively.

Several biotic and environmental factors act in concert to limit the geographical distribution patterns of plant pathogens. Climate and specificity of the interactions between microbial pathogens and host strains are two of the most important factors.

When conditions permit, plant pathogenesis can have devastating effects on local agricultural productivity, and the societal impacts can be far reaching. For example, the rapid spread of *Phytophthora infestans*, which caused a successive and sustained epidemic of late blight potato disease in Ireland in the 1840s, had considerable lasting impacts on society in both Ireland and the United States. It was estimated that approximately one million people, nearly 12% of the population, died of starvation and related causes between 1846 and 1851, and that another two million Irish citizens emigrated within the decade 1845–55 (Purdon, 2000). The famine destroyed more human life in proportion to regional population than the vast majority of famines in modern times, many of which occur on the African continent. Species belonging to the genus *Phytophthora* are widely recognized as potent plant pathogens, but rapid developments in the chemical and biotechnological industries and in the storage and global distribution of food resources have provided powerful new strategies for mitigating the societal impacts of these organisms (Garbelotto *et al.*, 2003). Consequently, catastrophic famines on the scale of the Irish incident are unlikely to occur in any part of the world. However, many developing national economies depend on the capacity to produce food crops for consumption and export. In this regard, geographical placement remains a powerful factor in determining the prevalence of detrimental host–pathogen interactions.

Fig. 10.1 The microbial geographical information system (MGIS), which is being developed jointly by the Idaho National Engineering Laboratory, the Yellowstone National Park Service (YNP), and Montana State University, features interactive web-based capability covering several sampling locations at the Yellowstone National Park (a), particularly around the Yellowstone lake located at the southwest corner of YNP (b). The Yellowstone area is one of the most geologically unique regions in the world, and the satellite photograph of the site was taken between September 30 and October 11, 1994 through the Shuttle *Endeavour's* flight windows. At 2,320 m above sea level, Yellowstone lake is the largest high altitude lake in North America. YNP was established in 1872, and it covers an area of 9,000 km^2 with an average altitude of 2,440 m above sea level. The major plateau was created by lava flows which occurred more than 100,000 years ago. The evidence of contemporary volcanic activity is apparent through the concentration of approximately 10,000 geysers and 200 hot springs. The extensive layers of data collected by MGIS include the presence of specific phylogenetic groups, pH, temperature, and other geochemical information. A query was submitted for the occurrence of *Chloroflexus* species. The results revealed their occurrence in microbial mats of five aquatic ecosystems, including Octopus Spring with average water temperature > 90°C (c). The hot water surges approximately every 5 minutes from the dark shadowed hole at the bottom of the pool at the southeast corner of the spring. The white deposits surrounding the spring are silica sinter (spring deposits). Layered microbial mats in Octopus Spring can be seen in (d) and (e). The mats contain *Aquifex* and *Thermotoga* species. The mats also harbor *Synechococcus* species (cyanobacterium) and *Chloroflexus* species (green non-sulfur bacteria). Most of the organisms are aerobic chemolithotrophs. The cyanobacteria sometimes form mushroom-like blobs that resemble stromatolites in calm regions of the spring (e). The MGIS interactive map server can be accessed at the following web address: http://remus.inel.gov/ynphome. The map of YNP is by courtesy of the United States Geological Survey. The satellite image of Yellowstone lake is by courtesy of NASA (Curator: Kim Dismukes; Responsible NASA Official: John I. Petty). Photographs of Octopus Spring and the microbial mats are by courtesy of Allan Treiman and the Lunar and Planetary Institute.

Table 10.1 Genetic, structural, and symptomatic categories of viruses infecting major plants.

Microbial phylogenetic category	Genus, species, or other technical name	Susceptible hosts	Plant disease
Single-stranded RNA non-enveloped monopartite genomes	*Poyvirus* 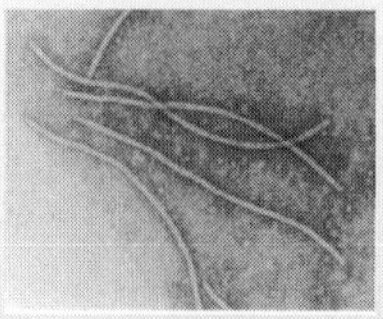 × *300,000*	Potato	Potato Virus "Y". Shortening of the stem internodes, spearing and malformation of the upper leaves, defoliation, and early plant death.
	Tobamovirus 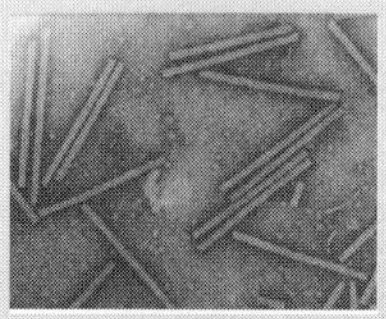 × *300,000*	Tobacco	Tobacco Mosaic Virus. Systemic necrosis on tobacco.
	Maize Chlorotic Dwarf Virus 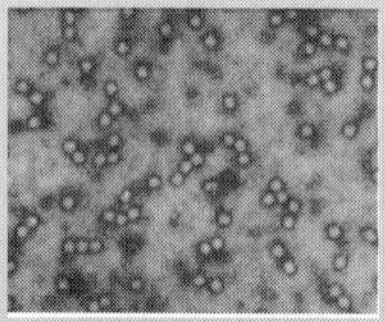 × *280,000*	Maize	Yellowing of youngest leaves in the whorl and a distinct fine yellow striping. Stunted growth.
Single-stranded RNA non-enveloped bipartite genomes	*Eurovirus*  × *300,000*	Oats	Oats Golden Stripe Virus. Symptoms vary seasonally, including yellow striping on tip leaves showing systemic infection.
	Campovirus 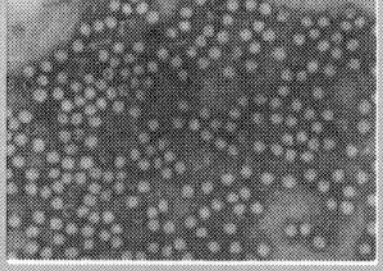 × *300,000*	Cowpea	Cowpea Mosaic Virus. Symptoms include "green on green" mosaic on primary leaves. Leaf deformation. The virus is transmitted by aphids and seeds.
	Fabavirus 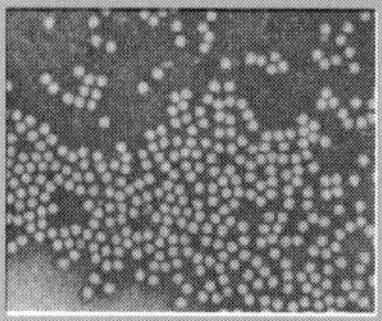 × *280,000*	Beans	Broad Bean Wilt Virus Spread by several species of aphid. Blackened growing tip and stems near ground level. Wilting Symptoms are worse in cold weather.

Table 10.1 *Continued*

Microbial phylogenetic category	Genus, species, or other technical name	Susceptible hosts	Plant disease
Single-stranded RNA non-enveloped tripartite genomes	*Hordeivirus* 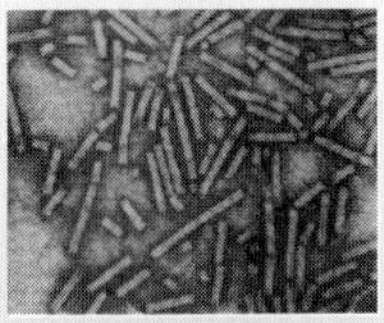 × *300,000*	Barley, wheat	Barley Stripe Virus. Found worldwide in its only natural hosts, wheat and barley. Mild stripe mosaic to lethal necrosis. No vectors, but high seed infection rate.
	Cucumovirus 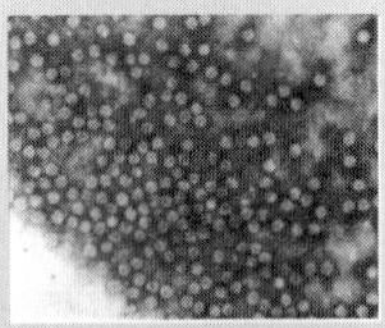 × *300,000*	Cucumber	Cucumber Mosaic Virus. Affects more than 400 plants. Symptoms include mosaic forms, flecking, dwarfing, and fern leaf. Transmitted by aphids.
	Alfalfa Mosaic Virus 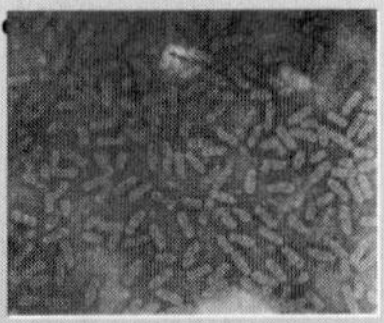 × *300,000*	Alfalfa	Alfalfa Mosaic Virus. Transmitted by aphids non-persistently; seed transmitted in some hosts. Present in most countries. Symptoms include mosaic, mottle, necrosis, and stunting.
Single-stranded RNA non-enveloped quadripartite genomes	*Temivirus*  × *280,000*	Maize	Maize Stripe Virus. Progressively severe chlorosis. Spreads in Australia, Botswana, Guadaloupe, India, Kenya, Mauritius, Nigeria, Peru, the Philippines, Réunion, Sao Tome & Principe, USA, and Venezuela.
Double-stranded DNA non-enveloped	*Baculovirus* 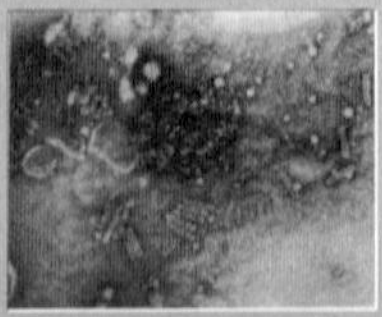 × *150,000*	Cocoa	Cocoa Swollen Shoot Virus. Limited to Ghana and Nigeria, and transmitted by mealybug vectors. Swelling of the root and stems follows discoloration of leaves and vein system.
	Caulimovirus 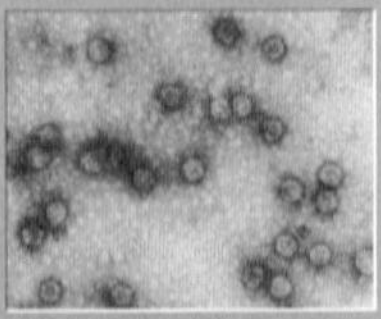 × *300,000*	Cauliflower, cabbage	Cauliflower Mosaic Virus. Affects many cruciferous plants. Causes mosaic symptoms on cauliflower, or black stippling of cabbage head and veins. Stunted growth.

Table 10.1 *Continued*

Microbial phylogenetic category	Genus, species, or other technical name	Susceptible hosts	Plant disease
Single-stranded DNA, non-enveloped	*Geminivirus* 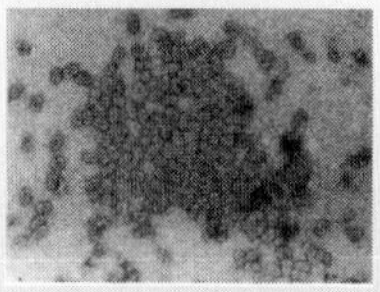 × 300,000	Infects many species of *Gramineae,* including maize, millet, wheat, barley, oats, rye, rice, sugarcane, and many grasses	Maize Streak Virus. Systemic chlorotic streaking on plant leaves. In maize, small, round whitish spots on young leaves. Spots elongate and coalesce to form long discontinuous chlorotic streaks distributed uniformly over all leaf surfaces.
Double-stranded RNA, non-enveloped	*Cryptovirus* 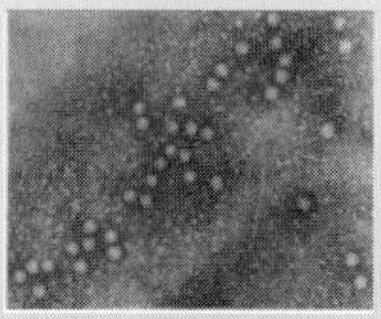 × 300,000	Beet	Beet Cryptic Virus. Present, but with no evidence of further spreading in Japan, Italy and the UK. Asymptomatic infection transmitted by seeds.
Single-stranded RNA, enveloped	*Tospovirus* 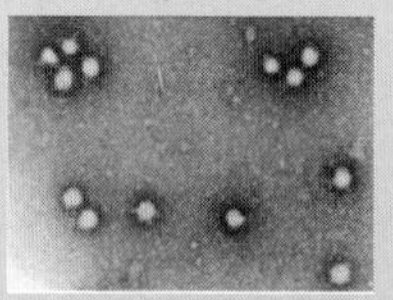 × 150,000	Tomato	Tomato Spotted Wilt Virus. Wide host range includes economically important plants representing 35 plant families, including dicots and monocots. Geographical range has recently expanded beyond tropical and subtropical regions to the global scale. Symptoms include yellow or brown ringspots, black streaks on petioles or stems, necrotic leaf spots, and tip dieback.
	Rhabdovirus 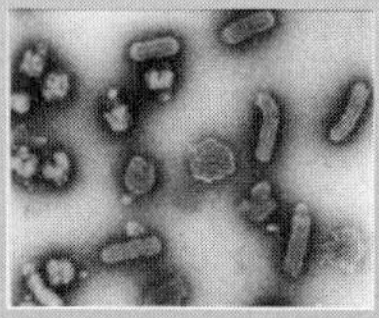 × 300,000	Maize	Maize Mosaic Virus. May have been responsible for the decline of the Maya civilization in central America. Transmitted by the leafhopper *Peregrinus maidis* in a persistent manner. Widely distributed in tropical countries. Symptoms include yellow spots and stripes in leaves, and stunting of maize and other members of the *Gramineae.*

Geographical location strongly influences the relative success or failure of schemes designed to control microbial plant pathogens through the overwhelming application of chemicals and biological control agents in regions of the world where these strategies are affordable. For example, the geographical distribution of microbial diseases affecting alfalfa

Table 10.2 Examples of fungal plant pathogens.

Microbial phylogenetic category	Genus. species or other technical name	Susceptible hosts	Plant disease
Filamentous fungus	*Pythium ultimum*	Most plants	Damping-off, rot. 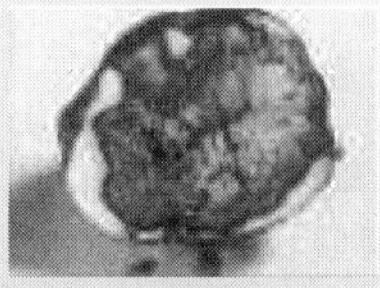
Filamentous fungus	*Phytophthora infestans*	Potato and tomato	Late blight 
Filamentous fungus	*Crinipellis perniciosa*	Cocoa	Witches' Broom 

Images by courtesy of Scott Bauer, USDA.

in the United States shows a remarkable pattern associated with climatic conditions (Table 10.4). This phenomenon raises the question of how projected climate change will affect the diversity, distribution, and activities of microbial plant pathogens. Finding answers to this question is central to the effectiveness of strategies being developed to mitigate the societal impacts of climate change, especially in countries with limited food reserves.

Impacts of global environmental change on microbial pathogens and plant diseases

The reports of climate assessments conducted by the Intergovernmental Panel on Climate Change (IPCC), under the auspices of the United Nations Framework Convention on Climate Change (UNFCCC), include scenarios representing the impacts of projected global climate change attributed to the increasing concentration of greenhouse gases in the atmosphere (IPCC, 2001a–d). There are three major ways through which the implied outcomes of these projections can affect the interactions between plants and their microbial pathogens:

1 Changes in the geographical distribution pattern of plants, resulting from shifts in climate-induced biomes, may create new opportunities for host–pathogen encounters.

Table 10.3 Examples of bacterial plant pathogens.

Microbial phylogenetic category	Genus, species, or other technical name	Susceptible hosts	Plant disease
Bacteria Proteobacteria; Alpha subdivision. Gram-negative non-sporing motile, rod-shaped, soil organism.	*Agrobacterium tumefaciens*	Attacks most dicotyledonous plants	Crown gall tumors 
Bacteria Proteobacteria; Gamma subdivision Gram-negative, aerobic, motile, rod-shaped, polar flagella, oxidase negative, arginine dihydrolase negative, DNA 58–60 mol% GC	*Pseudomonas syringae*	Tomato, olive	Frost damage through ice nucleation: bacterial canker 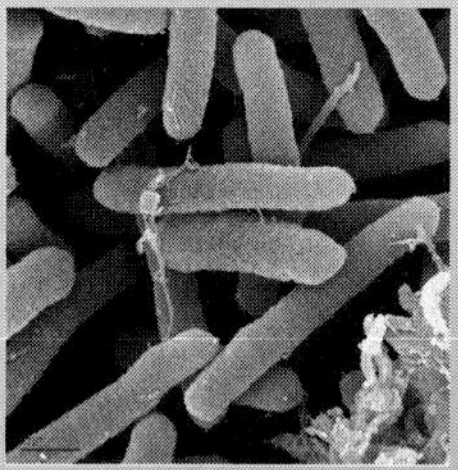 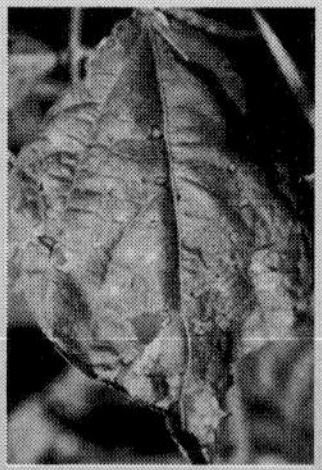
Bacteria Gram-negative, fastidious, xylem-inhabiting microorganism	*Xyllela fastidiosa*	Almond, grape, alfalfa, oleander	Pierce disease in grapevines, leaf scorch in almonds 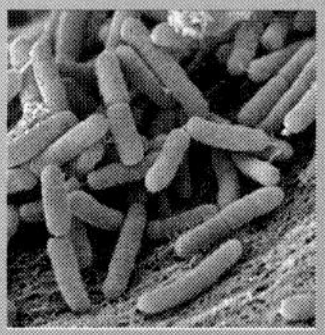 

Scanning electron micrograph of *Agrobacterium tumefaciens* cells as they begin to infect a carrot cell by courtesy of A. G. Matthysse, K.V. Holmes, R.H.G. Gurlitz (www.ostp.gov/Science/html/infection.html). Image of tomato gall by courtesy of Thorsten Kraska (www.keinfer.net/~tomato/Tomato/d_outl.html); *Pseudomonas syringae* cells by courtesy of Gordon Vrdoljak, Electron Microscope Lab, UC Berkeley. Image of Tomato leaf infected by *P. syringae* by courtesy of Susan Hirano, University of Wisconsin. Image of *Xyllela fastidiosa* by courtesy of the Joint Genome Institute www.jgi.doe.gov/JGI_microbial/html/ and http://www.cnr.berkeley.edu/xylella/page2.html.

2 Changes in regional average temperatures may produce changes in the geographical distribution of insect and bird populations that serve as vectors for the transmission of microbial agents of plant diseases.
3 Changes in climate and in particular rainfall patterns and incident ultraviolet radiation may directly affect the survival of microbial pathogens in the environment during transient phases between susceptible hosts.

Changes in the atmospheric concentration of carbon dioxide and the average global temperature will potentially affect the distribution pattern of plants having different photosynthetic strategies, known as the C3 and the C4 pathways. The protein responsible for photosynthesis, ribulose bisphosphate carboxylase/oxygenase (RuBisCO), is the most abundant enzyme on Earth, arguably because of its inefficiencies (Ellis, 1979). In some circumstances, RuBisCO recognizes molecular oxygen as a substrate instead of carbon dioxide,

Table 10.4 Geographical influence on the distribution pattern of microbial diseases in the United States. Incidence of disease in shaded areas, with high disease transmission in the darker shades. Courtesy of ABI Alfalfa, Inc.

Diseases of alfalfa and the environmental conditions that support their occurrence	Pathogenic microorganisms	Geographical distribution within the United States
***Phytophthora* root rot** Occurs in soils with poor water drainage.	*Phytophthora megasperma f. sp. medicaginis*	
Bacterial wilt Occurs in most soil types but damage can be more severe in the presence of nematodes or root-feeding insects that create sites for entry into root system. More common in cold climates.	*Clavibacter michiganense* subspecies *insidiosum*	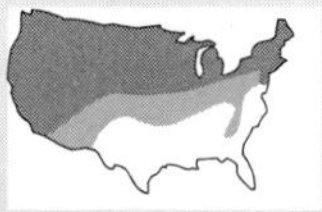
Fusarium wilt Occurs in most soil types but damage can be more severe in the presence of nematodes or root-feeding insects that create sites for entry into root system. More common in warm climates.	*Fusarium oxysporum* subspecies *medicaginis*	
Anthracnose Occurs mostly in spring or fall and spreads rapidly under warm wet conditions from spores produced on lower stems of infected plants.	*Fusarium oxysporum f.sp. medicaginis*	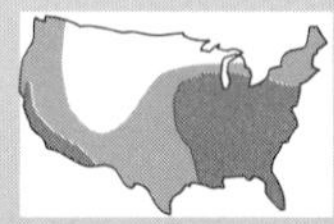
Rhizoctonia Wet humid conditions; root damage generally occurs in warm soils or those conditions that favor high-temperature flooding injury.	*Rhizoctonia solani*	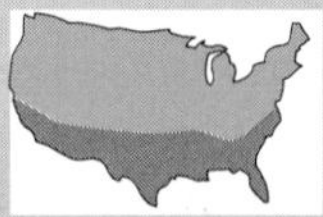
***Aphanomyces* root rot** Occurs in soils with poor drainage where water stands for an extended amount of time, or when there is an extended wet period at planting.	*Aphanomyces euteiches*	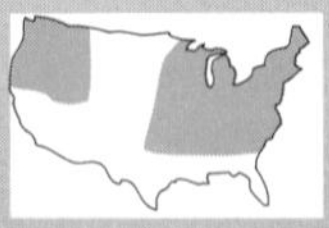
***Verticillium* wilt** Thought to occur only in cooler northern climates until it was identified in the late 1980s in parts of Southern California. Disseminated by dry or fresh plant material on harvest equipment. Cutter bar blades of mowing equipment are extremely effective at spreading the pathogen spores.	*Verticillium albo-atrum*	

Source: http://www.americasalfalfa.com/major_diseases.htm.

leading to the synthesis of one molecule of 3-phosphoglycerate and one molecule of glycolate instead of two molecules of 3-phosphoglycerate (Fig. 10.2). Glycolate is largely useless for energetic functions in plants. The oxygen–substrate reaction (oxygenase function of RuBisCO) competes with the carbon dioxide–substrate reaction (carboxylase function of RuBisCO), and overproduction of RuBisCO is the strategy adopted by some plants to optimize the rate of photosynthesis. However, at 25°C, the rate of the oxygenase function is 25% of the rate of the carboxylase function. This creates a serious inefficiency for plants, and as ambient temperature increases, the difference in the solubility of atmospheric oxygen and carbon dioxide creates a shift in the balance of these two substrates and their availability for terrestrial plant photosynthesis. Therefore, near the equator where ambient temperatures are relatively high, the carboxylase reaction would become less effective. In addition, plants

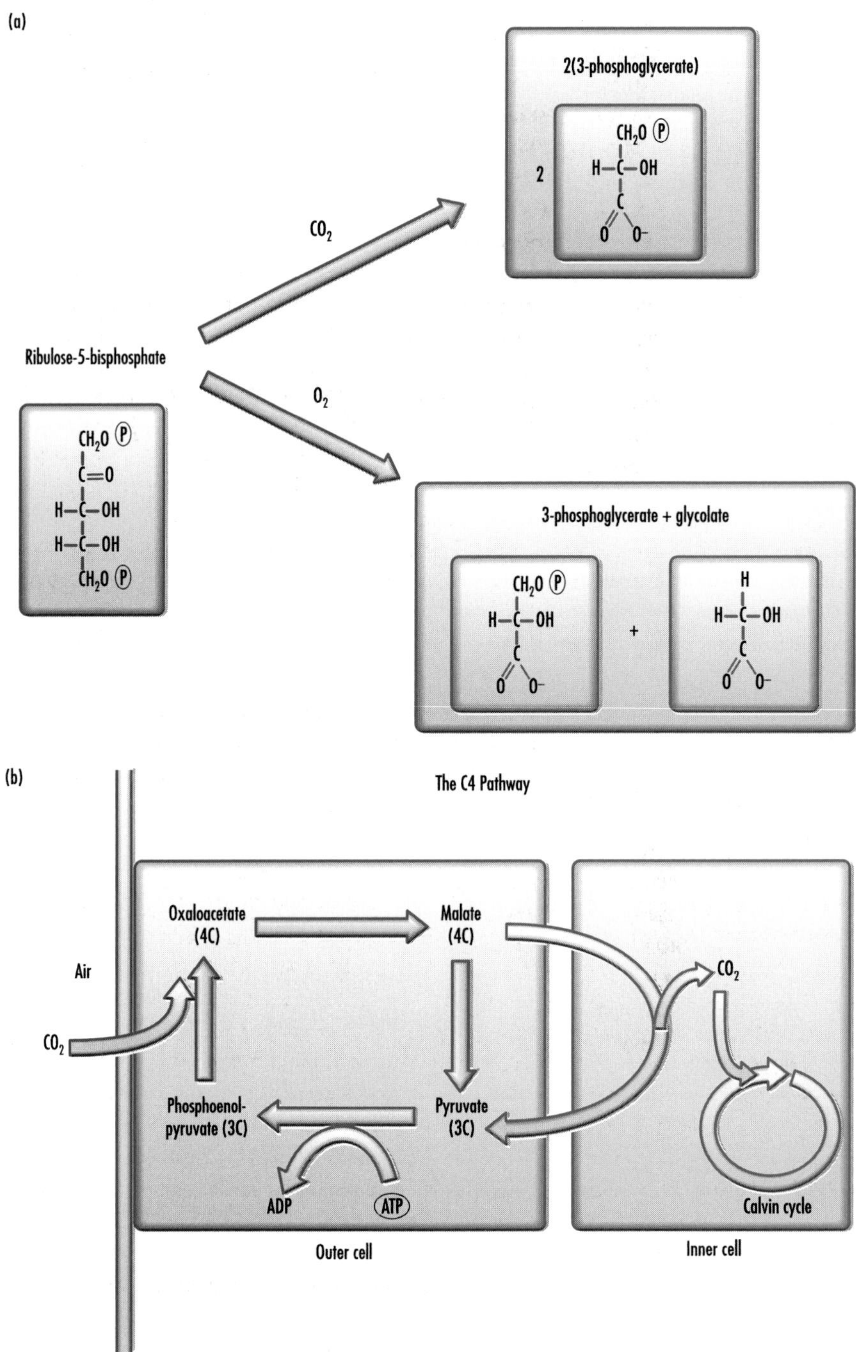

Fig. 10.2 Competitive oxygenase and carboxylase functions of the photosynthesis enzyme, ribulose-5-bisphosphate carboxylase, common in C3 plants (a), and the C4 pathway (b). By courtesy of MIT Biology Hypertext (www.mit.edu/esgbio/ www/ps/other.html).

inhabiting dry habitats regulate pore openings to minimize water loss, however, this also restricts access to carbon dioxide, thereby favoring the oxygenase function of RuBisCO. Consequently, plants in hot and/or dry environments have adapted to the laxity in RuBisCO function by evolving a different pathway for fixing carbon dioxide. The alternative pathway is referred to as "C4" because it involves the production of four-carbon compound intermediates (oxaloacetate and malate) that are effective in trafficking carbon dioxide directly to the general biochemical pathway for carbon fixation (Calvin cycle). These C4 intermediate compounds are recycled back to C3 compounds (pyruvate and phosphenol pyruvate) through an energy-requiring process. Plants that do not have the capability to utilize the C4 pathway are referred to as C3 plants. In general, C3 plants predominate in temperate regions where they do not need to expend the extra energy required to override the oxygenase function of RuBisCO, whereas C4 plants, including several grasses, sugarcane, and maize, dominate in tropical zones, where the high temperatures would favor the oxygenase function of RuBisCO (Fig. 10.3).

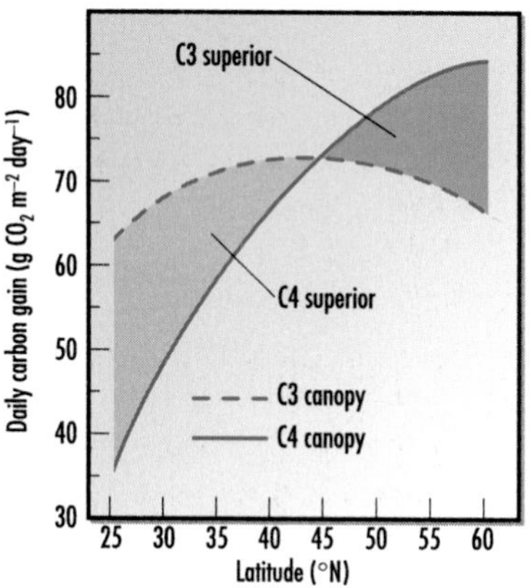

Fig. 10.3 Comparative geographical distribution and the carbon-fixing activities of C3 and C4 plants. By courtesy of University of California, Santa Barbara, Global Environmental Change program (http://real.geog.ucsb.edu/).

Under a global warming scenario where the plant-growing season in temperate regions is lengthened according to increases in precipitation and average temperatures, the geographical range of C4 plants will likely increase. Consequently, the geographical range of pathogens associated with these plants will likely increase to the same extent. It is impossible to predict the exact details of other environmental parameters that will accompany shifts in climatic conditions. Furthermore, many biotic factors affect the survival of plant pathogens, including grazing by protozoa and competition with the microorganisms. Therefore, it is difficult to be certain about the stability of new geographical ranges for microbial pathogens predicted solely on the basis of average temperature and precipitation. It is also not prudent to rule out other adaptive strategies that could emerge among C3 plants to retain their current geographical ranges, in spite of increases in average global temperatures. Therefore, C4 plants are not likely to simply invade niches "vacated" by C3 plants, along with associated pathogens.

Several microbial plant pathogens are disseminated through insect vectors, including aphids, bees, and other plant-feeding insects. Much research has been conducted on the potential impacts of global climate change on the population dynamics and geographical distributions of insect vectors of disease, although most of the work has focused on human diseases such as malaria, dengue fever, and the natural geographical range of their insect vectors (Aron and Patz, 2001; Hales *et al.*, 2002; McMichael, 2001; Patz *et al.*, 1998). In principle, the same set of environmental parameters that influence the incidence of vector-borne animal and human diseases should be important for plant diseases, except that the mobility of infected plants is rather limited. Another important difference between insect vectors of animal/human diseases compared to the insect vectors of plant diseases is that in many cases, microbial pathogens have an important segment of their life cycle in the vector, in which they may reproduce to large numbers before infecting a new host. In contrast, it is unlikely for microbial agents of plant diseases to also have a replicating life cycle in insects due to physiological barriers that limit viral infection across major phylogenetic groups. However, the general lack of specificity of pathogen–vector relationships in the case of plant diseases suggests that microbial pathogens can be transmitted effectively by many different kinds of insects, birds, or animals that (a) feed on infected plants, (b) have a wide range of mobility, and (c) exhibit features that support the transient storage and dissemination of microbial propagules. These vectors may also provide opportunities for genetic exchange mechanisms that contribute to the diversity of microbial pathogens, thereby circumventing host resistance mechanisms and acquiring new genetic determinants of pathogenesis.

ANIMAL DISEASES

Nearly all terrestrial ecosystems inhabited by animals are influenced by human actions. Humans are also responsible for nurturing a staggering number of domesticated animals either through agricultural practices or as home pets. The close proximity of humans and animals and the similarities between their physiological systems strongly suggest cross-susceptibility to microbial disease agents. In many cases, animals are purported to be the source of human pathogens that have wreaked havoc on the public health system (e.g. AIDS), and in other cases, genetic exchange between microorganisms has broadened the host ranges of both animal and human pathogens resulting in the emergence of highly potent strains (e.g. the flu virus and SARS). Nevertheless, certain quintessential animal diseases warrant particular attention because of their linkage to evolutionary events that contribute to the diversity of microbial pathogens, and to landscape factors that facilitate their dissemination and transmission across phylogenetic boundaries (Muller *et al.*, 2000). Two examples are discussed to illustrate this point, namely mad cow disease, and foot and mouth disease, both of which concern agricultural animals, and have large impacts on human welfare.

Mad cow disease

During the final years of the twentieth century and early in the twenty-first century, at least two major events related to animal diseases caused by microbial pathogens created global-level scares that cut across several spheres of society, while raising new interests in research on microbial diversity and the geography of infectious diseases. The first was the epidemic of "mad cow" disease in England, and the second was the near pandemic of "foot and mouth" disease with the epicenter also in Europe. These two events engendered different lessons for the understanding of microbial diversity in the contexts of molecular and environmental determinants of interactions between microorganisms and potential hosts. Mad cow disease, or bovine or transmissible spongiform encephalopathy (BSE or TSE) is unique among animal diseases in the sense that its causative agent is a "prion" or protein molecule that has so far not been associated with any organized independent cellular life form (Fig. 10.4). The only related disease is the Cruetzfeld–Jacob Disease (CJD), a human neurodegenerative disorder (Aguzzi, 1997; Prusiner, 1995). BSE has been responsible for the sacrifice of hundreds of thousands of potentially infected animals and for the implementation of restrictions on international trade in animal products. The exact character of the misfolded prion protein or its structural isoform with respect to disease causation has been researched extensively, although several important questions remain (Bounias and Purdey, 2003; Van Everbroeck *et al.*, 2002).

Two major aspects of BSE are of particular interest to the discussion of microbial diversity and the geography of animal diseases. The first is the possibility that insect vectors are involved in the transmission of prion proteins, which raises additional concerns about the apparent environmental stability of the agent and the likelihood that the range of susceptible hosts will expand. The pattern of infection of farm animals by the prion agent follows a horizontal transmission model, which suggests the involvement of ectoparasites in the infection process. Additionally, the exposure of fly larvae and mites to infected tissue has been used to show that these insects can be effective vectors for the prion protein, although this phenomenon has not been established under natural field conditions (Lupi, 2003; Matthews, 2003). The second aspect concerns the ongoing debate about the evolutionary origins of the prion protein and the myriad hypotheses linking the development of BSE to many biotic and environmental factors. These hypotheses include the speculation that the prion protein is a product of the retrograde evolution of an extinct microorganism; that the transformation of the normal prion protein to the isoform is caused by a bacterial toxin; that exposure to organophosphate pesticides is implicated in the development of BSE; and that the ecological balance of trace metals is the most important factor in the etiology of BSE. The interaction between the prion protein and metal ions is the most thoroughly developed of these hypotheses, and its connection to the biogeochemistry of trace metals warrants further discussion (Purdey, 1998, 2000, 2001, 2003; see also Chapter 8).

The normal version of the prion molecule is a copper-containing glycoprotein that is expressed through the circadian-mediated pathway (Purdey, 2003). In the absence of copper, manganese can bind to the prion protein, producing an intermediate variant form of the protein that is resistant to protease. As yet unclear steps are required to transform this intermediate molecule into a fully-fledged pathogenic variety. There appears to be some correlation between the incidences of BSE and the balance of trace metal concentrations in animal diets, where ecosystems that support clusters of sporadic BSE are also characterized by high concentrations of manganese and low concentrations of copper, zinc, and selenium. Substi-

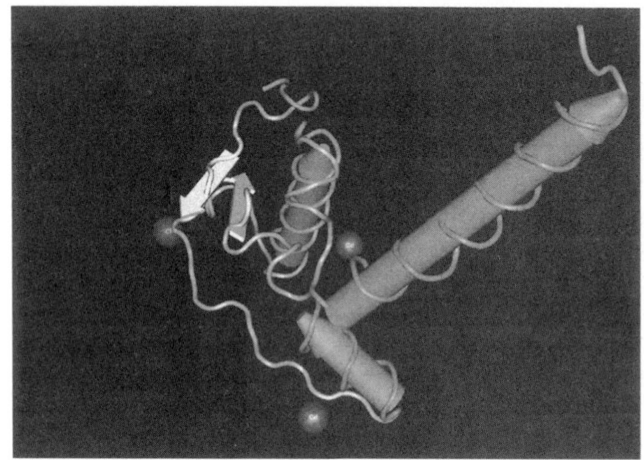
(a)

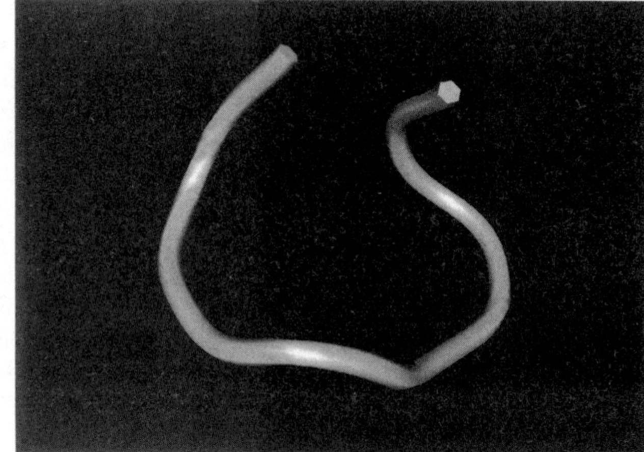
(b)

PrP^C + PrP^Sc ↔ Heterodimer

Homodimer

Dissociation

(c)

Fig. 10.4 The normal version of the prion protein is found in healthy brain tissue, and it is referred to as the cellular prion protein (PrP^C). The three-dimensional structure of the protein shown in (a) is derived from the sequence of approximately 210 amino acids and nuclear magnetic resonance (NMR) analysis (Zahn *et al.*, 2001). An octapeptide repeat sequence consisting of proline-histidine-glycine-glycine-tryptophan-glycine-glutamine is among the most conserved segments of all mammalian prion proteins (Zahn, 2003). The NMR solution structure of the repeat sequence shows a structural motif (loop conformation and a β turn) that is associated with reversible PrP^C oligomerization at neutral pH values (b). An enigmatic morphological change in the normal prion protein leads to a disease-associated variant named after scrapie (PrP^{Sc}). Exposure to PrP^{Sc} leads to the accumulation of filamentous polymeric versions of the protein that are found in diseased tissue. Furthermore, PrP^{Sc} can lead to increased conversion of PrP^C to more PrP^{Sc} according to the heterodimer model presented in (c).

tution of manganese for copper in the metal-binding domain causes the prion protein to misfold into a stable manganese protein isoform that cannot perform the normal antioxidant function of the natural protein. Differences in the trace metal concentrations of locally grown animal feed have been suggested as the main reason for the narrow geographical range of BSE epidemics (Purdey, 2001).

Foot and mouth disease

European countries had barely recovered from the BSE epidemic when the epidemic of foot and mouth disease (FMD) in farm animals emerged in 2001–2. FMD is a contagious viral disease affecting cloven-hoofed animals such as cows. Infected animals show painful blisters on their hooves and discolored tongues (Fig. 10.5). Although records of the disease date back to at least the 1600s, the causative agent was not identified until 1897 (Alexandersen *et al.*, 2003). The disease is extremely contagious, but its geographical range shows a strong dependence on climatic conditions (Fig. 10.6). Urashima and colleagues (2003) used seasonal models to explore the potential that global warming could lead to increases in the incidence of FMD. These investigators demonstrated that 64% of the variation in observed FMD cases between 1999 and 2002 could be explained according to a linear relationship between seasonal parameters and disease incidence. Their models predicted that warmer climatic conditions would lead to an increase in the number of FMD cases. However, it is unclear from these results whether the linkage of FMD to climate is influenced by the genetic variety of the pathogen.

There are seven genetically distinct strains of the virus that causes FMD, and they differ in their occurrence across geographical regions (Mason *et al.*, 2003; Fig. 10.7). The viruses are members of the genus *Aphthovirus*, belonging to the family *Picornaviridae*. This is a highly diverse family of viruses that are implicated in several human and animal diseases (Fig. 10.8). The viruses can survive for up to two weeks in the environment, and even longer when associated with body fluids such as urine; surviving up to six months in animal waste slurry. Environmental survival is prolonged during cold seasons (Alexandersen *et al.*, 2003). The fact that the disease is endemic in several tropical countries is not easily interpreted as an indication of adaptation to climatic conditions because socioeconomic factors in these countries confound any simplistic interpretation. Eradication of FMD usually requires the disposal of all animals that may have come into contact with the virus—a very expensive proposition.

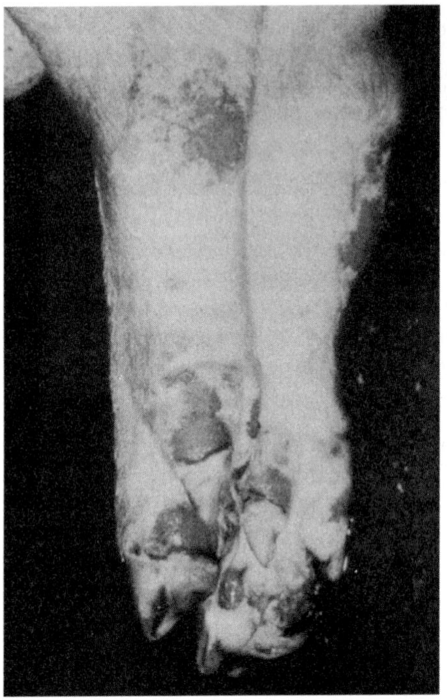

(a)

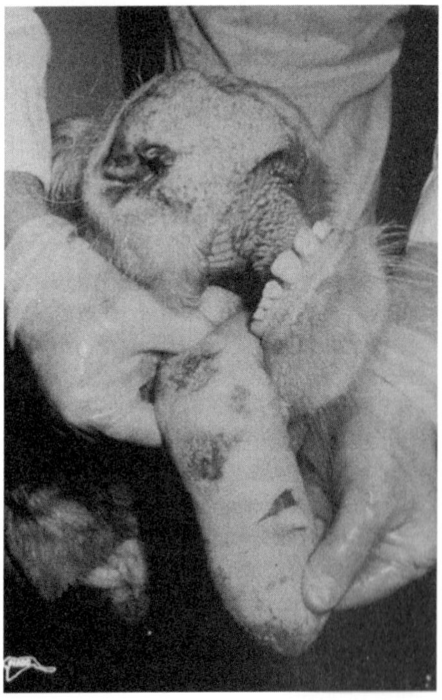

(b)

Fig. 10.5 Symptoms of viral foot and mouth disease show painful blisters on hooves (a) and infected tongues (b). Mortality is low, but its extreme contagiousness makes the disease extremely costly to eradicate from an infected herd. Images from the United States Animal and Plant Health Inspection Service (APHIS) of the United States Department of Agriculture.

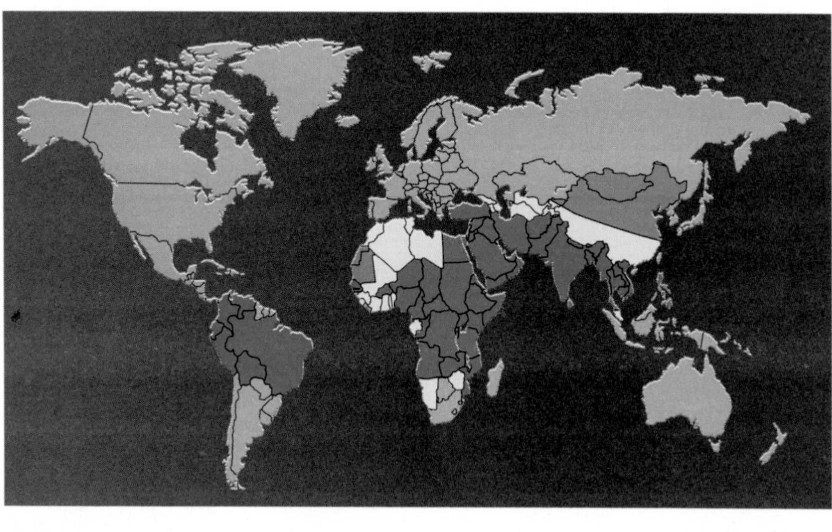

Fig. 10.6 Geographical distribution of foot and mouth disease according to endemicity in early 2002. Map by courtesy of Advanced Veterinary Information System (AVIS).

HUMAN DISEASES

Modern human populations are able to colonize nearly every terrestrial environment on Earth, partly because of their technological prowess for manipulating landscapes, but also due to a remarkable ability to prevent or cure microbial infections that could otherwise result in unsustainable levels of mortality. It is arguable that all human diseases demonstrate a geographical pattern, but this phenomenon is more potent in the consideration of infectious diseases caused by viruses and bacteria. Despite the unprecedented success of antibiotic therapy and vaccinations, most of the human mortality on Earth is caused by microbial pathogens. A rigorous surveillance of microbial diversity is essential to maintain control over the emergence of new infectious diseases, the colonization of new habitats by recognized pathogens, and the acquisition of new genetic determinants of pathogenesis and drug resistance. Two cases are discussed as examples to illustrate the importance of a global perspective on microbial diversity and its implications for human diseases. These cases are tuberculosis, caused by a soil-borne bacterium, but exacerbated by virus infection; and cholera, caused by a water-borne bacterium, and disseminated through pathways that are influenced by changes in global climate.

Tuberculosis

Mycobacterium tuberculosis, an obligately aerobic acid-fast Gram-positive bacterium, is the causative agent of tuberculosis, the leading cause of human mortality attributable to bacterial infection worldwide. The annual number of deaths from tuberculosis is only exceeded by the number of deaths from Human Immunodeficiency Virus/Acquired Immune Deficiency Syndrome (HIV/AIDS). The epidemiology of these two human diseases interacts in complex ways, since tuberculosis is the leading cause of death among people living with AIDS (Ebrahim *et al.*, 1997; WHO, 2003; Williams and Dye 2003). Every year, at least two million

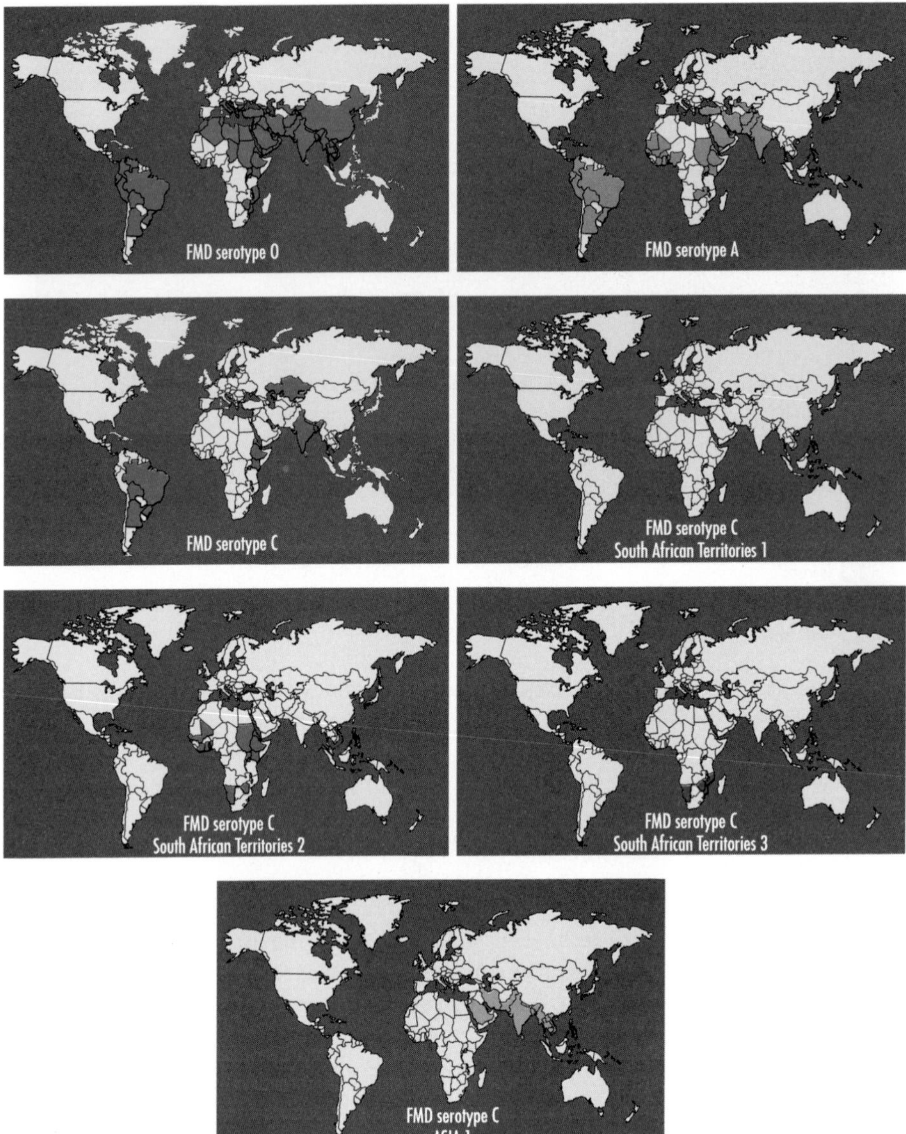

Fig. 10.7 The seven viral strains associated with FMD (serotypes O, A, C, SAT1, SAT2, SAT3, and Asia1) have different geographical ranges, raising the possibility that their stability and transmission are affected by different environmental conditions. Maps by courtesy of Advanced Veterinary Information System (AVIS).

people die from tuberculosis and eight million new cases are reported, despite the general agreement that it is a curable and largely preventable disease (WHO, 2002).

The global incidence of tuberculosis shows a strong geographical pattern, with more cases occurring in the tropical zones. However, the geographical pattern is being diffused by increasing global travel and the importation of cases into regions where the environmental survival of the bacterium is not stable (Figs. 10.9 and 10.10). Rios and colleagues (2000) used a mathematical model to explore the linkage between seasonality and tuberculosis infection. Their results demonstrated an annual seasonal pattern with higher incidence during summer and autumn, although this observation can be interpreted in different ways. The transmission of tuberculosis from person to person is enhanced during cold seasons in temperate zones when most people are indoors, or during very hot seasons in the tropics, also when most people are indoors. In addition, conversion of tuberculosis carriers to clinical-case

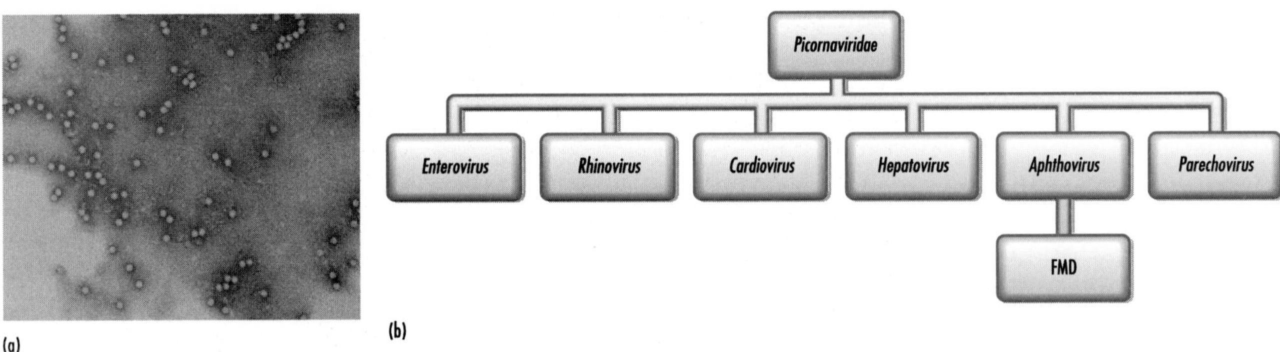

(a)
(b)
Picornaviridae
Enterovirus
Rhinovirus
Cardiovirus
Hepatovirus
Aphthovirus
Parechovirus
FMD

Fig. 10.8 The causative agent of foot and mouth disease (FMD) is a small Picornavirus (a) (family: *Picornaviridae*; genus: *Aphthovirus*). There are seven serotypes of the pathogen: 0, A, C, Asia1, SAT1, SAT2, and SAT3, and there are several subtypes associated with these serotypes. The viruses that cause the disease belong to a versatile family of viruses (b) that includes many human and animal pathogens such as polio (genus: *Enterovirus*); colds (genus: *Rhinovirus*); heart disease (encephalomyocarditis, genus: *Cardiovirus*); liver disease (hepatitis, genus: *Hepatovirus*); and Echovirus disease (genus: *Echovirus 22*). Electron micrograph of the virus by courtesy of Dr. Bhella, University of Glasgow.

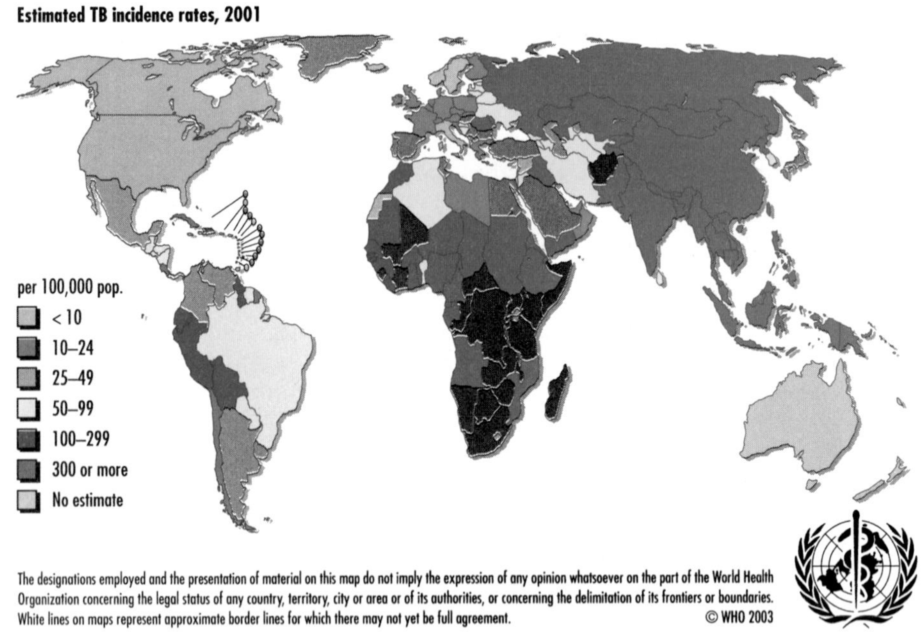

Estimated TB incidence rates, 2001
per 100,000 pop.
< 10
10–24
25–49
50–99
100–299
300 or more
No estimate
The designations employed and the presentation of material on this map do not imply the expression of any opinion whatsoever on the part of the World Health Organization concerning the legal status of any country, territory, city or area or of its authorities, or concerning the delimitation of its frontiers or boundaries. White lines on maps represent approximate border lines for which there may not yet be full agreement.
© WHO 2003

Fig. 10.9 The global incidence of tuberculosis shows an interaction between geography and regional socioeconomic characteristics. Map by courtesy of the World Health Organization, Geneva.

patients may be precipitated by superinfection with pathogens that impact the immune system, including the cold and flu viruses.

Several factors have complicated efforts to control the spread of tuberculosis worldwide. The genus *Mycobacterium* represents a very diverse, metabolically versatile, soil-inhabiting group of bacteria (Cheung and Kinkle, 2001). Resistance to multiple antibiotics is rapidly spreading among *M. tuberculosis* isolates, which makes treatment with contemporary antibiotics difficult to predict (Friedberg and Fischhaber, 2003; Smyth and Emmerson, 2000). The mechanisms of genetic exchange in this group of bacteria have not been thoroughly researched. Therefore, the routes through which multiple antibiotic resistance determinants are acquired are not presently clear. In addition, related organisms such as the *M. avium–M. intracellulare* complex, which were considered to be of little risk to human health, are now

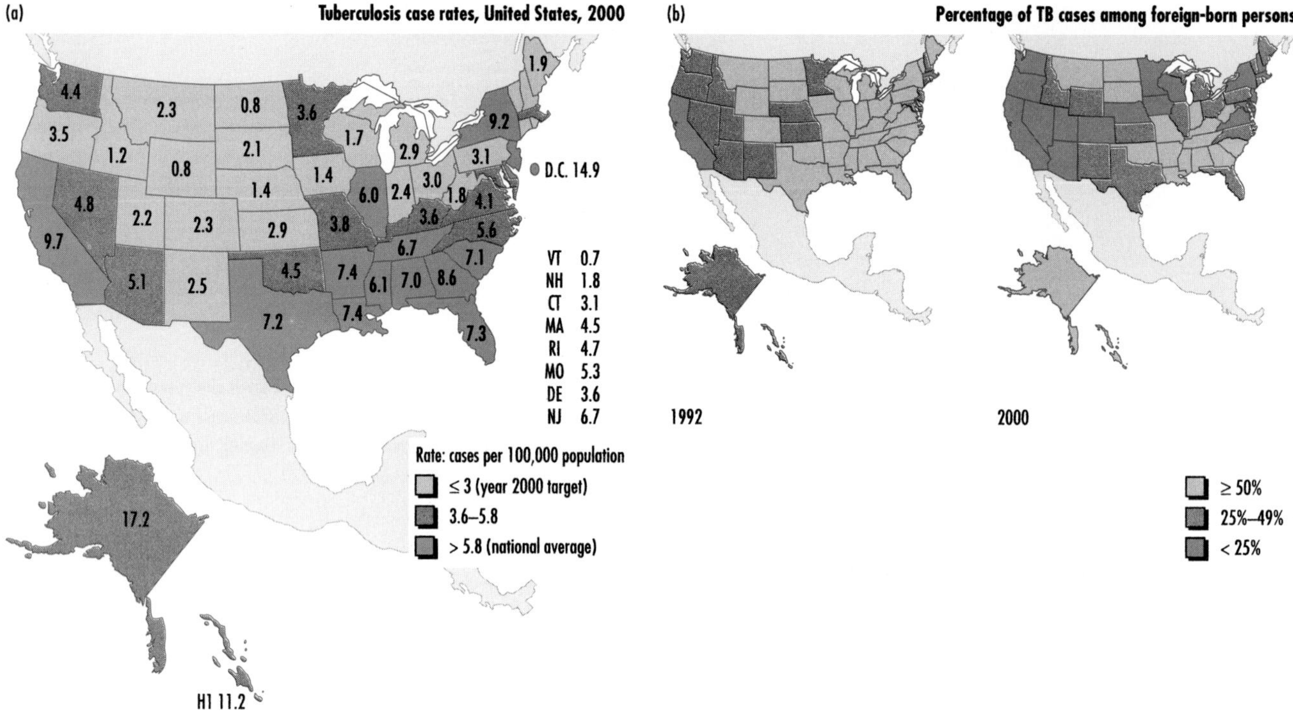

Fig. 10.10 The national incidence of tuberculosis in the United States shows an interaction between geographical factors (a) and immigration (b) (Davidow *et al.*, 2003). Maps by courtesy of the U.S. Center for Disease Control, Atlanta (www.cdc.gov/nchstp/tb/ worldtb2002/NIT.htm).

recognized as major opportunistic pathogens among those living with AIDS. It is possible that interspecies interactions among these organisms can exacerbate the difficulties of dealing with tuberculosis.

Cholera

Water-borne human diseases are among the leading causes of morbidity and mortality in children worldwide. *Vibrio cholerae*, the causative agent of cholera, is one of the most deadly agents of enteric disease acquired from contaminated water and food. The diversity of *V. cholerae* is extensive, including serogroups and serotypes based on the O antigen of the cell wall, and biotypes based on phenotypic differences exhibited in the bacterial population. A thorough understanding of this diversity is extremely important for models being developed to control cholera. For example, more than 150 serogroups of *V. cholerae* are recognized, but only two of these, O1 and O139, have been linked to cholera epidemics (Huq *et al.*, 2001). Although no serotypes have been identified for group O139, group O1 has three, namely, Inaba, Ogawa, and Hikojima. Two biotypes, classical and El Tor, are associated with serogroups Inaba and Ogawa. The distribution of cholera cases does not show a distinct geographical pattern at the global level, although socioeconomic factors influencing access to clean water is probably the most important determinant of cholera epidemics (Fig. 10.11). However, the global distribution of cholera has been unequivocally linked to geographic patterns at the regional level, and to major climatic conditions, including global phenomena such as El Niño (Borroto and Martinez-Piedra, 2000; Collins, 1996; Colwell, 1996). In 2001, 58 countries reported 184,311 cases and 2,728 deaths from cholera infection (approximately 33% more than in the previous year). The case-fatality rate (CFR) decreased to 1.48% from the 3.6% reported in 2000, although the CFR shows considerable geographical variation

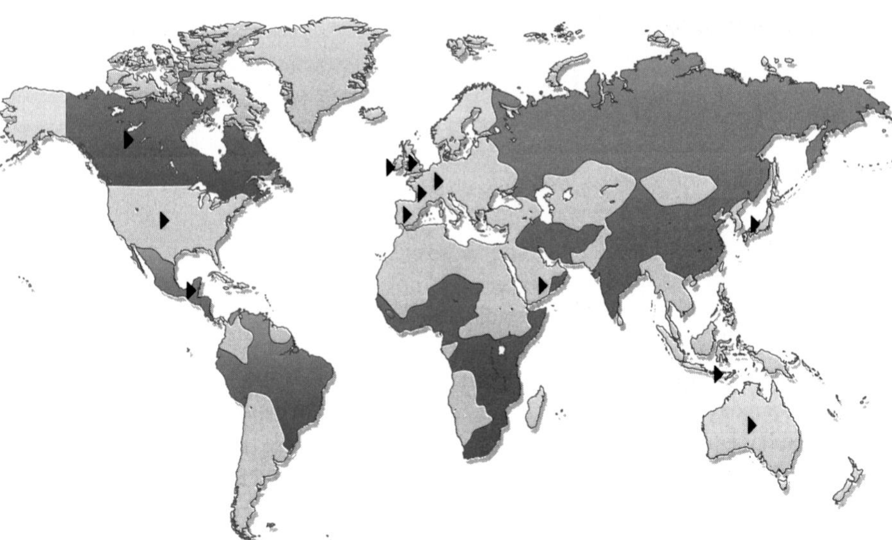

Fig. 10.11 Geographical distribution of cholera during 2000–2001. Shaded areas represent reported cholera cases and triangles represent reported imported cases of cholera. Data from the World Health Organization, Geneva.

from a low value of 0.22% in South Africa to approximately 30% in other African countries (WHO, 2003a).

Recent efforts to control the global spread of regional cholera epidemics have focused on immigration policies, behavioral change, and on the prediction of parameters that link the survival and transportation of *Vibrio cholerae* in marine waters to phytoplankton blooms and sea-surface temperatures (Bouma and Pascual, 2001). There is a historical record of seasonality and interannual variations in the incidence of cholera epidemics. This record was analyzed for the period between 1891 and 1940 by Bouma and Pascual (2001), and they discovered that during post El Niño events when sea-surface temperature rises, epidemics with high mortality occur. In addition, there is a significant correlation between sea-surface temperatures in the Pacific Ocean and the number of deaths that occur from cholera infections in the spring season at coastal areas of the Bay of Bengal. Although causal relationships between climate and cholera epidemics have not been established, several plausible mechanisms have been proposed to explain the interactions. One of these explanations invokes the prolonged survival of *V. cholerae* for up to 15 months on the mucilaginous sheaths of phytoplankton, including *Anabaena variabilis*, which grow to large population densities in warm waters (Islam *et al.*, 1990 and 1994). The plankton thus serves as aquatic reservoirs that facilitate long-range transport of infectious bacteria. This scenario has led to the proposal to develop a forecasting model on cholera epidemics based partially on remote-sensing data on marine chlorophyll concentrations and water temperature (Huq *et al.*, 2001).

Given the abundance and diversity of *Vibrio* species in the marine environment, it is reasonable to suppose that environmental factors act on a finer scale in addition to episodic climate events to support the survival and dissemination of the specific pathogenic serogroups of *V. cholerae*. Differentiating among these sorts of fine-scale organism–environment interactions is the subject of intense research (Benediktsdottir *et al.*, 2000).

Understanding the interactions between pathogenic varieties of microorganisms, environmental factors, and human-host characteristics, is increasingly recognized as a major pathway for controlling the burden of diseases associated with pathogens in developing countries, and for modeling the constellation of factors that contribute to the emergence of new diseases (Martin, 2003). Such understanding can aid in the development of broad-spectrum public health strategies, including but not limited to more effective vaccination and innovative methods of biological control. For example, understanding the geographical dimension of the distribution of pathogenic forms of *Helicobacter pylori* has contributed to the clinical and epidemiological management of gastrointestinal diseases associated with common *H. pylori* exposures and the environmental influences on the genetic exchange mechanisms that support the acquisition of the "pathogenicity island" in the genome of different strains of this organism (Covacci *et al.*, 1999; Graham, 2003). Other examples include the understanding of time–space clustering of diarrhea disease associated with *Clostridium difficile*, and the evolution of hepatitis B viruses from the perspective of the geographical selection of genetic varieties (Kroker *et al.*, 2001; Robertson and Margolis, 2002). Advances in the geographical dimensions of microbial diversity and pathogenesis are fueled by the development of innovative approaches in molecular epidemiology and in spatial mapping techniques, including satellite-enabled remote sensing and geographical information systems (Balmelli and Piffaretti, 1996; Bithell, 2000; Glass, 2000; Kistemann *et al.* 2002; Scott *et al.*, 1996).

DISEASES OF MARINE ORGANISMS

Global marine ecosystems contain a major proportion of the biological diversity on Earth, but relative to terrestrial systems, there is a paucity of knowledge concerning the adverse interactions existing across phylogenetic groups in this domain. In general, marine mammals are sensitive to the same categories of pathogens as terrestrial mammals. However, many findings that several organisms in the marine environment are succumbing to pathogens that may have "crossed over" from domestic and agricultural environments have raised serious concerns. The compromise of immune systems by exposure to toxic chemical pollutants has been implicated in such cross-over events. Additional stress factors have been attributed to global environmental change and habitat destruction (Harvell *et al.*, 1999). For example, there are documented cases of seals contracting canine distemper virus, which was typically limited to dogs; sardines contracting herpes virus disseminated from pilchards; and corals killed by a soil-borne fungus (Harvell *et al.*, 2002).

These challenges to conventional thinking on ecosystem health exemplify the need for a vigorous integrative approach toward the understanding of the co-evolution of pathogenic microorganisms and host organisms in environments that are undergoing rapid change. Recent problems associated with coral ecosystems have been studied extensively, but not much has resulted in terms of attributions of cause and effect. For example, coral bleaching has been attributed, separately and in various combinations, to several environmental factors including exposure to ultraviolet radiation, bacterial infection, protracted exposure to light, high temperatures, changing salinity, high turbidity, oceanic deposition of desert dust, and dysfunctional symbiosis (Garrison *et al.*, 2003; Rowan *et al.*, 1997). In fact, coral bleaching may result from a generalized stress response, which after exceeding a hypothetical threshold leads to irreversible damage (Hoegh-Guldberg, 1999; Marshall and Baird, 2000). At least 18 different diseases of corals are thought to be linked to microbial infections (Kushmaro *et al.*, 1996; Ritchie and Smith, 1995). This include the "black-band disease" where a band of bacterial consortium, consisting of the photosynthetic gliding cyanobacterium *Phormidium corallyticum*, sulfate-reducing bacteria, sulfur-oxidizing bacteria, and other heterotrophic bacteria, radiates outward at a rate of up to 1 cm per day during warm-water conditions (Carlton and Richardson, 1995; Frias-Lopez *et al.*, 2003; Richardson and Kuta, 2003). Figure 10.12 shows an infected coral, and Fig. 10.13 demonstrates the irreversible damage imposed on corals after a protracted infection period. The white area in the center is dead tissue-free coral skeleton. The coral tissue is killed by the anoxic conditions within the microbial band and by exposure to hydrogen sulfide produced by the sulfate-reducing microorganisms. Organic compounds secreted by the affected coral cells are used as nutrients by the microorganisms, but clearly, this is not a sustainable situation. Ultimately, the skeleton will be attacked by boring algae, sponges, clams, and parrot fish. These scavenger

Fig. 10.12 Telltale ringworm symptom of black-band disease of coral caused by bacterial infection. By courtesy of the United States Geological Survey.

(a)

(b)

(c)

Fig. 10.13 A large star coral (*Montastrea annularis*) attacked by black band disease in 1988 (a) and mostly dead by 1998 (b). By 2001, the coral mortality was complete and irreversible (c). By courtesy of the United States Geological Survey.

organisms reduce the size of the coral by approximately 1 cm annually (Kuta and Richardson, 2002; USGS, 2003).

Black-band disease of coral is infectious, putting many healthy geographically separated corals at risk. However, some studies indicate that a predisposition to infection is required for the establishment of bacterial pathogens on coral (Richardson and Kuta, 2003). Environmental conditions that predispose coral to infection include stress factors such as sedimentation, nutrient loading, toxic chemicals, and higher than normal temperatures. In this context, the hypothesis proposed by scientists at the United States Geological Survey (USGS) that dust from the Sahara desert is partly responsible for the declining health of coral reefs in the Caribbean ocean warrants close attention from an Earth system perspective (Garrison *et al.*, 2003). Biogeochemical cycling and ocean fertilization with trace metals, nutrients, and microbes are at the core of the "Sahara dust" hypothesis. There is ample satellite-based evidence that dust from the Sahara desert settles as far as the Amazon forest where it is a major source of phosphate nutrients, and as far as the Caribbean islands, where it contributes to the red iron and clay rich soils. Historical records of effluent dust from increasing desertification in northern Africa also correlate with the decline of coral reefs.

According to the authors of the "Sahara dust" hypothesis, direct fertilization of benthic algae by iron and other nutrients interacting with the ammonium and nitrogen oxides content of submarine water, produces an optimum growth environment for potential pathogens including bacteria, viruses, and fungal spores that are also transported by the dust. For example, the causative agent of Aspergillosis in the Caribbean population of sea fans has been associated with a terrestrial strain of *Aspergillus* species that do not ordinarily inhabit marine environments (Garrison *et al.*, 2003). Although the "Sahara dust" hypothesis remains to be verified in detail, there are already sufficient data to support the growing realization that the impacts of global environmental change phenomena such as desertification have strong implications for regional microbial diversity and the complex ecological functions in which microorganisms are intricately involved.

THE BENEFICIAL EFFECTS OF MICROBE–EUKARYOTE INTERACTIONS

This chapter has focused on the detrimental repercussions of interactions between microorganisms and large multicellular eukaryotes, plants, animals, and humans. It is worth emphasizing that these interactions are classified as detrimental, only if the observer adopts the perspective of the eukaryotic hosts that suffer morbidity and mortality as a result of encounters with pathogenic microorganisms. In fact, pathways to disease causation are full of many different interactions among microorganisms that co-inhabit a host, and between the host cells and the invading microbial species. It is only when intrinsic defense mechanisms are overwhelmed and the potential pathogen population reaches a critical level (infectious dose) that the disease state can progress. Prominent among intrinsic host defense mechanisms are the beneficial presence and action of many other microorganisms that are also present on the host. For example, in plants, the recognition that many pathogens can be controlled by the application of other microorganisms has been exploited as strategies for biological control of plant diseases (see Box 10.1). A detailed understanding of the microbial diversity in the phytosphere is required to take full advantage of such strategies (Zhou *et al.*, 2002). Many other beneficial interactions between microorganisms and plants have been discussed in previous chapters. These include the symbiotic relationship between *Rhizobium* species and leguminous plants that is essential for nitrogen fixation (Mhamdi *et al.*, 2002). Many rhizosphere-inhabiting fungi and bacteria also produce extracellular enzymes that make nutrients available for plants. The understanding of soil microbial diversity is also needed for improved management of agricultural practices such as the application of fertilizers and pesticides that have broad ecosystem impacts.

BOX 10.1

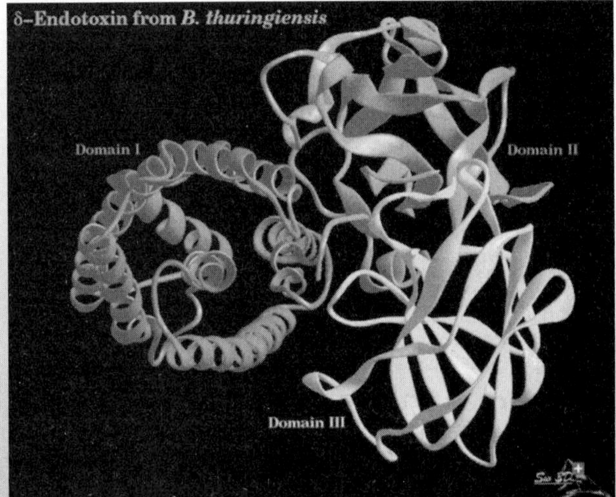

(a)

(b)

(c)

The subjective nature of the judgment about whether or not a particular interaction between prokaryotes and eukaryotes is beneficial is exemplified here by the biotechnological control of plant pests using the spores of *Bacillus thuringiensis* (Bt-toxin). *B. thuringiensis* is a soil bacterium that is selectively pathogenic against many strains of insects. As such, the interaction between the prokaryote and eukaryote in this regard is not beneficial. However, many insects that are sensitive to Bt-toxin are pests for food crops, and the biological control of such pests is beneficial for the human agricultural enterprise. The spores produced as part of the natural life cycle of *B. thuringiensis* contain a crystalline protein (Cry) which breaks down under the alkaline condition of insect guts to release a toxin that paralyses the digestive system of insects which ingest the spores as part of their feeding habits. Although all Cry proteins show the three domains presented in (a), there is considerable molecular diversity in the genes encoding Cry, and the diversity results in different levels of effectiveness of the Cry protein in inducing morbidity and mortality against a wide range of insect orders. Five main categories of Cry protein are recognized as described in the accompanying table.

Traditional use of the Bt-toxin involves the deliberate release of *B. thuringiensis* spores into the environment where crops are to be protected. However, the spores are sensitive to various environmental factors, including natural ultraviolet light radiation; therefore their effectiveness as pesticides is limited in certain ecosystems. This limitation has led to the successful cloning of the Bt gene directly into crop plants that express the Cry protein, and are therefore resistant to certain crop pests. The large-scale cultivation of such genetically engineered crops, for example, corn plants resistant to the European corn borer (b) and cotton plants resistant to the cotton bollworm (c) has become controversial due to public concerns about genetically engineered foods. However, there is no credible scientific evidence for potentially detrimental effects of the Bt-toxin on humans. On the other hand, the potential effects on non-target wildlife populations have not been ruled out (Webber *et al.*, 2003). Protein diagram and photographs are by courtesy of the Agricultural Research Services of the United States Department of Agriculture (Keith Weller, (b); and Scott Bauer, (c)).

There are also many beneficial interactions between microorganisms and animals. Perhaps the most widely known example is the intestinal microorganisms that are essential for digestive processes in ruminants and other organisms. Herbivores consume large amounts of cellulosic and other plant materials, which are fermented in part by the complex microbial community present in the intestines (White *et al.*, 1999). There are two major reasons for the considerable and growing research interest in understanding the microbial diversity of the rumen in agricultural animals. The first is to facilitate more efficient utilization of animal feed in order to maximize the cost-benefit ratio. Secondly, the product of microbial fermentation in the rumen is methane, a potent greenhouse gas. For example, it has been estimated that annual emission of methane from cows (live weight 650 kg, annual milk production: 7,550 kg, fat content 3.8%) is approximately 183 kg CH_4/cow or 137 kg C-CH_4/cow (diet with corn silage) (Rossi *et al.*, 2001). This is a large amount of methane released to the atmosphere, given that the number of cows in the global agricultural industry exceeds five billion. It might be possible to reduce this level of greenhouse gas emissions by manipulating the microbial diversity of cow rumen. Such manipulations may be achieved by introducing "beneficial" microorganisms that might displace closely related naturally occurring strains (Chan and Dehority, 1999).

Humans also benefit from close physical association with microorganisms. As in the case of other animals, the human digestive system is populated by a highly diverse community of microorganisms that assist in the digestive process, and in the synthesis of trace nutrients (Marsh, 1995). The human skin serves not only as a physical barrier against the entry of many potentially pathogenic microorganisms, but also as a habitat for a complex microbial community, many of whose members are beneficial through their interactions with exogenous microorganisms (Papapetropoulou and Sotiracopoulou, 1994). Understanding the diversity of the human microflora has implications for strategies being developed to control traditional and emerging infectious diseases. For example, the emergence of resistance to multiple antibiotics in pathogenic microorganisms has led to renewed interest in phage therapy, or the control of bacteria by lytic viruses (Godany *et al.*, 2003; Joerger, 2003). Several obstacles must be overcome before phage therapy can be implemented reproducibly, and its success will depend on better knowledge of the diversity of potential receptors and evolutionary selection factors that influence the activity of bacteriophage particles in complex microbial communities inhabiting an even more complex internal environment of multicellular eukaryotes.

CONCLUSION

Biogeography, defined as the study of the geographical distribution of organisms, their habitats, and the historical and biological factors which produce them (Lincoln *et al.*, 1982), is a major component of how we understand and catalog biodiversity. Ecological and historical biogeographies have enjoyed relatively little attention in microbiology when compared with other disciplines of biology, but the situation is changing. The driving forces behind this change are attributable to the recognized impacts of interactions between microorganisms (viruses, prokaryotes, and eukaryotes) and the large multicellular eukaryotes that are important to human affairs. The distributions of plant, human, and animal pathogens exhibit strong geographical patterns which reflect, in part, the interactions between local microbial diversity, environmental parameters, and host-susceptibility factors. Various attempts by humans to control diseases and to improve general welfare through industrial activities also influence microbial diversity. For example, the wide distribution of antibiotics reinforces the molecular determinants of microbial diversity through selection for drug-resistant varieties. Similarly, the deployment of pesticides, fertilizers, and other chemicals affects local microbial diversity with implications for the integrity of global ecosystem functioning. Finally, the association of special microbial physiological properties with particular geological factors and geographical landscapes provides essential resources for biotechnology, including access to enzymes which can function under high temperature and pressure conditions. Clearly, a more intensified research investment in exploring the geographical dimensions of microbial diversity will benefit public health, ecosystem integrity, and major industrial activities.

QUESTIONS FOR FURTHER INVESTIGATION

1 Plant pathogens are responsible for large economic losses in the global agricultural industry. The solution to this problem has relied heavily on the application of toxic chemical pesticides, some of which have detrimental ecosystem effects because of their action on non-target species. Alternative strategies for dealing with microbial plant pathogens include biological (biotechnological) control measures, where pathogens are targeted with antagonistic organisms or their products. Search the printed literature and the World-Wide-Web for examples of chemical pesticides used to treat at least two of the plant diseases discussed in this chapter. In addition, find examples of biological control strategies for at least two plant pathogens described in the chapter.

(a) Discuss how factors contributing to the evolution of diversity in the pathogen population might lead to the development of resistance to these two kinds of pathogen control agents.

(b) Compare and contrast the ecological risks and benefits of using the chemical and biological control agents.

2 Many microbial pathogens that infect animals are transmitted by mechanisms that are also relevant to the transmission of human infectious diseases. As such, rigorous quarantine strategies are used to separate infected animals from healthy populations. The success of quarantine strategies depends on the availability of molecular screening techniques that can be used to differentiate individual animals according to infection status. In some cases, microbial pathogens evolve rapidly to the extent that a diverse population of potential pathogens may emerge quickly from a single clone. For the two animal diseases discussed in this chapter, search the literature to find and discuss examples of molecular screening techniques that are used to discriminate on the basis of infection. In addition, find and discuss examples of how the recognition of diversity in the pathogen population has confounded disease status screening. In your examples, is there a geographical dimension to where variants of recognized pathogens are likely to emerge? If so, what are the likely reasons for this geographical dimension?

3 The reports of the Intergovernmental Panel on Climate Change (IPCC) include substantial discussion of the possible impacts of projected climate change on infectious human diseases. In this context, much emphasis has been placed on insect-borne diseases such as Lyme disease, and on water-born diseases such as cholera. In general, there have been very little analysis and discussion of microbial pathogens, such as the flu virus, known to be strongly influenced by seasonal changes, and by rapid evolutionary process generating tremendous molecular diversity on an annual basis. Develop arguments for and against the hypothesis that climate change accompanying a 2°C increase in average global temperature will reduce the global burden of flu disease on the basis of annual mortality. You may wish to consult IPCC reports and published journal articles on scenarios of climate change and their implications on public health. The website of the World Health Organization includes data on the geographical distribution of annual morbidity and mortality attributable to the flu virus.

4 The integrity of coral reefs depends on complex and sensitive symbiotic relations that cut across phylogenetic groups, and the health state of the world's coral reefs has been regarded as the essential "pulse" for marine ecosystem health. Discuss the evidence for and against at least two separate hypotheses that have been proposed as the cause of coral bleaching. In what ways does the diversity of organisms involved in coral symbiosis contribute to the discussion of cause and effect regarding the etiological agents of bleaching?

SUGGESTED READINGS

Alexandersen, S., Z. Zhang, A.I. Donaldson, and A.J.M. Garland. 2003. The pathogenesis and diagnosis of foot-and-mouth disease. *Journal of Comparative Pathology*, **129**: 1–36.

Aron, J.L. and J.A. Patz (eds.) 2001. *Ecosystem Change and Public Health: A global Perspective.* Baltimore, Johns Hopkins University Press.

Balmelli, T. and J.-C. Piffaretti. 1996. Analysis of the genetic polymorphism of *Borrelia burgdorferi* sensu lato by multilocus enzyme electrophoresis. *International Journal of Systematic Bacteriology*, **46**: 167–72.

Bouma, M.J. and M. Pascual. 2001. Seasonal and interannual cycles of endemic cholera in Bengal 1891–1940 in relation to climate and geography. *Hydrobiologia*, **460**: 147–56.

Carlton, R.G. and L.L. Richardson. 1995. Oxygen and sulfide dynamics in a horizontally migrating cyanobacterial mat: Black-band disease of corals. *FEMS Microbiology Ecology*, **18**: 155–62.

Colwell, R.R. 1996. Global climate and infectious disease: The cholera paradigm. *Science*, **274**: 2025–31.

Covacci, A., J.L. Telford, G.G. Del, J. Parsonnet, and R. Rappuoli. 1999. *Helicobacter pylori* virulence and genetic geography. *Science*, **284**: 1328—33.

Crisci, J.V., L. Katrinas, and P. Posadas. 2003. *Historical Biogeography: An Introduction.* Cambridge, MA: Harvard University Press.

Desbiez, C., C. Wipf-Scheibel, and H. Lecoq. 2002. Biological and serological variability, evolution and molecular epidemiology of Zucchini yellow mosaic virus (ZYMV, Potyvirus) with special reference to Caribbean islands. *Virus Research*, **85**: 5–16.

Garrison, V.H., E.A. Shinn, W.T. Foreman, D.W. Griffin, C.W. Holmes, C.A. Kellogg, M.S. Majewski, L.L. Richardson, K.B. Ritchie, and G.W. Smith. 2003. African and Asian dust: From desert soils to coral reefs. *Bioscience*, **53**: 469–80.

Glass, G.E. 2000. Update: Spatial aspects of epidemiology: The interface with medical geography. *Epidemiologic Reviews*, **22**: 136–9.

Harvell, C.D., C.E. Mitchell, J.R. Ward, S. Altizer, A.P. Dobson, R.S. Ostfeld, and M.D. Samuel. 2002. Climate warming and disease risks for terrestrial and marine biota. *Science*, **296**: 2158–62.

Hoegh-Guldberg, O. 1999. Climate change, coral bleaching and the future of the world's coral reefs. *Marine and Freshwater Ecology*, **50**: 839–66.

IPCC. 2001. *Climate Change 2001: Synthesis Report. Contribution of Working Groups I, II and III to the Third Assessment Report of the Intergovernmental Panel on Climate Change.* R.T. Watson and the Core Writing Team (eds.) Cambridge, England: Cambridge University Press.

Kistemann, T., F. Dangendorf, and J. Schweikart. 2002. New perspectives on the use of Geographical Information Systems (GIS) in environmental health sciences. *International Journal of Hygiene and Environmental Health*, **205**: 169–81.

Kushmaro, A., Y. Loya, M. Fine, and E. Rosenberg. 1996. Bacterial infection and coral bleaching. *Nature*, **380**: 396.

Kuta, K.G. and L.L. Richardson. 2002. Ecological aspects of black-band disease of corals: Relationships between disease incidence and environmental factors. *Coral Reefs*, **21**: 393–8.

Narayanasamy, P. 2002. *Microbial Plant Pathogens and Crop Disease Management.* Enfield, New Hampshire: Science Publishers.

Prusiner, S.B. 1995. The prion diseases. *Scientific American*, **272**: 30–7.

Purdey, M. 1998. High dose exposure to systemic phosmet insecticide modifies the phosphatidylinositol anchor on the prion protein: The origins of new variant transmissible spongiform encephalopathies? *Medical Hypotheses*, **50**: 91–111.

Purdey M. 2000. Ecosystems supporting clusters of sporadic TSEs demonstrate excesses of the radical generating divalent cation manganese. *Medical Hypotheses*, **54**: 278–306.

Smyth, E.T. and A.M. Emmerson. 2000. Geography is destiny: Global nosocomial infection control. *Current Opinion in Infectious Diseases*, **13**: 371–5.

Zhou, J., B. Xia, D.S. Treves, L.-Y. Wu, T.L. Marsh, R.V. O'Neill, A.V. Palumbo, and J.M. Tiedje. 2002. Spatial and resource factors influencing high microbial diversity in soil. *Applied and Environmental Microbiology*, **68**: 326–34.

MICROBIAL DIVERSITY AND GLOBAL ENVIRONMENTAL ISSUES

Chapter contents

INTRODUCTION

There is a meta-cultural conventional wisdom that the intrinsic values of tangible endowments are never fully appreciated until their demise occurs. In view of the fact that no microorganism has ever been proven to become extinct, it follows that we may never truly appreciate the significance of their individual contributions to the global ecosystem (Pimm and Raven, 2000). There is no shortage of answers to the question: **Of what use are microorganisms?** Many consumer products resulting from the multibillion dollar biotechnology industry demonstrate the contributions of specific microorganisms to the global economy. However, it is more challenging to find answers to the question: **Of what use is microbial diversity?** The easy but shortsighted response is that the existing diversity of microorganisms provides a resource reservoir from which individual species with special traits can be selected to serve biotechnological purposes. This response is inadequate because it ignores the functional *gestalt* of microbial communities, where multiple and complex interactions define the sustainability of critical ecological processes. Therefore, it is important to make a distinction here between **microbiological resources** and **microbiological diversity**, which is the variability of genes, species, and community composition. A compendium of resource-driven interests in microbial diversity is now available (Bull, 2003). The emerging conundrum regarding the appreciation of microbial diversity in the context of mainstream discussions on environmental conservation is an extension of the difficulty faced by investigators attempting to quantitatively value global biodiversity as an integrated community of organisms. These attempts are sometimes rejected for being too subjective or too frequently plagued by uncertainties and wide margins of error (Milon and Shogren, 1995; OECD 2002).

A long stretch of the early history of environmental microbiology was necessarily dominated by observations of microbial forms and functions in pure cultures. The construction of "microbial diversity" as a distinct concept emerged from classical microbial systematics, which was based on user-defined phenotypic differences along a continuum of anatomical and physiological traits. The concept of microbial diversity also relied on inferences about possible interactions among organisms isolated from similar ecosystems. These inferences were based on the characteristics of individual isolates under well-defined, unchanging growth conditions. It is now possible to observe microbial forms and functions within various ecological communities with unprecedented clarity using new technical capabilities engendered by progress in molecular biology. Comparative assessments of molecular sequence data have revolutionized the traditional concepts of microbial diversity, and these comparisons bring new opportunities for understanding and appreciating the concerted influence of microbial communities to the integrity of ecosystems. It is fairly straightforward to glimpse the ecological influences of microorganisms through long-term habitat monitoring programs at the local level, but a global ecosystem perspective is essential for building robust information databases that link microbial diversity to the dynamic nature of environmental parameters. This chapter focuses on strategies for making such linkages, beginning with the relationship between indexes of microbial diversity and the indicators related to changes in global environmental conditions. Thereafter, specific global environmental change events are considered as examples to demonstrate the potential for using assessments of microbial diversity to detect environmental change and to use such knowledge in the best service of ecosystem restoration.

MICROBIAL DIVERSITY AND INDEXES OF ENVIRONMENTAL CHANGE

The international Convention on Biological Diversity (CBD) defines biodiversity as "the variability among living organisms from all sources including, *inter alia*, terrestrial, marine, and other aquatic ecosystems and the ecological complexes of which they are part" (UNEP, 1992). For those concerned with quantitative assessments of biological diversity, the key issue in the CBD definition is how to measure **variability**. The development of robust indicators of biodiversity has been fraught with problems for several reasons pertaining to the different scales at which biodiversity is usually presented. For example, there are no universally accepted discrete boundaries between functional genes, between species, or between ecosystems. Therefore, it is not possible to quantify the absolute numbers of varieties within each of these three categories across phylogenetic domains. Perlman and Adelson (1997) have alleged that the current definitions of biodiversity on the basis of genes, species, and ecosystems fail both in theory and practice because they do not recognize the conceptual difficulties inherent in these constituent terms; they ignore practical and technical problems involved in making real-world inventories; they do not account for the lack of commensurability between different scales; and they make no distinctions in the worth of biodiversity units within each scale of assessment.

Despite these shortcomings, attempts to develop quantitative indexes of biological diversity that are sensitive to environmental change have relied on three major concepts, namely scale, component, and viewpoint (Van Kooten, 1998). The **scale** aspect focuses on the criteria of species richness and the geographical distributions of individuals among the species (evenness). Species richness within a local ecosystem is referred to as **alpha** diversity. The variation in alpha diversity among ecosystems within the same landscape is referred to as **beta** diversity, and, when measurable, **gamma** diversity represents species richness at the regional and global scales (Fig. 11.1). Gamma diversity is sensitive primarily to phenomena that impact the environment at the global level (e.g. major climate shifts) as opposed to local-scale impacts such as erosion, fires, the introduction of new invasive species, and exposure to toxic chemical pollutants, all of which influence alpha and beta diversities. Applications of the **component** concept in assessing biodiversity are aimed at empirically determining the minimum viable number of species necessary to maintain ecosystem functions. Finally, the **viewpoint** concept, perhaps the vaguest and most subjective of the three criteria, refers to the necessity of juxtaposing alternative viewpoints in assessing biodiversity, including but not limited to utilitarian, practical, and aesthetic viewpoints (NRC, 1999; OECD, 2002).

Not all these criteria apply uniformly to the assessments of microbial diversity without major refinements. For example, in estimating microbial species richness, it is frequently not possible to account for all species in an environmental sample. Therefore, carefully defined subsets are designated as indicator or surrogate species, and these are monitored to repre-

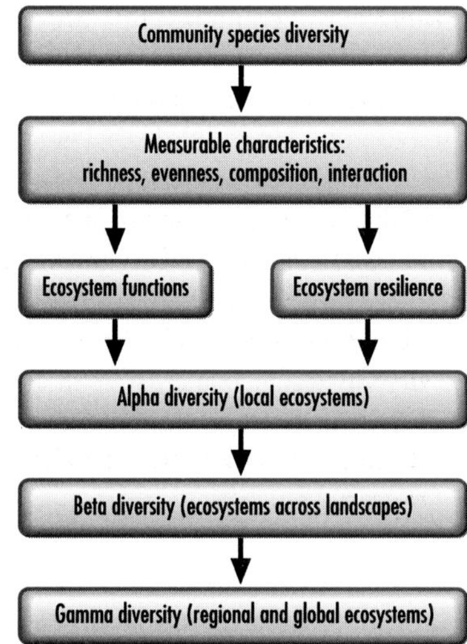

Fig. 11.1 Indexes of species diversity and ecosystem function are based on some quantitatively measurable characteristics such as species richness, evenness of distribution, and composition. Some measures are more difficult to quantify, such as interactions among different species. Indexes of species diversity and parameters of environmental change are linked through the roles performed by species in sustaining ecosystem functions and ecosystem resilience. These roles are integrated into measures of biodiversity at different geographical scales, namely alpha, beta, and gamma diversities, which are sensitive to perturbations at the habitat, ecosystem, and global scales, respectively.

sent overall diversity (Fjeldsa, 2000; Williams and Humphries, 1996). Species "weighting" can also be used as a shortcut to arrive at a quantitative estimate of changing diversity in microbial communities where only a fraction of existing species have been characterized. For example, the presence and relative population densities of organisms known to have unique ecological functions (e.g. nitrogen fixation or biodegradation of a particular toxic compound) are assigned larger weights than the presence of organisms whose ecological functions are either unknown or not immediately relevant for the purpose of the biodiversity assessment. Weighting is facilitated by the construction of phylogenetic trees describing the relationships between microorganisms present in a particular habitat (see Chapter 5). Phylogenetic trees provide a visual representation of the level of redundancy that may exist within groups and among distantly related species. Changes in environmental conditions that are likely to affect most members of phylogenetic groups can also be weighted more highly than changes that affect a fraction of members belonging to phylogenetic groups with several representatives.

Quantitative measures of species diversity

No single quantitative index of biological diversity can capture the magnitude and direction of environmental change. However, there is no shortage of ecological techniques for monitoring changes in the biological diversity of specific ecosystems, although most of the available field methods were designed for macroscopic organisms with fairly well-known spatial distributions and physiological characteristics (Purvis and Hector, 2000). There are several reasons why the tools that have been developed for assessing biodiversity in plant and animal communities are not appropriate for measuring microbial diversity. Ecologists have traditionally relied on integrating multiple biotic and abiotic indexes to explain past trends in environmental parameters and to account for trajectories of change. The simplest measures of species diversity rely only on the number of species (s) and the total number of individuals representing all the species (N). For example, the Margalef index (D_m) computes the species diversity according to the following equation (Margalef, 1958 and 1963):

$$D_m = (s-1) \div \sqrt{\log N}$$

The Margalef index and similar indexes do not support the differentiation of communities that have identical s and N values because the evenness of the distribution of individuals within the communities is not considered. To correct this shortcoming, some investigators have introduced the concept of species dominance into measures of diversity. For example, Simpson (1949) demonstrated that if two individuals are selected randomly from a community, the probability (P_d) that the two individuals belong to the same species is a measure of dominance (d_s), and it is given by the following equation:

$$d_s = 1 - P_d = 1 - \sum n_i(n_i - 1) \div N(N-1)$$

where n_i is the abundance of individual members belonging to species i. Simpson's measure of dominance has been modified to compute species diversity as follows:

$$D_s = 1 \div P_d = N(N-1) \div \sum n_i(n_i - 1)$$

Simpson's diversity index has been referred to as the probability of an interspecific encounter, which expresses the number of times required to select two independent individuals at random from the community before both are found to belong to the same species (Hurlburt, 1971; Brower *et al.*, 1998). Estimates of D_s are based on the assumption that the data on the number of species and abundance of individual members are derived from randomly collected environmental samples. However, in cases where it is possible to conduct an exhaustive sampling of a community (e.g. microcosm experiments), where direct molecular methods that can capture the complete spectrum of diversity are used (see Figs. 11.2 and 11.3), or where other non-random methods are applied, modified Simpson's indexes of dominance (λ) and diversity (Δ_s) are represented as follows:

$$\Delta_s = 1 \div \lambda = N^2 \div \sum n_i^2$$

These probability-based measures of species diversity have not found as much use as measures based on the concept of uncertainty as defined by information-theoretic indexes. In a community with relatively low species diversity, there is a relatively high level of certainty that the identity of a species selected at random can be predicted. Conversely, in a relatively diverse community, the level of certainty in predicting the species identity of a randomly selected individual is low. The Shannon diversity index (H') is perhaps the best known of diversity measures rooted in information theory (Perkins, 1982; Shannon, 1948):

$$H' = -\sum p_i \log p_i$$

where p_i is the fraction of the total number of individuals in the community that belong to species i. The equation can be rewritten to facilitate the calculation of H' without the need to convert abundances (n_i) to proportions (p_i) as follows:

$$H' = (N \log N - \sum (n_i \log n_i)) \div N$$

Skujins and Klubek (1982) used Shannon's index to investigate the correlations between the diversity of microbial populations, the nitrification potential, changes in organic carbon content, and nitrogen content pools of soils along a montane sere consisting of meadow–aspen–fir–spruce (a sere is the sequence of communities that develops through the process of ecological succession). These investigators demonstrated that when organic

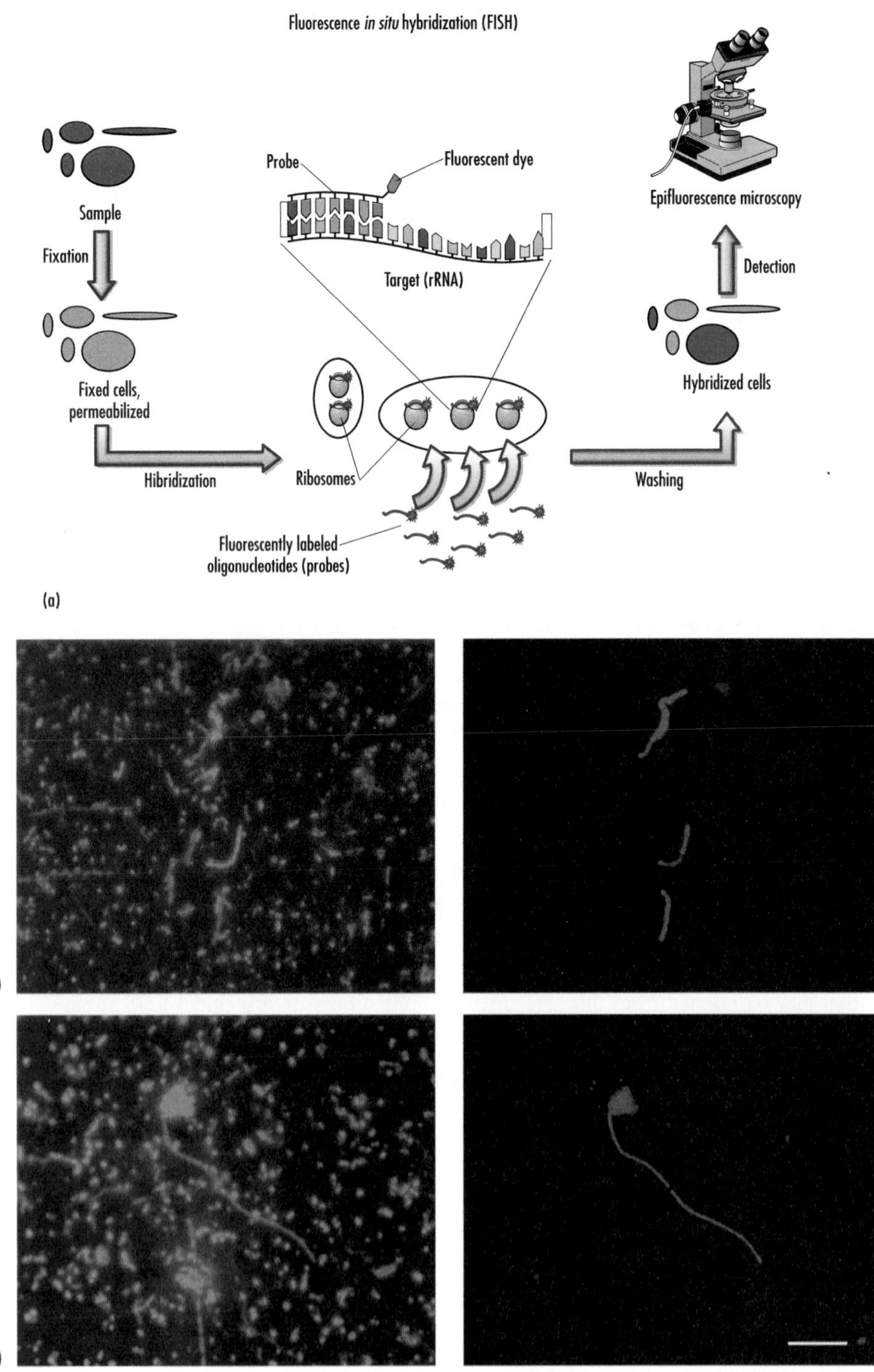

Fig. 11.2 Direct molecular methods aimed at collecting data appropriate for computing indexes of microbial diversity include fluorescence *in situ* hybridization (FISH). (a) Diagrammatic scheme for conducting FISH with oligonucleotide rRNA gene probes. This technique is particularly suitable for directly assessing microbial diversity in environmental samples where many of the organisms are known and their molecular signatures have been identified. (b) and (c) *In situ* identification of bacteria in a water sample from Piburger See in Germany by a combination of hybridization with CY3-labeled, rRNA-targeted oligonucleotide probes and DAPI staining. (b) Hybridization with probe EUB338 specific for Bacteria. (c) Hybridization with probe BET42a specific for beta-subclass *Proteobacteria.* Identical microscopic fields are shown by epifluorescence microscopy using filter sets specific for DAPI (left) and CY3 (right). By courtesy of Dr. Frank Oliver Glöckner, MPI for Marine Microbiology, Bremen, Germany.

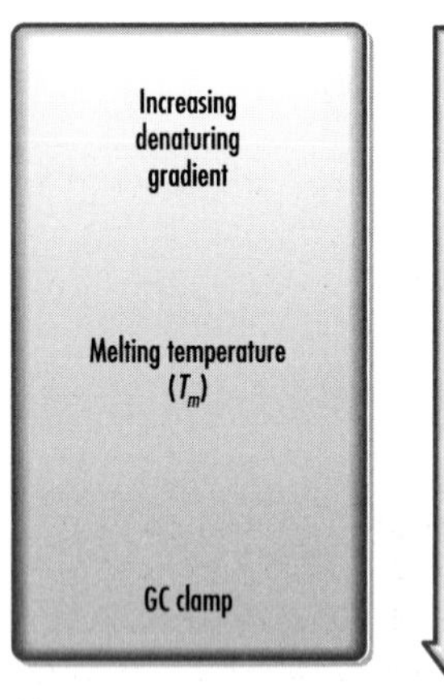

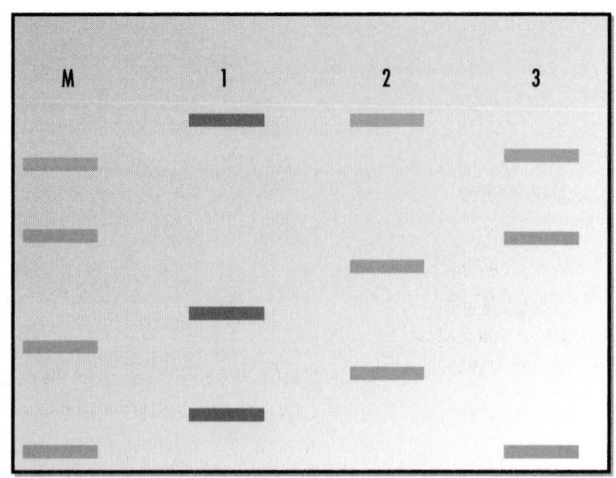

(a)

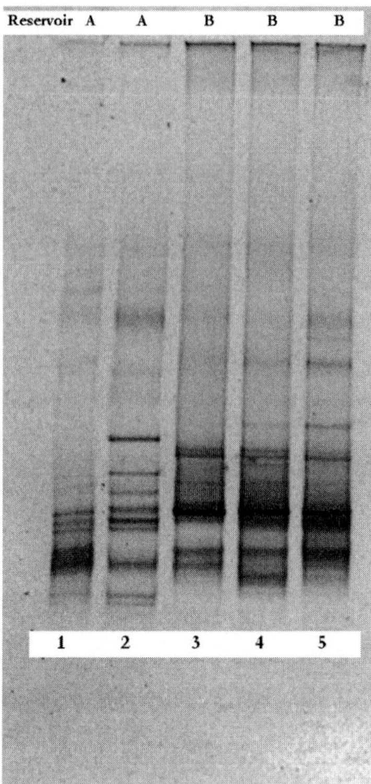

(b)

Fig. 11.3 Polymerase chain reaction and denaturing gradient gel electrophoresis (PCR-DGGE) based on unique 16S rDNA sequences are particularly suitable for assessing overall diversity in microbial communities where many of the organisms have not been cultured. (a) depicts the scheme for DGGE. Lane M represents a molecular size marker, and each band in lanes 1, 2, and 3 represents microbial phylotypes after separation of 16S rDNA PCR products. (b) shows an electrophoresis gel result of DGGE used to resolve the variety and differences of bacterial communities in groundwater well samples originating from two different hot oil reservoirs (A and B). The replications demonstrate considerable within-site variability, which has to be considered when determining how much band differences contribute to differences in between-site microbial diversity. Strategies to quantify diversity assessments based on DGGE are not yet fully developed, but the technique is invaluable for comparative assessments. By courtesy of Odd Gunnar Brakstad, Norway.

carbon increased along the sere from 2.15 to 26.8%, and total nitrogen increased from 0.13 to 0.98%, the index of microbial diversity according to Shannon's formula (H') increased from 0.87 to 1.28. Furthermore, there was a statistically significant correlation between H' and organic carbon and total nitrogen (r^2 = 0.99 and 0.98, respectively). The conclusion that may be reached from this assessment is limited because it includes both species richness and evenness in the same index of diversity (see Pielou 1966a and b). It is possible to express evenness as a separate measure by determining the closeness between observed species abundances and the abundance associated with a hypothetical situation with maximum possible species diversity. For example, the maximum possible species diversity (H_{max}) is observed only when the set of N individuals is distributed evenly among s number of species, or

$$n_i = N \div s$$

The maximum possible Shannon's diversity index can thus be expressed as:

$$H_{max} = (\log N! - (s - r)\log c! - r\log(c + 1)!) \div N$$

where c is the integer value of $N \div s$, and r is the remainder. The index of evenness (E') or relative diversity is expressed by the ratio of observed diversity (H') and maximum diversity (H_{max}):

$$E' = H' \div H_{max}$$

See Box 11.1 for hypothetical examples of how these measures may be applied to studies in the assessment of microbial diversity.

BOX 11.1

Hypothetical example demonstrating the derivation of the diversity index for unicellular microbial populations.

Imagine that there are 3.5×10^2 microbial cells (N), determined by direct microscopic counting, in a 1 ml of sample obtained from an oligotrophic lake. Imagine further that molecular staining through fluorescent in situ hybridization targeted against organisms in this sample revealed the presence of three species (s). See Figs. 11.2 and 11.3 for technique and visual representations of microscopic and molecular methods for assessing microbial diversity. If the three species are present in the following numbers:

Species 1 (S1) = 120 cells (34.28%)
Species 2 (S2) = 80 cells (22.86%)
Species 3 (S3) = 150 cells (42.86%)

Then, Shannon's diversity index (H') is computed according to the following equation:

$$H' = -\Sigma p_i \log p_i$$

$$\begin{aligned} H' &= -(0.343\log 0.343 + 0.229\log 0.229 + 0.429\log 0.429) \\ &= -(0.343(-0.465) + 0.220(-0.640) + 0.429(-0.368)) \\ &= -(-0.159 - 0.147 - 0.158) \\ &= 0.464 \end{aligned}$$

To compute the maximum possible diversity, H_{max}, we use the following equation:

$$H_{max} = (\log N! - (s - r)\log c! - r\log(c+1)!) \div N$$

And we need to calculate c and r:

$$N \div s = 350 \div 3 = 116.67$$

Therefore, $c = 116$, and $r = 2$

$$\begin{aligned} H_{max} &= (\log 350! - (3-2)\log 116! - 2\log(116+1)!) \div 350 \\ &= (740.02 - (1)(190.531) - (2)(192.599)) \div 350 \\ &= 164.363 \div 350 = 0.469 \end{aligned}$$

The index of evenness (E') is a measure of how close the observed H' is to the maximum possible H_{max}:

$$\begin{aligned} E' &= H' \div H_{max} \\ &= 0.464 \div 0.469 \\ &= 0.989 \end{aligned}$$

The logical conclusion is that the three species are very nearly evenly distributed in the water sample. Note: For very large values of N, the common hand-held calculator may not have sufficient computing capacity. In such cases, the following approximation may be used for computing log $N!$:

$$\log N! = (n_i + 0.5)\log n_i - 0.4343 n_i + 0.3991$$

There is, however, considerable theoretical analyses suggesting that the highly dynamic and random growth pattern of most prokaryotes leads to uneven distribution of species in microbial communities. For example, Curtis and colleagues (2001) have used a different statistical approach based on log-normal **species abundance curves** (SAC) to estimate the diversity of prokaryotes on a small scale (70 per ml in sewage; 160 per ml in oceans, and 6400–38,000 per gm in soils). On a larger scale, they estimate that the entire prokaryote diversity in oceans is unlikely to exceed two million, whereas a metric ton of soil could contain four million different taxonomic groups. The assumption of a log-normal distribution of species in prokaryotic communities means that very few species dominate communities in terms of large numbers of individuals, and a small number of species have relatively few individual members, whereas, most species have an intermediate number of individuals. The SAC approach is based on two measurable variables: (i) the total number of individuals in a prokaryotic community (N_T); and (ii) the abundance of the most abundant members of that community (N_{max}). It is assumed that the least abundant taxonomic group has an abundance of 1 (N_{min}). The relationship between these variables is defined by the following equation:

$$N_T = \int_{N_{min}}^{N_{max}} NS(N)\,dN$$

Where the area under the curve $S(N)$ is the number of taxa that contain N individuals in a log-normal community. The function $NS(N)$ is defined as the "individuals curve". In a commentary on the SAC approach, Ward (2001) noted that efforts to quantify the number of species in a prokaryotic community presupposes consensus on the definition of "species" among prokaryotes (see Chapter 1 for a discussion on challenges surrounding species concepts). Partly because of this challenge, Curtis and colleagues (2001) concluded that experimental approaches to solve the conundrum of estimating prokaryote diversity will be fruitless. However, innovative research approaches that define prokaryotic diversity in terms of metabolic capacity are yielding interesting results (for example, see Tyson *et al.*, 2004 and Venter *et al.*, 2004). Furthermore, in the case of viruses where genomic fingerprints are potentially reliable indicators of diversity, the combination of experimental and statistical approaches is providing long-awaited information on marine viral communities (Breitbart *et al.*, 2002).

In a bid to develop a conceptual model for evaluating ecological indicators, the US National Research Council (NRC) identified **ecosystem productivity**, the capacity to capture solar energy and store it as carbon-based molecules, as the foundation upon which specific categories of ecological indexes should be erected. Since ecosystem productivity is influenced by biodiversity, temperature, moisture, and soil fertility, indexes developed to monitor changes in each of these categories must be clearly related to changes in other categories,

Table 11.1 The major categories of national ecological indicators recommended by the National Research Council (NRC, 2000). Although not explicitly stated, the numbers, forms, and functions of microorganisms are intricately connected to the larger scale indicators that might seem more directly connected to the desired target information.

Target information	Indicators
Extent and status of ecosystems	Land cover and land use
Ecological capital	Total species diversity
	Native species diversity
	Nutrient run-off
	Soil organic matter
Ecological functioning	Carbon storage
	Production capacity
	Net primary production
	Lake trophic status
	Stream oxygen
	Nutrient use efficiency
	Nutrient balance

and more directly to measures of productivity (NRC, 2000). Although the specific coverage of microbial diversity is slim in the NRC report, it is arguable that microorganisms are active participants in all the recommended ecological indicators (Table 11.1). For example, changes in land-use patterns are probably the largest anthropogenic impact on ecosystems. These changes result in the modification of several environmental factors that affect microbial diversity. The conversion of native soil biological communities to agricultural practices dominated by single plant species, supported by periodical application of nitrogen and phosphorus-based fertilizers and pesticides, is bound to change the microbial diversity of soils. The larger the proportion of Earth's terrestrial surface is devoted to monoculture-type farming, the larger still will be the impact on the abundance and distribution of specific types of soil microorganisms. Similarly, changes in the indicators of ecological capital such as nutrient run-off, which can lead to the eutrophication of receiving water systems, are typically accompanied by changes in the population diversity of photosynthetic and heterotrophic microorganisms (Paerl *et al.*, 2003). Even more profoundly, changes in any indicator of ecosystem functioning are likely to be rooted in microbial activities through their involvement in the biogeochemical cycles.

The large ratio of surface area to mass that is characteristic of many microorganisms renders them highly sensitive to changing environmental conditions, particularly in the aquatic environment. In addition, the situation of many microorganisms at the base of important ecological processes suggests that indicators of microbial community diversity and function can serve as early warning systems for environmental changes that may result in major ecological catastrophes. Therefore, interest in developing biochemical and molecular indicators or biosensors useful for monitoring microbial communities is intensifying, and the results of this concerted effort have produced robust tools, although in many cases, these tools are adapted for specific contemporary environmental issues. The following paragraphs address case studies focusing on the relevance of microbial diversity to these environmental issues, however, the need for cross-cutting indicators of microbial diversity and function is paramount, and should be acknowledged as a major direction for future research.

GLOBAL CLIMATE CHANGE

Discussions focused on the relevance of microbial diversity to major shifts in the global climate should be preceded by a declaration that in the history of the Earth, microorganisms have indeed survived several major climate changes; and that the current projected scenarios of climate change are not likely to precipitate novel irreversible configurations of the microbial world. This declaration is predicated not only on the natural historical record of microorganisms, but also on the recognition of their prolific and extremely effective adaptive strategies. Concern for the unpredictability of the direction of microbial adaptation to major environmental changes, and the necessity for predicting the trajectory of change in order to mitigate adverse impacts on human settlements and other organisms, are driving the efforts to seek indicators of pace and impacts of climate change at all levels of ecological organization. Therefore, it is important that microbial indicators of large-scale ecological change are consistent with indicators based on large multicellular organisms and on abiotic environmental factors including global mean temperature, sea surface temperatures, and atmospheric carbon dioxide concentrations (Butterworth *et al.*, 2001). For example, the elevated atmospheric concentration of carbon dioxide is likely to result in changes in plant growth characteristics, affecting root systems, exudates, and litter production. Changes in vegetation cover will in turn affect the growth and distribution of free-living fungi, mychorrhizal relationships, soil bacterial diversity, and the occurrence of plant diseases. These simple interactions may then cascade to modify the activities of fungivores, bacterivores, and omnivores. All these changes are likely to be accompanied by dramatic fluctuations in local nitrogen cycling, and in the efficiency of other biogeochemical cycles (Carter, 1986; Wall *et al.*, 2001).

In aquatic ecosystems, certain microbial indicators have proven to be very useful in understanding the ecological impacts of climate-linked events such as the increasing frequency of hurricanes. For example, Paerl and colleagues (2003) have demonstrated for estuaries in North Carolina that Hurricane Floyd-induced hydrological changes in river discharge and flushing rates are correlated with changes in phytoplankton taxonomic groups as a function of contrasting growth characteristics and photosynthetic pigment signatures. Elevated freshwater flows were linked to an increase in the contributions of organisms with relatively high nutrient uptake rates such as diatoms, cryptophytes, and chlorophytes to the chlorophyll-a pool of the water system. Furthermore, *Cyanobacteria* predominated under conditions of minimal flushing and longer residence times for freshwater in the estuary.

The episodic nature of extreme events such as hurricanes and other climate change-linked impacts makes it difficult to generate consistent reproducible data on microbial indicators. With increasing frequency of these events, it should be possible to test various hypotheses that have been proposed to explain the changes in local microbial diversity in response to global scale events. However, the investigation of relatively more stable aquatic ecosystems may provide a natural microcosm where some of these hypotheses can be explored on a smaller scale. For example, a glimpse of how microbial diversity responds to environmental parameters in aquatic ecosystems is given by James Hollibaugh and colleagues at the University of Georgia through their Microbial Observatory project on Mono Lake, a meromictic soda lake in California (Box 11.2). The study involved simultaneous time series and cross-sectional assessment of abiotic parameters including changes in temperature and incident light by depth, correlated with biotic factors such as molecular-level diversity, dominant species, and chlorophyll concentration. Shifts in the microbial diversity index, measured by PCR-DGGE, coincided with marked gradients from the surface level in the diffusion of oxygen (oxycline), dissolved salt (chemocline), and temperature inversion (thermocline) (see Fig. 11.3 for general DGGE protocol). Of particular note is the distribution of *Picocystis*, a recently discovered unicellular chlorophytic alga, with respect to geochemical gradients. The increase in chlorophyll fluorescence at the depth of 15–20 m, where light intensity and oxygen concentration are relatively low, was attributed to *Picocystis* cells that have colonized this unique ecological zone.

BOX 11.2

(a)

Mono Lake (a) in central California provides an excellent opportunity for microbiologists investigating the adaptation of microbial diversity to changing environmental conditions within a manageable geographical scale. The lake is a natural laboratory where the interactions between hydrological parameters and biogeochemical processes can be examined as a model system for understanding broad-ranging topics in microbial diversity and ecology. The salt concentration in the lake (84 to 94 grams per liter) makes it a hypersaline ecosystem. The lake is also home to alkalinophiles (pH 9.8); although it receives freshwater from the environment, it is not subject to the kinds of mixing reactions that characterize estuaries. Therefore, the lake exhibits strong gradients in physicochemical conditions according to depth (Humayoun *et al.*, 2003). Being anoxic at the bottom, reduced compounds such as sulfides accumulate to levels that may affect the biochemistry of metals ions, thereby exerting major influences on the structure of microbial communities along the lake's depth. The photograph in (a) shows the lake's characteristic mineral encrusted tufa towers. Photograph by courtesy of NASA (credit: Tony Phillips).

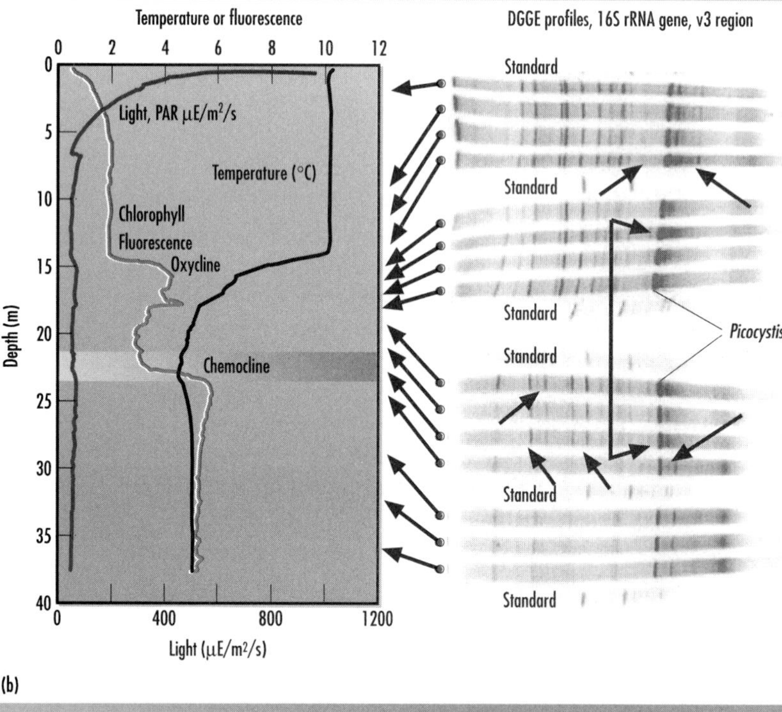

(b)

(b) (left): Graph showing the distribution of physical properties at a location near the deepest part of the lake, southeast of Paoha Island, on May 17, 2000. The data were collected by lowering a Self-Contained Autonomous MicroProfiler (SCAMP; Precision Measuring Engineering, Carlsbad, CA), a portable, lightweight microstructure profiler designed to measure extremely small-scale (~1 mm) fluctuations of electrical conductivity, temperature, and oxygen concentration in lakes, reservoirs, estuaries, and the oceans. SCAMP records at 0.1 second intervals showing water temperature; light intensity; chlorophyll fluorescence. The map also shows the depth below the lake surface at which the other measurements were taken. The data in the panel on the right were generated by analyzing samples of water collected at the depths indicated by the arrows.

(b) (right): Results of a DGGE experiment used to resolve the microbial diversity of a Mono Lake sample site according to depth. The arrows point to examples of places where there is a shift in the composition of the microbial community. Most shifts coincide with the chemocline and the oxycline. The dark band in the middle of the lanes continuing across all lanes is from the chloroplast of *Picocystis*. Data by courtesy of James Hollibaugh, University of Georgia at Athens.

Picocystis species, strain ML, accounts for approximately 25% of the primary production during the winter bloom, and its contribution exceeds 50% during other seasons (Roesler *et al.*, 2002). Therefore, environmental factors, which are not conducive to the growth of this organism will be reflected in the observation of a decline in primary production, and in all other biogeochemical processes depending on primary productivity. As such, *Picocystis* is a good indicator organism for ecosystem function in this seemingly low-diversity environment. However, the robustness of biological indicators is subject to strong biotic influences, and indeed Humayoun and colleagues (2003) demonstrated a complex bacterial diversity in the proximity of the habitat colonized by *Picocystis.* By employing protocols based on direct DNA analysis to reveal the presence of phylogenetic groups, these investigators showed that samples collected from the oxycline zone had low bacterial diversity with only nine phylotypes, whereas the chemocline contained 27 phylotypes. These bacteria belong to the five major lineages of the domain Bacteria, namely alpha- and gamma-*Proteobacteria*; *Cytophaga-Flexibacter- Bacteroides*; *Actinobacteria*; *Bacillus*; and *Clostridium.* It is not yet clear from these findings what is the nature of the interactions that occur between the dominant primary producer, *Picocystis*, and the diverse heterotrophic bacteria with which it co-exists. There are strong prospects that additional broad-spectrum indicators of extreme environmental change will be recovered from ecosystems typified by Mono Lake.

STRATOSPHERIC OZONE DEPLETION

As in the case of humans, microbial activities contribute ozone-depleting chemicals to the atmosphere, and they are also sensitive to the impacts of stratospheric ozone depletion through increased exposure to ultraviolet radiation. A sophisticated real-time information database is available for monitoring the size of the ozone layer in the stratosphere, and there is a well-developed index of specific health risks to human populations in many part of the world (Fig. 11.4) (de Gruijl *et al.*, 1994; Setlow *et al.*, 1993). In contrast, indexes of ecological impacts of increased UV exposure are either non-existent or poorly developed. This lack of information is due in part to the very wide range of exposure and responses exhibited by wildlife organisms and ecosystem processes to increased UV exposures (Blumthaler and Ambach, 1990; Fleischmann, 1989). In addition, ecological effects of UV exposure are likely to cascade hierarchically, rather than to remain with affected individuals or processes (Herndl *et al.*, 1993; Lyons *et al.*, 1998). Therefore, attempts to produce generalized response models that could serve as the foundation for the development of reliable indicators are fraught with uncertainties.

Biological responses to UV radiation can occur at the level of DNA damage or systemically through a reduction in the growth rate of organisms. Theoretically, biological response is modeled after the **action spectrum** of UV radiation, the weighting parameter that describes the variation in energy at different wavelengths. The relationship between the action spectrum and biological response produces an index referred to as the biologically effective dose (BED), which is described by the following equation:

$$\mathrm{BED} = \int_{\lambda} UV(\lambda)A(\lambda)\mathrm{d}\lambda$$

Where *UV* and *A* represent the ultraviolet irradiance and action spectrum values at a given wavelength (λ), respectively. The total UV dose is then computed by integrating over the entire wavelength range. Figure 11.5 shows four examples of action spectra.

The first repercussion of UV exposure on microbial populations is damage to DNA through the production of pyrimidine dimers and cyclobutane. These compounds are capable of generating mutations that vary in their lethality. In view of the fact that UV-protective compounds are not common in the microbial community, and their relatively small size may preclude the development of protective pigmentation, microorganisms are frequently considered as highly sensitive to the impact of increasing UV radiation (Aas

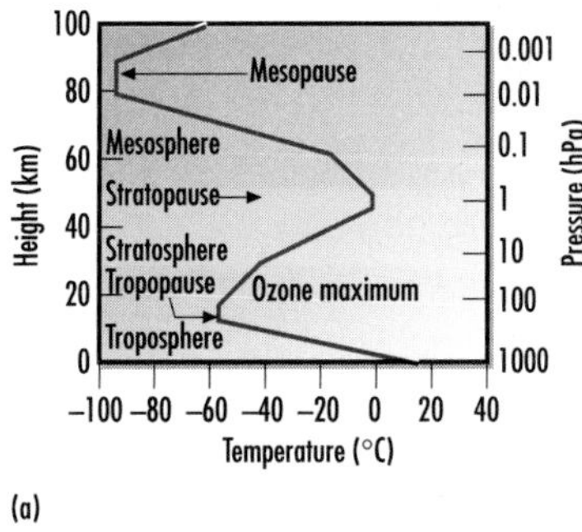

(a)

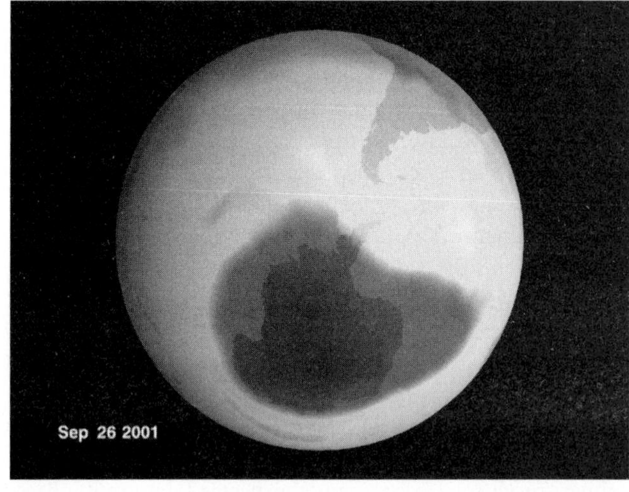

(b)

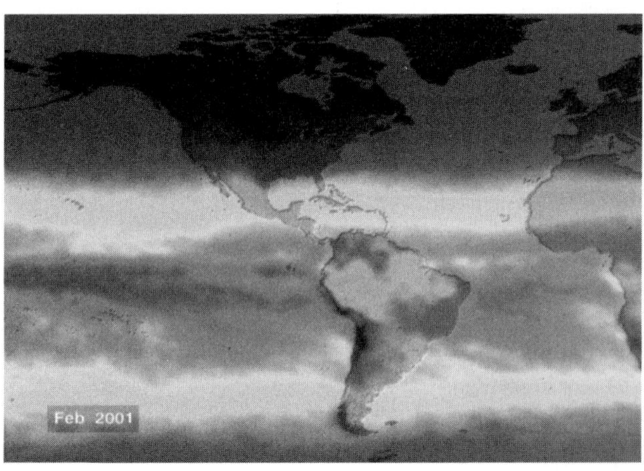

(c)

(d)

Fig. 11.4 (a) The size of the stratospheric ozone layer is continuously monitored by the Total Ozone Mapping Spectrometer (TOMS) satellite operated by NASA. The global maps are updated daily. Ozone concentrations are measured in Dobson units. (b) shows a substantial depletion of the ozone layer in the Antarctic region. The size of the ozone layer is inversely proportional to the incident UV dose on the Earth's surface, although the relationship is complex because it is influenced by several other environmental factors. Incident UV irradiance is used to compute the biologically effective dose for human risk factors. (c) and (d) show the computed erythemal UV dose for the Americas and globally. For a complete discussion of these ecological indicators, see the website for TOMS at http://jwocky.gsfc.nasa.gov/. Data used in (a) are from the British Met Office: http://www.met-office.gov.uk/research/stratosphere/. Maps on Ozone and UV dose are by courtesy of NASA/Goddard Space Flight Center Scientific Visualization Studio.

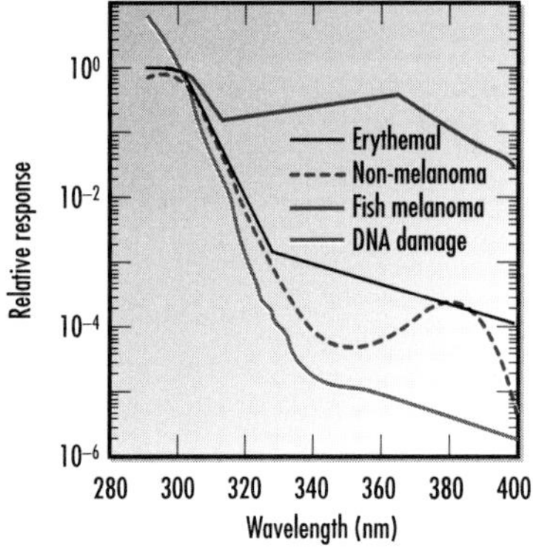

Fig. 11.5 Graph showing four action spectra for UV-related biological responses, namely microbiologically relevant DNA damage action spectrum (Setlow, 1974), the erythemal action spectrum (McKinlay and Diffey, 1987), the mammalian non-melanoma skin cancer action spectrum (de Gruijl and van der Leun, 1994), and the action spectrum for melanoma induction in a species of fish (Setlow *et al.*, 1993). The spectra for DNA damage, mammalian non-melanoma, and erythema decline rapidly with increasing wavelength, suggesting that the UV-B portion (280–320 nm) of the spectrum is more effective at eliciting a response than UV-A (360–400 nm). Theoretically, these biological responses are more sensitive to ozone depletion, which increases irradiance levels in the UV-B region. In contrast, the fish melanoma action spectrum shows a significant dependence on UV-A, suggesting that ozone depletion would have a relatively low effect on melanoma induction in certain kinds of fish. Direct comparisons of UV dose required for an induced effect should not be made on the basis of action spectra alone. The spectra only indicate the relative effectiveness of particular wavelengths, not the actual dose required to produce a biological response. Data from the Center for International Earth Science Information Network (http://sedac.ciesin.columbia.edu/ozone/docs/AS.html).

et al., 1996; Garcia-Pichel, 1994; Winter *et al.*, 2001). It is difficult to generalize from this reasoning because exposure to UV radiation (360–400 nm) in certain microbial populations can have beneficial effects on photo-enzymatic mechanisms which are required for repairing damage to DNA (Kaiser and Herndl, 1999). In addition, exposure of complex organic compounds to UV radiation increases their breakdown, thereby increasing the pool of nutrients that are biologically available to support microbial growth (Wetzel *et al.*, 1995; Tam *et al.*, 2003). Most microorganisms are expected to respond to UV exposure, but there is a wide variation in responses (Arrieta *et al.*, 2000; Jeffrey *et al.*, 1996; Joux *et al.*, 1999).

Attempts to quantify the net impact of UV radiation on natural microbial communities have focused on developing models for estimating the effect of ozone depletion on primary ecosystem production. With respect to evaluating the impact of UV irradiance on production in microbial communities, several investigators have used DGGE analysis of 16S rRNA and 16S rDNA sequences to monitor changes in microbial diversity across geographical zones that vary in incident UV. For example, Winter and colleagues (2001) demonstrated, through field measurements and laboratory experiments, minimal changes in the bacterioplankton community structure in different geographical locations identified by different UV radiation regimes. Approximately less than or equal to 10% of the operational taxonomic units recovered from the North Sea were determined to show adverse sensitivity to UV radiation. The conclusion that the composition of microbial communities is not strictly sensitive to excess UV radiation might suggest the futility of basing indexes of ozone depletion and UV exposure on microbial diversity. However, it is worth noting that most of the studies in this direction have focused on average responses, while recognizing the tremendous diversity of both adverse and potentially beneficial responses to UV among microorganisms. In such cases, a more effective approach would be to select specific sentinel species to indicate the direction of changes in microbial population dynamics and primary production. Much more research is needed before such sentinel approaches can be sufficiently developed for integration into ecosystem indicator models that include, for example, prophage induction and mutation assays.

TOXIC CHEMICAL POLLUTION

The distribution of toxic chemicals from industrial and domestic origins into aquatic and terrestrial ecosystems is now clearly recognized as a global phenomenon in terms of their sources, sinks, and ecological impacts. To address this problem, the international community has attempted to restrict the manufacture, sale, and use of several chemicals. For example, the United Nations Environment Program (UNEP, 1992; see http://www.chem.unep.ch/pops/) coordinated an international initiative to ban persistent organic pollutants (POPs). Different categories of chemicals are included in the POPs list, namely, the pesticides hexachlorobenzene, toxaphene, chlordane, aldrin, DDT, mirex, Dieldrin, pentachlorophenol, Endrin; industrial byproducts of polyaromatic hydrocarbon compounds (PAHs), dioxins, and furans; and the industrial chemicals polychlorinated biphenyls (PCBs) and hexabromobiphenyl. These chemicals persist in the environment, bioaccumulate through the food web, and pose a risk of causing adverse effects to human health and the ecosystem, including precipitating the emergence of new diseases (King, 2000). In addition, evidence of long-range transport of these chemicals to regions where they have never been used or produced and the consequent threats they pose to the global ecosystem have commanded the urgent attention of the research community. In general, POPs are sparingly soluble in water, relatively highly soluble in fatty tissue, and volatile. These characteristics contribute to their long environmental half-lives, and to their ubiquitous distribution in nature. In addition to POPs, ecosystem contamination by heavy metals is widely considered to be a global environmental problem. Unlike POPs, most heavy metals of concern are natural constituents of the Earth's crust, but anthropogenic activities have drastically altered the geochemical cycles that regulate their distribution. Of particular importance for ecosystem health are lead, mercury, and chromium. These metals are never eliminated from ecosystems, but they are capable of forming organic and inorganic compounds that alter their compartmentalization into different ecological zones (Doelman *et al.*, 1994).

There is a two-prong connection between microbial diversity and toxic chemicals in the environment. First, microorganisms are presumed to be particularly sensitive to the toxic effects of pollutants, and as such, ecosystem functions that depend on microbial activities will likely suffer adverse impacts from repeated exposures (Domsch *et al.*, 1983; Ogunseitan *et al.*, 2000). Secondly, much research has been dedicated toward exploiting the physiological diversity that exists in the microbial community for the purpose of environmental remediation and ecosystem restoration after pollution incidents (Pritchard *et al.*, 1995; Ogunseitan, 1994 and 2002). In both cases, qualitative and quantitative knowledge of microbial diversity in target ecosystems is essential for predicting impacts on ecosystem functions, or the direction and outcome of toxic chemical biotransformation (Kennedy and Papendick, 1995; Tebbe *et al.*, 1992).

Both biochemical and molecular genetic techniques have been used to explore the impacts of chemical exposures on microbial diversity (Bakermans and Madsen, 2002). For example, Fig. 11.6 shows the results of DGGE used to analyze the variation in microbial communities transecting a petrochemical-contaminated aquifer environment. The results of several such studies are consistent with the observation that toxic chemical exposures tend to narrow the spectrum of microbial diversity because organisms that are not capable of resisting the toxic effects either die or enter a static metabolic phase, leaving those that have evolved resistance mechanisms to proliferate and become dominant members of the impacted ecosystem (Ogunseitan, 1994 and 2000). Biotechnological approaches to environmental remediation have employed this kind of adaptation/selection process for identifying microorganisms with enhanced capacity to detoxify pollutants.

CONSERVATION OF GLOBAL BIODIVERSITY

The World Health Organization (WHO) declared in 1980 that smallpox had been eradicated from the Earth. This of course only meant that the human disease condition has been extinguished from human societies, and the news was greeted with much jubilation around the world. About 20 years later, on June 30, 1999, WHO announced plans to destroy the last remaining stock of smallpox virus particles. At the time, the news was met with controversy from various sectors (Ogunseitan, 2002a). The debates were framed by two very different schools of thought. At one end were those who believed that stockpiles of smallpox virus remain undeclared by rogue nations for the clandestine purpose of developing biological weapons. Therefore, as a precaution, the known stock of the virus should not be destroyed in case vaccines would need to be produced quickly in the event of biological warfare or other emergencies. At the other end were those individuals who opposed any deliberate actions that humans might take to cause the extinction of any other organism, including viruses. The latter argument was embedded in a growing appreciation of biological diversity at all levels of organization. In addition to human health-related concerns about microbial diversity, the rapid expansion of the biotechnology industry, fueled in part through the discovery of novel properties among microbial populations, provided the rationale for the integration of microorganisms into the 1992 Convention on Biological Diversity (Kelly, 1995).

The establishment of the global Microbial Resources Centers (MIRCEN) network by the United Nations Educational, Social, and Cultural Organization (UNESCO) is one of the major results of renewed interests in microbial diversity. The objectives of MIRCEN include the provision of a global infrastructure that incorporates national, regional, and interregional cooperating laboratories geared to the management, distribution, and utilization of the microbial gene pools; reinforcing the conservation of microorganisms, with emphasis

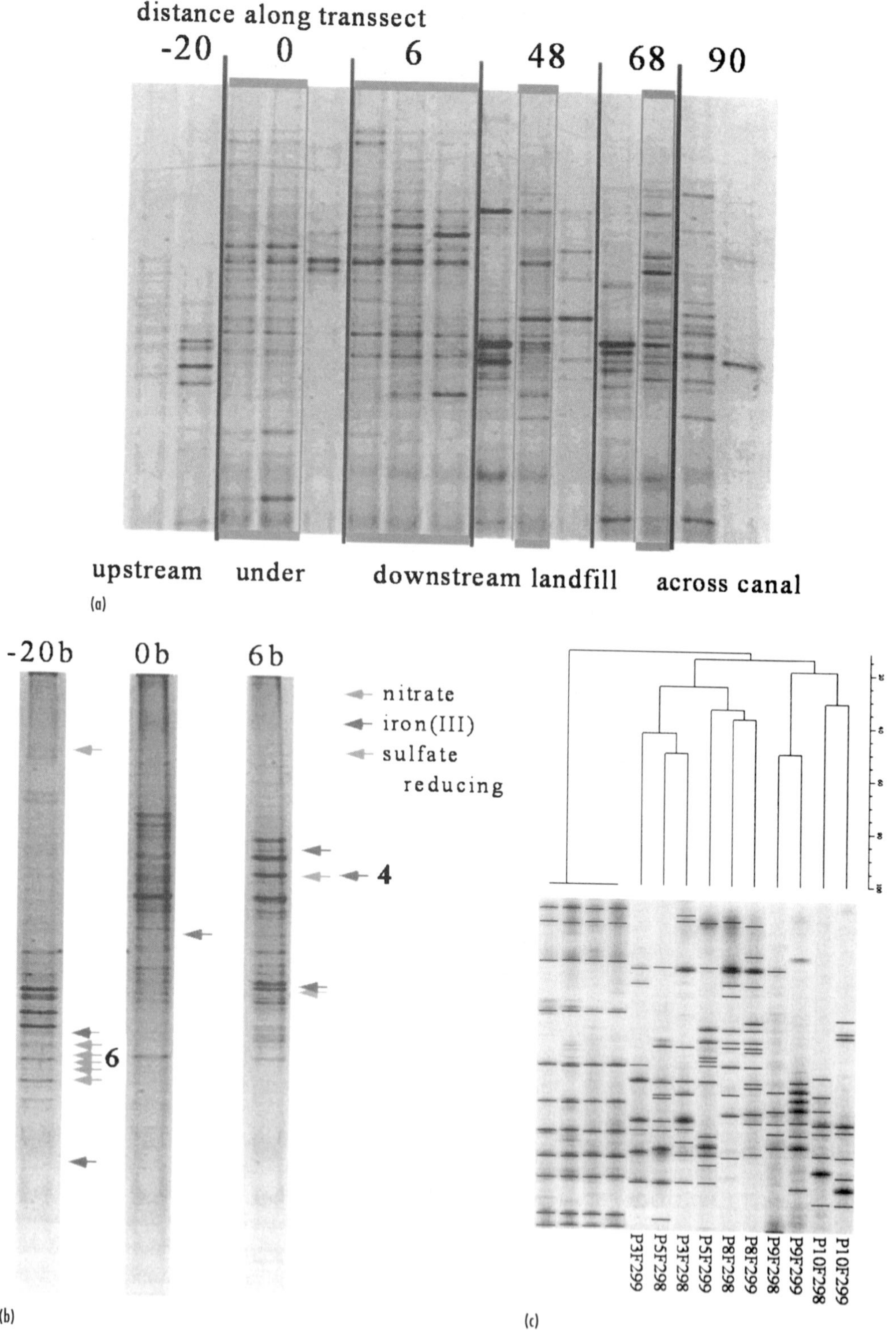

Fig. 11.6 Exposure of microbial communities to potentially toxic environmental pollutants tends to reduce the spectrum of microbial diversity, as demonstrated in (a) with DGGE conducted on samples collected along a transect at various distances from the center of a petrochemical pollution plume (0 m, and "under"). A high level of variation in band position occurs outside the plume, and the band positions are relatively similar within the plume zones (boxes). (b) shows the results of phylogenetic identification of specific bands resolved from samples collected at −20, 0, and 6 m along the transect. The solid, hatched, and hollow arrows represent organisms that use nitrate, iron (III), and sulfate as terminal electron acceptors, respectively. All the iron (III) reducers belong to the family *Geobacteriaceae*, and these are the only organisms in these systems that were shown to degrade the petrochemicals benzene, toluene, ethylbenzene, and xylene. Members of the *Geobacteriaceae* within the plume are extremely diverse as shown by fine-scale DGGE results for 14 isolates in (c). The dendogram on the right side of the panel shows that the 14 isolates can be placed into only seven phylogenetic groups. Data credit to Wilfred Roling and Komang Ralebitso-Senior at Vrije Universiteit, Amsterdam.

Fig. 11.7 The global network of UNESCO's Microbial Resources Centers (MIRCENS) is associated with existing academic and/or research institutes in various countries throughout the world. Source: UNESCO (see the program webpage at http://www.ejbiotechnology.info/content/mircen/).

on *Rhizobium* gene pools, in developing countries with an agrarian base; fostering the development of new inexpensive technologies native to specific regions of the world; promoting the economic and environmental applications of microbiology; and establishing focal centers in the network for the training of microbiology research staff. There are 32 MIRCENs distributed globally (Fig. 11.7).

Several countries also maintain microbial conservation centers. The largest and best known of these is the American Type Culture Collection (ATCC; http://www.atcc.org/SearchCatalogs/tasc2.cfm). The idea for ATCC was proposed in 1925 by a committee of scientists who recognized the need for a central collection of microorganisms that would serve scientists all over the world. The current ATCC facility is equipped with Biosafety Levels 2 and 3 containment stations for processing extremely pathogenic microorganisms. The ATCC collection of bacteria is the most diversified assemblage of prokaryotes in the world, containing nearly 18,000 strains in more than 750 genera. The bacteria collection represents every important prokaryotic physiological group, including more than 3,600 type cultures of validly described species, and approximately 500 bacteriophages. ATCC's mycology collection includes more than 27,000 strains of filamentous fungi and yeasts distributed among 1,500 genera and 7,000 species. The yeast collection includes more than 2,000 genetic strains of the biotechnology "work-horse" *Saccharomyces cerevisiae* and other yeasts. In addition, the ATCC protistology collection is the only general service collection of protozoa in the United States and one of two in the world. ATCC has the only service collection of parasitic protozoa in the world. Approximately 200 of the protozoans are *Tetrahymena* strains. ATCC's animal virus collection includes more than 2,000 viruses. Other collections include *chlamydiae* and *rickettsiae*.

In addition to the live culture collections maintained by MIRCEN and ATCC, several microbial conservatories focus on sub-organism collections to facilitate research on molecular diversity in the microbial world. The molecular databases include the complete genomic sequences maintained by The Institute for Genomic Research (TIGR) Microbial Database (http://www.tigr.org/tdb/mdb/mdbcomplete.html) and the Genomes Online Database (GOLD) that provides a World-Wide-Web resource for accessing complete and ongoing genome projects around the world (Bernal *et al.*, 2001) (http://wit.integratedgenomics.com/GOLD/). Finally, the database on ribosomal RNA and DNA sequences has been extremely influential in providing molecular tools for assessing microbial diversity in different ecosystems all over the world (Maidak *et al.*, 2001). The Ribosomal Database Project (RDP) is housed at the Center for Microbial Ecology at Michigan State University (http://www.cme.msu.edu/RDP).

The microbial conservatories have been instrumental in elevating support for microbial diversity research to levels that approach those dedicated to botanical and zoological species. However, deeper insights into the nature of speciation and processes that generate molecular and physiological diversity among microorganisms are needed to facilitate the establishment of more comprehensive, evidence-based global inventories of microbial diversity.

CONCLUSION

In the history of planet Earth, microorganisms have been responsible for the most profound forms of global environmental change. The emergence of new phylogenetic lineages accompanied by innovative physiological processes has continuously modified aquatic, terrestrial, and atmospheric systems in ways that reinvent niches and create opportunities for speciation. It is difficult to compare the global impacts of contemporary changes attributed to human industrial activities with the "natural" changes that are engendered by microbial activities. Perhaps attempts to compare these impacts are misguided because human activities can be viewed as a protracted extension of microbial evolution. Instead, an all-encompassing model of the interactions between all life forms and the environment (co-evolution) may serve better to place recent changes in a broader context that facilitates deeper understanding of the relationship between causes, effects, and remedial strategies. Problems such as reduction of biodiversity, toxic chemical pollution, climate change, and the ozone hole have assumed a status of urgency in human affairs primarily because we "like" the seemingly hierarchical organization of the biosphere with humans at the top. Therefore, we wish to maintain the *status quo* and to restrict the pace of co-evolution when in fact, organism extinctions, displacements, and replacements have dominated the Earth's history. In this context, the appreciation of microbial diversity hinges on utilitarian values: the microbial world is extremely rich in resources with which we can manipulate environmental conditions, and detect subtle changes in ecosystem integrity. So far, we have only tapped into the "tip of the iceberg" of microbial resources.

If there is a "silver lining" in the "cloudy" prospects of global environmental change impacts, it is that research on integrating various indexes of environmental change is intensifying. The need to incorporate microbial indicators into existing indexes is now unquestionable. The reliability of microbial indicators depends on how well we can capture the variety of responses which characterize the inherent diversity of microbial communities. The additional benefit of developing robust microbial indicators of environmental change is the probable discovery of novel capabilities that will contribute to the repertoire of biotechnological strategies for addressing environmental problems. The global nature of these problems warrants far-reaching solutions that are accessible to scientists all over the world. Therefore, carefully annotated conservatories of microorganisms located in strategic geographical regions should facilitate rapid technology transfers. A few of these conservatories already exist, but they need to be strengthened and made more uniformly accessible, although the advisable caution about restricting access to pathogens needs to be considered seriously. In the end, the study of microbial diversity and its implications is by necessity a multidisciplinary endeavor. With the rapid progress in the exploration of Space, microbial diversity will soon likely become also a multiplanetary endeavor. Indeed, it can be argued that we are already at the *inter*planetary stage. Hopefully, we will have mastered microbial solutions to the global environmental problems on Earth before the multiplanetary stage of microbial diversity is with us.

QUESTIONS FOR FURTHER INVESTIGATION

1 Find and discuss examples of microbial products and microbial processes that best illustrate your understanding of subtle differences in answers to the two questions: "Of what use are microorganisms?" and "Of what use is microbial diversity?"

2 The integration of redundancy into ecological indicators is likely to provide a robust information network on the size and status of ecosystems, ecological capital, and ecosystem functioning. For each of the indicators listed in Table 11.1 find in the literature at least two examples of microorganisms, microbial communities, or microbial processes that can serve the role of "ecological indicators".

3 Describe two methods for producing data on the composition of microbial communities that can be used for generating single indexes of diversity. Search the literature for at least one application of each of the methods that you described.

4 Access the interactive website dedicated to NASA's Total Ozone Monitoring Spectrometer (http://jwocky.gsfc.nasa.gov/). Follow the instructions on the website to produce a global map of today's incident ultraviolet radiation. Identify the approximate UV dose around your residence. Discuss the potential impacts of that dose for the structure and function of microbial communities.

5 Search the literature for specific examples of microorganisms that are capable of degrading or detoxifying each of the chemical categories included in the list of persistent organic pollutants (POPs) established by the United Nations Environment Program (UNEP). Do these degradative organisms cluster around any specific phylogenetic branch?

6 Compare and contrast the challenges facing the conservation of global microbial diversity as opposed to the challenges facing the conservation of global animal and/or plant diversity.

SUGGESTED READINGS

Arrieta, J.M., M.G. Weinbauer, and G.J. Herndl. 2000. Interspecific variability in sensitivity to UV radiation and subsequent recovery in selected isolates of marine bacteria. *Applied and Environmental Microbiology*, **66**: 1468–73.

Bernal, A., U. Ear, and N. Kyrpides. 2001. Genomes OnLine Database (GOLD): A monitor of genome projects world-wide. *Nucleic Acids Research*, **29**: 126–7.

Bull, A.T. (ed.) 2003. *Microbial Diversity and Bioprospecting*. Washington, DC: American Society for Microbiology.

Kelly, J. 1995. Microorganisms, indigenous intellectual property rights and the Convention on Biological Diversity. In D. Allsop, R.R. Colwell, and D.L. Hawksworth (eds.) *Microbial Diversity and Ecosystem Function*, pp. 415–26. Wallingford, England: United Nations Environment Program and Center for Agriculture and Biotechnology International.

King, J. 2000. Environmental pollution and the emergence of new diseases. In L. Margulis, C. Matthews, and A. Haselton (eds.) *Environmental Evolution* (second edition), pp. 249–62. Cambridge, MA: MIT Press.

Maidak, B., J. Cole, T. Lilburn, C. Parker, P. Saxman, R. Farris, G. Garrity, G. Olsen, T. Schmidt, and J. Tiedje. 2001. The RDP-II (Ribosomal Database Project). *Nucleic Acids Research*, **29**: 171–3.

NRC. 1999. *Perspectives on Biodiversity: Valuing its Role in an Ever Changing World*. Washington, DC: National Research Council. National Academy Press.

NRC. 2000. *Ecological Indicators for the Nation 2000*. Washington, DC: National Research Council. National Academy Press.

Ogunseitan, O.A. 2000. Microbial proteins as biomarkers of ecosystem health. In K. Scow, G.E. Fogg, D. Hinton, and M.L. Johnson (eds.) *Integrated Assessment of Ecosystem Health*, pp. 207–22. Boca Raton, FL: CRC Press.

Ogunseitan, O.A., S. Yang, and J.E. Ericson. 2000. Microbial delta-aminolevulinate dehydratase as a biosensor of lead (Pb) bioavailability in contaminated environments. *Soil Biology and Biochemistry*, **32**: 1899–906.

Paerl, H.W., J. Dyble, P.H. Moisander, R.T. Noble, M.F. Piehler, J.L. Pinckney, T.F. Steppe, L. Twomey, and L.M. Valdes. 2003. Microbial indicators of aquatic ecosystem change: Current applications to eutrophication studies. *FEMS Microbiology Ecology*, **1561**: 1–14.

Pimm, S.L. 1984. The complexity and stability of ecosystems. *Nature*, **307**: 321–6.

Pimm, S. and P. Raven. 2000. Extinction by numbers. *Nature*, **403**: 843–5.

Wall, D.H., G. Adams, and A.N. Parsons. 2001. Soil biodiversity. In F.S. Chapin, O.E. Sala, and E. Huber-Sannwald (eds.) *Global Biodiversity in a Changing Environment*, pp. 47–82. New York: Springer-Verlag.

PARTIAL LIST OF SEQUENCED MICROBIAL GENOMES

Organism and strain	Genome size (Mb)	Phylogenetic domain	Institution	Internet link
Aeropyrum pernix K1	1.67	*Archaea* Hyperthermophilic achaeon. Optimal growth temperature is 90–95°C and 3.5% salinity	National Institute for Technology and Evaluation, Japan	http://www.bio.nite.go.jp/cgi.bin/dogan/genome_top.cgi%27ape%27
Agrobacterium tumefaciens C58	5.30	*Bacteria* Crown gall bacterium, induces tumors in plants. Model for interdomain genetic exchange	University of Washington, Seattle Genome Center, and Cereon Corporation	http://www.agrobacterium.org/
Aquifex aerolicus VF5	1.50	*Bacteria*	Diversa Corporation	http://www.biocat.com
Archaeoglobus fulgidus DSM4304	2.18	*Archaea* Hyperthermophilic sulfate reducer, causes souring of oil deposits in subsurface due to formation of iron sulfides	Institute for Genomic Research	http://www.tigr.org/tigr-scripts/CMR2/GenomePage3.spl?database=gaf
Bacillus halodurans C-125	4.20	*Bacteria* Salt-tolerant Gram-positive bacterium	Japan Marine Science and Technology Center	http://www.jamstec.go.jp/jamstec-e/index-e.html
Bacillus subtilis 168	4.20	*Bacteria* Motile Gram-positive bacterium	International Consortium, European Commission	http://europa.eu.int/comm/index.htm.
Borrelia burgdoferi B31	1.44	*Bacteria* Causes Lyme disease. Commensal inhabitant of *Ixodes* ticks mid-gut	The Institute for Genomic Research	http://www.tigr.org/tigr-scripts/CMR2/GenomePage3.spl?database=gbb
Buchnera species APS	0.64	*Bacteria* Obligate endocellular bacterial symbiont of pea aphids, *Acyrthosiphon pisum*	University of Tokyo and RIKEN Genomic Science Center	http://buchnera.gsc.riken.go.jp/
Caulobacter crescentus CB15	4.01	*Bacteria* Asymmetrically differentiating Gram-negative bacterium that grows in dilute aquatic environments	The Institute for Genomic Research	http://www.tigr.org/CMR2/BackGround/gcc.html
Campylobacter jejuni NCTC 11168	1.64	*Bacteria* The leading cause of food poisoning. It is three times more commonly found as a food pathogen than *Salmonella*	Sanger Centre, United Kingdom	http://www.sanger.ac.uk/
Chlamydia pneumoniae CWL029	1.23	*Bacteria* Common source of community acquired pneumonia	University of California at Berkeley and Stanford	http://chlamydia-www.berkeley.edu:4231/

Continued

Organism and strain	Genome size (Mb)	Phylogenetic domain	Institution	Internet link
Chlamydia trachomatis serovar D (D/UW-3/Cx)	1.05	*Bacteria* Obligate intracellular human parasitic bacterium. It is transmitted sexually, causing infertility in women.	University of California at Berkeley and Stanford	http://chlamydia-www.berkeley.edu:4231/
Chlorobium tepidum TLS	2.10	*Bacteria* Photosynthetic green sulfur bacterium	The Institute for Genomic Research	http://www.tigr.org/tigr-scripts/CMR2/GenomePage3.spl?database=gct
Clostridium acetobutylicum ATCC 824	4.10	*Bacteria* Solventogenic sporulating anaerobe used industrially for fermentation between 1920s and 1950s before petrochemical processing	Genome Therapeutics Corporation	http://www.genomecorp.com/
Deinococcus radiodurans R1	3.28	*Bacteria* Radiation resistant organism exhibiting dual chromosomes and mega-plasmids that facilitate genetic redundancy as a mechanism for abating damage from extreme environmental stress	The Institute for Genomic Research	http://www.tigr.org/tigr.scripts CMR2/GenomePage3.spl? database=gdr
Escherichia coli O157:H7 strain EDL933	5.5	*Bacteria* Enteropathogenic inhabitant of animal intestines linked to meat contamination and poisoning	University of Wisconsin	http://www.genome.wisc.edu
Haemophilus influenza KW20	1.83	*Bacteria* Humans are the only known natural hosts for the commensal inhabitant of the upper respiratory mucosa. Causes otitis media and respiratory disease in children, whereas particularly invasive strains cause meningitis	The Institute for Genomic Research	
Halobacterium species NRC-1	2.57	*Archaea* Extremely halophilic. Requires 5–10 times the salinity of seawater to grow, resulting in high internal salt concentrations. Most of the encoded proteins are acidic	Halobacterium Genome Consortium	http://zdna2.umbi.umd.edu/~haloweb/
Helicobacter pylori 26695	1.66	*Bacteria* Intestinal inhabitant produces urease that facilitates survival at acidic pH of the gastric environment. One of the most common chronic bacterial infections of humans. Causes gastritis and potentiates gastric carcinomas	The Institute for Genomic Research	http://www.tigr.org/tigr-scripts/CMR2/GenomePage3.spl?database=ghp
Methanobacterium thermoautotrophicum Delta H	1.75	*Archaea* Moderate thermophile found in most anaerobic environments. Autotroph, requiring only CO_2, H_2, and salts for growth	Genome Therapeutics Corporation and Ohio State University	http://www.genomecorp.com//img/sequence_meth.gif
Lactococcus lactis IL1403	2.36	*Bacteria* One of the most important organisms employed in the dairy industry as leaven	Genoscope	http://www.genoscope.cns.fr/externe/English/Projets/Projet_AI/AI.html
Listeria monocytogenes EGD-e	2.94	*Bacteria* Human pathogen	European Commission Consortium Pasteur Institute	http://genolist.pasteur.fr/ListiList/

Organism and strain	Genome size (Mb)	Phylogenetic domain	Institution	Internet link
Methanococcus jannaschii DSM 2661	1.66	*Archaea* The first complete genome sequence for an *Archaea*, and the first complete sequence of an autotroph. Sea floor surface inhabitant grows at pressures of more than 200 atmospheres and at an optimum temperature of 85°C. Strict anaerobe and producer of the greenhouse gas methane	The Institute for Genomic Research	
Mesorhizobium loti MAFF303099	7.59	*Bacteria* Forms globular nodules and performs nitrogen fixation on several *Lotus* species	Kasuza DNA Research Institute, Japan	http://www.kazusa.or.jp/rhizobase/
Mycobacterium leprae	3.26	*Bacteria* Causative agent of leprosy	Sanger Centre, United Kingdom	http://www.sanger.ac.uk/Projects/M_leprae/
Mycobacterium tuberculosis H37Rv	4.40	*Bacteria* Causative agent of tuberculosis	Sanger Centre, United Kingdom	http://www.sanger.ac.uk/Projects/M_tuberculosis/
Mycoplasma genitalum G-37	0.58	*Bacteria* Lacks cell wall. Smallest known genome for an autonomously replicating organism. Implicated in urethritis	The Institute for Genomic Research	http://www.tigr.org/tigr-scripts/CMR2/GenomePage3.spl?database=gmg
Mycoplasma pneumoniae M129	0.81	*Bacteria* Lacks cell wall. Causes human respiratory infections	University of Heidelberg	http://www.zmbh.uni-heidelberg.de/M_pneumoniae/MP_Home.html
Neisseria meningitidis MC58 (ATCC BAA-335)	2.27	*Bacteria* Causative agent of meningitis and septicemia. Genetic phase variation controls differential gene expression and contributes to the evasion of the immune system	The Institute for Genomic Research	http://www.tigr.org/ igr-scripts/CMR2/GenomePage3.spl?database=gnm
Pasteurella multocida Pm70	2.40	*Bacteria* Animal pathogen, particularly domestic fowl	University of Minnesota	http://www.cbc.umn.edu/ResearchProjects/Pm/pmhome.html
Pseudomonas aeruginosa PAO1	6.30	*Bacteria* Opportunistic pathogen. Complicates burns and cystic fibrosis	University of Washington, Seattle *Pseudomonas aeruginosa* Community Annotation Project.	http://www.pseudomonas.com/
Pyrococcus horikoshii OT3	1.80	*Archaea* Hyperthermophile, discovered in the hot aquatic zones of the Okinawa Trench. Optimum growth temperature is near boiling at 98°C. Genome includes 2,061 genes encoding for heat-resistant proteins.	Japan Biotechnology Center (NITE)	http://www.bio.nite.go.jp/cgi-bin/dogan/genome_top.cgi?%27ot3%27
Rickettsia conorii Malish 7	1.27	*Bacteria* Obligate intracellular bacterial commensal inhabitants of arthropods. Occasionally causes human diseases such as Mediterranean spotted fever, which is transmitted by ticks	Unités des Rickettsies, CNRS, Marseille / Information Génétique et Structurale, CNRS, Marseille	http://polyc.cnrs-mrs.fr/RicBase/index.html
Saccharomyces cerevisiae S288C	13.00	*Eukarya* Versatile unicellular yeasts involved in food and beverage fermentation	International Consortium	http://genome-www.stanford.edu/Saccharomyces/

Continued

Organism and strain	Genome size (Mb)	Phylogenetic domain	Institution	Internet link
Salmonella typhi CT18	4.80	*Bacteria* Causative agent of typhoid fever	Sanger Centre, United Kingdom	http://www.sanger.ac.uk/Projects/S_typhi/
Sinorhizobium meliloti 1021	6.7	*Bacteria* Rhizosphere inhabitant, symbiotic association in nodules of *Medicago, Melilotus*, and *Trigonella* sp. Bacteroids fix atmospheric nitrogen by reducing N_2 to NH_3	European and Canadian Consortium and Stanford University	http://sequence.toulouse.inra.fr/meliloti.html
Staphylococcus aureus N315	2.81	*Bacteria* Common inhabitant of human skin. Implicated in pimples, boils, and acne	Juntendo University, Japan and NITE	http://www.ncbi.nlm.nih.gov/cgi-bin/Entrez/chrom?gi=179anddb=G
Streptococcus pyogenes MGAS8232	1.89	*Bacteria* Human pathogen implicated in upper respiratory tract infections and sore throat	Laboratory of Human-Bacterial Pathogenesis, Rocky Mountain Laboratories	http://www.ncbi.nlm.nih.gov/cgi-bin/Entrez/framik?db=Genomeandgi=234
Streptomyces coelicolor A3(2)	8.7	*Bacteria* The model representative of soil-dwelling bacteria with a complex life cycle involving fungal-like mycelial growth and spore formation. Produces pharmaceutical compounds such as 70% of all natural antibiotics, and anti-tumor drugs	Sanger Centre, United Kingdom	http://www.sanger.ac.uk/Projects/S_coelicolor/
Sulfolobus solfataricus P2	2.99	*Archaea* Most widely studied aerobic crenarchaeon. Grows optimally at 80°C and pH 2–4 metabolizing sulfur.	Canadian and European Consortium	http://www-archbac.u-psud.fr/projects/sulfolobus/sulfolobus.html
Synechocystis sp. PCC 6803	3.57	*Bacteria* Unicellular non-nitrogen-fixing cyanobacterium inhabiting fresh water	Kazusa DNA Research Institute, Japan	http://www.kazusa.or.jp/cyano/cyano.html
Thermoplasma acidophilum	1.56	*Archaea* Thermophilic heterotroph grows at 55–60°C and pH 0.5–4. Among the most acidophilic organisms known. Surprisingly, lacks protective outer shells (S-layer, cell wall), but maintains a near-neutral cytosolic pH	Max Planck Institute for Biochemistry, Germany	http://www.biochem.mpg.de/baumeister/genome/
Thermotoga maritime MSB8	1.80	*Bacteria* Inhabitant of geothermal marine sediment, originally isolated at Vulcano, Italy. Grows at 80°C. Conversion of cellulose to hydrogen renders it promising as a tool for renewable carbon and energy sources. Small-subunit rRNA phylogeny indicates that it is one of the deepest and most slowly evolving lineages in the *Bacteria*	The Institute for Genomic Research	http://www.tigr.org/tigr-scripts/CMR2/GenomePage3.spl?database=btm
Treponema pallidum Nichols	1.14	*Bacteria* Spirochaete. Causative agent of the sexually transmitted disease syphilis	The Institute for Genomic Research and The University of Texas	http://www.tigr.org/tigr-scripts/CMR2/GenomePage3.spl?database=gtp
Ureaplasma urealyticum Serovar 3	0.75	*Bacteria* Causative agent of sexually transmitted disease	University of Alabama, Applied Biosystems, and Eli Lilly	http://genome.microbio.uab.edu/uu/uugen.htm

Organism and strain	Genome size (Mb)	Phylogenetic domain	Institution	Internet link
Vibrio cholerae serotype 01, Biotype El Tor, strain N16961	4.00	*Bacteria* Causative agent of the devastating water-borne disease cholera. Genome contains two circular chromosomes, with the small chromosome containing genes that originated outside the Proteobacteria group and a gene capture system (integron island) suggesting ancestral capture of a megaplasmid by an ancestral *Vibrio* species	The Institute for Genomic Research	http://www.tigr.org/tigr-scripts/CMR2/GenomePage3.spl?database=gvc
Xanthomonas axonopodis pv. Citri 306	5.17	*Bacteria* Causative agent of citrus canker, affecting commercial citrus cultivars with significant economic impact on citrus production worldwide	Organization for Nucleotide Sequencing and Analysis – Consortium	http://cancer.lbi.ic.unicamp.br/xanthomonas/
Xylella fastidiosa 9a5c	2.68	*Bacteria* Plant pathogen that infects orange trees causing a disease known as "amarelinho" in Brazil	Organization for Nucleotide Sequencing and Analysis – Consortium	http://watson.fapesp.br/genoma.htm
Leishmania major Friedlin	0.27 (Chromosome Number 1)	*Eukarya* Causative agent of Leishmaniasis	Seattle Biomedical Research Institute	http://www.sbri.org/programs.htm
Plasmodium falciparum Isolate 3D7	1.00 (Chromosome Numbers 1–3)	*Eukarya* Causative agent of devastating mosquito-borne disease malaria	The Institute for Genomic Research, National Medical Research Institute, and Sanger Institute (Chromosome 3)	http://www.tigr.org/tdb/e2k1/pfa1/
Notable microbial genomes in progress				
Acidithiobacillus ferrooxidans (Thiobacillus ferrooxidans) ATCC 23270	2.90	*Bacteria* Iron-oxidizing bacteria important in the development of acid mine drainage	The Institute for Genomic Research	http://tigrblast.tigr.org/ufmg/index.cgi?database=a_ferrooxidans%7Cseq
Bacillus anthracis Ames	4.50	*Bacteria* Human pathogen. Causative agent of anthrax. Potential biological weapon	The Institute for Genomic Research	http://tigrblast.tigr.org/ufmg/index.cgi?database=b_anthracis%7Cseq
Burkholderia pseudomallei K96243	6.00	*Bacteria* Previously known as *Pseudomonas pseudomallei*. Causative agent of melioidosis, a tropical disease of humans and animals. Resistant to many antibiotics	Sanger Centre, Defense Evaluation Research Agency, and the Central Public Health Laboratory, United Kingdom	http://www.sanger.ac.uk/Projects/B_pseudomallei/
Candida albicans SC5314	16.00	*Eukarya* Human pathogen, implicated in mucosal infections in generally healthy persons to life-threatening systemic infections in immunodeficient individuals	Stanford University	http://sequence-www.stanford.edu/group/candida/index.html
Carboxydothermus hydrogenoformans	2.10	*Bacteria* Thermophilic organism, important as a potential source of hydrogen	The Institute for Genomic Research and Center of Marine Biotechnology	http://tigrblast.tigr.org/ufmg/index.cgi?database=c_hydrogenoformans%7Cseq

Continued

Organism and strain	Genome size (Mb)	Phylogenetic domain	Institution	Internet link
Chloroflexus aurantiacus J-10-fl	3.00	*Bacteria* Filamentous, thermophilic, anoxygenic, gliding phototroph. Forms prominent microbial mats in neutral to alkaline hot springs. Optimal growth temperature between 50°C and 60°C	Joint Genome Institute	http://www.jgi.doe.gov/JGI_microbial/html/chloroflexus/chloro_homepage.html
Chromobacterium violaceum CCT 3496 / JMC 3496	4.20	*Bacteria* Gram-negative aquatic and soil organisms. Produce violacein, a purple pigment which exhibits trypanocidal and antibiotic activity	The Virtual Institute of Genomic Research, Brazil	http://www.brgene.lncc.br/indexCV.html
Corynebacterium diphtheria NCTC13129	3.1	*Bacteria* Causative agent of diphtheria	Sanger Centre, United Kingdom, World Health Organization, and Public Health Laboratory, United Kingdom	http://www.sanger.ac.uk/Projects/C_diphtheriae/
Cytophaga hutchinsonii ATCC 33406	4.00	*Bacteria* Aerobic Gram-negative inhabitant of soils. Rapidly digests crystalline cellulose, the most abundant biopolymers on Earth	Joint Genome Institute	http://www.jgi.doe.gov/JGI_microbial/html/cytophaga/cytoph_homepage.html
Desulfobacterium autotrophicum HMR2	4.4	*Bacteria* Sulfate-reducing bacteria inhabiting marine sediments. Capable of lithoautotrophic growth using hydrogen plus sulfate and carbon dioxide as sole sources of energy and carbon	Real Environmental Genomix, Germany	http://www.regx.de/organisms/
Desulfotalea psychrophila LSv54	4.0	*Bacteria* Sulfate-reducing bacteria inhabiting marine sediments. Capable of lithoautotrophic growth at temperatures below 0°C	Real Environmental Genomix, Germany	http://www.regx.de/organisms/
Entamoeba histolytica	~20	*Eukarya* Water-borne causative agent of amoebic dysentery	The Institute for Genomic Research	http://www.lshtm.ac.uk/mp/bcu/enta/homef.htm
Erwinia carotovora subspecies atroseptica	4.0	*Bacteria* Plant pathogen causing soft rot of potato. Member of the Enterobacteriaceae, it is related to *Escherichia, Shigella, Salmonella*, and *Yersinia*	Sanger Centre, United Kingdom	http://www.sanger.ac.uk/Projects/E_carotovora/
Ferroplasma acidarmanus Fer 1	2.00	*Archaea* Able to transform the sulfide found in metal ores to sulfuric acid, the chemical pollutant that contaminates mining sites and drains into nearby rivers, streams and groundwater	Joint Genome Institute	http://www.jgi.doe.gov/JGI_microbial/html/ferroplasma/ferro_homepage.html
Geobacter sulfurreducens	2.50	*Bacteria* Sulfur-reducing soil inhabitant	The Institute for Genomic Research and University of Massachusetts, Amherst	http://tigrblast.tigr.org/ufmg/index.cgi?database=g_sulfurreducens%7Cseq
Giardia lamblia WB	12.00	*Eukarya* Water-borne pathogen. Model for ancient eukaryotic cell evolution	Marine Biological Laboratory, Woods Hole	http://jbpc.mbl.edu/Giardia-HTML/

Organism and strain	Genome size (Mb)	Phylogenetic domain	Institution	Internet link
Haloarcular morismortui ATCC 43049	~4.0	*Archaea* Salt-tolerant aquatic organism	University of Maryland Biotechnology Institute and the Institute for Systems Biology	http://zdna2.umbi.umd.edu/
Magnetospirillum magnetotacticum MS-1 (ATCC 31632)	4.5	*Bacteria* Gram-negative, motile, microaerophilic spiral cells, 0.5 × 5 microns, belonging to the alpha-subdivision of the Proteobacteria. Forms single-domain magnetosome crystals of the iron mineral magnetite (Fe_3O_4). Magnetotaxis following the magnetic field lines of the Earth towards regions favorable to growth	Joint Genome Institute	http://www.jgi.doe.gov/JGI_microbial/html/magnetospirillum/mag_spirill_homepage.html
Methanosarcina barkeri Fusario	2.8	*Archaea* Inhabits anaerobic sediments, mud, and in the rumen of cattle where it produces methane gas	Joint Genome Institute	http://www.jgi.doe.gov/JGI_microbial/html/methanosarcina/methano_homepage.html
Mycobacterium Tuberculosis CSU93	4.40	*Bacteria* Causative agent of tuberculosis, the leading cause of death due to infection worldwide. About 8 million new cases of active tuberculosis arise annually resulting in 3 million deaths	The Institute for Genomic Research	http://www.tigr.org/tigr-scripts/CMR2/GenomePage3.spl?database=gmt
Neisseria meningitidis Serogroup C Strain FAM18	2.20	*Bacteria* Causative agent of meningitis and septicemia in cyclical epidemics, particularly along the African meningitis geographical "belt"	Sanger Centre, United Kingdom	http://www.sanger.ac.uk/Projects/N_meningitidis/seroC.shtml
Nitrosomonas europaea	2.2	*Bacteria* Autotrophic nitrifying organism. Most studied of the ammonia-oxidizing bacteria that participate in the biogeochemical nitrogen cycle	Joint Genome Institute	http://www.jgi.doe.gov/JGI_microbial/html/nitrosomonas/nitro_homepage.html
Prochlorococcus marinus MIT9313	2.4	*Bacteria* Marine cyanobacterium, probably the most abundant photosynthetic organism on the planet, representing up to 50% of the total chlorophyll in certain parts of the ocean. Smallest known photosynthetic organism	Joint Genome Institute	http://www.jgi.doe.gov/JGI_microbial/html/Prochlorococcus_mit9313/prochlo_mit9313_homepage.html
Pseudomonas fluorescens PfO-1	5.5	*Bacteria* Nonpathogenic saprophyte inhabiting soils, aquatic systems, and plant surfaces. Produces a soluble, greenish fluorescent pigment under conditions of low iron availability	Joint Genome Institute	http://www.jgi.doe.gov/JGI_microbial/html/pseudomonas/pseudo_homepage.html
Pseudomonas putida	6.1	*Bacteria* Soil aerobe capable of degrading many aromatic organic pollutants	The institute for Genomic Research and QIAGEN German Consortium	http://www.qiagen.com/sequencing/psputida.html
Pseudomonas syringae pv. Tomato DC3000	6.0	*Bacteria* Plant opportunistic pathogen, causative agent of ice nucleation	The Institute for Genomic Research	http://tigrblast.tigr.org/ufmg/index.cgi?database=p_syringae%7Cseq

Continued

Organism and strain	Genome size (Mb)	Phylogenetic domain	Institution	Internet link
Pyrococcus abyssi GE5	1.8	*Archaea* Thermophile inhabiting 3500 meters deep in the south-east Pacific ocean. Optimal growth conditions are 103°C and 200 atmospheres pressure	Genoscope	http://www.genoscope.cns.fr/Pab/
Ruminococcus albus 8	4.00	*Bacteria* Rumen bacterium contributing to degradation of cellulosic materials and methane production	The Institute for Genomic Research	http://tigrblast.tigr.org/ufmg/index.cgi?database=r_albus%7Cseq
Rhodobacter sphaeroides 2.4.1	4.34	*Bacteria* Metabollically diverse, capable of photosynthesis, lithotrophy, aerobic and anaerobic respiration, and molecular nitrogen fixation. Harbors dual chromosome with evidence of sequence duplication. Possible model for the evolution of diploid systems found in eukaryotes	University of Texas, Houston Health Science Center	http://www-mmg.med.uth.tmc.edu/sphaeroides/
Rickettsia conori	1.2	*Bacteria* Causative agent of *fievre boutonneuse* (Mediterranean fever) in humans in contact with the brown dog tick	Genoscope	http://www.genoscope.cns.fr/externe/English/Projets/Projet_BU/BU.html
Sphingomonas aromaticivorans F199	3.60	*Bacteria* Yellow pigmented obligate aerobes. Lack lipopolysaccharide, but possess glycosphingolipid, with thermoreversible gel formation and viscosity useful as thermostable agar substitutes (Gelrite). Linked to the death of coral reefs	Joint Genome Institute	http://www.jgi.doe.gov/JGI_microbial/html/sphingomonas/sphingo_homepage.html
Yersinia pestis IM5P12	4.67	*Bacteria* Causative agent of bubonic plague	University of Wisconsin	http://www.genome.wisc.edu/
Yersinia pseudotuberculosis IP 32953	~5	*Bacteria* Closely related to *Yersinia pestis*. Human and animal pathogen. Widespread in soils and waters	Lawrence Livermore National Laboratory and The Pasteur Institute	http://bbrp.llnl.gov/bbrp/html/microbe.html

GLOSSARY

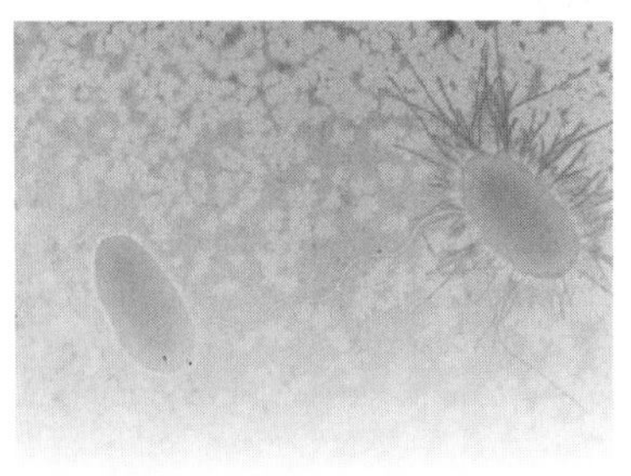

abiogenesis: The concept of spontaneous generation whereby it is held that living organisms or their essential components can emerge from non-living materials.

abiotic: Processes, conditions, or factors that do not require the involvement of biological organisms.

acetylene reduction assay: The enzymatic conversion of acetylene to ethylene, a process used to measure the rate of nitrogen fixation by microbial producers of nitrogenase.

acclimatization (acclimation): Process describing the capacity of organisms to adapt phenotypically to variable and usually unfavorable environmental conditions.

accretion: Term used in sedimentology or planetology to denote the increase in size by the addition of new materials from sources outside the system.

acid mine drainage: Ecosystem pollution resulting from mineral mining facilities where iron- and sulfur-oxidizing bacteria generate sulfuric acid through the leaching of sulfurylated ores. In addition to the acid, the drainage contains various combinations of toxic metals that may pose substantial hazards to ecosystem health.

acidophiles: Microorganisms that prefer to live in extremely acidic environments. For example, *Thiobacillus thiooxidans* grows at the minimum, optimum, and maximum pH values of 0.5, 2.0–2.8, and 4.0–6.0, respectively.

acrasin: 3′, 5′-cyclic adenosine monophosphate, which is produced by slime molds during the initiation of fruiting body formation.

actinobiology (actinology; radiology): The field of research focused on the effects of radiation on living organisms.

actinomycetes: Bacteria that sometimes resemble fungi because they form branching filaments.

adaptation: The process by which an organism copes with environmental stress. It can occur through evolutionary modification of physiological, morphological, or behavioral characteristics, resulting in the enhanced survival of the organism.

adaptive radiation: The evolutionary emergence of a wide variety of species that have adapted to distinct niches from a narrower spectrum of primitive species.

adhesins: Adhesion factors or chemical substances synthesized by biofilm-forming microorganisms to aid their attachment to solid surfaces.

aerobic: A term used to denote processes, conditions, or organisms that are associated with, or function best in the presence of, molecular oxygen.

aerosol: Small particulate materials dispersed in a gaseous medium, usually the atmosphere. Aerosolized particles are typically characterized by particle size (e.g. $PM_{2.5}$ refers to particulate matter less than or equal to 2.5 micrometers in diameter) and settling rate. Many microorganisms are dispersed in the environment as aerosols. Abiotic aerosols include smoke from natural fires and haze.

AFLP (amplified fragment length polymorphism): A technique for detecting small sequence differences among DNA fragments from different organisms. The "fingerprinting" technique detects DNA restriction fragments through PCR analysis according to the following steps: Restriction of the DNA with two restriction enzymes, preferably a hexa-cutter and a tetra-cutter recognizing unique 6- and 4-nucleotide sequences, respectively; ligation of double-stranded adapters to the ends of the restriction fragments; amplification of a subset of the restriction fragments using two primers complementary to the adapter and restriction site sequences, and extended at their 3′ ends by selective nucleotides; electrophoresis of the amplified restriction fragments on denatured polyacrylamide gels; visualization and comparative assessment of the DNA fingerprints.

algae: Photosynthetic and eukaryotic protoctists, some of which are microscopic and unicellular. Others have few cells (protists) but many more are large multicellular organisms. Algae are typically aquatic, contributing substantially to biogeochemical cycling in freshwater and marine ecosystems.

alkaliphiles: Microorganisms that prefer to live in extremely alkali environments. For example, *Nitrobacter* species grow at the minimum, optimum, and maximum pH values of 6.6, 7.6–8.6, and 10.0, respectively.

allochthonous: Not indigenous. Term used to describe opportunistic microorganisms that are either not normally present, or not major contributors to the ecology of a habitat, but are capable of rapidly proliferating following a major change in local environmental conditions, including the episodic introduction of nutrients.

amensalism: A form of ecological interaction in which the growth and activity of one organism is inhibited by another organism.

ammonification: The enzymatic formation of ammonium from nitrogen compounds carried out mostly by specialized bacteria.

anaerobic: Process, condition, or organism defined by the absence of oxygen. Anaerobic respiration is the metabolic process analogous to aerobic respiration, but where electrons are transferred from a reduced compound to an inorganic molecule that is different from oxygen. Sometimes used interchangeably with **anoxic**.

anagenesis: A theory of speciation that proposes the generation of higher levels of specialization through progressive evolution.

annealing temperature: The highest temperature at which two complementary strands of nucleic acid will form a stable double stranded structure. The annealing temperature is directly proportional to the proportion of guanine and cytosine residues, and indirectly proportional to the percentage of sequence mismatch.

anoxyphotobacteria: Photosynthetic bacteria in possession of photosystem I. These organisms conduct photosynthesis in anaerobic environments. Oxygen is not a product of this kind of photosynthesis.

antagonism (antibiosis or allelopathy): A form of ecological interaction in which one organism produces chemical substances that inhibit the growth and/or activities of one or more other organisms.

antibody: A proteinous immunoglobulin molecule produced by plasma cells in response to a foreign chemical or microbiological substance (antigen). A population of antibody molecules may be homogenous (monoclonal) or heterogeneous (polyclonal). Antibodies function as part of the disease prevention strategy of organisms to eliminate invading antigens from the body of vertebrates and certain marine animals. In the assessment of microbial diversity, the specificity of antibody–antigen reactions has been used to detect organisms in various environments and to distinguish among closely related varieties of organisms.

antigen: A substance that can elicit an immune response when introduced into the body of vertebrates or marine animals capable of synthesizing immunoglobulin molecules (antibodies). Haptens are molecules that need to be combined with other substances to become true antigens. Antigens may be soluble as in the case of several microbial toxins, or particulate as in the case of viruses. Proteins and polysaccharides are the most potent antigens. Antigens represented by specific molecules on the surface of microorganisms have been used extensively for taxonomic purposes. For example, the Kauffmann–White classification scheme for *Salmonella* species involves the notation of specific serotypes which are defined by the organism's "O", "H", and "Vi" antigens.

apoprotein: Polypeptide molecules isolated from non-polypeptide moieties with which they naturally associate to form functional protein complexes. For examples, ferritin without its ferric hydroxide core may be referred to as apoferritin, and apohemoglobin is the protein of hemoglobin without its heme group.

aquifer: Subsurface water systems (groundwater). Typically contains a diverse array of autochthonous microbial populations, but subject to the introduction of non-native microorganisms and chemical contaminants through pollution events.

Archaea: Prokaryotic unicellular microorganisms lacking murein in their cell walls, with ether bonds in their membrane phospholipids. According to a widely accepted view, one of three major phylogenetic domains of life, the other two being Bacteria and Eukarya.

Archaean: Geological time period from the formation of the Earth (approximately 4.5 billion years ago) to 2.4 billion years ago, during the Precambrian era.

ascomycetes: Filamentous fungi defined by the use of an ascus sac to contain ascospores.

assimilation: The metabolic integration of nutrient chemicals into organism biomass.

asthenosphere: The upper mantle of the Earth, or the layer beneath the lithosphere.

atmosphere: The gaseous layer that forms the interphase between the solid components of planets and outer Space. The chemical composition of atmospheres is influenced by the biogeochemical cycling of elements and is important for the maintenance of biodiversity. On Earth, the atmosphere consists of 79% nitrogen, 20% oxygen, 1% argon, approximately 0.03% carbon dioxide, and trace quantities of other gases including methane, nitrogen oxides, and ozone. Small changes in the concentration of the trace components such as methane and ozone can have strong influences on living organisms. Atmosphere is also the unit of air pressure. One atmospheric pressure is the pressure of air at sea level and it is equivalent to a force of approximately 10^5 Pascals (Newtons per square meter) or 14.7 pounds per square inch. Hyperbarophilic microorganisms thrive in the deep oceans at pressures of 400 atmospheres.

autochthonous: Indigenous inhabitants of a microbial ecosystem. The organisms are resilient in the sense that they maintain routine population densities and ecological functions despite variable environmental conditions.

autopoiesis: "Self-making" or the ability of organisms (cells) to independently maintain their own metabolism and reproduction.

autotrophs: Organisms that have the capacity to produce (fix) organic carbon compounds from carbon dioxide.

Bacteria: According to a widely accepted view, one of three major phylogenetic domains of life, the other two being Archaea and Eukarya. Bacteria are distinguished by unicellularity, lack of a nucleus, and cell walls containing murein.

bacteriochlorophyll: The light-gathering pigment found in green and purple anaerobic photosynthetic bacteria.

bacteriophage: Virus particles specialized in the infection of bacteria.

banded iron formation: A sedimentary rock with alternating iron-rich (Fe_2O_3) and iron-deficient bands. Banded iron formations (BIF) are typical of Precambrian rocks, and they are considered to be evidence for the emergence of oxygen-evolving photosynthesis approximately 3.5 billion years ago. BIF presumably resulted from the oxidation of Fe^{2+} by oxygen, and the resulting Fe_2O_3 was deposited in the primitive ocean. BIF deposits reached a peak occurrence in rocks dated from 2.5–3.0 billion years ago. Contemporary mining of iron ore taps ancient BIF deposits in Australia, South Africa, and the United States.

barophiles: Microorganisms typically found in deep seas, which grow best under conditions of high pressure.

benthos: Term used to describe organisms inhabiting the bottom section of aquatic environments.

biodegradation: Enzymatic conversion of complex chemical compounds into simpler compounds.

biogas: Primarily methane with traces of hydrogen sulfide and carbon dioxide produced by anaerobic microorganisms during the degradation of organic matter.

bioluminescence: The enzymatic action of luciferase to produce oxyluciferin from luciferin proteins. The reaction is associated with light production.

biomass: The total mass of living (micro) organisms in a particular habitat.

biome: A large geographical region characterized by a dominant climax ecological community. A biome typically has several communities at different successional stages.

biosensor: An artificial devise based on genetic, immunological, or other biochemical process used for the detection of specific types of microorganisms, or the occurrence of specific microbial processes in complex environmental systems.

biotype: A variety of organisms distinguished by metabolic and/or physiological properties.

blight: A plant disease resulting in plant death or withering without producing rot.

bloom: Visible microbial biomass in aquatic systems produced by large numbers of cyanobacteria. Blooms are also attributable to excessive algal growth resulting from eutrophication.

$C_0t_{1/2}$: The amount of time required for half the concentration of single-stranded nucleic acids in a reaction to form stable double-stranded structures. It is usually taken as an index of diversity in a microbial community gene pool.

carrying capacity: The ability of an ecosystem to sustain its biological population.

chemoautotroph (chemolithotroph): Microorganisms capable of growing on carbon dioxide or carbonates as the sole source of carbon while deriving energy through the oxidation of reduced inorganic compounds.

chemocline: In aquatic systems, the layer formed by a sharp gradient in chemical composition. For example, salt concentrations may change sharply with depth in lakes that periodically receive freshwater input.

chemoorganotroph (heterotroph): A microorganism capable of growing on organic compounds as the source of energy and carbon.

chemotaxis: The movement of microorganisms towards or away from a chemical gradient, which is either desirable or toxic, respectively.

chlorosis: Discoloration or yellowing of plant tissue due to loss of the photosynthetic pigment chlorophyll. Chlorosis may result from microbial infection of plants.

circirdian rhythm: Organism's intrinsic daily physiological rhythms independent of fluctuating environmental conditions.

cladogenesis: A theory of speciation that proposes the divergence of populations through progressive branching and splitting of early lineages.

cladogram: Also known as a "phylogenetic tree." Branching line diagrams of evolutionary relationships among species, strains, or molecular sequences. Independent lineages are represented by line segments. Branching points represent major periods of departure from the ancestral lineage, such as in the case of speciation events.

climax community: A microbial community in which all possible niches are occupied. The phrase is used to signify the terminal development stage of ecological succession.

co-metabolism: The enzymatic conversion of a chemical substrate into products by a microorganism that does not derive carbon, energy, or other forms of nutrients from the process. Typically, this is a fortuitous process as the enzyme is synthesized by the microorganism for a different purpose.

consortium: A community of two or more microbial species capable of combined metabolic activities that would otherwise proceed slowly or not at all without the presence of any one member.

Crenarchaeota: A kingdom within the Archaea domain characterized by extremely thermophilic members.

crown gall: Plant tumor, which is a symptom of infection by *Agrobacterium tumefaciens.* Many broad-leafed plants are susceptible to the disease.

cyanobacteria: Photosynthetic prokaryotes. They live mostly in aquatic environments and microbial mats where they generate oxygen by enzymatic cleavage of water.

dendogram: A diagram representing the relatedness among different microbial strains or species according to phylogenetic analysis.

denitrification: Enzymatic reduction of nitrate or nitrite to nitrogen or other more reduced forms of nitrogen oxides. In anaerobic or micro-aerobic respiration, nitrogen oxides are used as terminal electron acceptors.

desulfurization: The enzymatic cleavage of sulfur, usually as hydrogen sulfide, from organic sulfur compounds.

detritivores: Organisms characterized by consumption of detritus or the degradation of waste materials and dead organisms. They are important in global geochemical cycling.

DGGE (denaturing gradient gel electrophoresis): A genetic analysis technique suitable for detecting single base changes and polymorphisms in whole genomes or DNA fragments. Under a concentration gradient of denaturing agent, the mobilities of DNA fragments of the same length but different nucleotide sequence are different. The technique can also be used to determine DNA fragment melting points.

diazotroph: Organisms that are capable of fixing molecular nitrogen by enzymatic reduction of N_2 to the equivalent of ammonia.

diel: Occurring daily; 24-hour periods.

diversity: A qualitative or quantitative measure of the types of species present in a microbial community. Measures of diversity assess differences among organisms with respect to a given trait at the molecular, physiological, or anatomic levels. Microbial diversity is subsumed under the general term biodiversity, which defines the total of all species in any habitat.

dormancy: The viable state of metabolic inactivity in which many kinds of microorganisms reside to cope with adverse environmental conditions. For example, bacteria and fungi form spores. The viable-but-non-culturable state that characterizes several microorganisms present in the environment can be considered a state of dormancy.

dysphotic zone: The zone in a body of water that does not receive sufficient light to support photosynthetic activity by plants, but with sufficient light to support other types of phototrophism carried out by microorganisms.

dystrophic: Fresh water systems that have low concentrations of dissolved nutrients, and high amounts of suspended colloids and fragments from plant material.

ELISA (enzyme-linked immunosorbent assay): ELISA is an antigen-detection strategy where antibodies are tagged with a quantifiable marker that is expressed following attachment to a specific antigen. ELISA has been used extensively for detecting the presence of many microorganisms, including viruses in complex environmental systems.

epilimnion: The layer above the thermocline in an aquatic ecosystem.

epiterranean: Organisms inhabiting the interface between soil and air.

estuary: An intermediary ecosystem defined by the mixing of freshwater from a river outlet and marine water from ocean tidal activity.

Eubacteria: Now renamed simply bacteria, refers to the first proposed name for non-archaea prokaryotes.

Eukarya: According to one view, the third major phylogenetic domain of life, the other two being Bacteria and Archaea. Eukaryotes are traditionally defined by possession of a membrane-bound nucleus containing the genetic material organized as tightly packaged chromosomes. Microscopic eukaryotes include some species of algae, protozoa, and fungi.

euphotic zone: In aquatic ecosystems, the depth of water receiving light, and consequently, the zone that supports the growth of photosynthetic organisms.

eutrophication: The process of "self-feeding" by which ecosystems sustain sufficient nutrients to support the growth of inhabiting organisms. In aquatic environments, excessive input of nutrients has detrimental impacts when oxygen is depleted through respiration to sustain growth. Run-off containing nitrogen and phosphorus fertilizers can lead to the eutrophication of water systems.

evenness: A quantitative measure of the distribution of individual microorganisms or species within a microbial community.

exoenzymes (abiotic enzymes): Enzymes that are active outside the cell. They are either secreted by living cells into their environment or leaked out of the cell following cell lyses.

facultative: Microorganisms that have the metabolic flexibility to conduct either one of two mutually exclusive metabolic processes such as anaerobic and aerobic respiration.

FAME analysis: Fatty acid methyl ester analysis is a technique for identifying microorganisms through unique combinations of fatty acids in their cells.

fastidious: A term used to describe microorganisms with extremely stringent growth requirements.

feedback mechanism: The process by which the production of a biochemical material leads to the enhancement (positive feedback) or inhibition (negative feedback) of the same production process. Feedback mechanisms contribute to the overall economy of metabolism where, for example, after reaching a certain concentration, the end product of a metabolic pathway inhibits an enzyme within the pathway. Similarly, accumulation of a substrate can induce the production of catabolic enzymes. Feedback mechanisms may also act on large scales to regulate major biogeochemical cycles, but this requires further research to elucidate the contributions of many different organisms involved in most element cycles.

FISH (fluorescent *in situ* hybridization): A technique for identifying phylogenetic groups of microorganisms in natural microbial communities without prior isolation or cultivation. The technique relies on genomic hybridization to nucleotide probes labeled with fluorescent molecules. FISH is typically combined with microscopy.

flagellum (plural flagella): An extracellular proteinous appendage of some prokaryotes responsible for cell motility. The flagellum may occur singly (monotrichous) or in multiple forms distributed evenly over the cell surface (peritrichous) or in a polar region (amphitrichous). Flagella in most bacteria project from the cell surface except in the members of the phylogenetic order *Spirochaetales*, where the flagella occur between layers of the cell envelope.

fossil: The preserved remains of an organism, typically in rock. Body fossils represent transformations of bones, shells, or other structures, whereas trace fossils refer to imprints, tracks, or burrows.

gene: The basic unit of heredity. Genes are made up of nucleic acid sequences. Structural genes (DNA) are either transcribed into messenger RNA, and subsequently translated into protein molecules, or they are transcribed into ribosomal RNA or transfer RNA and used directly in cells. Regulatory genes function to mediate the transcription and translation of structural genes. All the genes in an organism are collectively referred to as the genome. A major distinction between prokaryotes and eukaryotes is the organization of the genome. The genome of prokaryotes exists as a covalently closed chromosome(s) and occasional plasmid(s) in the cytoplasm. The genome of eukaryotes is organized in a special compartment—the nucleus.

genomics: The structural and functional investigation of the entire genetic material of organisms. Genomics has been facilitated by the invention of rapid DNA and RNA sequencing techniques.

greenhouse effect: The climatic phenomenon where average temperatures increase in response to the increased concentration of "greenhouse gases" such as carbon dioxide and methane in the atmosphere.

halophiles: Microorganisms that thrive in a high salt environment, up to 30% NaCl. For example, *Halococcus* species.

hormesis: The seemingly paradoxical observation that exposure of organisms to sub-toxic concentrations of a toxic substance can stimulate the organism's growth.

hot spring: An aquatic ecosystem or thermal spring defined by temperatures exceeding 37°C.

humic acids: Abundant soil organic material characterized by irregular high molecular weight acidic polymers.

hybridization: The annealing of two complementary single strands of nucleic acids (DNA and/or RNA) to form double-stranded molecules whose stability depends on the extent of sequence similarity.

hypolimnion: In aquatic ecosystems, the deep, cold zone beneath the thermocline.

k-strategists: Species of microorganisms that maintain their population densities at the approximate carrying capacity of the environment. They are also known as the autochthonous organisms whose populations do not fluctuate to extremes in response to environmental changes.

lichen: A symbiotic association between fungi (mostly *Ascomycetes* or *Basidiomycetes*) and cyanobacteria. Most of the lichen tissue is made up of the fungal hyphae and cyanobacterial cells are distributed throughout the hyphal network. The cyanobacteria provide nutrients for the fungi, which protect the cyanobacteria from desiccation. Lichens grow in nutrient poor surfaces such as rocks and tree bark because of the photosynthetic activity of the cyanobacteria.

lignin: Abundant complex polymers that provide structural support for the woody part of plants.

lithosphere: The rigid crustal plates of the Earth, immediately above the asthenosphere, where most of the biological diversity resides.

littoral zone: The shore of a lake to a depth of 10 m or the shore of a continental shelf to a depth of 200 m.

lysogeny: The reversible latent condition in which a bacteriophage exists with its genome either integrated into the host chromosome or replicating autonomously in the host cytoplasm.

magnetotaxis: Locomotion of microorganisms towards a defined geomagnetic field. These kinds of organisms produce intracellular granules that aid their responses to magnetic fields.

melting temperature (T_m): The temperature at which double-stranded nucleic acid molecules dissociate ("melt") into two single-stranded molecules. The T_m depends on the proportion of guanine and cytosine bonds (G + C ratio) in the double-stranded molecule and the extent of sequence mismatch.

mesophiles: Microorganisms that grow optimally within the temperature range of 15°C to 35°C. Most prokaryotes that have been described belong to this group.

mesozoic: Geological era from 245 to 66 million years ago.

metabolism: The combination of all enzymatic processes that sustain an organism's physiology, including biochemical reactions resulting in the synthesis of new compounds (anabolism) and the degradation of compounds (catabolism).

metagenomics: A term coined to describe the analysis of genetic material extracted directly from environmental samples, in order to facilitate the characterization of microbial diversity that encompasses culturable and unculturable organisms. Metagenomics-based research involves the construction of composite libraries of microbial genomes that can be screened for particular functional traits, or for performing comparative assessments of genetic potential cross ecological systems.

metalloproteins: Protein molecules that require metal ions to function. Many microbial enzymes are metalloproteins, and the intracellular metal ion concentrations are under tight regulation to avoid potential toxicity.

meteorite: Fragments of asteroids that fall to Earth, surviving the disintegration by vaporization that usually occurs for materials passing through the atmosphere at great speed. Stony meteorites are most common, consisting of silicate minerals. Iron meteorites consist primarily of iron and some nickel. Carbonaceous chondrites are meteorites containing clay-type silicate materials and organic compounds that have been investigated as originating from biological sources, or as contributing materials to the origin of life on Earth.

methanogens: Microorganisms that produce methane through the enzymatic reduction of carbon dioxide or low molecular weight fatty acids. Most belong to the Archaea domain.

microaerophiles: Microorganisms that grow best under very low concentrations of oxygen. In aquatic environments, they are found beneath the water surface and above the sediment layer at a region where the diffusion gradient of oxygen reaches a sufficiently low concentration to support growth.

microbial loop: The concept that is used to explain the participation of microorganisms in nutrient cycling and food webs in marine ecosystems. The microbial loop is essentially a food chain that works within, or in parallel to, the classical food chain. In the microbial loop, heterotrophic bacteria obtain their carbon and energy sources directly from dissolved inorganic materials. These organisms, which are too small to be preyed on directly by copepods, are grazed by flagellates and ciliates. Ciliates are then consumed by copepods and the process continues up the classical food chain. The microbial loop concept is considered a major milestone in marine biology.

microbial mat: A densely structured heterogeneous microbial community in the benthos. Phototrophic bacteria are typically responsible for primary productivity in microbial mats. Stromatolites are fossilized microbial mats.

mineralization: The enzymatic transformation of an element from organic forms to its inorganic form. An example of mineralization is the total conversion of cellulose to carbon dioxide.

mitochondrion (plural: mitochondria): An organelle of eukaryotic cells occurring as a semi-autonomous structure with its own genome. Mitochondria function in cellular energy generation through respiration and the tri-carboxylic acid cycle. Mitochondria are absent from certain eukaryotes, including protozoa belonging to the phylogenetic orders *Pelobiontida* and *Diplomonadida*. The serial endosymbiotic theory posits that mitochondria originated through symbiotic interactions among ancestral prokaryotes. The prokaryotic origin of mitochondria is supported by nucleic acid sequence analysis.

mixotrophism: The capacity of some microorganisms (mixotrophs) to conduct both autotrophic and heterotrophic modes of metabolism.

molecular marker: Traceable genetic trait that can be used to identify specific microorganisms defined by phenotypic affiliation or ability to perform a certain ecological function in a complex heterogeneous community.

mutagen: A physical (e.g. electromagnetic radiation), chemical (e.g. nitrous acid and 5-bromo-deoxyuridine), or biological (e.g. transposable nucleic acid; transposons) agent that can cause mutations (nucleotide sequence alterations) in the genetic material.

mycorrhiza: Symbiotic interaction between filamentous fungi and plants in the rhizosphere.

neutralism: Used to define the absence of a relationship between two different species present in the same habitat. Neutralism is very difficult to prove.

niche: The concept used to define both the physical and functional situations of an organism relative to the biotic and abiotic components of its immediate environment. The niche concept encompasses all parameters that support or restrict the existence of species. Related concepts include **niche diversification**, or the total number of possible microhabitats and functions necessary to maintain ecosystem integrity; **niche breadth**, which defines the hyperspace in which organisms exist (upper and lower boundaries of physical and chemical parameters that support growth); **niche size**, which is an index of the number of species that can coexist in a community or habitat; and **niche overlap**, which defines direct competition between two or more species for the same ecological role, resources, or physical space.

nitrification: Part of the nitrogen cycle in which ammonium is enzymatically oxidized to nitrogen oxides, particularly nitrite and nitrate.

nitrogen fixation: The enzymatic conversion of molecular nitrogen to ammonia, which is then converted to organic nitrogen required by organisms for cellular metabolism. In nature, this is primarily a symbiotic process between plants and a selected group of microorganisms. Approximately 5–8% of the global amount of fixed nitrogen is attributed to lightning events which split molecular nitrogen, there enabling N atoms to combine with oxygen in the air to form nitrogen oxides. Industrially, nitrogen is fixed through Haber's process where under high pressure and temperature (600°C), atmospheric nitrogen and hydrogen are catalytically combined to form ammonia, which is either used directly or processed further to generate, for example, ammonium nitrate fertilizer.

numerical taxonomy: In microbial systematics, a strategy of positioning organisms in a phylogenetic tree by comparing large numbers of characteristics.

oligonucleotide: Short fragment of nucleic acid typically consisting of 30 or fewer nucleotides. Can be labeled and used as a hybridization probe.

oligotrophic: Limited supply of nutrients. Describes environments where the growth of organisms is retarded because of inadequate access to nutrients.

organelle: Small "organs" representing distinct intracellular features specialized in certain functions. For example, prokaryotic organelles include ribosomes, carboxysomes, and flagella. Examples of organelles for eukaryotes include the products of endosymbiosis, namely mitochondria and the plastids.

osmophiles: Organisms capable of living in environments with high concentrations of sugar.

ozone layer: The layer of ozone in the stratosphere. Ozone can be depleted by catalytic reactions involving halogenated compounds of industrial or biological origin.

Pangaea: The supercontinent which existed 200 million years ago.

Paleozoic: Geological era from 570 to 245 million years ago.

parasitism: Ecological interaction in which one organism depends entirely on another organism for nutrients and/or habitat.

PCR (polymerase chain reaction): A method originally developed by Kary Mullis for amplifying specific segments of DNA, through the annealing of oligonucleotide primers, and the extension of the primers based on the template DNA present in the target genetic material. Heat-stable DNA polymerase is essential for PCR. The amplification of target RNA is possible through the action of reverse transcriptase (RT-PCR).

pelagic zone: In marine environments, the zone between the sea floor and the edge of the continental shelf.

phanerozoic: The period from 570 million years ago to the present.

phosphobacteria: Phosphatase-producing bacteria capable of producing orthophosphate from organic phosphate compounds.

photolithotrophs: Microorganisms that can grow using light energy and carbon dioxide (or carbonates) as the sole source of carbon.

photosynthesis: The process through which certain organisms grow using light energy captured by photosynthetic pigments (e.g. chlorophyll) to drive reactions that produce organic matter from carbon dioxide and an electron donor (e.g. hydrogen, water, hydrogen sulfide).

photosystem I: The photosynthesis system that does not require external electron donors and does not produce reduced coenzyme.

photosystem II: In contrast to photosystem I, photosystem II is a photosynthetic system that requires external electron donors and produces reduced coenzyme.

phylogenetic: The evolutionary relationships within and between taxonomic groups. The phylogeny or "family tree" of a group is represented by diagrams tracing the relationships between ancestors and descendants.

phytoplankton: An aquatic community of unicellular photosynthetically active organisms.

plasmid: An autonomously replicating satellite DNA carried by some bacteria in addition to the chromosome.

population: All individuals representing a group of organisms belonging to the same species and occupying the same geographical location at the same time. Populations sharing a common gene pool as a local interbreeding group are referred to as "demes".

P/R ratio: A quantitative measure of the relationship between photosynthesis and respiration in an ecosystem.

Precambrian: The period from the formation of Earth to 570 million years ago.

prions: Causative agent of neurodegenerative diseases. Pathogenic prions are deformed versions of proteins that are normally required for cellular functioning.

prokaryotes: Unicellular microorganisms traditionally defined by their lack of a nucleus. Instead, the genetic material (chromosome) is loosely organized within the cell.

proteome: The total protein content of a cell or of a microbial community.

***Protoctista*:** A phylogenetic domain consisting of eukaryotic organisms defined by a multi-genomic nature and aquatic habitats, excepting true fungi, plants, and animals. Protoctists include algae, amebas, ciliates, foraminiferans, seaweeds, and water molds. Protists within this group are either unicellular or exhibit a small number of cells.

protozoa: Eukaryotic microorganisms without rigid cell walls. Protozoa are important in bacterial ecology because of their grazing activities.

psychrophiles: Microorganisms that thrive in cold environments. They are defined by the ability to proliferate at 0°C. Cold tolerance in psychrophilic bacteria is associated with plasma membranes which contain unsaturated fatty acids. Psychrophiles in the Antarctic contain polyunsaturated fatty acids, which were previously thought to be absent in prokaryotes.

protocooperation: A mutually beneficial but unnecessary interaction between two species.

quorum sensing: The ability of microorganisms to monitor and respond to population densities in complex microbial communities.

recalcitrant chemicals: Chemicals, usually anthropogenic, which are resistant to microbiological degradation.

respiratory quotient: An index of the efficiency of respiration defined as the number of carbon dioxide molecules produced from each molecule of oxygen.

r-strategists: Organisms which are optimized for maximizing growth rate under conditions in which exogenous nutrients become available, as opposed to organisms that keep growth rate in check while favoring maintaining population density around the environment's carrying capacity (k-strategists). In environmental microbiology, r-strategists are also known as allochthonous or zymogenous microorganisms.

RAPD (random amplification of polymorphic DNA): The technique invented for identifying molecular level differences among organisms through the conduction of PCR using random primers. The primers anneal to different parts of the genome in different organisms, leading to differently sized amplification products. A comparative assessment of the amplification products can provide an index of diversity in a microbial community.

redox potential (E_h in volts): The quantitative index of the electron exchange tendency of a system. It is a relative index based on the assumption that the redox potential of a standard hydrogen electrode is zero.

remote sensing: The use of Earth-orbiting satellites to investigate physical structures, ecosystem boundaries, and changes in environmental parameters without direct contact sampling. Extensive data processing, photogrametric image analysis, and ground-level verification are usually required to correctly interpret data acquired through remote sensing.

RFLP (restriction fragment length polymorphism): A technique for determining the level of nucleotide sequence differences among different genomes. A restriction enzyme with a strict nucleotide sequence recognition profile is used against each genome. If the frequency of occurrence of the recognition sequence differs among the genomes being compared, the resulting restriction fragment profiles will be different.

rRNA (ribosomal ribonucleic acid): A component of ribosomes, the location of protein synthesis in all cells. Prokaryotic ribosomes contain three major kinds of rRNA molecules: 5S, 16S, and 23S rRNA. More than 10,000 16S and 16S-like rRNA and 1,000 23S and 23S-like rRNA genes have been sequenced by laboratories trying to understand the phylogenetic relationships among organisms.

semaphoront: The basic unit of biological systematics, representing an individual at a particular phase of the life cycle.

serial endosymbiotic theory: Explains the origin of eukaryotes from ancestral prokaryotic cells that engaged in a series of symbiotic interactions occurring in a specific sequence, leading to the establishment of organelles such as mitochondria, plastids, and undulipodia.

serotype: A microorganism that has antigens (protein molecules that can induce immune response in mammals) that distinguish the organism from other closely related varieties.

siderophore: A microbial product secreted by some organisms, notably pseudomonads, to form stable coordination compounds with iron. The major types of siderophores are catecholate and hydroxamate, and they play critical roles in maintaining homeostatic control of metal bioavailability.

skotophiles: Microorganisms that prefer growth in dark environments.

species: The smallest formal unit in the hierarchy of microbiological classification, representing a group of organisms formally recognized as distinct from other groups. Members of a single species may be recognized as different strains when they exhibit characteristics that are temporary or easily selected in a population.

spores: Reproductive cells produced by some prokaryotes in response to adverse environmental conditions. Spores are important in the survival and dispersal of microorganisms in the environment.

SSCP (single-strand conformation polymorphism): A technique for detecting genetic diversity through the observation of small differences in nucleotide sequences. The technique is facilitated by PCR, and it takes advantage of differences in the mobility of single-stranded DNA according to sequence information, which allows the formation of secondary structures with variable stability.

strain: Within a taxon, organisms or populations of organisms that are genetically similar, but distinguishable from other organisms or populations within that taxon.

stratosphere: The region of the atmosphere that extends from the tropopause to the stratopause, at about 50 km. The stratosphere features the ozone layer, which absorbs much of the solar electromagnetic radiation in the ultraviolet range.

stromatolite: Biogenic sedimentary structure of laminated silicate or carbonate rocks. Stromatolites are fossilized microbial communities consisting of cyanobacteria and other microorganisms typical of living microbial mats.

symbiosis: The ecological relationship in which two different species occupy the same microhabitat with extensive obligatory interactions, which may not necessarily be mutually beneficial.

synergism: The ecological relationship in which two species engage in non-obligatory mutually beneficial interactions.

TGGE (temperature gradient gel electrophoresis): Technique for screening a population of microorganisms for mutation in a particular genetic locus. When a temperature gradient is applied during the electrophoretic separation of nucleic acids (DNA or RNA), fragments of identical length but different sequence can be separated.

thermocline: In aquatic ecosystems, the zone defined by drastic change in water temperature.

thermophiles: Organisms that thrive in extremely hot environments. For example, the optimum growth temperature for *Pyrococcus furiosus* is around 100°C. *Methanopyrus* and *Pyrodictium* species can grow at maximum temperatures of 110°C and 113°C, respectively. Thermophilic bacteria have saturated fatty acids in their membranes. Hyperthermophilic Archaea do not have fatty acids in their membranes. Instead, they have repeating subunits of phytane, a branched, saturated, isoprenoid. Protein molecules in thermophiles undergo post-translational modification, including dehydration, to render them heat stable.

tokogenetic: A term used to describe the genetic relationship between individuals.

transposon: Mobile genetic element found in many phylogenetic groups. The mechanism of mobility involves non-homologous recombination.

troposphere: The region of the atmosphere that extends from the Earth's surface to the tropopause, approximately 10–16 km in altitude, depending on the latitude at which the measurement is taken. An important feature of the troposphere is that most of the biological influences on atmospheric composition are limited to this range, except for the impact on the tropospheric ozone layer, which should be recognized as distinct from the tropospheric ozone that results from urban air pollution.

undulipodium (plural: undulipodia): An organelle of eukaryotes, including cilia, that functions in motility associated with sensory or feeding activity. The serial endosymbiotic theory posits that undulipodia originated through symbiosis among prokaryotes.

UV: Ultraviolet light representing the 100–400 nm segment of the electromagnetic spectrum.

viruses: Particulate obligately infectious genetic entities consisting primarily of a protein coat and genome made up of DNA or RNA.

visible light: The 400–800 nm section of the electromagnetic spectrum. Visible light is required for most photosynthetic activity.

xenobiotic: A chemical compound that is not a product of biological metabolism, but of industrial manufacture. Many xenobiotic compounds are recalcitrant and/or toxic to living systems.

xerophiles: Microorganisms that live in extremely dry environments lacking easily available water.

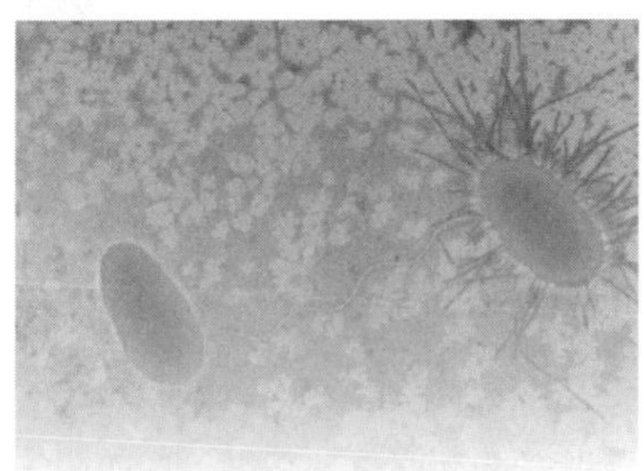

REFERENCES

Aas, P., M.M. Lyons, R. Pledger, D.L. Mitchell, and W.H. Jeffrey. 1996. Inhibition of bacterial activities by solar radiation in nearshore waters and the Gulf of Mexico. *Aquatic Microbial Ecology*, **11**: 229–38.

Acea, M.J., A. Prieto-Fernandez, and N. Diz-Cid. 2003. Cyanobacterial inoculation of heated soils: Effect on microorganisms of C and N cycles and on chemical composition in soil surface. *Soil Biology and Biochemistry*, **35**: 513–24.

Adams, M.B. 2003. Ecological issues related to N deposition to natural ecosystems: Research needs. *Environment International*, **29**: 189–99.

Adler, P.R. and P.L. Sibrell. 2003. Sequestration of phosphorus by acid mine drainage floc. *Journal of Environmental Quality*, **32**: 1122–9.

Agnihotri S., K. Kulshreshtha, and S.N. Singh. 1999. Mitigation strategy to contain methane emission from rice-fields. *Environmental Monitoring and Assessment*, **58**: 95–104.

Aguzzi, A. 1997. Prion research: the next frontier. *Nature*, **389**: 795–8.

Al-Awadhi, H., R.H. Al-Hasan, N.A. Sorkhoh, S. Salamah, and S.S. Radwan. 2003. Establishing oil-degrading biofilms on gravel particles and glass plates. *International Biodeterioration and Biodegradation*, **51**: 181–5.

Alexandersen, S., Z. Zhang, A.I. Donaldson, and A.J.M. Garland. 2003. The pathogenesis and diagnosis of foot-and-mouth disease. *Journal of Comparative Pathology*, **129**: 1–36.

Alfreider, A., C. Vogt, D. Hoffmann, and W. Babel. 2003. Diversity of ribulose-1,5-bisphosphate carboxylase/oxygenase large-subunit genes from groundwater and aquifer microorganisms. *Microbial Ecology*, **45**: 317–28.

Alp, B., G.M. Stephens, and G.H. Markx. 2002. Formation of artificial, structured microbial consortia (ASMC) by dielectrophoresis. *Enzyme and Microbial Technology*, **31**: 35–43.

Alper, J. 2003. Water splitting goes *au naturel*. *Science*, **299**: 1686–7.

Alperin, M.J. and W.S. Reeburgh. 1985. Inhibition experiments on anaerobic methane oxidation. *Applied and Environmental Microbiology*, **50**: 940–5.

Altschul, S.F., T.L. Madden, A.A. Schaffer, J. Zhang, Z. Zhang, W. Miller, and D.J. Lipman. 1997. Gapped BLAST and PST-BLAST: A new generation of protein database search programs. *Nucleic Acids Research*, **25**: 3389–402.

Amann, R.I., W. Ludwig, and K.H. Schleifer. 1995. Phylogenetic identification and *in situ* detection of individual microbial cells without cultivation. *Microbiology Reviews*, **59**: 143–69.

Amann, R.I., W. Ludwig, R. Schulze, S. Spring, E. Moore, and K-H. Schleifer. 1996b. rRNA-targeted oligonucleotide probes for the identification of genuine and former pseudomonads. *Systematic and Applied Microbiology*, **19**: 501–9.

Amann, R.I., J. Snaidr, M. Wagner, W. Ludwig, and K.H. Schleifer. 1996. *In situ* visualization of high genetic diversity in a natural microbial community. *Journal of Bacteriology*, **178**: 3496–500.

American Society for Microbiology. 2002. Microbe Library. http://www.microbelibrary.org/.

Andersson S.G. and C.G. Kurland. 1999. Origins of mitochondria and hydrogenosomes. *Current Opinion in Microbiology*, **2**: 535–41.

Andersson, S.G., A. Zomorodipour, J.O. Andersson, T. Sicheritz-Ponten, U.C. Alsmark, R.M. Podowski, A.K. Naslund, A.S. Eriksson, H.H. Winkler, and C.G. Kurland. 1998. The genome sequence of *Rickettsia prowazekii* and the origin of mitochondria. *Nature*, **396**: 133–40.

Andrews, J.A., R. Matamala, K.M. Westover, and W.H. Schlesinger. 2000. Temperature effects on the diversity of soil heterotrophs and the delta-^{13}C of soil-respired CO_2. *Soil Biology and Biochemistry*, **32**: 699–706.

Araujo, J.C., G. Brucha, J.R. Campos, and R.F. Vazoller. 2000. Monitoring the development of anaerobic biofilms using fluorescent in situ hybridization and confocal laser scanning microscopy. *Water Science and Technology*, **41**: 69–77.

Aravind, L., R.L. Tatusov, Y.I. Wolf, D.R. Walker, and E.V. Koonin. 1998. Evidence for massive gene exchange between archaeal and bacterial hyperthermophiles. *Trends in Genetics*, **14**: 442–4.

Arber, W. 1994. Bacteriophage transduction. In R.C. Webster and A. Granoff (eds.) *Encyclopedia of Virology*, pp. 107–13. London: Academic Press.

Archibald, J.M. and P.J. Keeling. 2002. Recycled plastids: A "green movement" in eukaryotic evolution. *Trends in Genetics*, **18**: 577–84.

Arias, Y.M. and B.M. Tebo. 2003. Cr (VI) reduction by sulfidogenic and nonsulfidogenic microbial consortia. *Applied and Environmental Microbiology*, **69**: 1847–53.

Arnold, F.H., P.L. Wintrode, K. Miyazaki, and A. Gershenson. 2001. How enzymes adapt: Lessons from directed evolution. *Trends in Biochemical Sciences*, **26**: 100–6.

Aron, J.L. and J.A. Patz (eds.) 2001. *Ecosystem Change and Public Health: A global Perspective.* Baltimore, Johns Hopkins University Press.

Arp, D.J. 2000. The nitrogen cycle. In E.W. Triplett (ed.) *Prokaryotic Nitrogen Fixation: A Model System for the Analysis of a Biological Process.* Norfolk, England: Horizon Scientific Press.

Arrhenius, S. 1903. The propagation of life in space. *Die Umschau*, **7**: 481.

Arrhenius, S. 1908. *Worlds in the Making: The Evolution of the Universe.* New York: HarperCollins.

Arrieta, J.M., M.G. Weinbauer, and G.J. Herndl. 2000. Interspecific variability in sensitivity to UV radiation and subsequent recovery in selected isolates of marine bacteria. *Applied and Environmental Microbiology*, **66**: 1468–73.

Ashelford, K.E., S.J. Norris, J.C. Fry, M.J. Bailey, and M.J. Day. 2000. Seasonal population dynamics and interactions of competing bacteriophages and their host in the rhizosphere. *Applied and Environmental Microbiology*, **66**: 4193–9.

Asner, G.P., A.R. Townsend, W.J. Riley, P.A. Matson, J.C. Neff, and C.C. Cleveland. 2001. Physical and biogeochemical controls over terrestrial ecosystem responses to nitrogen deposition. *Biogeochemistry (Dordrecht)*, **54**: 1–39.

Atlas, R.M. 1995. *Handbook of Media for Environmental Microbiology.* Boca Raton, FL: CRC Press.

Atlas, R.M. 1997. *Handbook of Microbiological Media.* Second Edition. Edited by L.C. Parks. Boca Raton, FL: CRC Press.

Atlas, R.M. and R. Bartha. 1998. *Microbial Ecology* (Fourth Edition). Menlo Park, CA: Benjamin Cumming.

Avery, O., C. MacLeod, and M. McCarty. 1944. Studies on the chemical nature of the substance inducing transformation of pneumococcal types. *Journal of Experimental Medicine*, **79**: 137–57.

Awramik, S.M. 1992. The oldest records of photosynthesis. *Photosynthesis Research*, **33**: 75–89.

Ayala, F.J. 1982. Gradualism versus punctuationalism in speciation: Reproductive isolation, morphology, genetics. In C. Barigozzi (ed.) *Mechanisms of Speciation. Proceedings from the International Meeting on Mechanisms of Speciation*, sponsored by the Academia Nazionale dei Lincei May 4–8, 1981, Rome, Italy, pp. 51–66; *Progress in Clinical and Biological Research*, volume 96. New York: Alan R. Liss.

Ayala, F.J. 1997. Vagaries of the molecular clock. *Proceedings of the National Academy of Sciences (USA)*, **94**: 7776–83.

Azam, F., T. Fenchel, J.G. Field, J.S. Gray, L.A. Meyer-Reil, and F. Thingstad. 1983. The ecological role of water-column microbes in the sea. *Marine Ecology Progress Series*, **10**: 257–63.

Azofsky, A.I. 2002. Size-dependent species-area relationships in benthos: Is the world more diverse for microbes? *Ecography*, **25**: 273–82.

Baker, J.O., C.I. Ehrman, W.S. Adney, S.R. Thomas, and M.E. Himmel. 1998. Hydrolysis of cellulose using ternary mixtures of purified cellulases. *Applied Biochemistry and Biotechnology*, **70**: 395–403.

Baker, N.R. 2001. *Curriculum Guidelines for Microbiology Majors: Recommendations.* American Society for Microbiology, http://www.asmusa.org.

Bakermans, C. and E.L. Madsen. 2002. Diversity of 16S rDNA and naphthalene dioxygenase genes from coal-tar-waste-contaminated aquifer waters. *Microbial Ecology*, **44**: 95–106.

Balkwill, D.L., F.R. Leach, J.T. Wilson, J.F. McNabband, and D.C. White. 1988. Equivalence of microbial biomass measures based on membrane lipid and cell wall components, adenosine triphosphate, and direct counts in sub-surface sediments. *Microbial Ecology*, **16**: 73–84.

Balmelli, T. and J.-C. Piffaretti. 1996. Analysis of the genetic polymorphism of *Borrelia burgdorferi* sensu lato by multilocus enzyme electrophoresis. *International Journal of Systematic Bacteriology*, **46**: 167–72.

Balows, A., H.G. Truper, M. Dworkin, W. Harder, and K.-H. Schleifer (eds.). 1991. *The Prokaryotes. A Handbook on the Biology of Bacteria: Ecophysiology, Isolation, Identification, Applications.* Second Edition. New York: Springer Verlag.

Barer, M.R., L.T. Gibbon, C.R. Harwood, and C.E. Nwoguh. 1993. The viable but not non-culturable hypothesis and medical microbiology. *Review of Medical Microbiology*, **4**: 183–91.

Barer, M.R. and C.R. Harwood. 1999. Bacterial viability and culturability. *Advances in Microbial Physiology*, **41**: 93–137.

Bartosch, S., C. Hartwig, E. Spieck, and E. Bock. 2002. Immunological detection of *Nitrospira*-like bacteria in various soils. *Microbial Ecology*, **43**: 26–33.

Bassler, B.L. 2002. Small talk: Cell-to-cell communication in bacteria. *Cell*, **109**:421–4.

Bates, T.S., B.K. Lamb, A. Guenther, J. Dignon, and R.E. Stoiber. 1992. Sulfur emissions to the atmosphere from natural sources. *Journal of Atmospheric Chemistry*, **14**: 315–37.

Bates, T.S. and P.K. Quinn. 1997. Dimethylsulfide (DMS) in the equatorial Pacific Ocean (1982 to 1996): Evidence of a climate feedback? *Geophysical Research Letters*, **24**: 861–4.

Baucher, M., J. Ralph, and W. Boerjan. 2003. Lignin biosynthesis review. *Annual Review of Plant Biology*, **54**: 519–46.

Baum, D.A. and M.J. Donoghue. 1995. Choosing among alternative "phylogenetic" species concepts. *Systematic Botany*, **20**: 560–73.

Beerling, D.J. 1999. Quantitative estimates of changes in marine and terrestrial primary productivity over the past 300 million years. *Proceedings of the Royal Society Biological Sciences (Series B)*, **266**: 1821–7.

Beijerinck, M. 1888. Die Bakterien de Papilionaceenknollchen. *Botanische Zeitung*, **46**: 725–804. In T.D. Brock (trans. and ed.) 1998. *Milestones in Microbiology: 1556 to 1940*, p. 220. Washington, DC: American Society for Microbiology Press.

Beja, O., E.V. Koonin, L. Aravind, L.T. Taylor, H. Seitz, J.L. Stein, D.C. Bensen, R.A. Feldman, R.V. Swanson, and E.F. DeLong. 2002. Comparative genomic analysis of archaeal genotypic variants in a single population and in two different oceanic provinces. *Applied and Environmental Microbiology*, **68**: 335–45.

Belay, N., R. Sparling, and L. Daniels. 1984. Dinitrogen fixation by a thermophilic methanogenic bacterium. *Nature*, **312**: 286–8.

Belviso, S. 2000. Methods for estimating the oceanic source of DMS on a global scale. *Oceanis*, **26**: 577–600.

Benediktsdottir, E., L. Verdonck, C. Sproer, S. Helgason, and J. Swings. 2000. Characterization of *Vibrio viscosus* and *Vibrio wodanis* isolated at different geographical locations: A proposal for reclassification of *Vibrio viscosus* as *Moritella viscosa* comb. nov. *International Journal of Systematic and Evolutionary Microbiology*, **50**: 479–88.

Benini, S., W.R. Rypniewski, K.S. Wilson, S. Miletti, S. Ciurli, and S. Mangani. 1999. A new proposal for urease mechanism based on the crystal structures of the native and inhibited enzyme from *Bacillus pasteurii*: Why urea hydrolysis costs two nickels. *Structure*, **7**: 205–16.

Benjamin, M.M. and B.D. Honeyman. 1992. Trace metals. In S.S. Butcher, R.J. Charlson, G.H. Orans, and G.V. Wolfe (eds.) *Global Biogeochemical Cycles*, pp. 318–52. New York: Academic Press.

Benlloch, S., F. Rodriguez-Valera, S.G. Acinas, and A.J. Martinez-Murcia. 1996. Heterotrophic bacteria, activity, and bacterial diversity in two coastal lagoons as detected by culture and 16S rRNA genes PCR amplification and partial sequencing. *Hydrobiologia*, **329**: 3–17.

Berbee, M.L. and J.W. Taylor. 1993. Dating the evolutionary radiations of the true fungi. *Canadian Journal of Botany*, **71**: 1114–27.

Berger, W.H. and G. Wefer. 1991. Productivity of the glacial ocean: Discussion of the iron hypothesis. *Limnology and Oceanography*, **36**: 1899–918.

Berman-Frank, I., P. Lundgren, and P. Falkowski. 2003. Nitrogen fixation and photosynthetic oxygen evolution in cyanobacteria. *Research in Microbiology*, **154**: 157–64.

Bernal, A., U. Ear, and N. Kyrpides. 2001. Genomes OnLine Database (GOLD): A monitor of genome projects world-wide. *Nucleic Acids Research*, **29**: 126–7.

Bettarel, Y., C. Amblard, T. Sime-Ngando, J.-F. Carrias, D. Sargos, F. Garabetian, and P. Lavandier. 2003. Viral lysis, flagellate grazing potential, and bacterial production in Lake Pavin. *Microbial Ecology*, **45**: 119–27.

Bettarel, Y., T. Sime-Ngando, C. Amblard, J.-F. Carrias, D. Sargos, F. Garabetian, and P. Lavandier. 2002. The functional importance of bacteriophages in the microbial loop of an oligomesotrophic lake over a diel cycle. *Annales de Limnologie*, **38**: 263–9.

Betts, R.P., P. Bankes, and J.G. Banks. 1989. Rapid enumeration of viable microorganisms by staining and direct microscopy. *Letters in Applied Microbiology*, **9**: 199–202.

Bhanu P.J. and J.K.H. Hörber (eds.). 2002. *Atomic Force Microscopy in Cell Biology*. San Diego: Academic Press.

Billings, S.A., S.M. Schaeffer, S. Zitzer, T. Charlet, S.D. Smith, and R.D. Evans. 2002. Alterations of nitrogen dynamics under elevated carbon dioxide in an intact Mojave Desert ecosystem: Evidence from nitrogen-15 natural abundance. *Oecologia*, **131**: 463–7.

Binnig, G., C.F. Quate, and C. Gerber. 1986. Atomic force microscope. *Physical Review Letters*, **56**: 930–3.

Bintrim, S.B., T.J. Donohue, J. Handelsman, G.P. Roberts, and R.M. Goodman. 1997. Molecular phylogeny of Archaea from soil. *Proceedings of the National Academy of Sciences (USA)*, **94**: 277–82.

Biotol (Biotechnology by Open Learning). 1992. *In Vitro Cultivation of Microorganisms*. Butterworth- Oxford, England: Heinemann.

Bishop, T.R., M.W. Miller, A. Wang, and P.M. Dierks. 1998. Multiple copies of the ALAD gene are located at the Lv locus in *Mus domesticus* mice. *Genomics*, **48**: 221–31.

Bithell, J.F. 2000. A classification of disease mapping methods. *Statistics in Medicine*, **19**: 2203–15.

Blodig W., A.T. Smith, W.A. Doyle, and K. Piontek. 2001. Crystal structures of pristine and oxidatively processed lignin peroxidase expressed in *Escherichia coli* and of the W171F variant that eliminates the redox active tryptophan 171. Implications for the reaction mechanism. *Journal of Molecular Biology*, **305**: 851–61.

Bloemberg, G.V., A.H.M. Wijfjes, G.E.M. Lamers, N. Stuurman, and B.J.J. Lugtenberg. 2000. Simultaneous imaging of *Pseudomonas fluorescens* WCS365 populations expressing three different autofluorescent proteins in the rhizosphere: New perspectives for studying microbial communities. *Molecular Plant–Microbe Interactions*, **13**: 1170–6.

Blom A., W. Harder, and A. Matin. 1992. Unique and overlapping stress proteins of *Escherichia coli*. *Applied and Environmental Microbiology*, **58**: 331–4.

Blumthaler, M. and W. Ambach. 1990. Indication of increasing solar ultraviolet-B radiation flux in Alpine regions. *Science*, **248**: 206–8.

Boerlin, P. 1997. Applications of multilocus enzyme electrophoresis in medical microbiology. *Journal of Microbiological Methods*, **28**: 221–31.

Boettcher, K.J. and E.G. Ruby. 1995. Detection and quantification of *Vibrio fischeri* autoinducer from the symbiotic squid light organs. *Journal of Bacteriology*, **177**: 1053–8.

Bohlke, J.K., G.E. Ericksen, and K. Revesz. 1997. Stable isotope evidence for an atmospheric origin of desert nitrate deposits in northern Chile and southern California, USA. *Chemical Geology*, **136**: 135–52.

Bohlool B.B. and E.L. Schmidt. 1980. The immunofluorescence approach in microbial ecology. *Advances in Microbial Ecology*, **4**: 203–41.

Bonam, D., L. Lehman, G.P. Roberts, and P.W. Ludden. 1989. Regulation of carbon monoxide dehydrogenase and hydrogenase in *Rhodospirillum rubrum*: Effects of CO and oxygen on synthesis and activity. *Journal of Bacteriology*, **171**: 3102–7.

Boone, D.R. and G.M. Maestrojuán. 1993. The methanogens. In N.P. Staley, R.G.E. Murray, and J.G. Holt (eds.) *Bergey's Manual of Determinative Bacteriology*, 9th Edition, pp. 719–36. Baltimore: Williams and Wilkins.

Boone, D.R., W.B. Whitman, and P. Rouviere. 1993. Diversity and taxonomy of methanogens. In J.G. Ferry (ed.) *Methanogenesis*, pp. 35–80. New York: Chapman and Hall.

Bormann, B.T., C.K. Keller, D. Wang, and F.H. Bormann. 2002. Lessons from the sandbox: Is unexplained nitrogen real? *Ecosystems*, **5**: 727–33.

Borroto, R. and R. Martinez-Piedra. 2000. Geographical patterns of cholera in Mexico, 1991–1996. *International Journal of Epidemiology*, **29**: 764–72.

Boschike, E. and T. Bley. 1998. Growth patterns of yeast colonies depending on nutrient supply. *Acta Biotechnologica*, **18**: 17–27.

Bosgelmez-Tinaz, G. 2003. Quorum sensing in Gram-negative bacteria. *Turkish Journal of Biology*, **27**: 85–93.

Boukhalfa, H. and A.L. Crumbliss. 2002. Chemical aspects of siderophore mediated iron transport. *BioMetals*, **15**: 325–39.

Bouma, M.J. and M. Pascual. 2001. Seasonal and interannual cycles of endemic cholera in Bengal 1891–1940 in relation to climate and geography. *Hydrobiologia*, **460**: 147–56.

Bounias, M., and M. Purdey. 2003. Transmissible spongiform encephalopathies: A family of etiologically complex diseases: A review. *Science of the Total Environment*, **297**: 1–19.

Bratbak, G., M. Heldal, S. Norland, and T.F. Thingstad. 1990. Viruses as partners in spring bloom trophodynamics. *Applied and Environmental Microbiology*, **56**: 1400–5.

Bratbak, G., M. Levasseur, S. Michaud, G. Cantin, E. Fernandez, B.R. Heimdal, and M. Heldal. 1995. Viral activity in relation to *Emiliania huxleyi* blooms: A mechanism of DMSP release? *Marine Ecology Progress Series*, **128**: 133–42.

Breitbart, M., P. Salamon, B. Andresen, J.M. Mahaffy, A.M. Segall, D. Mead, F. Azam, and F. Rohwer. 2002. Genomic analysis of uncultured marine viral communities. *Proceedings of the National Academy of Sciences USA*, **99**: 14250–5.

Brinkmann, H. and H. Philippe. 1999. Archaea: sister group of bacteria? Indications from tree reconstruction artifacts from ancient phylogenies. *Molecular Biology and Evolution*, **16**: 817–25.

Broda, E. and G.A. Preschek. 1983. Nitrogen fixation as evidence for the reducing nature of the early atmosphere. *Biosystems*, **16**: 1–8.

Bronson, E.C. and J.N. Anderson. 1994. Nucleotide composition as a driving force in the evolution of retroviruses. *Journal of Molecular Evolution*, **38**: 506–32.

Brosch, R., M. Lefevre, F. Grimont, and P.A.D. Grimont. 1996. Taxonomic diversity of pseudomonads revealed by computer interpretation of ribotyping data. *Systematic and Applied Microbiology*, **19**: 541–55.

Brower, J.E., J.H. Zar, and C.N. von Ende. 1998. *Field and Laboratory Methods for General Ecology*. Boston: McGraw-Hill.

Brown, J.R., C.J. Douady, M.J. Italia, W.E. Marshall, and M.J. Stanhope. 2001. Universal trees based on large combined protein sequence data sets. *Nature Genetics*, **28**: 281–5.

Brown, K., K. Djinovic-Carugo, T. Haltia, I. Cabrito, M. Saraste, J.J. Moura, I. Moura, M. Tegoni, and C. Cambillau. 2000b. Revisiting the catalytic CuZ cluster of nitrous oxide (N_2O) reductase. Evidence of a bridging inorganic sulfur. *Journal of Biological Chemistry*, **275**: 41133–6.

Brown, K., M. Tegnoi, M. Prudencio, A.S. Pereira, S. Besson, J.J. Moura, I. Moura, and C. Cambillau. 2000a. A novel type of catalytic copper cluster in nitrous oxide reductase. *Nature Structural Biology*, **7**: 191–5.

Bruggemann, R., L. Zelles, Q.Y. Bai, and A. Hartmann. 1995. Use of Hasse diagram technique for evaluation of phospholipid fatty acids distribution as biomarkers in selected soils. *Chemosphere*, **30**: 1209–28.

Brusseau, G.A., H.C. Tsien, R.S. Hanson, and L.P. Wackett. 1990. Optimization of trichloroethylene oxidation by methanotrophs and the use of a colorimetric assay to detect soluble methane monooxygenase activity. *Biodegradation*, **1**: 19–29.

Buchholz-Cleven, B.E.E., B. Rattunde, and K.L. Straub. 1997. Screening for genetic diversity of isolates of anaerobic Fe(II)-oxidizing bacteria using DGGE and whole-cell hybridization. *Systematic and Applied Microbiology*, **20**: 301–9.

Buesseler, K.O. and P.W. Boyd. 2003. Will ocean fertilization work? *Science*, **300**: 67–8.

Bui, E.T., P.J. Bradley, and P.J. Johnson. 1996. A common evolutionary origin for mitochondria and hydrogenosomes. *Proceedings of the National Academy of Sciences (USA)*, **93**: 9651–6.

Bull, A.T. (ed.) 2003. *Microbial Diversity and Bioprospecting*. Washington, DC: American Society for Microbiology.

Bult, C.J., O. White, G.J. Olsen, L. Zhou, R.D. Fleischmann, G.G. Sutton, J.A. Blake, L.M. FitzGerald, R.A. Clayton, J.D. Gocayne, A.R. Kerlavage, B.A. Dougherty, J.-F. Tomb, M.D. Adams, C.I. Reich, R. Overbeek, E.F. Kirkness, K.G. Weinstock, J.M. Merrick, A. Glodek, J.D. Scott, N.S. Geoghagen, J.F. Weidman, J.L. Fuhrmann, D.T. Nguyen, T. Utterback, J.M. Kelley, J.D. Peterson, P.W. Sadow, M.C. Hanna, M.D. Cotton, M.A. Hurst, K.M. Roberts, B.B. Kaine, M. Borodovsky, H.P. Klenk, C.M. Fraser, H.O. Smith, C.R. Woese, and J.C. Venter. 1996. Complete genome sequence of the methanogenic archaeon, *Methanococcus jannaschii*, *Science*, **273**: 1058–73.

Burford, J.R. and J.M. Bremner. 1972. Is phosphate reduced to phosphine in waterlogged soils? *Soil Biology and Biochemistry*, **4**: 489–95.

Burja, A.M. and R.T. Hill. 2001. Microbial symbionts of the Australian Great Barrier Reef Sponge, *Candidaspongia flabellate*. *Hydrobiologia*, **461**: 41–7.

Burke, D.H., J.E. Hearst, and A. Sidow. 1993. Early evolution of photosynthesis—clues from nitrogenase and chlorophyll iron proteins. *Proceedings of the National Academy of Sciences (USA)*, **90**: 7134–8.

Burlage, R.S., R. Atlas, D. Stahl, G. Geesey, and G. Sayler (eds.) 1998. *Techniques in Microbial Ecology*. Oxford, England: Oxford University Press.

Burner, R.V. and L.S. Moore. 1987. Microbialites: organosedimentary deposits of benthic microbial communities. *Palaios*, **2**: 241–54.

Butcher, S.S., R.J. Charlson, G.H. Orans, and G.V. Wolfe (eds.) 1992. *Global Biogeochemical Cycles*. New York: Academic Press.

Butterworth, F.M., A. Gunatilaka, and M.E. Gonsebatt (eds.) 2001. *Biomonitors and Biomarkers as Indicators of Environmental Change 2: A Handbook*. Dordrecht: Kluwer Academic.

Cahan, D. (ed.) 1994. *Hermann von Helmholtz and the Foundations of Nineteenth-Century Science. California Studies in the History of Science, Volume 10*. Berkeley, CA: The University of California Press.

Cain, A.J. and W.B. Provine. 1992. Genes and ecology in history. In T.J. Crawford and G.M. Hewitt (eds.) *Genes in Ecology*, pp. 3–28. Oxford, England: Blackwell Scientific.

Cain, C.C., D. Lee, R.H. Waldo III, A.T. Henry, E.J. Casida Jr., M.C. Wani, M.E. Wall, N.H. Oberlies, and J.O. Falkinham III. 2003. Synergistic antimicrobial activity of metabolites produced by a nonobligate bacterial predator. *Antimicrobial Agents and Chemotherapy*, **47**: 2113–17.

Caldwell, D.E., G.M. Wolfaardt, D.E. Korber, S. Karthikeyan, J.R. Lawrence, and D. Brannan. 2002. Cultivation of microbial consortia and communities. In C.J. Hurst, R.L. Crawford, G.R. Knudsen, M.J. McInerney, and L.D. Stetzenbach. *Manual of Environmental Microbiology*, pp. 92–100. Washington, DC: American Society for Microbiology Press.

Canfield, D.E. and A.Teske. 1996. Late Proterozoic rise in atmospheric oxygen concentration inferred from phylogenetic and sulphur-isotope studies. *Nature*, **382**: 127–32.

Cano, R.J. and M.K. Borucki. 1995. Revival and identification of bacterial spores in 25 to 40 million-year-old Dominican amber. *Science*, **268**: 1060–4.

Cao M., J.B. Dent, and O.W. Heal. 1995. Modeling methane emissions from rice paddies. *Global Biogeochemical Cycles*, **9**: 183–95.

Cao M., K. Gregson, and S. Marshall. 1998. Global methane emission from wetlands and its sensitivity to climate change. *Atmospheric Environment*, **32**: 3293–9.

Capone, D.G., A. Subramaniam, J.P. Montoya, M. Voss, C. Humborg, A.M. Johansen, R.L. Siefert, and E.J. Carpenter. 1998. An extensive bloom of the N_2-fixing cyanobacterium *Trichodesmium erythraeum* in the central Arabian Sea. *Marine Ecology Progress Series*, **172**: 281–92.

Carbonell, X., J.L. Corchero, R. Cubarsi, P. Vila, and A. Villaverde. 2002. Control of *Escherichia coli* growth rate through cell density. *Microbiological Research*, **157**: 257–65.

Carlsson, P. and D.A. Caron. 2001. Seasonal variation of phosphorus limitation of bacterial growth in a small lake. *Limnology and Oceanography*, **46**: 108–20.

Carlton, R.G. and L.L. Richardson. 1995. Oxygen and sulfide dynamics in a horizontally migrating cyanobacterial mat: Black-band disease of corals. *FEMS Microbiology Ecology*, **18**: 155–62.

Carroll, J.W., M.C. Mateescu, K. Chava, R.R. Colwell, and A.K. Bej. 2001. Response and tolerance of toxigenic *Vibrio cholerae* O1 to cold temperatures. *Antonie van Leeuwenhoek*, **79**: 377–84.

Carter, M.R. 1986. Microbial biomass as an index for tillage-induced changes in soil biological properties. *Soil and Tillage Research*, **7**: 29–40.

Casamayor, E.O., C. Pedros-Alio, G. Muyzer, and R. Amann. 2002. Microheterogeneity in 16S ribosomal DNA-defined bacterial populations from a stratified planktonic environment is related to temporal changes and to ecological adaptations. *Applied and Environmental Microbiology*, **68**: 1706–14.

Casamayor, E.O., H. Schafer, L. Baneras, C. Pedros-Alio, and G. Muzer.

2000. Identification of and spatio-temporal differences between microbial assemblages from two neighboring sulfurous lakes: Comparison by microscopy and denaturing gradient gel electrophoresis. *Applied and Environmental Microbiology*, **66**: 499–508.

Catling D., K. Zahnle, and C. McKay. 2001. Biogenic methane and the rise of oxygen. *Astrobiology*, **1**: 392–3.

Cerniglia, C.E. and K.L. Shuttleworth. 2002. Methods for isolating polycyclic aromatic hydrocarbon (PAH)-degrading microorganisms and procedures for determination of biodegradation intermediates and environmental monitoring of PAHs. In C.J. Hurst, R.L. Crawford, G.R. Knudsen, M.J. McInerney, and L.D. Stetzenbach (eds.) *Manual of Environmental Microbiology*, pp. 972–86. Washington, DC: American Society for Microbiology Press.

Chan, W.W. and B.A. Dehority. 1999. Production of *Ruminococcus flavefaciens* growth inhibitor(s) by *Ruminococcus albus. Animal Feed Science and Technology*, **77**: 61–71.

Chao, L. and D.E. Carr. 1993. The molecular clock and the relationship between population size and generation time. *Evolution*, **47**: 688–90.

Charlson, R.J. 1993. Gas-to-particle conversion and CNN production. In G. Restelli and G. Angeletti (eds.) *Dimethylsulfide: Oceans, Atmospheres, and Climate*, pp. 275–86. Dordrecht: Kluwer.

Charlson, R.J., T.L. Anderson, and R.E. McDuff. 1992. The sulfur cycle. In S.S. Butcher, R.J. Charlson, G.H. Orans, and G.V. Wolfe (eds.) *Global Biogeochemical Cycles*, pp. 285–300. New York: Academic Press.

Charlson, R.J., J.E. Lovelock, M.O. Andrea, and S.G. Warren. 1987. Oceanic phytoplankton, atmospheric sulfur, cloud albedo and climate. *Nature*, **326**: 655–61.

Chauhan, S. and M.R. O'Brian. 1995. A mutant *Bradyrhizobium japonicum* delta-aminolevulinic acid dehydratase with an altered metal requirement functions *in situ* for tetrapyrrole synthesis in soybean root nodules. *Journal of Biological Chemistry*, **270**: 19823–7.

Chauhan, S., D.E. Titus, and M.R. O'Brian. 1997. Metals control activity and expression of the heme biosynthesis enzyme delta-aminolevulinic acid dehydratase in *Bradyrhizobium japonicum. Journal of Bacteriology*, **179**: 5516–20.

Chen, X., S. Schauder, I. Pelczer, B.L. Bassler, F. M. Hughson, N. Potier, and A. Van Dorsselaer. 2002. Structural identification of a bacterial quorum-sensing signal containing boron. *Nature*, **415**: 545.

Cheng, C.-N. 1975. Extracting and desalting amino acids from soils and sediments: evaluation of methods. *Soil Biology and Biochemistry*, **7**: 319–22.

Cheung, P.-Y. and B.K. Kinkle. 2001. *Mycobacterium* diversity and pyrene mineralization in petroleum-contaminated soils. *Applied and Environmental Microbiology*, **67**: 2222–9.

Chien, Y-T., V. Auerbuch, A.D. Brabban, and S.H. Zinder. 2000. Analysis of genes encoding alternative nitrogenase in the archaeon *Methanosarcina barkeri* 227. *Journal of Bacteriology*, **182**: 3247–53.

Chisholm, S.W. 2000. Oceanography: Stirring times in the Southern Ocean. *Nature*, **407**: 685–7.

Chisholm S.W., P.G. Falkowski, and J.J. Cullen. 2001. Discrediting ocean fertilization. *Science*, **294**: 309–10.

Chopp, D.L., M.J. Kirisits, B. Moran, and M.R. Parsek. 2002. A mathematical model of quorum sensing in a growing bacterial biofilm. *Journal of Industrial Microbiology and Biotechnology*, **29**: 339–46.

Christensen, B.B., J.A.J. Haagensen, A. Heydorn, and S. Molin. 2002. Metabolic commensalism and competition in a two-species microbial consortium. *Applied and Environmental Microbiology*, **68**: 2495–502.

Chung, A.P., F. Rainey, M.F. Nobre, J. Burghardt, and M.S. Da Costa. 1997. *Meiothermus cerbereus* sp. nov., a new slightly thermophilic species with high levels of 3-hydroxy fatty acids. *International Journal of Systematic Bacteriology*, **47**: 1225–30.

Cicerone, R.J. and R.S. Oremland. 1988. Biogeochemical aspects of atmospheric methane. *Global Biogeochemical Cycles*, **2**: 299–327.

Cleveland, C.C., A.R. Townsend, and S.K. Schmidt. 2002. Phosphorus limitation of microbial processes in moist tropical forests: Evidence from short-term laboratory incubations and field studies. *Ecosystems*, **5**: 680–91.

Clewell, D.B. (ed.) 1993. *Bacterial Conjugation*. New York: Plenum Press.

Coale, K.H. 1991. Effects of iron, manganese, copper, and zinc enrichments on productivity and biomass in the subarctic Pacific. *Limnology and Oceanography*, **36**: 1851–64.

Cohan, F.M. 1994. Genetic exchange and evolutionary divergence in prokaryotes. *Trends in Ecology and Evolution*, **9**: 175–80.

Cohan, F.M. 1994a. The effects of rare but promiscuous genetic exchange on evolutionary divergencies in prokaryotes. *American Naturalist*, **143**: 965–86.

Cohan, F.M. 1995. Does recombination constrain neutral divergence among bacterial taxa? *Evolution*, **49**: 164–75.

Cohan, F.M. 2001. Bacterial species and speciation. *Systematic Biology*, **50**: 513–24.

Cohan, F.M. 2002. What are bacterial species? *Annual Review of Microbiology*, **56**: 457–87.

Colebatch, G., B. Trevaskis, and M. Udvardi. 2002. Symbiotic nitrogen fixation research in the postgenomics era. *New Phytologist*, **153**: 37–42.

Collins, A.E. 1996. The geography of cholera. In B.S. Drasar and B.D. Forrest (eds.) *Cholera and the Ecology of Vibrio cholerae*, pp. 255–94. New York: Chapman and Hall.

Colwell, R.R. 1996. Global climate and infectious disease: The cholera paradigm. *Science*, **274**: 2025–31.

Colwell, R.R. 1996a. Microbial diversity—global aspects. In R.R. Colwell, U. Simidu, and K. Ohwada (eds.) *Microbial Diversity in Time and Space*, pp. 1–11. New York: Plenum Press.

Colwell, R.R. 1997. Microbial diversity: The importance of exploration and conservation. *Journal of Industrial Microbiology and Biotechnology*, **18**: 302–7.

Colwell, R.R., R.A. Clayton, B.A. Ortiz-Condel, D. Jacobs, and E. Russek-Cohen. 1995. The microbial species concept and biodiversity. In D. Allsopp, R.R. Colwell, and D.L. Hawksworth (eds.) *Microbial Diversity and Ecosystem Function*, pp. 3–15. Oxford: CAB International.

Colwell, R.R. and D.J. Grimes. (eds.) 2000. *Nonculturable Microorganisms in the Environment.* Washington, DC: American Society for Microbiology.

Colwell, R.R., U. Simidu, and K. Ohwada (eds.) 1996. *Microbial Diversity in Time and Space.* New York: Plenum Press.

Cornelissen, J.H.C., T.V. Callaghan, J.M. Alatalo, A. Michelsen, E. Graglia, A.E. Hartley, D.S. Hik, S.E. Hobbie, M.C. Press, C.H. Robinson, G.H.R. Henry, G.R. Shaver, G.K. Phoenix, D.G. Jones, S. Jonasson, F.S. Chapin III, U. Molau, C. Neill, J.A. Lee, J.M. Melillo, B. Sveinbjornsson, and R. Aerts. 2001. Global change and arctic ecosystems: Is lichen decline a function of increases in vascular plant biomass? *Journal of Ecology*, **89**: 984–94.

Correia, I.J., C.M. Paquete, R.O. Louro, T. Catarino, D.L. Turner, and A.V. Xavier. 2002. Thermodynamic and kinetic characterization of trihaem cytochrome c3 from *Desulfuromonas acetoxidans*. *European Journal of Biochemistry*, **269**: 5722–30.

Corsetti, F.A., S.M. Awramik, and D.A. Pierce. 2003. A complex microbiota from snowball Earth times: Microfossils from the Neoproterozoic Kingston Peak Formation, Death Valley, USA. *Proceedings of the National Academy of Sciences (USA)*, **100**: 4399–404.

Costerton, J.W., Z. Lewandowski, D.E. Caldwell, D.R. Kober, and H.M. Lappin-Scott. 1995. Microbial biofilms. *Annual Review of Microbiology*, **49**: 711–46.

Covacci, A., J.L. Telford, G.G. Del, J. Parsonnet, and R. Rappuoli. 1999. *Helicobacter pylori* virulence and genetic geography. *Science*, **284**: 1328—33.

Cracraft, J. 1983. Species concepts and speciation analysis. In R.F. Johnston (ed.) *Current Ornithology*, Volume 1, pp. 159–87. New York: Plenum Press.

Cracraft, J. 1989. Speciation and its ontology: The empirical consequences of alternative species concepts for understanding patterns and processes of differentiation. In D. Otte and J.A. Endler (eds.) *Speciation and its Consequences*, pp. 28–37. Sunderland, MA: Sinauer.

Crick, F.H.C. 1981. *Life itself: Its origin and nature.* New York: Simon and Schuster.

Crick, F.H.C. and L.E. Orgel. 1973. Propose the idea of colonizing space by deliberately seeding space with genetic material: "Directed panspermia". *Icarus*, **19**: 341.

Crisci, J.V., L. Katrinas, and P. Posadas. 2003. *Historical Biogeography: An Introduction.* Cambridge, MA: Harvard University Press.

Cummings, C.A. and D.A. Relman. 2002. "Microbial forensics"—Cross-examining pathogens. *Science*, **296**: 1976–9.

Curtis, T.P., W.T. Sloan, and J.W. Scannell. 2002. Estimating prokaryotic diversity and its limits. *Proceedings of the National Academy of Sciences USA*, **99**: 10494–9.

Curutchet, G., E. Donati, C. Oliver, C. Pogliani, and M.R. Viera. 2001. Development of *Thiobacillus* biofilms for metal recovery. *Methods in Enzymology*, **337**: 171–86.

Danovaro, R. and C. Corinaldesi. 2003. Sunscreen products increase virus production through prophage induction in marine bacterioplankton. *Microbial Ecology*, **45**: 109–18.

Davidow, A.L., B.T. Mangura, E.C. Napolitano, and L.B. Reichman. 2003. Rethinking the socioeconomics and geography of tuberculosis among foreign-born residents of New Jersey, 1994–1999. *American Journal of Public Health*, **93**: 1007–12.

Davidson, E.A. 1991. Fluxes of nitrous oxide and nitric oxide from terrestrial ecosystems. In J.E. Rogers and W.B. Whitman (eds.) *Microbial Production and Consumption of Greenhouse Gases: Methane, Nitrogen Oxides, and Halomethanes*, pp. 219–35. Washington, DC: American Society for Microbiology.

Davies, J. 1994. Inactivation of antibiotics and the dissemination of resistance genes. *Science*, **264**: 375–82.

Davison, J. 1999. Genetic exchange between bacteria in the environment. *Plasmid*, **42**: 73–91.

de Gruijl, F.R. and J.C. van der Leun. 1994. Estimate of the wavelength dependency of ultraviolet carcinogenesis in humans and its relevance to the risk assessment of stratospheric ozone depletion. *Health Physics*, **67**: 319–25.

Decho A.W. and T. Kawaguchi. 1999. Confocal imaging of *in situ* natural microbial communities and their extracellular polymeric secretions using nanoplast (R) resin. *Biotechniques*, **27**: 1246–52.

Dedysh, S.N., P.F. Dunfield, M. Derakshani, S. Stubner, J. Heyer, and W. Liesack. 2003. Differential detection of type II methanotrophic bacteria in acidic peatlands using newly developed 16S rRNA-targeted fluorescent oligonucleotide probes. *FEMS Microbiology Ecology*, **43**: 299–308.

Deisenhofer, J. and H. Michel. 1992. High resolution crystal structures of bacterial photosynthetic reaction centers. In L. Ernster (ed.) *Molecular Mechanisms in Bioenergetics*, pp. 103–20. Amsterdam: Elsevier.

Delbes, C., J-J. Godon, and R. Moletta. 1998. 16S rDNA sequence diversity of a culture-accessible part of an anaerobic digestor bacterial community. *Anaerobe*, **4**: 267–75.

DeLong, E.F. and Pace, N.R. 2001. Environmental diversity of bacteria and archaea. *Systematic Biology*, **50**: 470–8.

Deming, J.W. and J.A. Baross. 2000. Survival, dormancy, and nonculturable cells in extreme deep-sea environments. In R.R. Colwell and D.J. Grimes (eds.) *Nonculturable Microorganisms in the Environment*, pp. 147–97. Washington, DC: American Society for Microbiology.

Dennis, P.C., B.E. Sleep, R.R. Fulthorpe, and S.N. Liss. 2003. Phylogenetic analysis of bacterial populations in an anaerobic microbial consortium capable of degrading saturation concentrations of tetrachloroethylene. *Canadian Journal of Microbiology*, **49**: 15–27.

Des Marais, D.J. 2001. Biogeochemical processes in microbial ecosystems. *Astrobiology*, **1**: 317–18.

Des Marais, D.J. 2003. Biogeochemistry of hypersaline microbial mats illustrates the dynamics of modern microbial ecosystems and the early evolution of the biosphere. *Biological Bulletin (Woods Hole)*, **204**: 160–7.

Des Marais, D.J., M.O. Harwit, K.W. Jucks, J.E. Kasting, D.N.C. Lin, J.I. Lunine, J. Schneider, S. Seager, W.A. Traub, and N.J. Woolf. 2002. Remote sensing of planetary properties and biosignatures on extrasolar terrestrial planets. *Astrobiology*, **2**: 153–81.

Desbiez, C., C. Wipf-Scheibel, and H. Lecoq. 2002. Biological and serological variability, evolution and molecular epidemiology of Zucchini yellow mosaic virus (ZYMV, Potyvirus) with special reference to Caribbean islands. *Virus Research*, **85**: 5–16.

Devai, I., L. Felfoldy, I. Wittner, and S. Plosz. 1988. Detection of phosphine: New aspects of the phosphorus cycle in the hydrosphere. *Nature*, **333**: 343–5.

Dewar, R.C. 1992. Inverse modeling and the global carbon cycle. *Trends in Ecology and Evolution*, **7**: 105–7.

Dias, J.M., M.E. Than, A. Humm, R. Huber, G.P. Bourenkov, H.D. Bartunik, S. Bursakov, J. Calvete, J. Caldeira, C. Carneiro, J.J. Moura, I. Moura, and M.J. Romao. 1999. Crystal structure of the first dissimilatory nitrate reductase at 1.9 A solved by MAD methods. *Structure*, **7**: 65–79.

Diaspro, A. (ed.) 2002. *Confocal and Two-Photon Microscopy: Foundations, Applications, and Advances.* New York: Wiley-Liss.

Dillon, L.S. 1978. *The Genetic Mechanism and the Origin of Life.* New York: Plenum Press.

Dilworth, M.J., R.R. Eady, R.L. Robson, and R.W. Miller. 1987. Ethane formation from acetylene as a potential test for vanadium nitrogenase *in vivo. Nature*, **327**: 167–8.

Dockendorff, T.C., A. Breen, O.A. Ogunseitan, J.G. Packard, and G.S. Sayler. 1992. Practical considerations of nucleic acid hybridization and reassociation techniques in environmental analysis. In M.A. Levin, R.J. Seidler, and M. Rogul (eds.) *Microbial Ecology: Principles, Methods, and Applications*, pp. 393–420. New York: McGraw-Hill.

Doelman, P., E. Jansen, M. Michels, and M.Van Til. 1994. Effects of heavy metals in soil on microbial diversity and activity as shown by the sensitivity-resistance index, an ecologically relevant parameter. *Biology and Fertility of Soils*, **17**: 177–84.

Domingues, M.R., J. Araujo, M.B.A. Varesche, and R.F. Vazoller. 2002. Evaluation of thermophilic anaerobic microbial consortia using fluorescence in situ hybridization (FISH). *Water Science and Technology*, **45**: 27–33.

Domsch, K.H., G. Jagnow, and T.H. Anderson. 1983. An ecological concept for the assessment of side effects of agrochemicals on soil microorganisms. *Residue Reviews*, **86**: 65–105.

Dong, Y.-H., J.-L. Xu, X.-Z. Li, and L.-H. Zhang. 2003. AiiA, an enzyme that inactivates the acylhomoserine lactone quorum-sensing signal and attenuates virulence of *Erwinia carotovora. Proceedings of the National Academy of Sciences (USA)*, **97**: 3526–31.

Donlan, R. 2002. Biofilms: Microbial life on surfaces. *Emerging Infectious Diseases*, **8**: 881–90.

Doolittle, W.F. 1999. Phylogenetic classification and the universal tree. *Science*, **284**: 2124–8.

Doudna, J.A. and T.R. Cech. 2002. The chemical repertoire of natural ribozymes. *Nature*, **418**: 222–8

Dowling, N.J.E., F. Widdel, and D.C. White. 1986. Phospholipid ester linked fatty acid biomarkers of acetate-oxidizing reducers and other sulfide forming bacteria. *Journal of General Microbiology*, **132**: 1815–25.

Dubiose, S.M., B.E. Moore, C.A. Sorber, and B.P. Sagik. 1979. Viruses in soil systems. In H.D. Isenberg (ed.) *CRC Critical Reviews in Microbiology* (Volume 7), pp. 245–85. Boca Raton, FL: CRC Press.

Ducklow, H. 1983. Production and fate of bacteria in the oceans. *Bioscience*, **33**: 494–501.

Dunbar, J., S. Takala, S. Barns, J.A. Davis, and C.R. Kuske. 1999. Levels of bacterial community diversity in four arid soils compared by cultivation and 16S rRNA gene cloning. *Applied and Environmental Microbiology*, **65**: 1662–9.

Dunn, A.K. and J. Handelsman. 2002. Toward an understanding of microbial communities through analysis of communication networks. *Antonie van Leeuwenhoek*, **81**: 565–74.

Dupin, H.J. and P.L. McCarty. 2000. Impact of colony morphologies and disinfection on biological clogging in porous media. *Environmental Science and Technology*, **34**: 1513–20.

Dworkin, M. 1996. Recent advances in the social and developmental biology of the myxobacteria. *Microbiological Reviews*, **60**: 70–102.

Dykhuizen D.E. and L. Green. 1991. Recombination in *Escherichia coli* and the definition of biological species. *Journal of Bacteriology*, **173**: 7257–68.

Ebrahim, S.H., T.A. Peterman, A.A. Zaidi, and F.F. Hamers. 1997. Geography of AIDS-associated Kaposi's sarcoma in Europe. *AIDS* (London), **11**: 1739–45.

Eder, W., L.L. Jahnke, M. Schmidt, and R. Huber. 2001. Microbial diversity of the brine-seawater interface of the Kebrit Deep, Red Sea, studied via 16S rRNA gene sequences and cultivation methods. *Applied and Environmental Microbiology*, **67**: 3077–85.

Edwards, C. (ed.) 1990. *Microbiology of Extreme Environments.* New York: McGraw-Hill.

Edwards, K.J., W. Bach, and D.R. Rogers. 2003. Geomicrobiology of the ocean crust: A role for chemoautotrophic Fe-bacteria. *Biological Bulletin (Woods Hole)*, **204**: 180–5.

Eguchi, T., Y. Nishimura, and K. Kakinuma. 2003. Importance of the isopropylidene terminal of geranylgeranyl group for the formation of tetraether lipid in methanogenic archaea. *Tetrahedron Letters*, **44**: 3275–9.

Eismann, F., G. Glindemann, A. Bergmann, and P. Kuschk. 1997. Soils as a source and sink of phosphine. *Chemosphere*, **35**: 523–33.

Eldredge, N. and S.J. Gould. 1988. Punctuated equilibrium prevails. *Nature*, **332**: 211–12.

Elena, S.F., V.S. Cooper, and R.E. Lenski. 1996. Punctuated evolution caused by selection of rare mutations. *Science*, **272**: 1802–4.

Elena, S.F. and R.E. Lenski. 2003. Evolution experiments with microorganisms: The dynamics and genetic bases of adaptation. *Nature Reviews Genetics*, **4**: 457–69.

Ellis, M.J., F.E. Dodd, G. Sawers, R.R. Eady, and S.S. Hasnain. 2003. Atomic resolution structures of native copper nitrite reductase from *Alcaligenes xylosoxidans* and the active site mutant Asp92Glu. *Journal of Molecular Biology*, **328**: 429–38.

Ellis, R.J. 1979. The most abundant protein in the world. *Trends in Biochemical Science*, **4**: 241–4.

Embley, T.M. and E. Stackebrandt. 1997. Species in practice: Exploring uncultured prokaryote diversity in natural samples. In M.F. Claridge, H.A. Dawah, and M.R. Wilson (eds.) *Species: The Units of Biodiversity*, pp. 66–81. London: Chapman and Hall.

Emelyanov, V.V. 2001. Rickettsiaceae, Rickettsia-like endosymbionts, and the origin of mitochondria. *Bioscience Reports*, **21**: 1–17.

Emiliani, C. 1992. *Planet Earth: Cosmology, Geology, and the Evolution of Life and Environment.* Cambridge, England: Cambridge University Press.

Ereshefsky, M. (ed.) 1992. *The Units of Evolution: Essays on the Nature of Species.* Cambridge, MA: MIT Press.

Ermler, U., W. Grabarse, S. Shima, M. Goubeaud, R.K. Thauer. 1997. Crystal structure of methyl-coenzyme M reductase: The key enzyme of biological methane formation. *Science*, **278**: 1457–62.

Erskine, P.T., R. Newbold, J. Roper, A. Coker, P.M. Shoolingin-Jordan, S.P. Wood, and J.R. Cooper. 1999. The Schiff base complex of yeast 5-aminolaevulinic acid dehydratase with laevulininc acid. *Protein Science*, **8**: 1250–6.

Evens R., J. Braven, and L. Brown. 1982. A high performance liquid chromatographic determination of free amino acids in natural waters in picomolar range suitable for shipboard use. *Chemical Ecology*, **1**: 99–106.

Falkowski, P.G. 1994. The role of phytoplankton photosynthesis in global biogeochemical cycles. *Photosynthesis Research*, **39**: 235–58.

Falkowski, P.G., R.T. Barker, and V. Smetacek. 1998. Biogeochemical controls and feedbacks on ocean primary production. *Science*, **281**: 200–6.

Falkowski, P.G., R.J. Scholes, E. Boyle, J. Canadell, D. Canfield, J. Elser, N. Gruber, K. Hibbard, P. Hogberg, S. Linder, F.T. Mackenzie, B. Moore III, T. Pedersen, Y. Rosenthal, S. Seitzinger, V. Smetacek, and W. Steffen. 2000. The global carbon cycle: A test of our knowledge of Earth as a system. *Science*, **290**: 291–6.

Farquhar, J., H.M. Bao, and M. Thiemens. 2000. Atmospheric influence of Earth's earliest sulfur cycle. *Science*, **289**: 756–8.

Feldman, R.A. and D.W. Harris. 2000. Beyond the human genome: High-throughput, fine scale, molecular dissection of earth's microbial biodiversity. *Journal of Clinical Ligand Assay*, **23**: 256–61.

Fenchel, T. 2002. Microbial behavior in a heterogeneous world. *Science*, **296**: 1068–71.

Fenchel, T. 2002a. *Origin and Early Evolution of Life.* Oxford: Oxford University Press.

Feray, C. and B. Montuelle. 2002. Competition between two nitrite-oxidizing bacterial populations: A model for studying the impact of wastewater treatment plant discharge on nitrification in sediment. *FEMS Microbiology Ecology*, **42**: 15–23.

Findlay, R.H. and F.C. Dobbs. 1993. Quantitative description of microbial communities using lipid analysis. In P.F. Kemp (ed.) *Handbook of Methods in Aquatic Microbial Ecology*, pp. 347–58. Boca Raton, FL: Lewis Publishers.

Finkel, S.E. and R. Kolter. 1999. Evolution of microbial diversity during prolonged starvation. *Proceedings of the National Academy of Sciences (USA)*, **96**: 4023–7.

Finlay, B.J. 2002. Global dispersal of free-living microbial eukaryote species. *Science*, **296**: 1061–3.

Finlay, B.J., J.O. Corliss, G. Esteban, and T. Fenchel. 1996. Biodiversity at the microbial level: The number of free-living ciliates in the biosphere. *Quarterly Review of Biology*, **71**: 221–37.

Fischer, U.R. and B. Velimirov. 2002. High control of bacterial production by viruses in a eutrophic oxbow lake. *Aquatic Microbial Ecology*, **27**: 1–12.

Fisher, M.M. and E.W. Triplett. 1999. Automated approach for ribosomal intergenic spacer analysis of microbial diversity and its application to freshwater bacterial communities. *Applied and Environmental Microbiology*, **65**: 4630–6.

Fjeldsa, J. 2000. The relevance of systematics in choosing priority areas for global conservation. *Environmental Conservation*, **27**: 67–75.

Fleischmann, E.M. 1989. The measurement and penetration of ultraviolet radiation into tropical marine water. *Limnology and Oceanography*, **34**: 1623–9.

Focht, D.D. and W. Verstraete. 1977. Biochemical ecology of nitrification and denitrification. *Advances in Microbial Ecology*, **1**: 135–214.

Frank, H.A. 1993. Carotenoids in photosynthetic bacterial reaction centers: Structure, spectroscopy, and photochemistry. In J. Deisenhofer and J.R. Norris (eds.) *The Photosynthetic Reaction Center*, pp. 221–37, Volume II. San Diego: Academic Press.

Fredrickson, A.G. and G. Stephanopoulos. 1981. Microbial competition. *Science*, **213**: 972–9.

Frias-Lopez, J., G.T. Bonheyo, Q. Jin, and B.W. Fouke. 2003. Cyanobacteria associated with coral black-band disease in Caribbean and Indo-Pacific reefs. *Applied and Environmental Microbiology*, **69**: 2409–13.

Friedberg, E.C. and P.L. Fischhaber. 2003. TB or not TB: How mycobacterium tuberculosis may evade drug treatment. *Cell*, **113**: 139–40.

Friedmann, E.I., J. Wierzchos, C. Ascaso, and M. Winklhofer. 2001. Chains of magnetite crystals in the meteorite ALH84001: Evidence of biological origin. *Proceedings of the National Academy of Sciences (USA)*, **98**: 2176–81.

Frigaard, N.-U., K.L. Larsen, and R.P. Cox. 1996. Spectrochromatography of photosynthetic pigments as a fingerprinting technique for microbial phototrophs. *FEMS Microbiology Ecology*, **20**: 69–77.

Fuhrman, J. 1992. Bacterioplankton roles in cycling of organic matter: The microbial food web. In P.G. Falkowski and A.D. Woodhead (eds.) *Primary Productivity and Biogeochemical Cycles in the Sea*, pp. 361–83. New York: Plenum Press.

Fuhrman, J.A. 1999. Marine viruses and their biogeochemical and ecological effects. *Nature*, **399**: 541–8.

Fuhrman, J.A. and D.G. Capone. 1991. Possible biogeochemical consequences of ocean fertilization. *Limnology and Oceanography*, **36**: 1951–9.

Fuhrman, J.A. and C.A. Suttle. 1993. Viruses in marine planktonic systems. *Oceanography*, **6**: 51–63.

Fuqua, W.C. and E.P. Greenberg. 2002. Listening on bacteria: Acyl-homoserine lactone signaling. *Nature Reviews Molecular Cell Biology*, **3**: 685–96.

Fuqua, W.C., S.C. Winans, and E.P. Greenberg. 1994. Quorum sensing in bacteria: The *LuxR-LuxI* family of cell density-responsive transcriptional regulators. *Journal of Bacteriology*, **176**: 269–75.

Galagan J.E., C. Nusbaum, A. Roy, M.G. Endrizzi, P. Macdonald, W. FitzHugh, S. Calvo, R. Engels, S. Smirnov, D. Atnoor, A. Brown, N. Allen, J. Naylor, N. Stange-Thomann, K. DeArellano, R. Johnson, L. Linton, P. McEwan, K. McKernan, J. Talamas, A. Tirrell, W. Ye, A. Zimmer, R.D. Barber Robert, I. Cann, D.E. Graham, D.A. Grahame, A.M. Guss, R. Hedderich, C. Ingram-Smith, H.C. Kuettner, J.A. Krzycki, J.A. Leigh, W. Li, J. Liu, B. Mukhopadhyay, J.N. Reeve, K. Smith, T.A. Springer, L.A. Umayam, O. White, R.H. White, E.C. de Macario, J.G. Ferry, K.F. Jarrell, H. Jing, A.J.L. Macario, I. Paulsen, M. Pritchett, K.R. Sowers, R.V. Swanson, S.H. Zinder, E. Lander, W.W. Metcalf, and B. Birren. 2002. The genome of *M. acetivorans* reveals extensive metabolic and physiological diversity. *Genome Research*, **12**: 532–42.

Galperin, M.Y. and E.V. Koonin. 1999. Functional genomics and enzyme evolution: Homologous and analogous enzymes encoded in microbial genomes. *Genetica (Dordrecht)*, **106**: 159–70.

Garbelotto, M., D.M. Rizzo, J.M. Davidson, K. Ivors, P.E. Maloney, D. Huberli, K. Hayden, T. Harnik, and S.T. Koike. 2003. *Phytophthora ramorum*: An emerging forest pathogen. *Phytopathology*, **93**: S28.

Garcia-Cantizano J., J.I. Calderon-Paz, and C. Pedros-Alio. 1994. Thymidine incorporation in Lake Ciso: Problems in estimating bacterial secondary production across oxic-anoxic interfaces. *FEMS Microbiology and Ecology*, **14**: 53–64.

Garcia-Pichel, F. 1994. A model for the internal self-shading in planktonic organisms and its implications for the usefulness of ultraviolet sunscreens. *Limnology and Oceanography*, **39**: 1704–17.

Garrison, V.H., E.A. Shinn, W.T. Foreman, D.W. Griffin, C.W. Holmes, C.A. Kellogg, M.S. Majewski, L.L. Richardson, K.B. Ritchie, and G.W. Smith. 2003. African and Asian dust: From desert soils to coral reefs. *Bioscience*, **53**: 469–80.

Gattinger, A., A. Guenthner, M. Schloter, and J.C. Munch. 2003. Characterization of *Archaea* in soils by polar lipid analysis. *Acta Biotechnologica*, **23**: 21–8.

Geesey, G.G., A.L. Neal, P.A. Suci, and B.M. Peyton. 2002. A review of spectroscopic methods for characterizing microbial transformations of minerals. *Journal of Microbiological Methods*, **51**: 125–39.

Gerdes, K., P.B. Rasmussen, and S. Molin. 1986. Unique type of plasmid maintenance function: Postsegregational killing of plasmid-free cells. *Proceedings of the National Academy of Sciences (USA)*, **83**: 3116–20.

Germot A., H. Philippe, and H. Guyader. 1996. Presence of a mitochondrial-type 70-kDa heat shock protein in *Trichomonas vaginalis* suggests a very early mitochondrial endosymbiosis in eukaryotes. *Proceedings of the National Academy of Sciences (USA)*, **93**: 14614–17.

Giancoli, D.C. 1991. Optical instruments. In D.C. Giancoli, *Physics* (Third Edition), pp. 654–84. Englewood Cliffs, NJ: Prentice Hall.

Gibbs, P.E.M., W.F. Witke, and A. Dugaiczyk. 1998. The molecular clock runs at different rates among closely related members of a gene family. *Journal of Molecular Evolution*, **46**: 552–61.

Gilbert, E.S., A.W. Walker, and J.D. Keasling. 2003. A constructed microbial consortium for biodegradation of the organophosphorus insecticide parathion. *Applied Microbiology and Biotechnology*, **61**: 77–81.

Gillespie, J.H. 1984. The molecular clock may be an episodic clock. *Proceedings of the National Academy of Sciences (USA)*, **81**: 8009–13.

Ginzburg, B., I. Chalifa, J. Gun, I. Dor, O. Hadas, and O. Lev. 1998. DMS formation by dimethylsulfoniopropionate route in freshwater. *Environmental Science and Technology*, **32**: 2130–6.

Giovannoni, S.J., T.B. Britschgi, C.L. Moyer, and K.G. Field. 1990.

Genetic diversity in Sargasso Sea bacterioplankton. *Nature*, **345**: 60–3.

Giovannoni, S.J., and M.S. Rappé, M.S. 2000. Evolution, diversity, and molecular ecology of marine prokaryotes. In D.L. Kirchman (ed.), *Microbial Ecology of the Oceans*, pp. 47–84. New York: Wiley-Liss.

Giraffa, G. 2001. Protein coding gene sequences: Alternative phylogenetic markers or possible tools to compare ecological diversity in bacteria? *Current Genomics*, **2**: 243–51.

Glass, G.E. 2000. Update: Spatial aspects of epidemiology: The interface with medical geography. *Epidemiologic Reviews*, **22**: 136–9.

Glauert, A.M. and P.R. Lewis. 1998. *Biological Specimen Preparation for Transmission Electron Microscopy*. Princeton, NJ: University Press.

Glindemann, D., R.M. De Graaf, and A.W. Schwartz. 1999. Chemical reduction of phosphate on the primitive earth. *Origins of Life and Evolution of the Biosphere*, **29**: 555–61.

Glindemann, D., M. Edwards, and P. Kuschk. 2003. Phosphine gas in the upper troposphere. *Atmospheric Environment*, **37**: 2429–33.

Glindemann, D., U. Stottmeister, and A. Bergmann. 1996. Free phosphine from the anaerobic biosphere. *Environmental Science and Pollution Research*, **3**: 17–19.

Glud, R.N., M. Kuhl, O. Kohls, and N.B. Ramsing. 1999. Heterogeneity of oxygen production and consumption in a photosynthetic microbial mat as studied by planar optodes. *Journal of Phycology*, **35**: 270–9.

Godany, A., G. Bukovska, J. Farkasovska, and I. Mikula. 2003. Phage therapy: Alternative approach to antibiotics. *Biologia*, **58**: 313–20.

Gold, T. 1999. *The Deep Hot Biosphere*. New York: Springer-Verlag.

Golubic, S. 2000. Microbial landscapes: Abu Dhabi and Shark Bay. In L. Margulis, C. Matthews, and A. Haselton (eds.) *Environmental Evolution*, pp. 117–40. Cambridge, MA: MIT Press.

González, J.M., R.P. Keene, and M.A. Moran. 1999. Transformation of sulfur compounds by an abundant lineage of marine bacteria in the alpha-subclass of the class *Proteobacteria*. *Applied and Environmental Microbiology*, **65**: 3810–19.

González, J.M., R. Simo, R. Massana, J.S. Covert, O. Casamayor, C. pedros-Alio, and M.A. Moran. 2000. Bacterial community structure associated to a DMSP producing North Atlantic algal bloom. *Applied and Environmental Microbiology*, **66**: 4237–46.

Gootschalk, G. 1989. Bioenergetics of methanogenic and acetogenic bacteria. In H.G. Schlegel and B. Bowien (eds.) *Autotrophic Bacteria*, pp. 383–413. Berlin: Springer-Verlag.

Gould, G.W. and J.E.L. Corry (eds.) 1980. *Microbial Growth and Survival in Extremes of Environment*. London: Academic Press.

Gould, S.J. 2002. *The Structure of Evolutionary Theory*, pp. 745–1022. Cambridge, MA: Belknap-Harvard.

Goward, S.N. and D.L. Williams. 1997. Landsat and Earth systems science: Development of terrestrial monitoring. *Photogrammetric Engineering and Remote Sensing*, **63**: 887–900.

Goward, T. and A. Arsenault. 2000. Cyanolichens and conifers: Implications for global conservation. *Forest Snow and Landscape Research*, **75**: 303–18.

Grabarse, W., F. Mahlert, E.C. Duin, M. Goubeaud, S. Shima, R.K. Thauer, V. Lamzin, and U. Ermler. 2001. On the mechanism of biological methane formation: Structural evidence for conformational changes in methyl-coenzyme M reductase upon substrate binding. *Journal of Molecular Biology*, **309**: 315–30.

Grabarse, W., F. Mahlert, S. Shima, R.K. Thauer, and U. Ermler. 2000. Comparison of three methyl-coenzyme M reductases from phylogenetically distant organisms: Unusual amino acid modification, conservation and adaptation. *Journal of Molecular Biology*, **303**: 329–44.

Graham, D.Y. 2003. The changing epidemiology of GERD: Geography and *Helicobacter pylori*. *American Journal of Gastroenterology*, **98**: 1462–70.

Graham, W.F. and R.A. Duce. 1979. Atmospheric pathways of the phosphorus cycle. *Geochimica et Cosmochimica Acta*, **43**: 1195–208.

Grant, R.F. 1999. Simulation of methanotrophy in the mathematical model ecosystems. *Soil Biology and Biochemistry*, **31**: 287–97.

Gray, M.W., G. Burger, and B.F. Lang. 1999. Mitochondrial evolution. *Science*, **283**: 1476–81.

Green, C.H., D.M. Heil, G.E. Cardon, G.L. Butters, and E.F. Kelly. 2003. Solubilization of man-ganese and trace metals in soils affected by acid mine runoff. *Journal of Environmental Quality*, **32**: 1323–34.

Greenblatt, C.L., A. Davis, B.G. Clement, C.L. Kitts, T. Cox, and R.J. Cano. 1999. Diversity of microorganisms isolated from amber. *Microbial Ecology*, **38**: 58–68.

Greenfield L.J., R.D. Hamilton, and C. Weiner. 1970. Nondestructive determination of protein, total amino acids, and ammonia in marine sediments. *Bulletin of Marine Science*, **20**: 289–304.

Griffith, F. 1928. The significance of pneumococcal types. *Journal of Hygiene*, **27**: 113–59.

Groffman, P.M. and C.L. Turner. 1995. Plant productivity and nitrogen gas fluxes in a tallgrass prairie landscape. *Landscape Ecology*, **10**: 255–66.

Grotzinger J.P. and N.P. James. 2000. Precambrian carbonates: Evolution of understanding. In J.P. Grotzinger and N.P. James (eds.) *Carbonate Sedimentation and Diagenesis in the Evolving Precambrian World*. Special Publication of the Society for Sedimentary Geology, **67**: 3–20.

Grotzinger J.P. and D.H. Rothman. 1996. An abiotic model for stromatolite morphogenesis. *Nature*, **383**: 423–5.

Guan, L.L. and K. Kamino. 2002. Bacterial response to siderophore and quorum-sensing chemical signals in the seawater microbial community. *BMC Microbiology*, **1**: 1–11.

Gubler, D.J. 1998. Resurgent vector-borne diseases as a global health problem. *Emerging Infectious Diseases*, **4**: 442–50.

Guckert, J.B., C.P. Antworth, P.D. Nichols, and D.C. White. 1985. Phospholipid, ester-linked fatty acid profiles as reproducible assays for changes in prokaryotic community structure of estuarine sediments. *FEMS Microbiology and Ecology*, **31**: 147–58.

Guckert, J.B., M.A. Hood, and D.C. White. 1986. Phospholipid, ester-linked fatty acid profile changes during nutrient deprivation of Vibrio cholera: Increases in the trans/cis ratio and proportions of cyclopropyl fatty acids. *Applied and Environmental Microbiology*, **52**: 794–801.

Gudmundsdottir, K., E. Sigurdsson, S.I. Thorbjarnardottir, and G. Eggertsson. 1999. Cloning and sequence analysis of the hemB gene of *Rhodothermus marinus*. *Current Microbiology*, **39**: 103–5.

Guerrero R., M. Piqueras, and M. Berlanga. 2002. Microbial mats and the search for minimal ecosystems. *International Microbiology*, **5**: 177–88.

Gupta, R.S. 1998a. What are archaebacteria: Life's third domain or monoderm prokaryotes related to Gram-positive bacteria? A new proposal for the classification of prokaryotic organisms. *Molecular Microbiology*, **29**: 695–707.

Gupta, R. 1998b. Protein phylogenies and signature sequences: A reappraisal of evolutionary relationships among archaebacteria, eubacteria, and eukaryotes. *Microbiology and Molecular Biology Reviews*, **62**: 1435–91.

Gupta, R.S. 2002. Phylogeny of bacteria: Are we now close to understanding it? *American Society for Microbiology News*, **68**: 284–91.

Guschin, D.Y., B.K. Mobarry, D. Proudnikov, D.A. Stahl, B.E. Rittmann, and A.D. Mirzabekov. 1997a. Oligonucleotids microchips as genosensors for determinative and environmental studies in microbiology. *Applied and Environmental Microbiology*, **63**: 2397–402.

Guschin, D.Y., G. Yerkov, A. Zaslavsky, A. Gemmell, V. Shick, D. Proudnikov, P. Arenkov, and A. Mirzabekov. 1997b. Manual manufacturing of oligonucleotide, DNA, and protein microchips. *Analytical Biochemistry*, **250**: 203–11.

Guttman, D.S. and D.E. Dykuizen. 1994. Clonal divergence in *Escherichia coli* as a result of recombination, not mutation. *Science*, **266**: 1380–3.

Haider, K. 1992. Problems related to the humidification processes in soils of temperate climates. In G. Stotzky and J.-M. Bollag (eds.) *Soil Biochemistry*, Volume 7, pp. 55–94. New York: Marcel Dekker.

Hales, B.A., D.A. Ritchie, G. Hall, R.W. Pickup, and J.R. Saunders. 1996. Isolation and identification of methanogen-specific DNA from blanket bog peat by PCR amplification and sequence analysis. *Applied and Environmental Microbiology*, **62**: 668–75.

Hales, S., N. de Wet, J. Maindonald, and A. Woodward. 2002. Potential effect of population and climate changes on global distribution of dengue fever: An empirical model. *Lancet* (North American Edition), **360**: 830–4.

Hall, B.G. 2001. *Phylogenetic Trees Made Easy. A How-to Manual for Molecular Biologists.* Sunderland, MA: Sinauer.

Hallam, S.J., P.R. Girguis, C.M. Preston, P.M. Richardson, and E.F. DeLong. 2003. Identification of methyl coenzyme M reductase A (*mcrA*) genes associated with methane-oxidizing archaea. *Applied and Environmental Microbiology*, **69**: 5483–91.

Han, S.H., Y.H. Zhuang, J.A. Liu and D. Glindemann. 2000. Phosphorus cycling through phosphine in paddy fields. *Science of the Total Environment*, **258**: 195–203.

Hantula J.A., T.K. Korhonen, and D.H. Bamford. 1990. Determination of taxonomic resolution power of SDS-polyacrylamide gel electrophoresis of total cellular proteins using *Enterobacteriaceae. FEMS Microbiology Letters*, **70**: 325–30.

Hantula J.A., A. Kurki, P. Vuoriranta, and D.H. Bamford. 1991. Rapid classification of bacterial strains by SDS-polyacrylamide gel electrophoresis: Population dynamics of the dominant dispersed phase bacteria of activated sludge. *Applied Microbiology and Biotechnology*, **34**: 551–5.

Hartman, H., J.G. Lawless, and P. Morrison (eds.) 1987. *Search for the Universal Ancestors: The Origin of Life.* Palo Alto, CA: Blackwell Scientific.

Harvell, C.D., K. Kim, J.M. Burkholder, R.R. Colwell, P.R. Epstein, D.J. Grimes, E.E. Hofmann, E.K. Lipp, A.D.M.E. Osterhaus, R.M. Overstreet, J.W. Porter, G.W. Smith, and G.R. Vasta. 1999. Emerging marine diseases—climate links and anthropogenic factors. *Science*, **285**: 1505–10.

Harvell, C.D., C.E. Mitchell, J.R. Ward, S. Altizer, A.P. Dobson, R.S. Ostfeld, and M.D. Samuel. 2002. Climate warming and disease risks for terrestrial and marine biota. *Science*, **296**: 2158–62.

Hayat, M.A. 2000. *Principles and Techniques of Electron Microscopy: Biological Applications.* Fourth Edition. New York: Cambridge University Press.

He, X., W. Chang, D.L. Pierce, L.O. Seib, J. Wagner, and C. Fuqua. 2003. Quorum sensing in *Rhizobium* sp. strain NGR234 regulates conjugal transfer (tra) gene expression and influences growth rate. *Journal of Bacteriology*, **185**: 809–22.

He, Y., T. Gaal, R. Karls, T.J. Donohue, R.L. Gourse, and G.P. Roberts. 1999. Transcription activation by CooA, the CO-sensing factor from *Rhodospirillum rubrum. Journal of Biological Chemistry*, **274**: 10840–5.

Head, I.M., W.D. Hiorns, T.M., Embley, A.J. McCarthy, and J.R. Saunders. 1993. The phylogeny of autotrophic ammonia-oxidizing bacteria as determined by analysis of 16S ribosomal RNA gene sequences. *Journal of General Microbiology*, **139**: 1147–53.

Heinrich, M.R. (ed.) 1976. *Extreme Environments: Mechanisms of Microbial Adaptation.* New York: Academic Press.

Heldal, M., S. Norland, K.M. Fagerbakke, F. Thingstad, and G. Bratbak. 1997. The elemental composition of bacteria: A signature of growth conditions? *Marine Pollution Bulletin*, **33**: 3–9.

Hennes, K.P. and M. Simon. 1995. Significance of bacteriophages for controlling bacterioplankton growth in a mesotrophic lake. *Applied and Environmental Microbiology*, **61**: 333–40.

Hennig, W. 1966. *Phylogenetic Systematics* (translated by D. D. Davis and R. Zangerl). Urbana, IL: University of Illinois Press.

Herndl, G.J., G. Mueller-Niklas, and J. Frick. 1993. Major role of ultraviolet-B in controlling bacterioplankton growth in the surface layer of the ocean. *Nature*, **361**: 717–19.

Herron, P.R. 1995. Phage ecology and genetic exchange in soil. In A.D.L. Akkermans, J.D. Van Elsas, and F.J. De Bruijn (eds.) *Molecular Microbial Ecology Manual*, pp. 1–12. Dordrecht: Kluwer Academic Press.

Higgins, D.G. and P.M. Sharp. 1988. Clustal. A package for performing multiple sequence alignment on a microcomputer. *Gene*, **73**: 237–44.

Hilario, E. and J.P. Gogarten. 1993. Horizontal transfer of ATPase genes: The tree of life becomes a net of life. *Biosystems*, **31**: 111–19.

Hines, M.E., P.T. Visscher, and R. Devereux. 2002. Sulfur cycling. In C.J. Hurst, R.L. Crawford, G.R. Knudsen, M.J. McInerney, and L.D. Stetzenbach (eds.) *Manual of Environmental Microbiology*, pp. 427–38. Washington, DC: American Society for Microbiology.

Hirsch, P., M. Bernhard, S.S. Cohen, J.C. Ensign, H.W. Jannasch, A.L. Koch, K.C. Marshall, A. Marin, J.S. Poindexter, S.C. Rittenberg, D.C. Smith, and H. Veldkamp. 1979. Life under conditions of low nutrient concentrations group report. In M. Shilo (ed.) *Strategies of Microbial Life in Extreme Environments, Dalem Konferenzen Life Sciences Research Report 13*, pp. 357–72. Weinheim, Germany: Verlag Chemie.

Hitchens, A.P. and M.C. Leikind. 1939. The introduction of agar into bacteriology. *Journal of Bacteriology*, **37**: 485–93.

Hoegh-Guldberg, O. 1999. Climate change, coral bleaching and the future of the world's coral reefs. *Marine and Freshwater Ecology*, **50**: 839–66.

Hoehler, T.M., B.M. Bebout, and D.J. Des Marais. 2001. The role of microbial mats in the production of reduced gases on the early Earth. *Nature*, **412**: 324–7.

Hoffmann, A.A. and P.A. Parsons. 1997. *Extreme Environmental Change and Evolution.* Cambridge, England: Cambridge University Press.

Hofle, M. 1988. Identification of bacteria by low molecular weight RNA profiles: A new chemotaxonomic approach. *Journal of Microbiological Methods*, **8**: 235–48.

Hoh, J.H. and P.K. Hansma. 1992. Atomic force microscopy for high-resolution imaging in cell biology. *Trends in Cell Biology*, **2**: 208–13.

Holmen, K. 1992. The global carbon cycle. In S.S. Butcher, R.J. Charlson, G.H. Orans, and G.V. Wolfe (eds.) *Global Biogeochemical Cycles*, pp. 239–62. New York: Academic Press.

Holmes, A.J., A. Costello, M.E. Lidstrom, and J.C. Murrell. 1995.

Evidence that particulate methane monooxygenase and ammonia monooxygenase may be evolutionarily related. *FEMS Microbiology Letters*, **132**: 203–8.

Honda, Y., R. Navarro-Gonzalez, and C. Ponnamperuma. 1989. A qualitative assay of biologically important compounds in simulated primitive Earth experiments. *Advances in Space Research*, **9**: 63–6.

Hoppert, M. and A. Holzenburg. 1998. *Electron Microscopy in Microbiology.* New York and Oxford: Springer in association with the Royal Microscopical Society.

Horikoshi, K. and T. Akiba. 1982. *Alkalophilic Microorganisms: A New Microbial World.* Berlin: Springer-Verlag.

Horikoshi, K., R. Aono, and S. Nakamura. 1993. The triangular halophilic archaebacterium *Haloarcula japonica* strain TR-1. *Experientia (Basel)*, **49**: 497–502.

Horikoshi, K. and W.D. Grant (eds.) 1998. *Extremophiles: Microbial Life in Extreme Environments.* New York: Wiley-Liss.

Horinouchi, M., T. Yoshida, H. Nojiri, H. Yamane, and T. Omori. 1999. Oxidation of dimethylsulfide by various aromatic compound oxygenases from bacteria. *Biotechnology Letters*, **21**: 929–33.

Horneck, G. 1998. Exobiology in earth orbit. In J. Chela-Flores and F. Raulin (eds.) *Exobiology: Matter, Energy, and Information in the Origin and Evolution of Life in the Universe*, pp. 205–12. Dordrecht, The Netherlands: Kluwer Academic.

Horneck, G. and A.C. Baumstark-Khan (eds.) 2002. *Astrobiology: The Quest for Conditions of Life.* Berlin: Springer-Verlag.

Horneck, G. and A. Brack (eds.) 2000. *Exobiology in the Solar System. Papers from the 24th General Assembly of the European Geophysical Society, at The Hague, April 19–23, 1999.* Oxford, UK: Pergamon.

Horneck, G., P. Rettberg, G. Reitz, J. Wehner, U. Eschweiler, K. Strauch, C. Panitz, V. Starke, and C. Baumstark-Khan. 2001. Protection of bacterial spores in space, a contribution to the discussion on panspermia. *Origins of Life and Evolution of the Biosphere*, **31**: 527–47.

Hoyle, F. 1980. *The Relation of Biology to Astronomy.* Cardiff, UK: University College Cardiff Press.

Hoyle, F. and C. Wickramasinghe. 1978. *Lifecloud, the Origin of Life in the Universe.* New York: Harper and Row.

Hoyle, F. and C. Wickramasinghe. 1979. *Diseases from Space.* New York: Harper and Row.

Hoyle, F. and C. Wickramasinghe. 1982. *Evolution from Space.* Hillside, NJ: Enslow.

Hoyle, F. and C. Wickramasinghe. 1984a. *From Grains to Bacteria.* Cardiff, UK: University College Cardiff Press.

Hoyle, F. and C. Wickramasinghe. 1984b. The availability of phosphorous in the bacterial model of the interstellar grains. *Astrophysics and Space Science*, **103**: 189.

Hoyle, F. and C. Wickramasinghe. 1986. The case for life as a cosmic phenomenon. *Nature*, **322**: 509.

Hoyle, F. and C. Wickramasinghe. 1990. Sunspots and influenza. *Nature*, **343**: 304.

Hoyle, F. and C. Wickramasinghe. 1990a. *The Theory of Cosmic Grains.* Dordrecht, The Netherlands: Kluwer Academic.

Hoyle, F. and C. Wickramasinghe. 1993. *Our Place in the Cosmos: The Unfinished Revolution.* London: J.M. Dent.

Hoyle, F. and C. Wickramasinghe. 1997. *Life on Mars? The Case for a Cosmic Heritage.* Redland: Clinical Press.

Hoyle, F., N.C. Wickramasinghe, and N.L. Jabir. 1983. 2.8–3.6 micron spectra of microorganisms with varying H_2O ice content. *Astrophysics and Space Science*, **92**: 439.

Hoyle, F., N.C. Wickramasinghe, and S. Al-Mufti. 1985b. The case for interstellar microorganisms. *Astrophysics and Space Science*, **110**: 401.

Hoyle F., N.C. Wickramasinghe, and S. Al-Mufti. 1985c. The measurement of the absorption properties of dry microorganisms and its relationship to astronomy. *Astrophysics and Space Science*, **113**: 413.

Hoyle, F., N.C. Wickramasinghe, and S. Al-Mufti. 1986. The viability with respect to temperature of microorganisms incident on the Earth's atmosphere. *Earth, Moon and Planets*, **35**: 79.

Hoyle, F., N.C. Wickramasinghe, and J. Watkins. 1985a. Legionnaires disease: Seeking a wider cause. *The Lancet*, 25 May: 1216.

Hubbell, D.H. and G. Kidder. 2003. *Biological nitrogen fixation.* Fact sheet document SL-16 of the Soil and Water Science Department, Florida Cooperative Extension Service, Institute of Food and Agricultural Sciences, University of Florida. http: //edis.ifas.ufl.edu.

Huddleston, A.S., N. Creswell, M.C.P. Neves, J.E. Beringer, S. Baumberg, D.I. Thomas, and E.M.H. Wellington. 1997. Molecular detection of streptomycin producing streptomycetes in Brazilian soils. *Applied and Environmental Microbiology*, **63**: 1288–97.

Hugenholtz P., C. Pitulle, K.L. Hershberger, and N.R. Pace. 1998b. Novel division level bacterial diversity in Yellowstone hot spring. *Journal of Bacteriology*, **180**: 366–76.

Hugenholtz, P., B.M. Goebel, and N.R. Pace. 1998a. Impact of culture-independent studies on the emerging phylogenetic view of bacterial diversity. *Journal of Bacteriology*, **180**: 4765–74.

Hughes, J.B., J.J. Hellmann, T.R. Ricketts, and B.J. Bohannan. 2001. Counting the uncountable: Statistical approaches to estimating microbial diversity. *Applied and Environmental Microbiology*, **67**: 4399–406.

Hull, D.L. 1997. The ideal species concept—and why we can't get it. In M.F. Claridge, H.A. Dawah, and M.R. Wilson (eds.) *Species: The Units of Biodiversity*, pp. 357–80. London: Chapman and Hall.

Humayoun, S.B., N. Bano, and J.T. Hollibaugh. 2003. Depth distribution of microbial diversity in Mono Lake, a meromictic soda lake in California. *Applied and Environmental Microbiology*, **69**: 1030–42.

Huq, A., R.B. Sack, and R.R. Colwell. 2001. Cholera and global ecosystems. In J.L. Aron and J.A. Patz (eds.) 2001. *Ecosystem Change and Public Health: A Global Perspective*, pp. 327–52. Baltimore: Johns Hopkins University Press.

Hurlburt, S.H. 1971. The nonconcept of species diversity: A critique and alternative parameters. *Ecology*, **52**: 577–86.

Hurst, C.J. and K.A. Reynolds. 2002. Sampling viruses from soil. In C.J. Hurst, R.L. Crawford, G.R. Knudsen, M.J. McInerney, and L.D. Stetzenbach (eds.) *Manual of Environmental Microbiology* (Second Edition), pp. 527–34. Washington, DC: American Society for Microbiology.

Imbert, M. and R. Blondeau. 1998. On the iron requirement of *lactobacilli* grown in a chemically defined medium. *Current Microbiology*, **37**: 64–6.

IPCC. 2001a. *Climate Change 2001: Synthesis Report. Contribution of Working Groups I, II and III to the Third Assessment Report of the Intergovernmental Panel on Climate Change.* R.T. Watson and the Core Writing Team (eds.) Cambridge, England: Cambridge University Press.

IPCC. 2001b. *Climate Change 2001: The Scientific Basis. Contribution of Working Group I to the Third Assessment Report of the Intergovernmental Panel on Climate Change.* J. T. Houghton, Y. Ding, D.J. Griggs, M. Noguer, P. J. van der Linden and D. Xiaosu (eds.) Cambridge, England: Cambridge University Press.

IPCC. 2001c. *Climate Change 2001: Impacts, Adaptation and Vulnerability. Contribution of Working Group II to the Third Assessment Report of the Intergovernmental Panel on Climate Change.* James J. McCarthy, Osvaldo F. Canziani, Neil A. Leary, David J. Dokken and Kasey S. White (eds.) Cambridge, England: Cambridge University Press.

IPCC. 2001d. *Climate Change 2001: Mitigation. Contribution of Working Group III to the Third Assessment Report of the Intergovernmental Panel on Climate Change.* Bert Metz, Ogunlade Davidson, Rob Swart and Jiahua Pan (eds.) Cambridge, England: Cambridge University Press.

Islam, M.S., B.S. Draser, and D.J. Bradley. 1990. Long-term persistence of toxigenic *V. cholerae* O1 in the mucilaginous sheath of a blue green alga, *Anabaena variabilis*. *Journal of Tropical Medicine and Hygiene*, **93**: 133–9.

Islam, M.S., M.A. Miah, M.K. Hasan, R.B. Sack, and M.J. Albert. 1994. Detection of non-culturable *V. cholerae* O1 associated with a cyanobacterium from an aquatic environment in Bangladesh. *Transactions of the Royal Society for Tropical Medicine and Hygiene*, **88**: 298–9.

Jaan A.J., B. Dahllof, and S. Kjelleberg. 1986. Changes in protein composition of three bacterial isolates from marine waters during short periods of energy and nutrient deprivation. *Applied and Environmental Microbiology*, **52**: 1419–21.

Jackman P.J.H. 1985. Bacterial taxonomy based on electrophoretic whole-cell protein patterns. In M. Goodfellow and D.E. Minnikin (eds.) *Chemical Methods in Bacterial Systematics*, pp. 115–29. London: Academic Press.

Jackson, R.W., K. Osborne, G. Barnes, C. Jolliff, D. Zamani, B. Roll, A. Stillings, D. Herzog, S. Cannon, and S. Loveland. 2000. Multiregional evaluation of the SimPlate heterotrophic plate count method compared to the standard plate count agar pour plate method in water. *Applied and Environmental Microbiology*, **66**: 453–4.

Jaffal, A.A., I.M. Banat, A.A. El Mogheth, H. Nsanze, A. Bener, and A.S. Ameen. 1997. Residential indoor airborne microbial populations in the United Arab Emirates. *Environment International*, **23**: 529–33.

Jaffe, D.A. 1992. The nitrogen cycle. In S.S. Butcher, R.J. Charlson, G.H. Orans, and G.V. Wolfe (eds.) *Global Biogeochemical Cycles*, pp. 263–84. New York: Academic Press.

Jaffe, E.K., M. Volin, C.R. Bronson-Mullins, R.L. Dunbrack Jr., J. Kervinen, J. Martins, J.F. Quinlan Jr., M.H. Sazinsky, E.M. Steinhouse, and A.T. Yeung. 2000. An artificial gene for human porphobilinogen synthase allows comparison of an allelic variation implicated in susceptibility to lead poisoning. *Journal of Biological Chemistry*, **275**: 2619–26.

Jana, B.B. 1994. Ammonification in aquatic environments: A brief review. *Limnologica*, **24**: 389–413.

Janakiraman, R.S. and Y.V. Brun. 1999. Cell cycle control of a holdfast attachment gene in *Caulobacter crescentus*. *Journal of Bacteriology*, **181**: 1118–25.

Janetos, A.C. and C.O. Justice. 2000. Land cover and global productivity: A measurement strategy for the NASA program. *International Journal of Remote Sensing*, **21**: 1491–512.

Janssen, A.J.H., R. Rutenberg, and C.J.N. Buisman. 2001. Industrial applications of new sulphur biotechnology. *Water Science and Technology*, **44**: 85–90.

Jeffrey, W.H., R.J. Pledger, P. Aas, S. Hager, R.B. Coffin, R.V. Haven, and D.L. Mitchell. 1996. Diel and depth profiles of DNA photodamage in bacterioplankton exposed to ambient solar ultraviolet radiation. *Marine Ecology Progress Series*, **137**: 283–91.

Jenkins, R.O., T-A. Morris, P.J. Craig, A.W. Ritchie, and N. Ostah. 2000. Phosphine generation by mixed- and monoseptic-cultures of anaerobic bacteria. *Science of the Total Environment*, **250**: 73–81.

Jensen, E.C., H.S. Schrader, B. Rieland, T.L. Thompson, K.W. Lee, K.W. Nickerson, and T.A. Kokjohn. 1998. Prevalence of broad-host-range lytic bacteriophages of *Sphaerotilus natans*, *Escherichia coli*, and *Pseudomonas aeruginosa*. *Applied and Environmental Microbiology*, **64**: 575–80.

Jensen, P.K., B. Aalbaek, R. Aslam, and A. Dalsgaard. 2001. Specificity for field enumeration of *Escherichia coli* in tropical surface waters. *Journal of Microbiological Methods*, **45**: 135–41.

Jeon, C.O., D.S. Lee, and J.M. Park. 2000. Morphological characteristics of microbial sludge performing enhanced biological phosphorus removal in a sequencing batch reactor fed with glucose as sole carbon source. *Water Science and Technology*, **41**: 79–84.

Jiang J.X. and X.P. Zhou. 2002. Maize dwarf mosaic disease in different regions of China is caused by sugarcane mosaic virus. *Archives of Virology*, **147**: 2437–43.

Joerger, R.D. 2003. Alternatives to antibiotics: Bacteriocins, antimicrobial peptides and bacteriophages. *Poultry Science*, **82**: 640–7.

Johnson, K.S. and D.M. Karl. 2002. Is ocean fertilization credible and creditable? *Science*, **296**: 467.

Joint, I., K. Tait, M.E. Callow, J.A. Callow, D. Milton, P. Williams, and M. Camara. 2002. Cell-to-cell communication across the prokaryote–eukaryote boundary. *Science*, **298**: 1207.

Jones, M.C., J.M. Jenkins, A.G. Smith, and C.J. Howe. 1994. Cloning and characterization of genes for tetrapyrrole biosynthesis from the cyanobacterium *Anacystis nidulans* R2. *Plant Molecular Biology*, **24**: 435–48.

Jones, T. 1997. *The History of the Light Microscope.* Third Edition. http://www.utmem.edu/~thjones/hist/hist_mic.htm.

Jones, W.J. 1991. Diversity and physiology of methanogens. In J.E. Rogers and W.B. Whitman (eds.) *Microbial Production and Consumption of Greenhouse Gases: Methane, Nitrogen Oxides, and Halomethanes*, pp. 39–55. Washington, DC: American Society for Microbiology.

Jonkers H.M. and R.M. Abed. 2003. Identification of aerobic heterotrophic bacteria from the photic zone of a hypersaline microbial mat. *Aquatic Microbial Ecology*, **30**: 127–33.

Jonkers, H.M., M.J.E. Van Der Maarel, H. Van Gemerden, and T.A. Hansen. 1996. Dimethylsulfoxide reduction by marine sulfate-reducing bacteria. *FEMS Microbiology Letters*, **136**: 283–7.

Jordan, D.B. and W.L. Ogren. 1981. Species variation in the specificity of ribulose bisphosphate carboxylase/oxygenase. *Nature*, **291**: 513–15.

Jorgensen, B.B. 1983. The microbial sulfur cycle. In W.E. Krumbein (ed.) *Microbial Geochemistry*, pp. 91–214. Oxford, England: Blackwell.

Joux, F., W.H. Jeffrey, P. LeBaron, and D.L. Mitchell. 1999. Marine bacterial isolates display diverse responses to UV-B radiation. *Applied and Environmental Microbiology*, **65**: 3820–7.

Joyce, G.F. 2002. The antiquity of RNA-based evolution. *Nature*, **418**: 214–21.

Juneau, P., J.E. Lawrence, C.A. Suttle, and P.J. Harrison. 2003. Effects of viral infection on photosynthetic processes in the bloom-forming alga *Heterosigma akashiwo*. *Aquatic Microbial Ecology*, **31**: 9–17.

Junge, K., F. Imhoff, T. Staley, and J.W. Deming. 2002. Phylogenetic diversity of numerically important arctic sea-ice bacteria cultured at sub-zero temperature. *Microbial Ecology*, **43**: 315–28.

Justice, C., E. Vermote, J. Townshend, R. Defries, D. Roy, D. Hall, V. Salomonson, J. Privette, G. Riggs, A. Strahler, W. Lucht, R. Myneni, Y. Knjazihhin, S. Running, R. Nemani, Z. Wan, A. Huete, W. van Leeuwen, R. Wolfe, L. Giglio, J. Muller, P. Lewis, and M. Barnsley. 1998. The Moderate Resolution Imaging Spectroradiometer (MODIS): Land remote sensing for global change research. *IEEE Transactions of Geoscience and Remote Sensing*, **36**: 1228–49.

Kaeberlein, T., K. Lewis, and S.S. Epstein. 2002. Isolating "uncultivable" microorganisms in pure culture in a simulated natural environment. *Science*, **296**: 1127–9.

Kaiser, E. and G.J. Herndl. 1997. Rapid recovery of marine bacterioplankton activity after inhibition by radiation in coastal waters. *Applied and Environmental Microbiology*, **63**: 4026–31.

Kamminga, H. 1982. Life from space—A history of panspermia. *Vistas in Astronomy*, **26**: 67.

Kaplan, H.B. and E.P. Greenberg. 1985. Diffusion of autoinducer is involved in regulation of the *Vibrio fischeri* luminescence system. *Journal of Bacteriology*, **163**: 1210–14.

Kappler, A. and S.B. Haderlein. 2003. Natural organic matter as reductant for chlorinated aliphatic pollutants. *Environmental Science and Technology*, **37**: 2714–19.

Kaprelyants A.S., J.C. Gottschal, and D.B. Kell. 1993. Dormancy in non-sporulating bacteria. *FEMS Microbiology Reviews*, **104**: 271–86.

Karthikeyan, S., G.M. Wolfaardt, D.R. Korber, and D.E. Caldwell. 1999. Functional and structural responses of a degradative microbial community to substrates with varying degrees of complexity in chemical structure. *Microbial Ecology*, **38**: 215–24.

Kashefi, K., D.E. Holmes, J.A. Baross, and D.R. Lovley. 2003. Thermophily in the Geobacteraceae: *Geothermobacter ehrlichii* gen. nov., sp. nov., a novel thermophilic member of the Geobacteraceae from the "Bag City" hydrothermal vent. *Applied and Environmental Microbiology*, **69**: 2985–93.

Kasting, J.F. 2001. The rise of atmospheric oxygen. *Science*, **293**: 819–20.

Kasting, J.F., A.A. Pavlov, and J.L. Siefert. 2001. A coupled ecosystem-climate model for predicting the methane concentration in the Archean atmosphere. *Origins of Life and Evolution of the Biosphere*, **31**: 271–85.

Kasting, J.F. and J.L. Siefert. 2002. Life and the evolution of earth's atmosphere. *Science*, **296**: 1066–8.

Kataoka, N., Y. Tokiwa, Y. Tanaka, K. Takeda, and T. Suzuki. 1996. Enrichment culture and isolation of slow-growing bacteria. *Applied Microbiology and Biotechnology*, **45**: 771–7.

Keim, C.N., U. Lins, and M. Farina. 2001. Elemental analysis of uncultured magnetotactic bacteria exposed to heavy metals. *Canadian Journal of Microbiology*, **47**: 1132–6.

Kell, D.B., A.S. Kaprelyants, D.H. Weichart, C.R. Harwood, and M.R. Barer. 1998. Viability and activity in readily culturable bacteria: A review and discussion of practical issues. *Antonie van Leeuwenhoek*, **73**: 169–87.

Kell, D.B. and M. Young. 2000. Bacterial dormancy and culturability: The role of autocrine growth factors: Commentary. *Current Opinion in Microbiology*, **3**: 238–43.

Keller, M. and R. Dirmeier. 2001. Hydrogen-sulfur oxidoreductase complex from *Pyrodictium abyssi. Methods in Enzymology*, **331**: 442–51.

Kelly, D. and J.C. Murrell. 1999. Microbial metabolism of methanesulfonic acid. *Archives of Microbiology*, **172**: 341–8.

Kelly, J. 1995. Microorganisms, indigenous intellectual property rights and the Convention on Biological Diversity. In D. Allsop, R.R. Colwell, and D.L. Hawksworth (eds.) *Microbial Diversity and Ecosystem Function*, pp. 415–26. Wallingford, England: United Nations Environment Program and Center for Agriculture and Biotechnology International.

Kemp, P.F. 1994. A philosophy of methods development: The assimilation of new methods and information into aquatic microbial ecology. *Microbial Ecology*, **28**: 159–62.

Kennedy, A.C. and R.I. Papendick. 1995. Microbial characteristics of soil quality. *Journal of Soil and Water Conservation*, **6–7**: 243–8.

Kersters K. and J. De Ley. 1980. Classification and identification of bacteria by electrophoresis of their proteins. In M. Goodfellow and R.G. Board (eds.) *Microbiological Classification and Identification*, p. 273. London: Academic Press.

Kessel, M., Y. Cohen, and A.E. Walsby. 1985. Structure and physiology of square-shaped and other halophilic bacteria from the Gavish Sabkha. In G.M. Friedman and W.E. Krumbein (eds.) *Ecological Studies: Analysis and Synthesis. Volume 53. Hypersaline Ecosystems: The Gavish Sabkha*, pp. 267–87. New York: Springer-Verlag.

Kessler, P.S., J. McLarnan, and J.A. Leigh. 1997. Nitrogenase phylogeny and the molybdenum dependence of nitrogen fixation in *Methanococcus maripaludis. Journal of Bacteriology*, **179**: 541–3.

Kho, D.-H., J.-H. Jang, H.-S. Kim, K.-S. Kim, and J.K. Lee. 2003. Quorum sensing of *Rhodobacter sphaeroides* negatively regulates cellular poly-beta-hydroxybutyrate content under aerobic growth conditions. *Journal of Microbiology and Biotechnology*, **13**: 477–81.

Kim, E.E. and H.W. Wyckoff. 1991. Reaction mechanism of alkaline phosphatase based on crystal structures. Two-metal ion catalysis. *Journal of Molecular Biology*, **218**: 449–64.

Kim, S.Y., S.E. Lee, Y.R. Kim, C.M. Kim, P.Y. Ryu, H.E. Choy, S.S. Chung, and J.H. Rhee. 2003. Regulation of *Vibrio vulnificus* virulence by the LuxS quorum-sensing system. *Molecular Microbiology*, **48**: 1647–64.

Kimura, M. 1980. A simple method for estimating evolutionary rates of base substitutions through comparative studies of nucleotide sequences. *Journal of Molecular Evolution*, **16**: 111–20.

King, G.M. 1992. Ecological aspects of methane oxidation, a key determinant of global methane dynamics. *Advances in Microbiology and Ecology*, **12**: 431–68.

King, J. 2000. Environmental pollution and the emergence of new diseases. In L. Margulis, C. Matthews, and A. Haselton (eds.) *Environmental Evolution* (second edition), pp. 249–62. Cambridge, MA: MIT Press.

Kistemann, T., F. Dangendorf, and J. Schweikart. 2002. New perspectives on the use of Geographical Information Systems (GIS) in environmental health sciences. *International Journal of Hygiene and Environmental Health*, **205**: 169–81.

Kitano, K., N. Maeda, T. Fukui, H. Atomi, T. Imanaka, and K. Miki. 2001. Crystal structure of a novel-type archaeal rubisco with pentagon symmetry. *Structure (Cambridge)*, **9**: 473–81.

Kjellerup, B.V., B.H. Olesen, J.L. Nielsen, B. Frolund, S. Odum, and P.H. Nielsen. 2003. Monitoring and characterization of bacteria in corroding district heating systems using fluorescence *in situ* hybridization and microautoradiography. *Water Science and Technology*, **47**: 117–22.

Klingmuller, W. 1991. Plasmid transfer in soil: A case by case study with nitrogen-fixing *Enterobacter. FEMS Microbial Ecology*, **85**: 107–15.

Kloos, D.U., M. Stratz, A. Guttler, R.J. Steffan, and K.N. Timmis. 1994. Inducible cell lysis system for the study of natural transformation and environmental fate of DNA released by cell death. *Journal of Bacteriology*, **176**: 7352–61.

Klotz, M.G. and J.M. Norton. 1998. Multiple copies of ammonia monooxygenase (amo) operons have evolved under biased AT/GC mutational pressure in ammonia-oxidizing autotrophic bacteria. *FEMS Microbiology Letters*, **168**: 303–11.

Knight, I.T., W.E. Holben, J.M. Tiedje, and R.R. Colwell. 1992. Nucleic acid hybridization techniques for detection, identification, and enumeration of microorganisms in the environment. In M.A. Levin, R.J. Seidler, and M. Rogul (eds.) *Microbial Ecology: Principles, Methods, and Applications*, pp. 65–91. New York: McGraw-Hill.

Knorr, M.A., R.E.J. Boerner, and M.C. Rillig. 2003. Glomalin content of forest soils in relation to fire frequency and landscape position. *Mycorrhiza*, **13**: 205–10.

Koizumi, Y., J.J. Kelly, T. Nakagawa, H. Urakawa, S. El-Fantroussi, S. Al-Muzaini, M. Fukui, Y. Urushigawa, and D.A. Stahl. 2002. Parallel characterization of anaerobic toluene- and ethylbenzene-degrading microbial consortia by PCR-denaturing gradient gel electrophoresis, RNA-DNA membrane hybridization, and DNA microarray technology. *Applied and Environmental Microbiology*, **68**: 3215–25.

Kolb S., C. Knief, S. Stubner, and R. Conrad. 2003. Quantitative detection of methanotrophs in soil by novel *pmoA*-targeted real-time PCR assays. *Applied and Environmental Microbiology*, **69**: 2423–9.

Kolber, Z.S., F.G. Plumley, A.S. Lang, J.T. Beatty, R.E. Blankenship, C.L. VanDover, C. Vetriani, M. Koblizek, C. Rathgeber, and P.G. Falkowski. 2001. Contribution of aerobic photoheterotrophic bacteria to the carbon cycle in the ocean. *Science*, **292**: 2492–5.

Konings, W.N., S-V. Albers, S. Koning, and A.J.M. Driessen. 2002. The cell membrane plays a crucial role in survival of bacteria and archaea in extreme environments. *Antonie van Leeuwenhoek*, **81**:61–72.

Koonin, E.V., A.R. Mushegian, M.Y. Galperin, and D.R. Walker. 1997. Comparison of archaeal and bacterial genomes: Computer analysis of protein sequences predicts novel functions and suggests a chimeric origin for the archaea. *Molecular Microbiology*, **25**: 619–37.

Korem, M., A.S. Sheoran, Y. Gov, S. Tzipori, I. Borovok, and N. Balaban. 2003. Characterization of RAP, a quorum sensing activator of *Staphylococcus aureus*. *FEMS Microbiology Letters*, **223**: 167–75.

Korner, C. 2003. Slow in, rapid out: Carbon flux studies and Kyoto targets. *Science*, **300**: 1242–3.

Kowalchuk, G.A., A.W. Stienstra, G.H.J. Heilig, J.R. Stephen, and J.W. Woldendorp. 2000. Composition of communities of ammonium-oxidising bacteria in wet, slightly acid grassland soils using 16S rDNA-analysis. *FEMS Microbiology Ecology*, **31**: 207–15.

Kozdroj, J. and J.D. van Elsas. 2001. Structural diversity of microbial communities in arable soils of a heavily industrialized area determined by PCR-DGGE fingerprinting and FAME profiling. *Applied Soil Ecology*, **17**: 31–42.

Kramer, S. and D. Kunkel. 2001. *Hidden Worlds: Looking Through a Scientist's Microscope*. Boston, MA: Houghton/Mifflin.

Krawiec, S. 1985. Concept of bacterial species. *International Journal of Systematic Bacteriology*, **35**: 217–20.

Krishnamoorthy, M. 2000. The World of ALAD. Web-based information resource. http://www.arches.uga.edu/~malini/source.htm

Kroker, P.B., M. Bower, and B. Azadian. 2001. *Clostridium difficile* infection, hospital geography and time-space clustering. *Quarterly Journal of Medicine*, **94**: 223–5.

Krug, E.C. and D. Winstanley. 2002. The need for comprehensive and consistent treatment of the nitrogen cycle in nitrogen cycling and mass balance studies: I. Terrestrial nitrogen cycle. *Science of the Total Environment*, **293**: 1–29.

Kundu, S. and M.S. Hargrove. 2003. Distal heme pocket regulation of ligand binding and stability in soybean leghemoglobin. *Proteins*, **50**: 239–48.

Kushmaro, A., Y. Loya, M. Fine, and E. Rosenberg. 1996. Bacterial infection and coral bleaching. *Nature*, **380**: 396.

Kushner, D.J. (ed.) 1978. *Microbial Life in Extreme Environments*. New York: Academic Press.

Kuta, K.G. and L.L. Richardson. 2002. Ecological aspects of black-band disease of corals: Relationships between disease incidence and environmental factors. *Coral Reefs*, **21**: 393–8.

Kwaasi, A.A.A., R.S. Parhar, F.A.A. Al-Mohanna, H.A. Harfi, K.S. Collison, and S.T. Al-Sedairy. 1998. Aeroallergens and viable microbes in sandstorm dust: Potential triggers of allergic and non-allergic respiratory ailments. *Allergy (Copenhagen)*, **53**: 255–65.

Lacey, A.J. (ed.) 1999. *Light Microscopy in Biology: A Practical Approach*. Oxford: Oxford University Press.

Lackner, K. 2003. A guide to CO_2 sequestration. *Science*, **300**: 1677–8.

Lake, J.A., M.W. Clark, E. Hendeson, S.P. Fay, M. Oakes, A. Scheinman, J.P. Thornber, and R.A. Mah. 1985. Eubacteria, halobacteria, and the origin of photosynthesis: The photocytes. *Proceedings of the National Academy of Sciences (USA)*, **82**: 3716–20.

Lawrence, J. 2001. Catalyzing bacterial speciation: Correlating lateral transfer with genetic headroom. *Systematic Biology*, **50**: 479–96.

Lawrence, J. and H. Ochman. 1998. Molecular archaeology of the *Escherichia coli* genome. *Proceedings of the National Academy of Sciences (USA)*, **95**: 9431–7.

Lawrence, J.R., G. Kopf, J.V. Headley, and T.R. Neu. 2001. Sorption and metabolism of selected herbicides in river biofilm communities. *Canadian Journal of Microbiology*, **47**: 634–41.

Lawrence, J.R., T.R. Neu, and K.C. Marshall. 2002. Colonization, adhesion, aggregation, and biofilms. In C.J. Hurst, R.L. Crawford, G.R. Knudsen, M.J. McInerney, and L.D. Stetzenbach (eds.) *Manual of Environmental Microbiology*, pp. 466–77. Washington, DC: American Society for Microbiology.

Lazcano, A. 2000. Origins of life: history of ideas. In L. Margulis, C. Matthews and A. Haselton (eds.) *Environmental Evolution*, pp. 83–96. Cambridge, MA: MIT Press.

Lazcano, A. and S.L. Miller. 1994. How long did it take for life to begin and evolve to cyanobacteria? *Journal of Molecular Evolution*, **39**: 546–54.

Le Du, M.H., C. Lamoure, B.H. Muller, O.V. Bulgakov, E. Lajeunesse, A. Menez, and J.C. Boulain. 2002. Artificial evolution of an enzyme active site: Structural studies of three highly active mutants of *Escherichia coli* alkaline phosphatase. *Journal of Molecular Biology*, **316**: 941–53.

Le Mer, J. and P. Roger. 2001. Production, oxidation, emission and consumption of methane by soils: A review. *European Journal of Soil Biology*, **37**: 25–50.

Leadbetter, J.R. and E.P. Greenberg. 2000. Metabolism of acyl homoserine lactone quorum-sensing signals by *Variovorax paradoxus*. *Journal of Bacteriology*, **182**: 6921–6.

Lechevalier, H. and M.P. Lechevalier. 1988. Chemotaxonomic use of lipids – an overview. In C. Ratledge and S.G. Wilkinson (eds.) *Microbial Lipids*, pp. 869–902. London: Academic Press.

LeChevallier, M.W., R.J. Seidler, and T.M. Evans. 1980. Enumeration and characterization of standard plate count bacteria in chlorinated and raw water supplies. *Applied and Environmental Microbiology*, **40**: 922–30.

Ledgham, F., I. Ventre, C. Soscia, M. Foglino, J.N. Sturgis, and A. Lazdunski. 2003. Interactions of the quorum sensing regulator QscR: Interaction with itself and the other regulators of *Pseudomonas aeruginosa* LasR and RhlR. *Molecular Microbiology*, **48**: 199–210.

Lee, G.-H. and G. Stotzky. 1999. Transformation and survival of donor, recipient, and transformants of *Bacillus subtilis in vitro* and in soil. *Soil Biology and Biochemistry*, **31**: 1499–508.

Lenski, R.E., C. Ofria, R.T. Pennock, and C. Adami. 2003a. The evolutionary origin of complex features. *Nature*, **423**: 139–44.

Lenski, R.E. and M. Travisano. 1994. Dynamics of adaptation and diversification in a 10,000-generation experiment with a bacterial population. *Proceedings of the National Academy of Sciences (USA)*, **91**: 6808–14.

Lenski, R.E., C.L. Winkworth, and M.A. Riley. 2003b. Rates of DNA sequence evolution in experimental populations of *Escherichia coli* during 20,000 generations. *Journal of Molecular Evolution*, **56**: 498–508.

Lesprit, P., F. Faurisson, O. Join-Lambert, F. Roudot-Thoraval, M. Foglino, C. Vissuzaine, and C. Carbon. 2003. Role of the quorum-sensing system in experimental pneumonia due to *Pseudomonas aeruginosa* in rats. *American Journal of Respiratory and Critical Care Medicine*, **167**: 1478–82.

Letelier, R.M. and D.M. Karl. 1998. *Trichodesmium* spp. physiology and nutrient fluxes in the North Pacific subtropical gyre. *Aquatic Microbial Ecology*, **15**: 265–76.

Levia, D. and E.E. Frost. 2003. A review and evaluation of stem flow literature in the hydrologic and biogeochemical cycles of forested and agricultural ecosystems. *Journal of Hydrology*, **274**: 1–29.

Levin, B.R. 1981. Periodic selection, infectious genetic exchange and genetic structure of *Escherichia coli* population. *Genetics*, **99**: 1–23.

Levin, B.R. and C.T. Bergstrom. 2000. Bacteria are different: Observations, interpretations, speculations, and opinions about the mechanisms of adaptive evolution in bacteria. *Proceedings of the National Academy of Sciences (USA)*, **97**: 6981–5.

Levy, S.B. 1998. The challenge of antibiotic resistance. *Scientific American*, **218**: 46–53.

Levy, S.B. 2002. *The Antibiotic Paradox.* Boston: Perseus.

Lewin, R.A. 2002. Prochlorophyta: A matter of class distinctions. *Photosynthesis Research*, **73**: 59–61.

Lewontin, R.C. 1995. Genes, environment, and organisms. In R.B. Silvers (ed.). *Hidden Histories of Science*, pp. 115–39. New York: New York Review.

Lewis Jr., W.M., M.C. Grant, and S.K. Hamilton. 1985. Evidence that filterable phosphorus is a significant atmospheric link in the phosphorus cycle. *Oikos*, **45**: 428–32.

Lewis, D.F.V. and G. Sheridan. 2001. Cytochromes P450, oxygen, and evolution. *The Scientific World*, **1**: 151–67.

Lewis, P.J., C.E. Nwoguh, M.R. Barer, C.R. Harwood, and J. Errington. 1994. Use of digitized video microscopy with a fluorogenic enzyme substrate to demonstrate cell- and compartment-specific gene expression in *Salmonella enteritidis* and *Bacillus subtilis. Molecular Microbiology*, **13**: 655–62.

Li, Y.-H., N. Tang, M.B. Aspiras, P.C. Lau, J.H. Lee, R.P. Ellen, and D.G. Cvitkovitch. 2002. A quorum-sensing signaling system essential for genetic competence in *Streptococcus mutans* is involved in biofilm formation. *Journal of Bacteriology*, **184**: 2699–708.

Lieberman R.L., D.B. Shrestha, P.E. Doan, B.M. Hoffman, T.L. Stemmler, and A.C. Rosenzweig. 2003. Purified particulate methane monooxygenase from *Methylococcus capsulatus* (Bath) is a dimer with both mononuclear copper and a copper-containing cluster. *Proceedings of the National Academy of Sciences (USA)*, **100**: 3820–5.

Lin, B-L., A. Sakoda, R. Shibasaki, N. Goto, and M. Suzuki. 2000. Modeling a global biogeochemical nitrogen cycle in terrestrial ecosystems. *Ecological Modeling*, **135**: 89–110.

Lincoln, R.J., G.A. Boxhall, and P.F. Clark. 1989. *A Dictionary of Ecology, Evolution, and Systematics.* Cambridge, England: Cambridge University Press.

Lins, U. and M. Farina. 1999. Phosphorus-rich granules in uncultured magnetotactic bacteria. *FEMS Microbiology Letters*, **172**: 23–8.

Lipski, A., E. Spieck, A. Makolla, and K. Altendorf. 2001. Fatty acid profiles of nitrite-oxidizing bacteria reflect their phylogenetic heterogeneity. *Systematic and Applied Microbiology*, **24**: 377–84.

Liu, J., F.B. Dazzo, O. Glagoleva, B. Yu, and A.K. Jain. 2001. CMEIAS: A computer-aided system for the image analysis of bacterial morphotypes in microbial communities. *Microbial Ecology*, **41**: 173–94.

Liu, W-T. and D.A. Stahl. 2002. Molecular approaches for the measurement of density, diversity, and phylogeny. In C.J. Hurst, R.L. Crawford, R. Knudsen, M.J. McInerney, and L.D. Stetzenbach (eds.) *Manual of Environmental Microbiology* (Second Edition), pp. 114–34. Washington, DC: ASM Press.

Lloyd, D., K.L. Thomas, A. Hayes, B. Hill, B.A. Hales, C. Edwards, J.R. Saunders, D.A. Ritchie, and M. Upton. 1998. Micro-ecology of peat: Minimally invasive analysis using confocal laser scanning microscopy, membrane inlet mass spectrometry, and PCR amplification of methanogen-specific gene sequences. *FEMS Microbiology and Ecology*, **25**: 179–88.

Lomans, B.P., C. van der Drift, A. Pol, and H.J.M. Op den Camp. 2002. Microbial cycling of volatile organic sulfur compounds. *Cellular and Molecular Life Sciences*, **59**: 575–88.

Lorenz, M.G., D. Gerjets, and W. Wackernagel. 1994. Release of transforming plasmid and chromosomal DNA from two cultured soil bacteria. *Archives of Microbiology*, **156**: 319–26.

Lorenz, P., K. Liebeton, F. Niehaus, and J. Eck. 2002. Screening for novel enzymes for biocatalytic processes: Accessing the metagenome as a resource of novel functional sequence space. *Current Opinion in Biotechnology*, **13**: 572–7.

Lovelock, J.E. 1995. New statements on the Gaia theory. *Microbiologia (Madrid)*, **11**: 295–304.

Lovelock, J.E. 2000. The Gaia hypothesis. In L. Margulis, C. Matthews and A. Haselton (eds.) *Environmental Evolution*, pp. 1–28. Cambridge, MA: MIT Press.

Lovley, D.R. 1993. Dissimilatory metal reduction. *Annual Review of Microbiology*, **47**: 263–90.

Lupi, O. 2003. Could ectoparasites act as vectors for prion diseases? *International Journal of Epidemiology*, **32**: 425–9.

Luton, P.E., J.M. Wayne, R.J. Sharp, and P.W. Riley. 2002. The *mcrA* gene as an alternative to 16S rRNA in the phylogenetic analysis of methanogen populations in landfills. *Microbiology*, **148**: 3521–30.

Lyons, M. M., P. Aas, J.D. Pakulski, L.V. Waasbergen, R.V. Miller, D.L. Mitchell, and W.H. Jeffrey. 1998. DNA damage induced by ultraviolet radiation in coral-reef microbial communities. *Marine Biology*, **130**: 537–43.

MacNaughton, S.J., T. Booth, T.M. Embley, and A.G. O'Donnell. 1996. Physical stabilization and confocal microscopy of bacteria on roots using 16S ribosomal RNA-targeted, fluorescent-labeled oligonucleotide probes. *Journal of Microbiological Methods*, **26**: 279–85.

Maeda, N., K. Kitano, T. Fukui, S. Ezaki, H. Atomi, and T. Imanaka. 1999. Ribulose bisphosphate carboxylase/oxygenase from the hyperthermophilic archaeon *Pyrococcus kodakaraensis* KOD1 is composed solely of large subunits and forms a pentagonal structure. *Journal of Molecular Biology*, **293**: 57–66.

Maidak, B., J. Cole, T. Lilburn, C. Parker, P. Saxman, R. Farris, G. Garrity, G. Olsen, T. Schmidt, and Tiedje, J. 2001. The RDP-II (Ribosomal Database Project). *Nucleic Acids Research*, **29**: 171–3.

Maier, R.M. 2000. Biogeochemical cycling. In R.M. Maier, I.L. Pepper, and C.P. Gerba (eds.) *Environmental Microbiology*, pp. 319–46. San Diego: Academic Press.

Majewski, J. and F.M. Cohan. 1999. Adapt globally, act locally: The effect of selective sweeps on bacterial sequence diversity. *Genetics*, **152**: 1459–74.

Mallet, M. 1995. A species definition for the modern synthesis. *Trends in Ecology and Evolution*, **10**: 294–9.

Mamet, R., R. Scharf, Y. Zimmels, S. Kimchie, and N. Schoenfeld. 1996. Mechanism of aluminum-induced porphyrin synthesis in bacteria. *Biometals*, **9**: 73–7.

Mancinelli, R.L., M.R. White, and L.J. Rothschild. 1995. DNA integrity and survival of *Synechococcus* (Nagli) exposed to solar irradiation and vacuum in earth orbit. *Journal of Phycology*, **31**(Supplement 3): 11.

Mandeel, Q.A., J.A. Abbas, and A.M. Saeed. 1995. Survey of *Fusarium* species in an arid environment of Bahrain: II. Spectrum of species on five isolation media. *Sydowia*, **47**: 223–39.

Manefield, M. and S.L. Turner. 2002. Quorum sensing in context: Out of molecular biology and into microbial ecology. *Microbiology*, **148**: 3762–4.

Maness, P.-C., S. Smolinski, A.C. Dillon, M.J. Heben, and P.F. Weaver. 2002. Characterization of the oxygen tolerance of a hydrogenase linked to a carbon monoxide oxidation pathway in *Rubrivivax gelatinosus*. *Applied and Environmental Microbiology*, **68**: 2633–6.

Maness, P.-C. and P.F. Weaver. 2001. Evidence for three distinct hydrogenase activities in *Rhodospirillum rubrum*. *Applied Microbiology and Biotechnology*, **57**: 751–6.

Mann, N.H. 2003. Phages of the marine cyanobacterial picophytoplankton. *FEMS Microbiology Reviews*, **27**: 17–34.

Marchesi, J.R., A.J. Weightman, B.A. Cragg, R.J. Parkes, and J.C. Fry. 2001. Methanogen and bacterial diversity and distribution in deep gas hydrate sediments from the Cascadia Margin as revealed by 16S rRNA molecular analysis. *FEMS Microbiology and Ecology*, **34**: 221–8.

Margalef, R. 1958. Information theory in ecology. *General Systems*, **3**: 36–71.

Margalef, R. 1963. On certain unifying principles in ecology. *American Naturalist*, **97**: 357–74.

Margulis, L. 1996. Archaeal–eubacterial mergers in the origin of *Eukarya*: Phylogenetic classification of life. *Proceedings of the National Academy of Sciences (USA)*, **93**: 1071–6.

Margulis, L. 2000. Symbiosis and the origin of protists. In L. Margulis, C. Matthews and A. Haselton (eds.) *Environmental Evolution*, pp. 141–58. Cambridge, MA: MIT Press.

Margulis, L., M.J. Chapman, D.S. Roos, and C.F. Delwiche. 1998. Endosymbioses: Cyclical and permanent in evolution (and reply). *Trends in Microbiology*, **6**: 342–6.

Margulis, L., J.O. Corliss, M. Melkonian, and D.J. Chapman (eds.). 2000. *Handbook of Protoctista: The Structure, Cultivation, Habitats and Life Histories of Eukaryotic Microorganisms and their Descendants Exclusive of Animals, Plants, and Fungi.* New York: Academic Press.

Margulis, L., M.F. Dolan, and R. Guerrero. 2000. The chimeric eukaryote: Origin of the nucleus from the karyomastigont in amitochondriate protests. *Proceedings of the National Academy of Sciences (USA)*, **97**: 6954–9.

Margulis, L. and R. Fester (eds.) 1991. *Symbiosis as a Source of Evolutionary Innovation, Speciation, and Morphogenesis.* Cambridge, MA: MIT Press.

Margulis, L., C. Matthews, and A. Haselton (eds.) 2000. *Environmental Evolution.* Cambridge, MA: MIT Press.

Margulis, L. and D. Sagan. 1991. Microcosmos: Four Billion Years of Evolution from our Microbial Ancestors. New York: Simon and Schuster.

Margulis, L. and D. Sagan. 2002. *Acquiring Genomes: A Theory of the Origin of Species.* New York: Basic Books.

Margulis, L., and K.V. Schwartz. 1998. *Five Kingdoms* (Third Edition). 448 pp. New York: W.H. Freeman & Co.

Marita, J.M., J. Ralph, C. Lapierre, L. Jouanin, and W. Boerjan. 2001. NMR characterization of lignins from transgenic poplars with suppressed caffeinc acid O-methyltransferase activity. *Journal of the Chemical Society Perkin Transactions*, **1**: 2939–45.

Maron P.-A., C. Coeur, C. Pink, A. Clays-Josserand, R. Lensi, A. Richaume, and P. Potier. 2003. Use of polyclonal antibodies to detect and quantify the NOR protein of nitrite oxidizers in complex environments. *Journal of Microbiological Methods*, **53**: 87–95.

Marsh, P.D. 1995. The role of continuous culture in modelling the human microflora. *Journal of Chemical Technology and Biotechnology*, **64**: 1–9.

Marsh, T.L., P. Saxman, J. Cole, and J. Tiedje. 2000. Terminal restriction fragment length polymorphism analysis program, a web-based tool for microbial community analysis. *Applied and Environmental Microbiology*, **66**: 3616–20.

Marshall, P.A. and A.H. Baird. 2000. Bleaching of corals in the Central Great Barrier Reef: Variation in assemblage response and taxa susceptibilities. *Coral Reefs*, **19**: 155–63.

Martin, D.R., L.L. Lundie, R. Kellum, and H.L. Drake. 1983. Carbon monoxide-dependent evolution of hydrogen by the homoacetate-fermenting bacterium *Clostridium thermoaceticum*. *Current Microbiology*, **8**: 337–40.

Martin, L.D. 2003. Earth history, disease, and the evolution of primates. In C.L. Greenblatt and M. Spigelman (eds.) *Emerging Pathogens: Archaeology, Ecology and Evolution of Infectious Disease*, pp. 13–24. New York: Oxford University Press.

Martin, W. 1999. A briefly argued case that mitochondria and plastids are descendants of endosymbionts but that the nuclear compartment is not. *Proceedings of the Royal Society Biological Sciences Series B*, **266**: 1387–95.

Martin, W. and M. Müller. 1998. The hydrogen hypothesis for the first eukaryote. *Nature*, **392**: 37–41.

Mason, P.W., J.M. Pacheco, Q.-Z. Zhao, and N.J. Knowles. 2003. Comparisons of the complete genomes of Asian, African and European isolates of a recent foot-and-mouth disease virus type O pandemic strain (PanAsia). *Journal of General Virology*, **84**: 1583–93.

Matheson, V.G., J. Munakata-Marr, G.D. Hopkins, P.L. McCarty, J.M. Tiedje, and L.J. Forney. 1997. A novel means to develop strain-specific DNA probes for detecting bacteria in the environment. *Applied and Environmental Microbiology*, **63**: 2863–9.

Matson, P.A. and P.M. Vitousek. 1990a. Remote sensing and trace gas fluxes. In R.J. Hobbs and H.A. Mooney (eds.) *Remote Sensing of Biosphere Functioning*, pp. 157–67. New York: Springer Verlag.

Matson, P.A. and P.M. Vitousek. 1990b. Ecosystem approach to a global nitrous oxide budget. *Bioscience*, **40**: 677–82.

Matthews, C. 2000. Chemical evolution in a hydrogen cyanide world. In L. Margulis, C. Matthews, and A. Haselton (eds.) *Environmental Evolution*, pp. 47–66. Cambridge, MA: MIT Press.

Matthews, D. 2003. BSE: A global update. *Symposium Series/The Society for Applied Microbiology*, **32**: 120S–25S.

Matthies, C., H.-P. Erhard, and H.L. Drake. 1997. Effects of pH on the comparative culturability of fungi and bacteria from acidic and less acidic forest soils. *Journal of Basic Microbiology*, **37**: 335–43.

Matz, C., J. Boenigk, H. Arndt, and K. Juergens. 2002. Role of bacterial phenotypic traits in selective feeding of the heterotrophic nanoflagellate *Spumella* sp. *Aquatic Microbial Ecology*, **27**: 137–48.

Mau, B., M. Newton, and B. Larget. 1999. Bayesian phylogenetic inference via Markov chain Monte Carlo methods. *Biometrics*, **55**: 1–12.

Mayden, R.L. 1997. A hierarchy of species concepts: the denouement in the saga of the species problem. In M.F. Claridge, H.A. Dawah, and M.R. Wilson (eds.) *Species: The Units of Biodiversity*, pp. 318–24. London: Chapman and Hall.

Mayer, F. and M. Hoppert. 2001. *Microscopic Techniques in Biotechnology*. Chichester: Wiley.

Maynard Smith, J. 1999. *The Origins of Life: From the Birth of Life to the Origin of Language*. New York: Oxford University Press.

Maynard-Smith, J., N.H. Smith, M. Orourke, and B.G. Spratt. 1993. How clonal are bacteria? *Proceedings of the National Academy of Sciences (USA)*, **90**: 4348–88.

Mayr, E. 1963. *Animal Species and Evolution*. Cambridge, MA: Belknap Press of Harvard University Press.

Mayr, E. 1970. *Populations, Species, and Evolution*, pp. 256–77. Cambridge, MA: Belknap Press of Harvard University Press.

Mayr, E. 1987. The ontological status of species: Scientific progress and philosophical terminology. *Biology and Philosophy*, **2**: 145–66.

Mayr, E. 1998. Two empires, or three? *Proceedings of the National Academy of Sciences (USA)*, **95**: 9720–3.

Mayr, E. 2001. *What Evolution Is*. New York: Basic Books.

McDougald, D., S.A. Rice, and S. Kjelleberg. 1999. New perspectives on the viable but nonculturable response. *Biologia (Bratislava)*, **54**: 617–23.

McDougald, D., S.A. Rice, D. Weichardt, and S. Kjellerberg. 1998. Nonculturability: Adaptation or debilitation? *FEMS Microbial Ecology*, **25**: 1–9.

McElroy, M. 2000. Comparison of planetary atmospheres on Mars, Venus, and Earth. In L. Margulis, C. Matthews and A. Haselton (eds.) *Environmental Evolution*, pp. 29–46. Cambridge, MA: MIT Press.

McFetters, G.A., A. Singh, S. Byun, P.R. Callis, and S. Williams. 1991. Acridine orange staining reaction as an index of physiological activity in *Escherichia coli*. *Journal of Microbiological Methods*, **13**: 87–97.

McIsaac, G. and M.B. David. 2003. On the need for consistent and comprehensive treatment of the N cycle. *Science of the Total Environment*, **305**: 249–55.

McKay, D.S., E.K. Gibson Jr., K.L. Thomas-Keprta, H. Vali, C.S. Romanek, S.J. Clemett, X.D.F. Chillier, C.R. Maechling, and R.N. Zare. 1996. Search for past life on Mars: Possible relic biogenic activity in Martian meteorite ALH84001. *Science*, **273**: 924–30.

McKinlay, A.F. and B.L. Diffey. 1987. A reference action spectrum for ultra-violet induced erythema in human skin. In W.F. Passchier and B.F.M. Bosnjakovich (eds.) *Human Exposure to Ultraviolet Radiation: Risks and Regulations*, pp. 83–7. International Congress Series.

McLean, R.J.C., J.M. Cassanto, M.B. Barnes, and J.H. Koo. 2001. Bacterial biofilm formation under microgravity conditions. *FEMS Microbiology Letters*, **195**: 115–19.

McMahon, K.D., M.A. Dojka, N.R. Pace, D. Jenkins, and J.D. Keasling. 2002. Polyphosphate kinase from activated sludge performing enhanced biological phosphorus removal. *Applied and Environmental Microbiology*, **68**: 4971–8.

McMichael, T. 2001. *Human Frontiers: Environments and Disease*. Cambridge, England: Cambridge University Press.

Mester, T. and J.A. Field. 1997. Optimization of manganese peroxidase production by the white rot fungus *Bjerkandera* sp. strain BOS55. *FEMS Microbiology Letters*, **155**: 161–8.

Meyer, J.-M. 2000. Pyoverdins, pigments, siderophores, and potential taxonomic markers of fluorescent *Pseudomonas* species. *Archives of Microbiology*, **174**: 135–42.

Mhamdi, R., G. Laguerre, M.E. Aouani, M. Mars, and N. Amarger. 2002. Different species and symbiotic genotypes of field rhizobia can nodulate *Phaseolus vulgaris* in Tunisian soils. *FEMS Microbiology Ecology*, **41**: 77–84.

Michaelis, W., R. Seifert, K. Nauhaus, T. Treude, V. Thiel, M. Blumenberg, K. Knittel, A. Gieseke, K. Peterknecht, T. Pape, A. Boetius, R. Amann, B.B. Jorgensen, F. Widdel, J. Peckmann, N.V. Pimenov, and M.B. Gulin. 2002. Microbial reefs in the Black Sea fueled by anaerobic oxidation of methane. *Science*, **297**: 1013–15.

Middledoe, M., N.O.G. Jorgensen, and N. Kroer. 1996. Effects of viruses on nutrient turnover and growth efficiency of non-infected marine bacterioplankton. *Applied and Environmental Microbiology*, **62**: 1991–7.

Miklausen, A.J. 1997. *The Brown Algal Origin of Land Plants and the Algal Origin of Life on Earth and in the Universe*. Shippensburg, PA: Ragged Edge Press.

Miller, R.V., T.A. Kokjohn, D.J. Saye, S. Ripp, S. Kidambi, J. Replicon, and O.A. Ogunseitan. 1993. Monitoring horizontal gene transfer in environmental microbial communities. *Journal of Cellular Biochemistry* (Supplement), **17C**: 185.

Miller, R.V., S. Ripp, J. Replicon, O.A. Ogunseitan, and T.A. Kokjohn. 1992. Virus-mediated gene transfer in freshwater environments. In M.J. Gauthier (ed.) *Gene Transfers and the Environment*. Berlin: Springer-Verlag.

Miller, S.L. 1953. A production of amino acids under possible primitive conditions. *Science*, **117**: 528–9.

Milon, J.W. and J.F. Shogren (eds.) 1995. Integrating economic and ecological indicators: Practical methods for environmental policy analysis. Westport, Connecticut: Praeger.

Mizohata, E., H. Matsumura, Y. Okano, M. Kumei, H. Takuma, J. Onodera, K. Kato, N. Shibata, T. Inoue, A. Yokota, and Y. Kai. 1999. Crystal structure of activated ribulose-1,5-bisphosphate carboxylase/oxygenase from green alga *Chlamydomonas reinhardtii* complexed with 2-carboxyarabinitol-1,5-bisphosphate. *Journal of Molecular Biology*, **316**: 679–91.

Molina, L., F. Constantinescu, L. Michel, C. Reimmann, B. Duffy, and G. Defago. 2003. Degradation of pathogen quorum-sensing molecules by soil bacteria: A preventive and curative biological control mechanism. *FEMS Microbiology Ecology*, **45**: 71–81.

Montoya, J.P., E.J. Carpenter, and D.G. Capone. 2002. Nitrogen fixation and nitrogen isotope abundances in zooplankton of the oligotrophic North Atlantic. *Limnology and Oceanography*, **47**: 1617–28.

Moon-van der Staay, S.Y., R. De Wachter, and D. Vaulot. 2001. Oceanic 18S rDNA sequences from picoplankton reveal unsuspected eukaryotic diversity. *Nature*, **409**: 607–10.

Moorbath, S. 1995. Age of the oldest rocks with biogenic components: An estimate for the age of the origin of life. *Journal of Biological Physics*, **20**: 85–94.

Moran, N.A., M.A. Munson, P. Baumann, and H. Ishikawa. 1993. A molecular clock in endosymbiotic bacteria is calibrated using the insect hosts. *Proceedings of the Royal Society of London, Series B*, **253**: 167–71.

Moreira, D. and P. López-García. 2002. The molecular ecology of microbial eukaryotes unveils a hidden world. *Trends in Microbiology*, **10**: 31–8.

Morel, F.M.M., R.J.M. Hudson, and N.M. Price. 1991. Limitation of productivity by trace metals in the sea. *Limnology and Oceanography*, **36**: 1742–55.

Morel F.M.M. and N.M. Price. 2003. The biogeochemical cycles of trace metals in the oceans. *Science*, **300**: 944–7.

Morel, G. and A. Cavalier. 2001. *In Situ Hybridization in Light Microscopy*. Boca Raton, FL: CRC Press.

Morgan, P., S.A. Lee, S.T. Lewis, A.N. Sheppard, and R.J. Watkinson. 1993. Growth and biodegradation by white-rot fungi inoculated into soil. *Soil Biology and Biochemistry*, **25**: 279–87.

Morris P.A., S.J. Wentworth, M. Byrne, M. Nelman, T. Longazo, C.C. Allen, D.S. McKay, and C. Sams. 2002. Biosignatures for interpreting life (past or present) from ancient terrestrial and Mars rocks. *Astrobiology*, **2**: 540–1.

Morris, S.A., R. Radajewski, and J.C. Murrell. 2002. Identification of the functionally active methanotroph populations in a peat soil by stable-isotope probing. *Applied and Environmental Microbiology*, **68**: 1446–53.

Morris, V.J. and A.P. Kirby. 1999. *Gunning Atomic Force Microscopy for Biologists*. London: Imperial College Press.

Moutin, T. 2000. Biogeochemical phosphate cycle as a key factor in the control of planktonic production and carbon export from the photic to the aphotic zone. *Oceanis*, **26**: 643–60.

Moyer, C.L., J.M. Tiedje, F.C. Dobbs, and D.M. Karl. 1996. A computer-simulated restriction fragment length polymorphism analysis of bacterial small-subunit rRNA genes: Efficacy of selected tetrameric restriction enzymes for studies of microbial diversity in nature. *Applied and Environmental Microbiology*, **60**: 871–9.

Muller, A.K., K. Westergaard, S. Christensen, and S.J. Sorensen. 2001. The effect of long-term mercury pollution on the soil microbial community. *FEMS Microbiology Ecology*, **36**: 11–19.

Muller, U., A. Kappeler, R.G. Zanoni, and U. Breitenmoser. 2000. The development of rabies in Switzerland: Landscape determines the course of wild animal epidemic. *Schweizer Archiv Fuer Tierheilkunde*, **142**: 431–8.

Mura, C., J.E. Katz, S.G. Clarke, and D. Eisenberg. 2003. Structure and function of an archaeal homolog of survival protein E (*SurEalpha*): An acid phosphatase with purine nucleotide specificity. *Journal of Molecular Biology*, **7**: 1559–75.

Murphy, D.B. 2001. *Fundamentals of Light Microscopy and Electronic Imaging*. New York: Wiley-Liss.

Murray, A. 1995. Phytoplankton exudation: Exploitation of the microbial loop as a defense against algal viruses. *Journal of Plankton Research*, **17**: 1079–94.

Murray, R.G.E. and C.F. Robinow. 1994. Light microscopy. In P. Gerhardt, R.G.E. Murray, W.A. Wood, and N.R. Kreig (eds.) *Methods for General and Molecular Bacteriology*, pp. 8–20. Washington, DC: American Society for Microbiology.

Muyzer, G. and K. Smalla. 2000. Application of denaturing gradient gel electrophoresis (DGGE) and temperature gradient gel electrophoresis (TGGE) in microbial ecology. *Antonie van Leeuwenhoek*, **73**: 127–41.

Nagahama, T., M. Hamamoto, T. Nakase, and K. Horikoshi. 2001. *Rhodotorula lamellibrachii* sp. nov., a new yeast species from a tubeworm collected at the deep-sea floor in Sagami Bay and its phylogenetic analysis. *Antonie van Leeuwenhoek*, **80**: 317–23.

Nakamura, M., K. Tsumoto, K. Ishimura, and I. Kumagai. 2001. A visualization method of filamentous phage infection and phage-derived proteins in *Escherichia coli* using biotinylated phages. *Biochemical and Biophysical Research Communications*, **289**: 252–6.

Nakayama, J., A.D.L. Akkermans, and W.M. de Vos. 2003. High-throughput PCR screening of genes for three-component regulatory system putatively involved in quorum sensing from low-G + C Gram-positive bacteria. *Bioscience Biotechnology and Biochemistry*, **67**: 480–9.

Narayanasamy, P. 2002. *Microbial Plant Pathogens and Crop Disease Management*. Enfield, New Hampshire: Science Publishers.

National Science Foundation. 2000. Microbial Observatories. http://www.nsf.gov/pubs/2000/nsf0021/nsf0021.htm (accessed August 8, 2003).

National Science Foundation. 2001. Biocomplexity in the environment: Integrated research and education in environmental systems. http://www.nsf.gov/pubs/2001/nsf0134/nsf0134.htm (accessed August 8, 2003).

National Science Foundation. 2003. *Complex Environmental Systems: Synthesis for Earth, Life, and Society in the 21st Century*. Arlington, Virginia: National Science Foundation (http://www.nsf.gov).

Navarrete, A., A. Peacock, S.J. Macnaughton, J. Urmeneta, J. Mas-Castella, D.C. White, and R. Guerrero. 2000. Physiological status and community composition of microbial mats of the Ebro Delta, Spain, by signature lipid biomarkers. *Microbial Ecology*, **39**: 92–9.

Nealson, K.H. 1999. Post-viking microbiology: New approaches, new data, new insights. *Origins of Life and Evolution of the Biosphere*, **29**: 73–93.

Nealson, K.H. and C.R. Myers. 1992. Microbial reduction of manganese and iron: New approaches to carbon cycling. *Applied and Environmental Microbiology*, **58**: 439–43.

Nealson, K.H., T. Platt, and J.W. Hastings. 1970. Cellular control of the synthesis and activity of the bacterial luminescent system. *Journal of Bacteriology*, **104**: 313–22.

Nealson, K.H. and D. Saffarini. 1994. Iron and manganese in anaerobic respiration: Environmental significance, physiology, and regulation. *Annual Review of Microbiology*, **48**: 311–43.

Neff, J.C., E.A. Holland, F.J. Dentener, W.H. McDowell, and K.M. Russell. 2000. The origin, composition and rates of organic nitrogen deposition: A missing piece of the nitrogen cycle? *Biogeochemistry (Dordrecht)*, **57–8**: 99–136.

Neilan, B.A. 1996. Detection and identification of cyanobacteria associated with toxic blooms: DNA amplification protocols. *Phycologia*, **35**(6): 147–55.

Neilan B.A., B.P. Burns, D.A. Relman, and D.R. Lowe. 2002. Molecular identification of cyanobacteria associated with stromatolites from distinct geographical locations. *Astrobiology*, **2**: 271–80.

Newman D.K. and J.F. Banfield. 2002. Geomicrobiology: How molecular-scale interactions underpin biogeochemical systems. *Science*, **296**: 1071–7.

Nishioka, K. 1998. *Report on cosmic dust capture research and development for the exobiology program*, Report CR-97–207698. Washington,

DC: National Aeronautics and Space Administration. Distributed by National Technical Information Service, Springfield, VA.

Nishiyama, Y., S. Nakamura, R. Aono, and K. Horikoshi. 1995. Electron microscopy of halophilic *Archaea*. In S. DasSarma and E.M. Fleischmann (eds.). *Archaea: A Laboratory Manual: Halophiles*, pp. 29–33. New York: Cold Spring Harbor Laboratory Press.

Nixon, K.C. and Q.D. Wheeler. 1990. An amplification of the phylogenetic species concept. *Cladistics*, **6**: 211–23.

Noack, D. 1986. A regulatory model for steady state conditions in populations of lysogenic bacteria. *Journal of Theoretical Biology*, **18**: 1–18.

Noble, P.A., R.W. Citek, and O.A. Ogunseitan. 1998. Tetranucleotide frequencies in microbial genomes. *Electrophoresis*, **19**: 528–35.

Norton, J.M., J.J. Alzerreca, Y. Suwa, and M.G. Klotz. 2002. Diversity of ammonia monooxygenase operon in autotrophic ammonia-oxidizing bacteria. *Archives of Microbiology*, **177**: 139–49.

Norton, J.M., J.M. Low, and M.G. Klotz. 1996. The gene encoding ammonia monooxygenase subunit A exists in three nearly identical copies in *Nitrosospira sp.* NpAV. *FEMS Microbiology Letters*, **139**: 181–8.

Notomista E., A. Lahm, A. Di Donato, and A. Tramontano. 2003. Evolution of bacterial and archaeal multicomponent monooxygenases. *Journal of Molecular Evolution*, **56**: 435–45.

Nouwens, A.S., S.A. Beatson, C.B. Whitchurch, B.J. Walsh, H.P. Schweizer, J.S. Mattick, and S.J. Cordwell. 2003. Proteome analysis of extracellular proteins regulated by the *las* and *rhl* quorum-sensing systems in *Pseudomonas aeruginosa* PAO1. *Microbiology*, **149**: 1311–22.

NRC. 1999. *Perspectives on Biodiversity: Valuing its Role in an Ever Changing World.* Washington, DC: National Research Council. National Academy Press.

NRC. 2000. *Ecological Indicators for the Nation 2000.* Washington, DC: National Research Council. National Academy Press.

Nriagu, J.O. 1996. The history of global metal pollution. *Science*, **272**: 223–4.

Nuebel U., M.M. Bateson, V. Vandieken, A.Wieland, M. Kuhl, and D.M. Ward. 2002. Microscopic examination of distribution and phenotypic properties of phylogenetically diverse Chloroflexaceae-related bacteria in hot spring microbial mats. *Applied and Environmental Microbiology*, **68**: 4593–603.

Nybroe O., A. Johansen, and M. Laake. 1990. Enzyme-linked immunosorbent assays for detection of Pseudomonas fluorescence in sediment samples. *Letters in Applied Microbiology*, **11**: 293–6.

O'Donnell, A.G., C. Falconer, M. Goodfellow, A.C. Ward, and E. Williams. 1993. Biosystematics and diversity amongst novel carboxydotrophic actinomycetes. *Antonie van Leeuwenhoek*, **64**: 325–40.

O'Morchoea S., O.A. Ogunseitan, G.S. Sayler G.S. and R.V. Miller. 1988. Conjugal transfer of R68.45 and FP5 between *Pseudomonas aeruginosa* in a freshwater environment. *Applied and Environmental Microbiology*, **54**: 1923–9.

Oberg, G. 2002. The natural chlorine cycle: Fitting the scattered pieces. *Applied Microbiology and Biotechnology*, **58**: 565–81.

Ochman, H., S. Elwyn, and N.A. Moran. 1999. Calibrating bacterial evolution. *Proceedings of the National Academy of Sciences (USA)*, **96**: 12638–43.

Ochman, H., J.G. Lawrence, and E.A. Groisman. 2000. Lateral gene transfer and the nature of bacterial innovations. *Nature*, **405**: 299–304.

O'Connor, S.M. and J.D. Coates. 2002. Universal immunoprobe for (per)chlorate-reducing bacteria. *Applied and Environmental Microbiology*, **68**: 3108–13.

OECD. 2002. *Handbook of Biodiversity Valuation.* Paris, France: Organization for Economic Cooperation and Development.

Ogram, A.V. 2000. Soil molecular ecology at age 20: methodological challenges for the future. *Soil Biology and Biochemistry*, **32**: 1499–504.

Ogunseitan, O.A. 1993. Direct extraction of proteins from environmental samples. *Journal of Microbiological Methods*, **17**: 273–81.

Ogunseitan, O.A. 1994. Biochemical, genetic, and ecological approaches to problem solving during in situ and off-site bioremediation. In D.L. Wise and D.J. Trantolo (eds.) *Process Engineering for Pollution Control and Waste Minimization*, pp. 171–92. New York: Marcell-Dekker.

Ogunseitan, O.A. 1995. Bacterial genetic exchange in nature. *Science Progress*, **78**: 183–204.

Ogunseitan, O.A. 1996. Protein profile variation in cultivated and native freshwater microorganisms exposed to chemical environmental pollutants. *Microbial Ecology*, **31**: 291–304.

Ogunseitan, O.A. 1997. Direct extraction of catalytic proteins from natural microbial communities. *Journal of Microbiological Methods*, **28**: 55–63.

Ogunseitan, O.A. 1998. Protein method for investigating mercuric reductase gene expression in aquatic environments. *Applied and Environmental Microbiology*, **64**: 695–702.

Ogunseitan, O.A. 1998a. Protein profile analysis for investigating genetic functions in microbial communities. In K. Cooksey (ed.) *Molecular Approaches to the Study of the Ocean.* London: Chapman and Hall.

Ogunseitan, O.A. 1998b. Extraction of proteins from aquatic and soil sources. In A.D.L. Ackkermans, J.D. van Elsas, and F.J. De Bruijn (eds.) *Molecular Microbial Ecology*, Chapter 4.1.6. The Netherlands: Kluwer Academic.

Ogunseitan, O.A. 2000. Microbial proteins as biomarkers of ecosystem health. In K. Scow, G.E. Fogg, D. Hinton, and M.L. Johnson (eds.) *Integrated Assessment of Ecosystem Health*, pp. 207–22. Boca Raton, FL: CRC Press.

Ogunseitan, O.A. 2002. Assessing microbial proteomes in the environment. In G. Bitton (ed.) *Encyclopedia of Environmental Microbiology.* New York: Wiley.

Ogunseitan, O.A. 2002a. Global eradication of smallpox. In K.R. Rasmussen (ed.) *Great Events of the Twentieth Century*, pp. 1974–5. Pasadena, CA: Salem Press.

Ogunseitan, O.A. 2002b. Episodic bioavailability of environmental mercury: Implications for the biotechnological control of mercury pollution. *African Journal of Biotechnology*, **1**: 1–9.

Ogunseitan, O.A. 2004. Assessing microbial proteomes in the environment. In G. Bitton (ed.) *Encyclopedia of Environmental Microbiology*, pp. 305–12. New York: Wiley.

Ogunseitan, O.A., J. LeBlanc, and E. Dalmasso. 2001. Microbial community proteomics. In P.A. Rochelle (ed.) *Environmental Molecular Microbiology*, pp. 125–40. Norfolk, England: Horizon Scientific Press.

Ogunseitan, O.A., J.F. LeBlanc, and P.A. Noble. 2002. Ecological dimensions of microbial proteomics. *Recent Research Developments in Microbiology*, **6**: 487–501.

Ogunseitan, O.A., G.S. Sayler, and R.V. Miller. 1990. Dynamic interactions between *Pseudomonas aeruginosa* and bacteriophages in freshwater. *Microbial Ecology*, **19**: 171–85.

Ogunseitan, O.A., G.S. Sayler, and R.V. Miller. 1992. Application of

DNA probes to analysis of bacteriophage distribution patterns in the environment. *Applied and Environmental Microbiology*, **58**: 2046–52.

Ogunseitan, O.A., S. Yang, and J. Ericson. 2000. Microbial delta-aminolevulinate dehydrogenase activity as a biosensor for lead (Pb) bioavailability in polluted environments. *Soil Biology and Biochemistry*, **32**: 1899–906.

Ohkuma, M., S. Noda, and K. Toshiaki. 1999. Phylogenetic diversity of nitrogen fixation genes in the symbiotic microbial community in the gut of diverse termites. *Applied and Environmental Microbiology*, **65**: 4926–34.

Olendzenski, L., O. Zhaxybayeva, and J.P. Gogarten. 2001. How much did horizontal gene transfer contribute to early evolution? Quantifying Archaeal genes in two bacterial lineages. *Astrobiology*, **1**: 404–5.

Olivier N., P. Hantzpergue, C. Gaillard, B. Pittet, R.R. Leinfelder, D.U. Schmid, and W. Werner. 2003. Microbialite morphology, structure and growth: A model of the Upper Jurassic reefs of the Chay Peninsula (Western France). *Palaeogeography Palaeoclimatology Palaeoecology*, **193**: 383–404.

Olson, B.H., O.A. Ogunseitan, P.A. Rochelle, C. Tebbe, and Y.-L. Tsai. 1991. The implications of horizontal gene transfer in the environmental impact of genetically engineered microorganisms. In M. Levin and H. Strauss (eds.) *Risk Assessment in Genetic Engineering: Environmental Release of Organisms*. New York: McGraw-Hill.

Omland, K.E. 1997. Correlated rates of molecular and morphological evolution. *Evolution*, **51**: 1381–93.

Onda, S. and S. Takii. 2002. Isolation and characterization of a Gram-positive polyphosphate-accumulating bacterium. *Journal of General and Applied Microbiology*, **48**: 125–33.

Ong, C., M.L.Y. Wong, and J. Smit. 1990. Attachment of the adhesive holdfast organelle to the cellular stalk of *Caulobacter crescentus*. *Journal of Bacteriology*, **172**: 1448–56.

Oparin, A.I. 1953. *The Origin of Life*. Second Edition. New York: Dover.

Oren, A., A. Ventosa, M.C. Gutierrez, and M. Kamekura. 1999. *Haloarcula quadrata* sp. *nov*., a square, motile archaeon isolated from a brine pool in Sinai (Egypt). *International Journal of Systematic Bacteriology*, **49**: 1149–55.

Oro, J., S.L. Miller, and A. Lazcano. 1990. The origin and early evolution of life on Earth. *Annual Review of Earth and Planetary Sciences*, **18**: 317–63.

Orphan, V.J., C.H. House, K.-U. Hinrichs, K.D. McKeegan, and E.F. DeLong. 2002. Multiple archaeal groups mediate methane oxidation in anoxic cold seep sediments. *Proceedings of the National Academy of Sciences (USA)*, **99**: 7663–8.

Otsuka, J., G. Terai, and T. Nakano. 1999. Phylogeny of organisms investigated by the base-pair changes in the stem regions of small and large ribosomal subunit RNAs. *Journal of Molecular Evolution*, **48**: 218–35.

Ouverney, C.C. and J.A. Fuhrman, J.A. 1999. Combined microautoradiography-16S rRNA probe technique for determination of radioisotope uptake by specific microbial cell types *in situ*. *Applied and Environmental Microbiology*, **65**: 1746–52.

Pace, N.R. 1997. A molecular view of microbial diversity and the biosphere. *Science*, **276**: 734–40.

Paddock, S.W. (ed.) 1999. *Confocal Microscopy Methods and Protocols*. Totowa, NJ: Humana Press.

Paerl, H.W., J. Dyble, P.H. Moisander, R.T. Noble, M.F. Piehler, J.L. Pinckney, T.F. Steppe, L. Twomey, and L.M. Valdes. 2003. Microbial indicators of aquatic ecosystem change: Current applications to eutrophication studies. *FEMS Microbiology Ecology*, **1561**: 1–14.

Paerl, H.W. and J.L. Pinkney. 1996. A mini-review of microbial consortia: Their roles in aquatic production and biogeochemical cycling. *Microbial Ecology*, **31**: 225–47.

Paget, E. and P. Simonet. 1994. On the track of natural transformation in soil. *FEMS Microbiology Ecology*, **15**: 109–17.

Paggi, R.A., C.B. Martone, C. Fuqua, and R.E. De Castro. 2003. Detection of quorum sensing signals in the haloalkaliphilic archaeon *Natronococcus occultus*. *FEMS Microbiology Letters*, **221**: 49–52.

Palleroni, N.J. 1997. Prokaryotic diversity and the importance of culturing. *Antonie van Leeuwenhoek*, **72**: 3–19.

Palumbo, A.V., C. Zhang, S. Liu, S.P. Scarborough, S.M. Pfiffner, and T.J. Phelps. 1996. Influence of media on measurement of bacterial populations in the subsurface: Numbers and diversity. *Applied Biochemistry and Biotechnology*, **57–8**: 905–14.

Palys, T., L.K. Nakamura, and F.M. Cohan. 1997. Discovery and classification of ecological diversity in the bacterial world: The role of DNA sequence data. *International Journal of Systematic Bacteriology*, **47**: 1145–56.

Pancost R.D., D.J.S. Sinninghe, S. de Lint, M.J.E. van der Maarel, J.C. Gottschal, and the Medinaut Shipboard Scientific Party. 2000. Biomarker evidence for widespread anaerobic methane oxidation in Mediterranean sediments by a consortium of methanogenic archaea and bacteria. *Applied and Environmental Microbiology*, **66**: 1126–32.

Panikov N.S. 1999. Fluxes of CO_2 and CH_4 in high latitude wetlands: Measuring, modeling, and predicting response to climate change. *Polar Research*, **18**: 237–44.

Pankhurst, C.E., S. Yu, B.G. Hawke, and B.D. Harch. 2001. Capacity of fatty acid profiles and substrate utilization patterns to describe differences in soil microbial communities associated with increased salinity or alkalinity at three locations in South Australia. *Biology and Fertility of Soils*, **33**: 204–17.

Pantastica-Caldas, M., K.E. Duncan, C.A. Istock, and J.A. Bell. 1992. Population dynamics of bacteriophage and *Bacillus subtilis* in soil. *Ecology*, **73**: 1888–902.

Papapetropoulou, M. and S. Sotiracopoulou. 1994. Effect of sea bathing on human skin microorganisms. *Microbial Ecology in Health and Disease*, **7**: 105–9.

Park, J.C., J.C. Lee, J.Y. Oh, Y.W. Jeong, J.W. Cho, H.S. Joo, W.K. Lee, and W.B. Lee. 2003a. Antibiotic selective pressure for the maintenance of antibiotic resistant genes in coliform bacteria isolated from the aquatic environment. *Water Science and Technology*, **47**: 249–53.

Park, S., P.M. Wolanin, E.A. Yuzbashyan, P. Silberzan, J.B. Stock, and R.H. Austin. 2003b. Motion to form a quorum. *Science*, **301**: 188.

Park, S.W., E.H. Hwang, H. Park, J.A. Kim, J. Heo, K.H. Lee, T. Song, E. Kim, Y.T. Ro, S.W. Kim, and Y.M. Kim. 2003. Growth of *mycobacteria* on carbon monoxide and methanol. *Journal of Bacteriology*, **185**: 142–7.

Pastor J., J. Solin, S.D. Bridgham, K. Updegraff, C. Harth, P. Weishampel, and B. Dewey. 2003. Global warming and the export of dissolved organic carbon from boreal peatlands. *Oikos*, **100**: 380–6.

Patil, G.P., R.P. Brooks, W.L. Myers, D.J. Rapport, and C. Taillie. 2001. Ecosystem health and its measurement at landscape scale: Toward the next generation of quantitative assessments. *Ecosystem Health*, **7**: 307–16.

Patz, J., W.J.M. Martens, D.A. Focks, and T.H. Jetten. 1998. Dengue fever epidemic potential as projected by general circulation models of global climate change. *Environmental Health Perspectives*, **106**: 147–53.

Paul, J.H. 1993. The advances and limitations of methodology. In T.E. Ford (ed.) *Aquatic Microbiology*, pp. 15–46. Boston, MA: Blackwell Scientific.

Paul, J.H., M.B. Sullivan, A.M. Segall, and F. Rohwer. 2002. Marine phage genomics. *Comparative Biochemistry and Physiology. Part B, Biochemistry and Molecular Biology*, **133B**: 463–76.

Pauli, A.S.L. and S. Kaitala. 1997. Phosphate uptake kinetics by *Acinetobacter* isolates. *Biotechnology and Bioengineering*, **53**: 304–9.

Perezgasga, L. and E. Silva. 2003. Alfonso L. Herrera and plasmogenesis: A Mexican theory on the origin of life. *Origins of Life and Evolution of the Biosphere*, **33**: 253.

Perkins, J.L. 1982. Shannon–Weaver or Shannon–Weiner? *Journal of Water Pollution Control Federation*, **54**: 1049–50.

Perlman, D. and G. Adelson. 1997. *Biodiversity: Exploring Values and Priorities in Conservation.* Oxford, England: Blackwell.

Petri, R.J. 1887. Eine kleine Modification des Koch'schen Plattenverfahrens. *Centralblatt Fur Bakteriologie*, **1**: 279–80. In T.D. Brock (trans. and ed.) 1998. *Milestones in Microbiology: 1556 to 1940*, p. 218. Washington, DC: American Society for Microbiology Press.

Petursdottir, S.K., G.O. Hreggvidsson, M.S. Da Costa, and J.K. Kristjansson. 2000. Genetic diversity analysis of *Rhodothermus* reflects geographical origin of the isolates. *Extremophiles*, **4**: 267–74.

Pfeiffer, T. and S. Bonhoeffer. 2003. An evolutionary scenario for the transition to undifferentiated multicellularity. *Proceedings of the National Academy of Sciences (USA)*, **100**: 1095–8.

Phillips, R.L., S.C. Whalen, and W.H. Schlesinger. 2001. Response of soil methanotrophic activity to carbon dioxide enrichment in a North Carolina coniferous forest. *Soil Biology and Biochemistry*, **33**: 793–800.

Pielou, E.C. 1966a. Shannon's formula as a measure of species diversity: Its use and misuse. *American Naturalist*, **100**: 463–5.

Pielou, E.C. 1966b. The measurement of diversity in different types of biological collections. *Journal of Theoretical Biology*, **13**: 131–44.

Pierson, B.K. and J.M. Olson. 1987. Photosynthetic bacteria. In J. Amesz (ed.) *Photosynthesis*, pp. 21–42. Amsterdam: Elsevier.

Pierson B.K. and M.N. Parenteau. 2000. Phototrophs in high iron microbial mats: Microstructure of mats in iron-depositing hot springs. *FEMS Microbiology Ecology*, **32**: 181–96.

Pimm, S. and P. Raven. 2000. Extinction by numbers. *Nature*, **403**: 843–5.

Pinkart, H.C., D.B. Ringelberg, Y.M. Piceno, S.J. MacNaughton, and D.C. White. 2002. Biochemical approaches to biomass measurements and community structure analysis. In C.J. Hurst, R.L. Crawford, R. Knudsen, M.J. McInerney, and L.D. Stetzenbach (eds.) *Manual of Environmental Microbiology* (Second Edition), pp. 101–13. Washington, DC: ASM Press.

Platt, U. and G. Honninger. 2003. The role of halogen species in the troposphere. *Chemosphere*, **52**: 325–38.

Poinar, G.O., B.J. Waggoner, and U-C. Bauer. 1993. Terrestrial soft-bodied protists and other microorganisms in Triassic amber. *Science*, **259**: 222–4.

Ponnamperuma, C. 1992. Cosmochemical evolution and the origins of life. In L. Margulis and L. Olendzenski (eds.) *Environmental Evolution*, pp. 17–27. Cambridge, MA: MIT Press.

Porter, K.G. and Y.S. Feig. 1980. The use of DAPI for identifying and counting aquatic microflora. *Limnology and Oceanography*, **25**: 943–8.

Pritchard, P.H., J.G. Mueller, S.E. Lantz, and D.L. Santavy. 1995. The potential importance of biodiversity in environmental biotechnology applications: Bioremediation of PAH-contaminated soils and sediments. In D. Allsop, R.R. Colwell, and D.L. Hawksworth (eds.) *Microbial Diversity and Ecosystem Function*, pp. 161–82. Wallingford, England: United Nations Environment Program and Center for Agriculture and Biotechnology International.

Prusiner, S.B. 1995. The prion diseases. *Scientific American*, **272**: 30–7.

Purdey, M. 1998. High dose exposure to systemic phosmet insecticide modifies the phosphatidy-linositol anchor on the prion protein: The origins of new variant transmissible spongiform encephalopathies? *Medical Hypotheses*, **50**: 91–111.

Purdey, M. 2000. Ecosystems supporting clusters of sporadic TSEs demonstrate excesses of the radical generating divalent cation manganese. *Medical Hypotheses*, **54**: 278–306.

Purdey, M. 2001. Does an ultraviolet photooxidation of manganese loaded copper depleted prion protein in the retina initiate the pathogenesis of TSE? *Medical Hypothesis*, **57**: 29–45.

Purdey, M. 2003. Does an infrasonic acoustic shock wave resonance of the manganese 3 + loaded/copper depleted prion protein initiate the pathogenesis of TSE? *Medical Hypotheses*, **60**: 797–820.

Purdon, E. 2000. *The Irish Famine, 1845–52.* Cork, Ireland: Mercier Press.

Purdy, K.J., D.B. Nedwell, and T.M. Embley. 2003. Analysis of the sulfate-reducing bacterial and methanogenic archaeal populations in contrasting Antarctic sediments. *Applied and Environmental Microbiology*, **69**: 3181–91.

Purkhold, U., A. Pommerening-Röser, S. Juretschko, M.C. Schmid, H.-P. Koops, and M. Wagner. 2000. Phylogeny of all recognized species of ammonia oxidizers based on comparative 16S rRNA and *amoA* sequence analysis: Implications for molecular diversity surveys. *Applied and Environmental Microbiology*, **66**: 5368–82.

Purvis, A. and A. Hector. 2000. Getting the measure of biodiversity. *Nature*, **405**: 212–19.

Raaijmaker, J.M., D.M. Weller, and L.S. Thomashow. 1997. Frequency of antibiotic-producing *Pseudomonas* spp. in natural environments. *Applied and Environmental Microbiology*, **63**: 881–7.

Radajewski, S., P. Ineson, N. Parekh, and J.C. Murrell. 2000. Stable-isotope probing as a tool in microbial ecology. *Nature*, **403**: 646–9.

Rambaut, A. and L. Bromham. 1998. Estimating divergence dates from molecular sequences. *Molecular Biology and Evolution*, **15**: 442–8.

Rannala, B. and Z.H. Yang. 1996. Probability distribution of molecular evolutionary trees: A new method of phylogenetic inference. *Journal of Molecular Evolution*, **43**: 304–11.

Raskin, L., J.M. Stromley, B.E. Rittmann, and D.A. Stahl. 1994. Group-specific 16S rRNA hybridization probes to describe natural communities of methanogens. *Applied and Environmental Microbiology*, **60**: 1232–40.

Rasmussen, B. 2000. Filamentous microfossils in a 3,235-million-year-old volcanogenic massive sulphide deposit. *Nature*, **405**: 676–9.

Ravin, A.W. 1960. The origin of bacterial species: Genetic recombination and factors limiting it between bacterial populations. *Bacteriological Review*, **24**: 201–9.

Ravin, A.W. 1963. Experimental approaches to the study of bacterial phylogeny. *American Naturalist*, **97**: 307–18.

Reeburgh, W.S. 1980. Anaerobic methane oxidation: Rate versus depth distribution in San Bay sediments. *Earth and Planetary Science Letters*, **47**: 345–52.

Reeburgh, W.S. 1982. A major sink and flux control for methane in marine sediments: Anaerobic consumption. In K. Fanning and F.T. Manheim (eds.) *The Dynamic Environment of the Ocean Floor*, pp. 203–17. Lexington, MA: Heath.

Rees, G.N., G. Vasiliadis, J.W. May, and R.C. Bayly. 1993. Production of poly-beta-hydroxybutyrate in *Acinetobacter* spp. isolated from activated sludge. *Applied Microbiology and Biotechnology*, **38**: 734–7.

Regan, J.M., G.W. Harrington, H. Baribeau, R. De Leon, and D.R. Noguera. 2003. Diversity of nitrifying bacteria in full-scale chloraminated distribution systems. *Water Research*, **37**: 197–205.

Reid, S.D., N.M. Green, J.K. Buss, B. Lei, and J.M. Musser. 2001. Multilocus analysis of extracellular putative virulence proteins made by group A Streptococcus: Population genetics, human serologic response, and gene transcription. *Proceedings of the National Academy of Sciences (USA)*, **98**: 7552–7.

Reid, S.D., C.J. Herbelin, A.C. Bumbaugh, R.K. Selander, and T.S. Whittam. 2000. Parallel evolution of virulence in pathogenic *Escherichia coli*. *Nature*, **406**: 64–7.

Relman, D.A. 1999. The search for unrecognized pathogens. *Science*, **284**: 1308.

Reysenbach, A.-L. and E. Shock. 2002. Merging genomes with geochemistry in hydrothermal ecosystems. *Science*, **296**: 1077–82.

Rhee, S.-K., D.E. Fennell, M.M. Haggblom, and L.J. Kerkhof. 2003. Detection by PCR of reductive dehalogenase motifs in a sulfidogenic 2-bromophenol-degrading consortium enriched from estuarine sediment. *FEMS Microbiology Ecology*, **43**: 317–24.

Rhie, G.-E., Y.J. Avissar, and S.I. Beale, 1996. Structure and expression of the *Chlorobium vibrioforme* hemB gene and characterization of its encoded enzyme, porphobilinogen synthase. *Journal of Biological Chemistry*, **271**: 8176–82.

Ribbe, M., D. Gadkari, and O. Meyer. 1997. N_2 fixation by *Streptomyces thermoautotrophicus* involves a molybdenum-dinitrogenase and a manganese-superoxide oxidoreductase that couple N_2 reduction to the oxidation of superoxide produced from O_2 by a molybdenum-CO dehydrogenase. *Journal of Biological Chemistry*, **272**: 26627–33.

Rice, S.A., M. Givskov, P. Steinberg, and S. Kjelleberg. 1999. Bacterial signals and antagonists: The interaction between bacteria and higher organisms. *Journal of Molecular Microbiology and Biotechnology*, **1**: 23–31.

Richardson, L. and K.G. Kuta. 2003. Ecological physiology of the black-band disease cyanobacterium *Phormidium corallyticum*. *FEMS Microbiology Ecology*, **43**: 287–98.

Richie, D.A., C. Edwards, I.R. McDonald, and J.C. Murrell. 1997. Detection of methanogens and methanotrophs in natural environments. *Global Change Biology*, **3**: 339–50.

Riding R. 1999. The term stromatolite: Towards an essential definition. *Lethaia*, **32**: 321–30.

Rillig, M.C., P.W. Ramsey, S. Morris, and E.A. Paul. 2003. Glomalin, an arbuscular-mycorrhizal fungal soil protein, responds to land-use change. *Plant and Soil*, **253**: 293–9.

Rillig, M.C., S.F. Wright, and V.T. Eviner. 2002b. The role of arbuscular mycorrhizal fungi and glomalin in soil aggregation: Comparing effects of five plant species. *Plant and Soil*, **238**: 325–33.

Rillig, M.C., S.F. Wright, M.R. Shaw, and C.B. Field. 2002a. Artificial climate warming positively affects arbuscular mycorrhizae but decreases soil aggregate water stability in an annual grassland. *Oikos*, **97**: 52–8.

Rios, M., J.M. Garcia, J.A. Sanchez, and D. Perez. 2000. A statistical analysis of the seasonality in pulmonary tuberculosis. *European Journal of Epidemiology*, **16**: 483–8.

Ripp S., O.A. Ogunseitan, and R.V. Miller. 1994. Transduction of a freshwater microbial community by a new *Pseudomonas aeruginosa* generalized transducing phage, UT1. *Molecular Ecology*, **3**: 121–6.

Ritchie, K.B. and G.B. Smith. 1995. Preferential carbon utilization by surface bacterial communities from water mass, normal, and white-band diseased *Acropora cervicornis*. *Journal of Marine Biotechnology*, **4**: 345–52.

Robbins, L. 2002. *Louis Pasteur and the Hidden World of Microbes*. Oxford, England: Oxford University Press.

Robertson, B.H. and H.S. Margolis. 2002. Primate hepatitis B viruses: Genetic diversity, geography and evolution. *Reviews in Medical Virology*, **12**: 133–41.

Robertson, K., L. Klemedtsson, S. Axelsson, and T. Rosswall. 1993. Estimates of denitrification in soil by remote sensing of thermal infrared emission at different moisture levels. *Biology and Fertility of Soils*, **16**: 93–197.

Robertson, L.A. and J.G. Kuenen. 1991. Physiology of nitrifying and denitrifying bacteria. In J.E. Rogers and W.B. Whitman (eds.) *Microbial Production and Consumption of Greenhouse Gases: Methane, Nitrogen Oxides, and Halomethanes*, pp. 189–99. Washington, DC: American Society for Microbiology.

Robinson, N.E. and A.B. Robinson. 2001. Molecular clocks. *Proceedings of the National Academy of Sciences (USA)*, **98**: 944–9.

Rocheleau, S., C.W. Greer, J.R. Lawrence, C. Cantin, L. Laramee, and S.R. Guiot. 1999. Differentiation of *Methanosaeta concilii* and *Methanosarcina barkeri* in anaerobic mesophilic granular sludge by fluorescent in situ hybridization and confocal laser microscopy. *Applied and Environmental Microbiology*, **65**: 2222–9.

Rodrigo, A.G., P.R. Bergquist, and P.L. Bergquist. 1994. Inadequate support for an evolutionary link between the metazoa and the fungi. *Systematic Biology*, **43**: 578–84.

Roelofs, G.J., P. Kasibhatla, L. Barrie, D. Bergmann, C. Bridgeman, M. Chin, J. Christensen, R. Easter, J. Feichter, A. Jeuken, E. Kjellstrom, D. Koch, C. Land, U. Lohmann, and P. Rasch. 2001. Analysis of regional budgets of sulfur species modeled for the COSAM exercise. *Tellus* (Series B: Chemical and Physical Meteorology), **53B**: 673–94.

Roels, J. and W. Verstraete. 2001. Biological formation of volatile phosphorus compounds (a review). *Bioresource Technology*, **79**: 243–50.

Roesler, C.S., C.W. Culbertson, S.M. Etheridge, R. Goericke, R.P. Kiene, L.G. Miller, and R.S. Oremland. 2002. Distribution, production, and ecophysiology of *Picocystis* strain ML in Mono Lake, California. *Limnology and Oceanography*, **47**: 440–52.

Rosenzweig, R.F., R.R. Sharp, D.S. Treves, and J. Adams. 1994. Microbial evolution in a simple unstructured environment: Genetic differentiation in *Escherichia coli*. *Genetics*, **137**: 903–17.

Rossello-Mora, R. and R. Amann. 2001. The species concept for prokaryotes. *FEMS Microbiology Reviews*, **25**: 39–67.

Rossi, F., P. Vecchia, and F. Masoero. 2001. Estimate of methane production from rumen fermentation. *Nutrient Cycling in Agroecosystems*, **60**: 89–92.

Rothermich, M.M., R. Guerrero, R.W. Lenz, and S. Goodwin. 2000. Characterization, seasonal occurrence, and diel fluctuation of poly(hydroxyalkanoate) in photosynthetic microbial mats. *Applied and Environmental Microbiology*, **66**: 4279–91.

Rowan, R., N. Knowlton, and J. Jara. 1997. Landscape ecology of algal symbionts creates variation in episodes of coral bleaching. *Nature*, **388**: 265–9.

Ruska, E. 1980. *The Early Development of Electron Lenses and Electron Microscopy*, pp. 113–16. Stuttgart: Verlag.

Sabater S. 2000. Structure and architecture of a stromatolite from a Mediterranean stream. *Aquatic Microbial Ecology*, **21**: 161–8.

Saha, U., and M. Sen. 1989. Methane oxidation by *Candida tropicalis. National Academy of Science Letters (India)*, **12**: 373–6.

Salyers, A. and D.D. Whitt. 2001. *Microbiology: Diversity, Disease, and the Environment.* Bethesda, MD: Fitzgerald Science Press.

Sarmiento, J.L. and M. Bender. 1994. Carbon biogeochemistry and climate change. *Photosynthesis Research*, **39**: 209–34.

Saye, D.J., O.A. Ogunseitan, G.S Sayler, and R.V. Miller. 1987. Potential for transfer of plasmids in a natural freshwater environment: Effect of plasmid donor concentration and natural freshwater community on transduction in *Pseudomonas aeruginosa. Applied and Environmental Microbiology*, **53**: 987–95.

Saye, D.J., O.A. Ogunseitan, G.S. Sayler, and R.V. Miller. 1990. Transduction of linked chromosomal genes between *Pseudomonas aeruginosa* during incubation in situ in a freshwater habitat. *Applied and Environmental Microbiology*, **56**: 140–5.

Schaule, G., H.C. Flemming, and H.F. Ridgway. 1993. Use of 5-cyano-2, 3-ditolyl chloride for quantifying planktonic and sessile respiring bacteria in drinking water. *Applied and Environmental Microbiology*, **59**: 3850–7.

Scheer, H. 1991. Structure and occurrence of chlorophylls. In H. Scheer (ed.) *Chlorophylls*, pp. 3–20. Boca Raton, FL: CRC Press.

Scherer, P. 1989. Vanadium and molybdenum requirement for the fixation of molecular nitrogen by two *Methanosarcina* strains. *Archives of Microbiology*, **151**: 44–8.

Scherer, S. 1990. The protein molecular clock: Time for a reevaluation. *Evolutionary Biology*, **24**: 83–106.

Schierup, M.H. and J. Hein. 2000. Consequences of recombination on traditional phylogenetic analysis. *Genetics*, **156**: 879–91.

Schlesinger, W.H. 1991. *Biogeochemistry: An Analysis of Global Change.* San Diego: Academic Press.

Schlessman, J.L., D. Woo, L. Joshua-Tor, J.B. Howard, and D.C. Rees. 1998. Conformational variability in structures of the nitrogenase iron proteins from *Azotobacter vinelandii* and *Clostridium pasteurianum. Journal of Molecular Biology*, **280**: 669–85.

Schmid, B., O. Einsle, H.J. Chiu, A. Willing, M. Yoshida, J.B. Howard, and D.C. Rees. 2002. Biochemical and structural characterization of the cross-linked complex of nitrogenase: comparison to the ADP-AlF4(-)-stabilized structure. *Biochemistry*, **41**: 15557–65.

Schmidt, A.R., H. von Eynatten, and M. Wagreich. 2001. The Mesozoic amber of Schliersee (southern Germany) is Cretaceous in age. *Cretaceous Research*, **22**: 423–8.

Schmidt, H.A., K. Strimmer, M. Vingron, and A. von Haeseler. 2002. TREE-PUZZLE: Maximum likelihood phylogenetic analysis using quartets and parallel computing. *Bioinformatics (Oxford)*, **18**: 502–4.

Schmidt, I. and E. Bock. 1997. Anaerobic ammonia oxidation with nitrogen dioxide by *Nitrosomonas eutropha. Archives of Microbiology*, **167**: 106–11.

Schmidt, I., C. Hermelink, K. van de Pas-Schoonen Katinka, M. Strous, H.J. op den Camp, J.G. Kuenen, and M.S.M. Jetten. 2002. Anaerobic ammonia oxidation in the presence of nitrogen oxides (NOx) by two different lithotrophs. *Applied and Environmental Microbiology*, **68**: 5351–7.

Scholten, J.C.M., J.C. Murrell, and D.P. Kelly. 2003. Growth of sulfate-reducing bacteria and methanogenic archaea with methylated sulfur compounds: A commentary on the thermodynamic aspects. *Archives of Microbiology*, **179**: 135–44.

Schuster, M., C.P. Lostroh, T. Ogi, and E.P. Greenberg. 2003. Identification, timing, and signal specificity of *Pseudomonas aeruginosa* quorum-controlled genes: A transcriptome analysis. *Journal of Bacteriology*, **185**: 2066–79.

Schweiger, F. and C.C. Tebbe. 1998. A new approach to utilize PCR-single-strand-conformation polymorphism for 16S rRNA gene-based microbial community analysis. *Applied and Environmental Microbiology*, **64**: 4870–6.

Scott J., G. Anthony, A.J. Hall, R. Dagan, J.M.S. Dixon, S.J. Eykyn, A. Fenoll, M. Hortal, L P. Jette, J.H. Jorgensen, F. Lamothe, C. Latorre, J.T. Macfarlane, D.M. Shlaes, L.E. Smart, and A. Taunay. 1996. Serogroup-specific epidemiology of *Streptococcus pneumoniae*: Associations with age, sex, and geography in 7,000 episodes of invasive disease. *Clinical Infectious Diseases*, **22**: 973–81.

Scott, W.G. 1999. RNA structure and catalysis. *Quarterly Reviews of Biophysics*, **32**: 241–84.

Scott, W.G., J.B. Murray, B.L. Stoddard, and A. Klug. 1996. Capturing the structure of a catalytic RNA intermediate: The hammerhead ribozyme. *Science*, **274**: 2065–9.

Seckbach, J. (ed.) 1999. *Enigmatic Microorganisms and Life in Extreme Environments.* Dordrecht: Kluwer Academic.

Seckbach, J. (ed.) 2002. *Symbiosis: Mechanisms and Model Systems.* Dordrecht: Kluwer Academic.

Selander, R.K., D.A. Caugant, H. Ochman, J.M. Musser, N. Gilmour, and T.S. Whittam. 1986. Methods of multilocus enzyme electrophoresis for bacterial population genetics and systematics. *Applied and Environmental Microbiology*, **51**: 873–84.

Sellers, P.J., R.E. Dickinson, D.A. Randall, A.K. Betts, F.G. Hall, J.A. Berry, G.J. Collatz, A.S. Denning, H.A. Mooney, C.A. Nobre, N. Sato, C.B. Field, and A. Henderson-Sellers. 1997. Modeling the exchanges of energy, water, and carbon between continents and the atmosphere. *Science*, **275**: 502–9.

Sepers, A.B.J. 1981. Diversity of ammonifying bacteria. *Hydrobiologia*, **83**: 343–50.

Setlow, R.B. 1974. The wavelengths in sunlight effective in producing cancer: A theoretical analysis. *Proceedings of the National Academy of Sciences (USA)*, **71**: 3363–6.

Setlow, R.B., E. Grist, K. Thompson, and A.D. Woodhead. 1993. Wavelengths effective in induction of malignant melanoma. *Proceedings of the National Academy of Sciences (USA)*, **90**: 6666–70.

Shannon, C.E. 1948. A mathematical theory of communication. *Bell System Technical Journal*, **27**: 379–424 and 623–56.

Shapiro, J.A. and M. Dworkin (eds.) 1997. *Bacteria as Multicellular Organisms.* New York: Oxford University Press.

Sheehan, J. and M.E. Himmel. 1999. Enzymes, energy, and the environment: Cellulase development in the emerging bioethanol industry. *Biotechnology Progress*, **15**: 817–27.

Sherr, E.B. and B.F. Sherr. 2002. Phagotrophy in aquatic microbial food webs. In C.J. Hurst, R.L. Crawford, G.R. Knudsen, M.J. McInerney, and L.D. Stetzenbach (eds.) *Manual of Environmental Microbiology* (Second Edition), pp. 409–18. Washington, DC: American Society for Microbiology.

Shima, S., E. Warkentin, R.K. Thauer, and U. Ermler. 2002. Structure and function of enzymes involved in the methanogenic pathway utilizing carbon dioxide and molecular hydrogen. *Journal of Bioscience and Bioengineering*, **93**: 519–30.

Shreeve, J. 1996. Are algae – not coral – reef's master builders? *Science*, **271**: 597–8.

Sicheritz-Ponten, T. and S.G.E. Andersson. 2001. A phylogenomic approach to microbial evolution. *Nucleic Acids Research*, **29**: 545–52.

Silva, S.R., C. Kendall, D.H. Wilkison, A.C. Ziegler, C.C.Y. Chang, and R. Avanzino. 2000. A new method for collection of nitrate from fresh water and the analysis of nitrogen and oxygen isotope ratios. *Journal of Hydrology*, **228**: 22–36.

Simo, R. 2001. Production of atmospheric sulfur by oceanic plankton: Biogeochemical, ecological, and evolutionary links. *Trends in Ecology and Evolution*, **16**: 287–94.

Simpson, G.G. 1951. The species concept. *Evolution*, **5**: 285–98.

Skujins, J. and B. Klubek. 1982. Soil biological properties of a montane forest sere: Corroboration of Odum's postulates. *Soil Biology and Biochemistry*, **14**: 505–14.

Slater, J.H., R. Whittenbury, and J.W.T. Wimpenny (eds.) 1983. *Microbes in their Natural Environments.* Thirty-fourth symposium of the Society for General Microbiology held at the University of Warwick. New York: Published for the Society for General Microbiology by Cambridge University Press.

Smit, J., C.S. Sherwood, and R.F.B. Turner. 2000. Characterization of high-density monolayers of the biofilm bacterium *Caulobacter crescentus*: Evaluating prospects for developing immobilized cell bioreactors. *Canadian Journal of Microbiology*, **46**: 339–49.

Smith, D.R., L.A. Doucette-Stamm, C. Deloughery, H.-M. Lee, J. Dubois, T. Aldredge, R. Bashirzadeh, D. Blakely, R. Cook, K. Gilbert, D. Harrison, L. Hoang, P. Keagle, W. Lumm, B. Pothier, D. Qiu, R. Spadafora, R. Vicare, Y. Wang, J. Wierzbowski, R. Gibson, N. Jiwani, A. Caruso, D. Bush, H. Safer, D. Patwell, S. Prabhakar, S. McDougall, G. Shimer, A. Goyal, S. Pietrovski, G.M. Church, C.J. Daniels, J.-I. Mao, P. Rice, J. Nolling, and J.N. Reeve. 2001. Complete genome sequence of *Methanobacterium thermoautotrophicum* delta H: Functional analysis and comparative genomics. *Journal of Bacteriology*, **179**: 7135–55.

Smith, T.J., S.E. Slade, N.P. Burton, J.C. Murrell, and H. Dalton. 2002. Improved system for protein engineering of the hydroxylase component of soluble methane monooxygenase. *Applied and Environmental Microbiology*, **68**: 5265–73.

Smyth, E.T. and A.M. Emmerson. 2000. Geography is destiny: Global nosocomial infection control. *Current Opinion in Infectious Diseases*, **13**: 371–5.

Snellinx, Z., S. Taghavi, J. Vangronsveld, and D. van der Lelie. 2003. Microbial consortia that degrade 2,4-DNT by interspecies metabolism: Isolation and characterization. *Biodegradation*, **14**: 19–29.

Sonea, S. 1988. The global organism: A new view of bacteria. *The Sciences*, **28**: 38–45.

Song, B. and B.B. Ward. 2003. Nitrite reductase genes in halobenzoate degrading denitrifying bacteria. *FEMS Microbiology Ecology*, **43**: 349–57.

Sorderlund, R. and B.H. Svensson. 1976. The global nitrogen cycle. *Global Cycles, Scope Report 7. Bulletin of the Ecological Research Commission (Stockholm)*, **22**: 23–73.

Souza, V., A. Castillo, and L.E. Eguiarte. 2002. The evolutionary ecology of *Escherichia coli. American Scientist*, **90**: 332–41.

Souza, V., P.E. Turner, and R.E. Lenski. 1997. Long-term experimental evolution in *Escherichia coli.* V. Effects of recombination with immigrant genotypes on the rate of bacterial evolution. *Journal of Evolutionary Biology*, **10**: 743–69.

Sowers, T., R.B. Alley, and J. Jubenville. 2003. Ice core records of atmospheric N_2O covering the last 106,000 years. *Science*, **301**: 945–8.

Srebotnik, E., K.A. Jensen, Jr, and K.E. Hammel. 1994. Fungal degradation of recalcitrant nonphenolic lignin structures without lignin peroxidase. *Proceedings of the National Academy of Sciences (USA)*, **20**: 12794–7.

Staley, J.T. 1997. Biodiversity: are microbial species threatened? *Current Opinion in Biotechnology*, **8**: 340–5.

Staley, J.T., R.W. Castenholz, R.R. Colwell, J.G. Holt, M.D. Kane, N.R. Pace, A.A. Salyers, and J.M. Tiedje. 1997. *The Microbial World: Foundation of the Biosphere. A Report from the American Academy of Microbiology.* Washington, DC: American Society for Microbiology.

Staley, J.T. and G.H. Orians. 1992. Evolution and the biosphere. In S.S. Butcher, R.J. Charlson, G.H. Orans, and G.V. Wolfe (eds.). *Global Biogeochemical Cycles*, pp. 21–54. New York: Academic Press.

Steinberg, P.D. and M.C. Rillig. 2003. Differential decomposition of arbuscular mycorrhizal fungal hyphae and glomalin. *Soil Biology and Biochemistry*, **35**: 191–4.

Stephens, J.R., G.A. Kowalchuk, M.V. Bruns, A.E. McCaig, C.J. Phillips, T.M. Embley, and J.I. Prosser. 1998. Analysis of β-subgroup Proteobacterial ammonia oxidizer populations in soil by denaturing gradient gel electrophoresis analysis and hierarchical phylogenetic probing. *Applied and Environmental Microbiology*, **64**: 2958–65.

Stephens, J.R., A.E. McCaig, Z. Smith, J.I. Prosser, and T.M. Embley. 1996. Molecular diversity of soil and marine 16S rRNA gene sequences related to β–subgroup ammonia-oxidizing bacteria. *Applied and Environmental Microbiology*, **62**: 4147–54.

Stephens, N.P. and D.Y. Sumner. 2003. Late Devonian carbon isotope stratigraphy and sea level fluctuations, Canning Basin, Western Australia. *Palaeogeography Palaeoclimatology Palaeoecology*, **191**: 203–19.

Stetter, K.O. 1996. Hyperthermophilic prokaryotes. *FEMS Microbiology Reviews*, **18**: 149–58.

Stewart, S.L., S.A. Grinshpun, K. Willeke, S. Terzieva, V. Ulevicius, and J. Donnelly. 1995. Effect of impact stress on microbial recovery on an agar surface. *Applied and Environmental Microbiology*, **61**: 1232–9.

Stoner, D.L., M.C. Geary, L.J. White, R.D. Lee, J.A. Brizzee, A.C. Rodman, and R.C. Rope. 2001. Mapping microbial biodiversity. *Applied and Environmental Microbiology*, **67**: 4324–8.

Storz, G. and R. Hengge-Aronis. 2000. *Bacterial Stress Responses.* Washington, DC: American Society for Microbiology.

Strauss, H. 1997. The isotopic composition of sedimentary sulfur through time. *Palaeogeography Palaeoclimatology Palaeoecology*, **132**: 97–118.

Sugawara, H., S. Miyazaki, J. Shimura, and Y. Ichiyanagi. 1996. Bioinformatics tools for the study of microbial diversity. *Journal of Industrial Microbiology and Biotechnology*, **17**: 490–7.

Sumner, D.Y. 1997. Late Archaean calcite–microbe interactions: Two morphologically distinct microbial communities that affected calcite nucleation differently. *Palaios*, **12**: 302–18.

Sumner, D.Y. 2001. Microbial influences on local carbon isotopic ratios and their preservation in carbonate. *Astrobiology*, **1**: 57–70.

Sunda, W., D.J. Kieber, R.P. Kiene, and S. Huntsman. 2002. An antioxidant function for DMSP and DMS in marine algae. *Nature*, **418**: 317–20.

Sundareshwar, P.V., J.T. Morris, P.J. Pellechia, H.J. Cohen, D.E. Porter, and B.C. Jones. 2001. Occurrence and ecological implications of pyrophosphate in estuaries. *Limnology and Oceanography*, **46**: 1570–7.

Suttle, C.A. 1994. The significance of viruses to mortality in aquatic microbial communities. *Microbial Ecology*, **28**: 237–43.

Svetlitchnyi, V., C. Peschel, G. Acker, and O. Meyer. 2001. Two membrane-associated NiFeS–carbon monoxide dehydrogenases from the anaerobic carbon-monoxide-utilizing eubacterium *Carboxydothermus hydrogenoformans*. *Journal of Bacteriology*, **183**: 5134–44.

Swofford, D.L. 2000. *PAUP*: Phylogenetic Analysis Using Parsimony and Other Methods* (software). Sunderland, MA: Sinauer Associates.

Tabita, F.R. 1995. Microbial ribulose bisphosphate carboxylase/oxygenase: A different perspective. *Photosynthesis Research*, **60**: 1–28.

Takai, K., H. Kobayashi, K.H. Nealson, and K. Horikoshi. 2003. *Deferribacter desulfuricans* sp. nov., a novel sulfur-, nitrate- and arsenate-reducing thermophile isolated from a deep-sea hydrothermal vent. *International Journal of Systematic and Evolutionary Microbiology*, **53**: 839–46.

Takashina, T., K. Otozati, T. Hamamoto, and K. Horikoshi. 1994. Isolation of halophilic and halotolerant bacteria from a Japanese salt field and comparison of the partial 16S rRNA gene sequence of an extremely halophilic isolate with those of other extreme halophiles. *Biodiversity and Conservation*, **3**: 632–42.

Tam, L., P.G. Kevan, and J.T. Trevors. 2003. Viable bacterial biomass and functional diversity in fresh and marine waters in the Canadian Arctic. *Polar Biology*, **26**: 287–94.

Tanaka, T., T. Kakizono, S. Nishikawa, K. Watanabe, K. Sasaki, N. Nishino, and S. Nagai. 1995. Screening of 5-aminolevulinic acid dehydratase inhibitors. *Seibutsu-Kogaku Kaishi*, **73**: 13–19.

Tanner, R.S. 2002. Cultivation of bacteria and fungi. In C.J. Hurst, R.L. Crawford, G.R. Knudsen, M.J. McInerney, and L.D. Stetzenbach (eds.) *Manual of Environmental Microbiology*, pp. 62–70. Washington, DC: American Society for Microbiology Press.

Taroncher-Oldenburg, G., E.M. Griner, C.A. Francis, and B.B. Ward. 2003. Oligonucleotide microarray for the study of functional gene diversity in the nitrogen cycle in the environment. *Applied and Environmental Microbiology*, **69**: 1159–71.

Taylor, F.J.R. 1974. Implications and extensions of the serial endosymbiosis theory of the origin of eukaryotes. *Taxon*, **23**: 229–58.

Taylor, F.J.R. 1976. Autogenous theories for the origin of eukaryotes. *Taxon*, **25**: 377–90.

Taylor, F.J.R. 1979. Symbioticism revisited: A discussion of the evolutionary impact of intracellular symbiosis. *Proceedings of the Royal Society of London*, **204**: 267–86.

Taylor, F.J.R. 2003. The collapse of the two-kingdom system, the rise of protistology and the founding of the International Society for Evolutionary Protistology (ISEP). *International Journal of Systematic and Evolutionary Microbiology*, **53**: 1707–14.

Tebbe, C.C., O.A. Ogunseitan, P.A. Rochelle, Y.L. Tsai, and B.H. Olson. 1992. Varied responses in the gene expression of heterotrophic bacteria isolated from the environment. *Applied Microbiology and Biotechnology*, **37**: 818–24.

Templeton, A.R. 1989. The meaning of species and speciation: A genetic perspective. In D. Otte and J.A. Endler (eds.) *Speciation and its Consequences*, pp. 3–27. Sunderland, MA: Sinauer.

Teplitski, M., J.B. Robinson, and W.D. Bauer. 2000. Plants secrete substances that mimic bacterial N-acyl homoserine lactone signal activities and affect population density-dependent behaviors in associated bacteria. *Molecular Plant–Microbe Interaction*, **13**: 637–48.

Teske, A., E. Alm, J.M. Regan, S. Toze, B.E. Rittmann, and D.A. Stahl. 1994. Evolutionary relationships among ammonia- and nitrite oxidizing bacteria. *Journal of Bacteriology*, **176**: 6623–30.

Teske, A., A. Dhillon, and M.L. Sogin. 2003. Genomic markers of ancient anaerobic microbial pathways: Sulfate reduction, methanogenesis, and methane oxidation. *Biological Bulletin (Woods Hole)*, **204**: 186–91.

Teske, A., K.-U. Hinrichs, V. Edgcomb, G.A. de Vera, D. Kysela, S.P. Sylva, M.L. Sogin, and H.W. Jannasch. 2002. Microbial diversity of hydrothermal sediments in the Guaymas Basin: Evidence for anaerobic methanotrophic communities. *Applied and Environmental Microbiology*, **68**: 1994–2007.

Thomashow, L.S., R.F. Bonsall, and D.M. Weller. 2002. Antibiotic production by soil and rhizosphere microbes *in situ*. In C.J. Hurst, R.L. Crawford, G.R. Knudsen, M.J. McInerney, and L.D. Stetzenbach. (eds.) *Manual of Environmental Microbiology* (Second Edition), pp. 638–59. Washington, DC: American Society for Microbiology.

Thompson, J.D., T.J. Gibson, F. Plewniak, F. Jeanmougin, and D.G. Higgins. 1997. The ClustalX-Windows interface: Flexible strategies for multiple sequence alignment aided by quality analysis tools. *Nucleic Acids Research*, **25**: 4876–82.

Thorpe, J.P. 1982. The molecular clock hypothesis: Biochemical evolution, genetic differentiation, and systematics. *Annual Review of Ecology and Systematics*, **13**: 139–68.

Timms-Wilson, T.M., L.S. van Overbeek, J.T. Trevors, M.J. Bailey, and J.D. van Elsas. 2002. Quantification of gene transfer in soil and the phytosphere. In C.J. Hurst, R.L. Crawford, G.R. Knudsen, M.J. McInerney, and L.D. Stetzenbach (eds.). *Manual of Environmental Microbiology* (Second Edition), pp. 648–59. Washington, DC: American Society for Microbiology.

Tomasz, A. 1965. Control of the competent state in *Pneumococcus* by a hormone-like cell product: An example of a new type of regulatory mechanism in bacteria. *Nature*, **208**: 155–9.

Tomb, H. and D. Kunkel. 1993. *Microaliens: Dazzling Journeys with an Electron Microscope*. New York: Farrar, Straus, and Giroux.

Topp, E. and R.S. Hanson. 1991. Metabolism of relatively important trace gases by methane-oxidizing bacteria. In J.E. Rogers and W.B. Whitman (eds.) *Microbial Production and Consumption of Greenhouse Gases: Methane, Nitrogen Oxides, and Halomethanes*, pp. 71–90. Washington, DC: American Society for Microbiology.

Torsvik, V., J. Gorksoyr, F.L. Daae, R. Sorheim, J. Michaelsen, and K. Salte. 1993. Diversity of microbial communities determined by DNA reassociation technique. In R. Guerrero and C. Pedros-Alio (eds.) *Trends in Microbial Ecology*, pp. 375–8. Madrid, Spain: Spanish Society for Microbiology.

Treguer, P., D.M. Nelson, A.J. Van Bennekom, D.J. DeMaster, A. Leynaert, and B. Queguiner. 1995. The silica balance in the world ocean: A re-estimate. *Science*, **268**: 375–9.

Trevors, J.T. 1999. Evolution of gene transfer in bacteria. *World Journal of Microbiology and Biotechnology*, **15**: 1–6.

Tromp, T.K., R.L. Shia, M. Allen, J.E. Eiler, and Y.L. Yung. 2003. Potential environmental impact of a hydrogen economy on the stratosphere. *Science*, **300**: 1740–2.

Trotsenko, Y. and V.N. Khmelenina. 2002. The biology and osmoadaptation of haloalkaliphilic methanotrophs. *Mikrobiologiya*, **71**: 149–59.

Tryland, I. and L. Fiksdal. 1998. Rapid enzymatic detection of heterotrophic activity of environmental bacteria. *Water Science and Technology*, **38**: 95–101.

Tunlid, A. and D.C. White. 1992. Biochemical analysis of biomass community structure nutritional status and metabolic activity of microbial communities in soil. In G. Stotzky and J.-M. Bollag (eds.) *Soil Biochemistry*, Volume 7. pp. 229–62. New York: Marcell Dekker.

Turner, A. and G.E. Millward. 2002. Suspended particles: Their role in estuarine biogeochemical cycles. *Estuarine Coastal and Shelf Science*, **55**: 857–83.

Tyler, S.C. 1991. The global methane budget. In J.E. Rogers and W.B. Whitman (eds.) *Microbial Production and Consumption of Greenhouse Gases: Methane, Nitrogen Oxides, and Halomethanes*, pp. 7–38. Washington, DC: American Society for Microbiology.

Tyson, G.W., J. Chapman, P. Hugenholtz, E.E. Allen, R.J. Ram, P.M. Richardson, V.V. Solovyev, E.M. Rubin, D.S. Rokhsar, and J. F. Banfield 2004. Community structure and metabolism through reconstruction of microbial genomes from the environment. *Nature*, **428**: 25–6.

UNEP. 1992. *Convention on Biological Diversity: Article 2. Secretariat of the Convention on Biological Diversity*. United Nations Environment Program, Nairobi, Kenya (http://www.biodiv.org/convention/articles.asp).

United States Geological Survey. 2003. Phosphate Rock. U.S. Geological Survey, Mineral Commodity Summaries, January 2003. By S.M. Jasinski (http://minerals.usgs.gov/minerals/pubs/commodity/phosphate_rock/).

Updegraff, K., S.D. Bridgham, J. Pastor, P. Weishampel, and C. Harth. 2001. Response of CO_2 and CH_4 emissions from peatlands to warming and water table manipulation. *Ecological Applications*, **11**: 311–26.

Upton, M., B. Hill, C. Edwards, J.R. Saunders, D.A. Ritchie, and D. Lloyd. 2000. Combined molecular ecological and confocal laser scanning microscopic analysis of peat bog methanogen populations. *FEMS Microbiology Letters*, **193**: 275–81.

Urashima, M., N. Shindo, and N. Okabe. 2003. Seasonal models of herpangina and hand-foot-mouth disease to simulate annual fluctuations in urban warming in Tokyo. *Japanese Journal of Infectious Diseases*, **56**: 48–53.

Urbach, E., and S.W. Chisholm. 1998. Genetic diversity in *Prochlorococcus* populations flow cytometrically sorted from the Sargasso Sea and Gulf Stream. *Limnology and Oceanography*, **43**: 1615–30.

Urey, H. and S.L. Miller. 1959. Organic compound synthesis on the primitive earth. *Science*, **130**: 245–51.

USEPA. 1992. *Biological Populations as Indicators of Environmental Change*. EPA/230/R-92/011—Volume 1. Washington, DC: Office of Policy, Planning, and Evaluation, U.S. Environmental Protection Agency.

USGS. 2003. Coral mortality and African dust: Black-band disease. Available online: http://coastal.er.usgs.gov/african_dust/blackband.html. Updated February 4, 2003.

Utåker, J.B., L. Baken, Q.Q. Jiang, and I. Nes. 1995. Phylogenetic analysis of seven new isolates of ammonia-oxidizing bacteria based on 16S rRNA sequences. *Systematics and Applied Microbiology*, **18**: 549–59.

Utsumi, M., S.E. Belova, G.M. King, and H. Uchiyama. 2003. Phylogenetic comparison of methanogen diversity in different wetland soils. *Journal of General and Applied Microbiology*, **49**: 75–84.

Valentine, D.L. 2002. Biogeochemistry and microbial ecology of methane oxidation in anoxic environments: A review. *Antonie van Leeuwenhoek*, **81**: 271–82.

Vallino, J. 2003. Modeling microbial consortiums as distributed metabolic networks. *Biological Bulletin (Woods Hole)*, **204**: 174–9.

Van den Berg, H.A. 2001. How microbes can achieve balanced growth in a fluctuating environment. *Acta Biotheoretica*, **49**: 1–21.

Van Everbroeck, B., P. Pals, J.-J. Martin, and P. Cras. 2002. Transmissible spongiform encephalopathies: The story of a pathogenic protein. *Peptides* (New York), **23**: 1351–9.

van Hannen, E.J., G. Zwart, M.P. van Agterveld, H.J. Gons, J. Ebert, and H.J. Laanbroek. 1999. Changes in bacterial and eukaryotic community structure after mass lysis of filamentous cyanobacteria associated with viruses. *Applied and Environmental Microbiology*, **65**: 795–801.

van Herk, C.M., A. Aptroot, and H.F. van Dobben. 2002. Long-term monitoring in the Netherlands suggests that lichens respond to global warming. *Lichenologist*, **34**: 141–54.

Van Kooten, K. 1998. Economics of conservation: A critical review. *Environmental Science and Policy*, **1**: 13–25.

Vance, R.E., J. Zhu, and J.J. Mekalanos. 2003. A constitutively active variant of the quorum-sensing regulator LuxO affects protease production and biofilm formation in *Vibrio cholerae*. *Infection and Immunity*, **71**: 2571–6.

Vanzin, G.F., J. Huang, S. Smolinski, K. Kronoveter, and P.-C. Maness. 2002. *Biological hydrogen from fuel gases. Proceedings of the 2002 U.S. DOE Hydrogen Program Review*. NERL/CP-610-32405. http://www.eere.energy.gov/hydrogenandfuelcells/pdfs/32405a1.pdf.

Varley, J.D. and P.T. Scott. 1998. Conservation of microbial diversity: A Yellowstone priority. *ASM News*, **64**: 147–51.

Vaulot, D., K. Romari, and F. Not. 2002. Are autotrophs less diverse than heterotrophs in marine picoplankton? *Trends in Microbiology*, **10**: 266–7.

Veldkamp, H. 1976. *Continuous Culture in Microbial Physiology and Ecology*. Shildon: Meadowfield Press.

Velicer, G.J. 2003. Social strife in the microbial world. *Trends in Microbiology*, **11**: 330–7.

Venter, J.C., K. Remington, J.F. Heidelberg, A.L. Halpern, D. Rusch, J.A. Eisen, D. Wu, I. Paulsen, K.E. Nelson, W. Nelson, D.E. Fouts, S. Levy, A.H. Knap, M.W. Lomas, K. Nealson, O. White, J. Peterson, J. Hoffman, R. Parsons, H. Baden-Tillson, C. Pfannkoch, Y.-H. Rogers, and H.O. Smith. 2004. Environmental genome shotgun sequencing of the Sargasso Sea. *Science*, **304**: 66–74.

Vesey, G., J. Narai, N. Ashbolt, K. Williams, and D. Veal. 1994. Detection of specific microorganisms in environmental samples using flow cytometry. *Methods in Cell Biology*, **42**: 489–522.

Vignais, P.M., B. Billoud, and J. Meyer. 2001. Classification and phylogeny of hydrogenases. *FEMS Microbiology Reviews*, **25**: 455–501.

Visca, P., G. Colotti, L. Serino, D. Verzili, N. Orsi, and E. Chiancone. 1992. Metal regulation of siderophore synthesis in *Pseudomonas aeruginosa* and functional effects of siderophore–metal complexes. *Applied and Environmental Microbiology*, **58**: 2686–93.

Visscher, P.T., R.F. Gritzer, and E.R. Leadbetter. 1999. Low-molecular-weight sulfonates: a major substrate for sulfate reducers in marine microbial mats. *Applied and Environmental Microbiology*, **65**: 3272–8.

Vogel, T.M. and M.V. Walter. 2002. Bioaugmentation. In C.J. Hurst, R.L. Crawford, G.R. Knudsen, M.J. McInerney, and L.D. Stetzenbach. (eds.) *Manual of Environmental Microbiology*, pp. 952–9. Washington, DC: American Society for Microbiology Press.

Von Bloh, W., S. Franck, C. Bounama, and H.-J. Schellnhuber. 2003. Maximum number of habitable planets at the time of Earth's origin: New hints for panspermia? *Origins of Life and Evolution of the Biosphere*, **33**: 219–31.

Von Wintzingerode, F., F.A. Rainey, R.M. Kroppenstedt, and E. Stackebrandt. 1997. Identification of environmental strains of *Bacillus mycoides* by fatty acid analysis and species-specific 16S rDNA oligonucleotide probe. *FEMS Microbiology Ecology*, **24**: 201–9.

Voordouw, G., Y. Shen, C.S. Harrington, A.J. Telang, T.R. Jack, and D.W.S. Westlake. 1993. Quantitative reverse sample genome probing of microbial communities and its application to oil field production waters. *Applied and Environmental Microbiology*, **59**: 4101–14.

Vreeland, R.H., W.D. Rosenzweig, and D.W. Powers. 2000. Isolation of a 250 million-year-old halotolerant bacterium from a primary salt crystal. *Nature*, **407**: 897–900.

Wagner, G.H. and D.C. Wolf. 1998. Carbon transformations and soil organic matter formation. In D.M. Sylvia, J.J. Fuhrmann, P.G. Hartel, and D.A. Zuberer (eds.) *Principles and Applications of Soil Microbiology*, pp. 259–94. Upper Saddle River, NJ: Prentice-Hall.

Wagner, M., A.J. Roger, J.L. Flax, G.A. Brusseau, and D.A. Stahl. 1998. Phylogeny of dissimilatory sulfite reductases supports an early origin of sulfate respiration. *Journal of Bacteriology*, **180**: 2975–82.

Wagner, V.E., D. Bushnell, L. Passador, A.I. Brooks, and B.H. Iglewski. 2003. Microarray analysis of *Pseudomonas aeruginosa* quorum-sensing regulons: Effects of growth phase and environment. *Journal of Bacteriology*, **185**: 2080–95.

Wall, D.H., G. Adams, and A.N. Parsons. 2001. Soil biodiversity. In F.S. Chapin, O.E. Sala, and E. Huber-Sannwald (eds.) *Global Biodiversity in a Changing Environment*, pp. 47–82. New York: Springer-Verlag.

Wallner, G., B. Fuchs, S. Spring, W. Beikser, and R. Amann. 1997. Flow sorting of microorganisms for molecular analysis. *Applied and Environmental Microbiology*, **63**: 4223–31.

Walsh, C. 2003. *Antibiotics: Actions, Origins, Resistance.* Washington, DC: American Society for Microbiology.

Ward, B.B. 2002. How many species of prokaryotes are there? *Proceedings of the National Academy of Sciences USA*, **99**: 10234–6.

Ward, B.B., K.A. Kirkpatrick, P.C. Novelli, and M.I. Scranton. 1987. Methane oxidation and methane fluxes in the ocean surface layer and deep anoxic waters. *Nature*, **327**: 226–8.

Ward, D.M. 1998. A natural species concept for prokaryotes. *Current Opinion in Microbiology*, **1**: 271–7.

Ward, D.M., R. Weller, and M.M. Bateson. 1990. 16S rRNA sequences reveal numerous uncultured microorganisms in a natural community. *Nature*, **345**: 63–5.

Watson, G.M. and F.R. Tabita. 1997. Microbial ribulose 1,5-bisphosphate carboxylase/oxygenase: A molecule for phylogenetic and enzymological investigation. *FEMS Microbiology Letters*, **146**: 13–22.

Watson, G.M., J.P. Yu, and F.R. Tabita. 1999. Unusual ribulose 1,5-bisphosphate carboxylase/oxygenase of anoxic *Archaea. Journal of Bacteriology*, **181**: 1569–75.

Watve, M.G. and R.M. Gangal. 1996. Problems in measuring bacterial diversity and a possible solution. *Applied and Environmental Microbiology*, **62**: 4299–301.

Webber, G., M. Zeller, and K. Hessler. 2003. *Bacillus thuringiensis: Sharing its natural talent with crops.* Iowa State University. Office of Biotechnology. Document Number 4H 949.

Weber, M., K. Taraz, H. Budzikiewicz, V. Geoffroy, and J.-M. Meyer. 2000. The structure of a pyoverdine from *Pseudomonas* sp. CFML 96.188 and its relation to other pyoverdins with a cyclic C-terminus. *BioMetals*, **13**: 301–9.

Weeks, S.J., B. Currie, and A. Bakun. 2002. Massive emissions of toxic gas in the Atlantic. *Nature*, **415**: 493–4.

Weinbauer, M.G., I. Brettar, and M.G. Hoefle. 2003. Lysogeny and virus-induced mortality of bacterioplankton in surface, deep, and anoxic marine waters. *Limnology and Oceanography*, **48**: 1457–65.

Weinbauer, M.G. and C.A. Suttle. 1999. Lysogeny and prophage induction in coastal and offshore bacterial communities. *Aquatic Microbial Ecology*, **18**: 217–25.

Weinbauer, M.G., S.W. Wilhelm, C.A. Suttle, R.J. Pledger, and D.L. Mitchell. 1999. Sunlight-induced DNA damage and resistance in natural viral communities. *Aquatic Microbial Ecology*, **17**: 111–20.

Weinberg, E.D. 1997. The *lactobacillus* anomaly: Total iron abstinence. *Perspectives in Biology and Medicine*, **40**: 578–83.

Weller, D.M. and L.S. Thomashow. 1990. Antibiotics: Evidence for their production and sites where they are produced. In R.R. Baker and P.E. Dunn (eds.) *New Directions in Biological Control: Alternatives for Suppression of Agricultural Pests and Diseases*, pp. 703–11. New York: Alan R. Liss.

Welsh, D. 2000. Ecological significance of compatible solute accumulation by micro-organisms: From single cells to global climate (Literature Review). *FEMS Microbiology Reviews*, **24**: 263–90.

Westall, F. 1998. The oldest fossil mineral bacteria from the Early Archaean of South Africa and Australia. In J. Chela-Flores and F. Raulin (eds.) *Exobiology: Matter, Energy, and Information in the Origin and Evolution of Life in the Universe*, pp. 181–6. Dordrecht, The Netherlands: Kluwer Academic.

Westherimer, F.H. 1987. Why nature chose phosphates. *Science*, **235**: 1173–8.

Wetmur, J.G. 1994. Influence of the common human delta-aminolevulinate dehydratase polymorphism on lead body burden. *Environmental Health Perspectives*, **102**(supplement 3): 215–19.

Wetzel, R.G., P.G. Hatcher, and T.S. Bianchi. 1995. Natural photolysis by ultraviolet irradiance of recalcitrant dissolved organic matter to simple substrates for rapid bacterial metabolism. *Limnology and Oceanography*, **40**: 1369–80.

Whalen, S.C. and W.S. Reeburgh. 1990. Consumption of atmospheric methane to sub ambient concentrations by tundra soils. *Nature*, **36**: 160–2.

Wheeler, Q.D. and R. Meier (eds.) 2000. *Species Concepts and Phylogenetic Theory: A Debate.* New York: Columbia University Press.

Whistler, C.A. and L.S. Pierson III. 2003. Repression of phenazine antibiotic production in *Pseudomonas aureofaciens* strain 30-84 by RpeA. *Journal of Bacteriology*, **185**: 3718–25.

White, B.A., I.K.O. Cann, S.A. Kocherginskaya, R.I. Aminov, L.A. Thill, R.I. Mackie, and R. Onodera. 1999. Molecular analysis of archaea, bacteria and eucarya communities in the rumen: Review. *Asian-Australasian Journal of Animal Sciences*, **12**: 129–38.

White, D.C. 1993. *In situ* measurement of microbial biomass, community structure, and nutritional status. *Philosophical Transactions of the Royal Society, London (Series A)*, **344**: 59–67.

White, D.C., D.B. Ringelberg, S. McNaughton, S.J. Alugupalli, and D. Schram. 1997. Signature lipid biomarker analysis for quantitative assessment in situ of environmental microbial ecology. In R.P. Eganhouse (ed.) *Molecular Markers in Environmental Geochemistry, Volume 2*, pp. 22–34. Published for the American Chemical Society by Oxford University Press, New York.

White, D.C., J.O. Stair, and D.B. Ringelberg. 1996. Quantitative comparisons of in situ microbial biodiversity by signature biomarker analysis. *Journal of Industrial Microbiology*, **17**: 185–96.

White, M.J.D. 1978. *Modes of Speciation.* San Francisco: W.H. Freeman.

Whitman, W.B. and J.E. Rogers. 1991. Research needs in the microbial production and consumption of radiatively important trace gases. In J.E. Rogers and W.B. Whitman (eds.) *Microbial Production and Consumption of Greenhouse Gases: Methane, Nitrogen Oxides, and Halomethanes*, pp. 287–91. Washington, DC: American Society for Microbiology.

WHO. 2001. *Cholera: 2001 Weekly Epidemiological Record.* Number 31. Geneva, Switzerland: World Health Organization (http://www.who.int/vaccine_research/diseases/cholera/documents/en/).

WHO. 2002. *Tuberculosis.* Fact Sheet Number 104. Geneva, Switzerland: World Health Organization (http://www.who.int/mediacentre/factsheets/who104/en/).

WHO. 2003. *Global health-sector strategy for HIV/AIDS 2003–2007.* Geneva, Switzerland: World Health Organization (http://www.who.int/hiv/pub/advocacy/ghss/en/).

Wickramasinghe, N.C. 1974. Formaldehyde polymers in interstellar space. *Nature,* **252**: 462–3.

Wiegel, J. and M.W.W. Adams (eds.) 1998. *Thermophiles: The Keys to Molecular Evolution and the Origin of Life?* Philadelphia, PA: Taylor and Francis.

Wierenga, R.K. 2001. The TIM-barrel fold: a versatile framework for efficient enzymes. *Federation of European Biochemical Societies (Letters),* **492**: 193–8.

Wiggins, B.A. and M. Alexander. 1987. Minimum bacterial density for bacteriophage replication: Implications for significance of bacteriophages in natural ecosystems. *Applied and Environmental Microbiology,* **49**: 19–23.

Wilhelm, S.W., W.H. Jeffrey, C.A. Suttle, and D.L. Mitchell. 2002. Estimation of biologically damaging UV levels in marine surface waters with DNA and viral dosimeters. *Photochemistry and Photobiology,* **76**: 268–73.

Wilhelm, S.W. and C.A. Suttle. 1999. Viruses and nutrient cycles in the sea. *Bioscience,* **49**: 781–8.

Wilhelm, S.W., M.G. Weinbauer, C.A. Suttle, and W.H. Jeffrey. 1998a. The role of sunlight in the removal and repair of viruses in the sea. *Limnology and Oceanography,* **43**: 586–92.

Wilhelm, S.W., M.G. Weinbauer, C.A. Suttle, R.J. Pledger, and D.L Mitchell. 1998b. Measurements of DNA damage and photoreactivation imply that most viruses in marine surface waters are infective. *Aquatic Microbial Ecology,* **14**: 215–22.

Wilkinson, M.H.F. and F. Schut. 1999. *Digital Image Analysis of Microbes: Imaging, Morphometry, Fluorometry, and Motility Techniques and Applications.* New York: Wiley.

Williams, B.G. and C. Dye. 2003. Antiretroviral drugs for tuberculosis control in the era of HIV/AIDS. *Science,* **301**: 1535–7.

Williams, P. and C. Humphries. 1996. Comparing character diversity among biotas. In K. Gaston (ed.) *Biodiversity: A Biology of Numbers and Difference,* pp. 54–76. Oxford, England: Blackwell Science.

Williams, S.T., A.M. Mortimer, and L. Manchester. 1987. Ecology of soil bacteriophages. In S.M. Goyar, C.P. Gerba, and G. Bitton (eds.) *Phage Ecology,* pp. 157–79. New York: Wiley.

Willke, B. and K.-D. Vorlop. 1994. Long-term observation of gel-entrapped microbial cells in a microscope reactor. *Biotechnology Techniques,* **8**: 619–22.

Wilson, E.O. 1994. *Naturalist.* Washington, DC: Island Press/Shearwater Books.

Wimpenny, J.W.T. (ed.) 1998. *CRC Handbook of Laboratory Model Systems for Microbial Ecosystems.* Boca Raton, FL: CRC Press.

Winogradsky, S. 1890. Recherches sur les Organismes de la Nitrification. *Compte Rendus,* **110**: 1013–16. In T.D. Brock (trans. and ed.) 1998. *Milestones in Microbiology: 1556 to 1940,* p. 231. Washington, DC: American Society for Microbiology Press.

Winter, C., M.M. Moeseneder, and G.J. Herndl. 2001. Impact of UV radiation on bacterioplankton community composition. *Applied and Environmental Microbiology,* **67**: 665–72.

Winter, M.J. 2002. The interactive periodic table. http://www.webelements.com.

Woese, C.R. 1998a. The universal ancestor. *Proceedings of the National Academy of Sciences (USA),* **95**: 6854–9.

Woese, C.R. 1998b. Default taxonomy: Ernst Mayr's view of the microbial world. *Proceedings of the National Academy of Sciences (USA),* **95**: 11043–6.

Woese, C.R., O. Kandler, and M.L. Wheelis. 1990. Towards a natural system of organisms: Proposal for the domains Archaea, Bacteria, and Eucarya. *Proceedings of the National Academy of Sciences (USA),* **87**: 4576–9.

Wolf, H.J. and R.S. Hanson. 1979. Isolation and characterization of methane-utilizing yeasts. *Journal of General Microbiology,* **114**: 187–94.

Wolin, M.J. and T.L. Miller. 1987. Bioconversion of organic carbon to CH_4 and CO_2. *Geomicrobiology Journal,* **5**: 239–59.

Wommack, K.E. and R.R. Colwell. 2000. Virioplankton: viruses in aquatic ecosystems. *Microbiology and Molecular Biology Review,* **64**: 69–114.

Woodruff, H.B. 1996. Impact of microbial diversity on antibiotic discovery, a personal history. *Journal of Industrial Microbiology and Biotechnology,* **17**: 323–7.

Wright, S.F. 1992. Immunological techniques for detection, identification, and enumeration of microorganisms in the environment. In M.A. Levin, Seidler, R.J. and M. Rogul (eds.) *Microbial Ecology Principles, Methods, and Applications* pp. 45–60, New York: McGraw-Hill.

Wright, S.F., M. Franke-Snyder, J.B. Morton, and A. Upadhyaya. 1996. Time-course study and partial characterization of a protein on hyphae of arbuscular mycorrhizal fungi during active colonization of roots. *Plant and Soil,* **181**: 193–203.

Wright, S.F. and A. Upadhyaya. 1996. Extraction of an abundant and unusual protein from soil and comparison with hyphal protein of arbuscular mycorrhizal fungi. *Soil Science,* **161**: 575–86.

Wright, S.F. and A. Upadhyaya. 1998. A survey of soils for aggregate stability and glomalin, a glycoprotein produced by hyphae of arbuscular mycorrhizal fungi. *Plant and Soil,* **198**: 97–107.

Wu, W.-S. and T.-W. Chen. 1999. Development of a new semiselective medium for detecting *Alternaria brassicicola* in cruciferous seeds. *Seed Science and Technology,* **27**: 397–409.

Xu, F., T. Byun, H.-J. Dussen, and K.R. Duke. 2003. Degradation of N-acylhomoserine lactones, the bacterial quorum-sensing molecules, by acylase. *Journal of Biotechnology,* **101**: 89–96.

Yamaguchi, K., N.J. Cosper, C. Stalhandske, R.A. Scott, M.A. Pearson, P.A. Karplus, and R.P. Hausinger. 1999. Characterization of metal-substituted *Klebsiella aerogenes* urease. *Journal of Biological and Inorganic Chemistry,* **4**: 468–77.

Yan, Z.-C., B. Wang, Y.-Z. Li, X. Gong, H.-Q. Zhang and P.-J. Gao. 2003. Morphologies and phylogenetic classification of cellulolytic myxobacteria. *Systematic and Applied Microbiology,* **26**: 104–9.

Yang, X., H. Beyenal, G. Harkin, and Z. Lewandowski. 2000. Quantifying biofilm structure using image analysis. *Journal of Microbiological Methods,* **39**: 109–19.

Yarmolinsky, M.B. 1995. Programmed cell death in bacterial populations. *Science,* **267**: 836–7.

Yoch, D.C. 2002. Dimethylsulfoniopropionate: Its sources, role in the marine food web, and biological degradation to dimethylsulfide. *Applied and Environmental Microbiology,* **68**: 5804–15.

Young, J.P.W. and B.R. Levin. 1992. Adaptation in bacteria: Unanswered ecological and evolutionary questions about well-studied molecules. In T.J. Crawford and G.M. Hewitt (eds.) *Genes in Ecology,* pp. 169–92. Oxford, England: Blackwell Scientific Publications.

Yu, X., X.-j. Zhang, X.-L. Liu, X.-D. Zhao, and Z.-S. Wang. 2003. Phosphorus limitation in biofiltration for drinking water treatment. *Journal of Environmental Sciences (China)*, **15**: 494–9.

Zahn, R. 2003. The octapeptide repeats in mammalian prion protein constitute a pH-dependent folding and aggregation site. *Journal of Molecular Biology*, **334**: 477–88.

Zahn, R., A. Liu, T. Luhrs, R. Riek, C. von Schroetter, F. Garcia, M. Billeter, L. Calzolai, G. Wider, and K. Wuthrich. 2000. NMR solution structure of the human prion protein. *Proceedings of the National Academy of Sciences USA*, **97**: 145–50.

Zamore, P.D. 2002. Ancient pathways programmed by small RNAs. *Science*, **296**: 1265–9.

Zaug, A.J. and T.R. Cech. 1986. The intervening sequence RNA of *Tetrahymena* is an enzyme. *Science*, **231**: 470–5.

Zavarzin, G.A. 1994. Microbial biogeography. *Zhurnal Obshchei Biologii*, **55**: 5–12.

Zawadzki, P. and F.M. Cohan. 1995. The size and continuity of DNA segments integrated in *Bacillus* transformation. *Genetics*, **141**: 1231–43.

Zdanowski, M.K. and P. Weglenski. 2001. Ecophysiology of soil bacteria in the vicinity of Henryk Arctowski Station, King George Island, Antarctica. *Soil Biology and Biochemistry*, **33**: 819–29.

Zelenev, V.V., A.H.C. van Bruggen, and A.M. Semenov. 2000. "BACWAVE," a spatial-temporal model for traveling waves of bacterial populations in response to a moving carbon source in soil. *Microbial Ecology*, **40**: 260–72.

Zelles, L., R. Rackwitz, Q.Y. Bai, T. Beck, and F. Beese. 1995. Discrimination of microbial diversity by fatty acid profiles of phospholipids and lipopolysaccharides in differently cultivated soils. *Plant and Soil*, **170**: 115–22.

Zengler, K., G. Toledo, M. Rappe, J. Elkins, E.J. Mathur, J.M. Short, and M. Keller. 2002. Cultivating the uncultured. *Proceedings of the National Academy of Sciences (USA)*, **99**: 15681–6.

Zeph, L.R., M.A. Onaga, and G. Stotzky. 1988. Transduction of *Escherichia coli* by bacteriophage P1 in soil. *Applied and Environmental Microbiology*, **54**: 1731–7.

Zhang, H., K. Ishige, and A. Kornberg. 2002. A polyphosphate kinase (PPK-2) widely conserved in bacteria. *Proceedings of the National Academy of Sciences (USA)*, **99**: 16678–83.

Zhang, L.-H. 2003. Quorum quenching and proactive host defense. *Trends in Plant Science*, **8**: 238–44.

Zhang, R.G., T. Skarina, J.E. Katz, S. Beasley, A. Khachatryan, S. Vyas, C.H. Arrowsmith, S. Clarke, A. Edwards, A. Joachimiak, and A. Savchenko. 2001. Structure of *Thermotoga maritima* stationary phase survival protein *SurE*: A novel acid phosphatase. *Structure (Cambridge)*, **9**: 1095–106.

Zhang, Y.-M., J.-K. Liu and T-Y. Wong. 2003. The DNA excision repair system of the highly radioresistant bacterium *Deinococcus radiodurans* is facilitated by the pentose phosphate pathway. *Molecular Microbiology*, **48**: 1317–23.

Zhong, Y., F. Chen, S.W. Wilhelm, L. Poorvin, and R.E. Hodson. 2002. Phylogenetic diversity of marine cyanophage isolates and natural virus communities as revealed by sequences of viral capsid assembly protein gene g20. *Applied and Environmental Microbiology*, **68**: 1576–84.

Zhou, J., B. Xia, D.S. Treves, L-Y. Wu, T.L. Marsh, R.V. O'Neill, A.V. Palumbo, and J.M. Tiedje. 2002. Spatial and resource factors influencing high microbial diversity in soil. *Applied and Environmental Microbiology*, **68**: 326–34.

Zhu, Y., S.S.K. Lee, and W. Xu. 2003. Crystallization and characterization of polyphosphate kinase from *Escherichia coli. Biochemical and Biophysical Research Communications*, **305**: 997–1001.

Zinder, N.D. and J. Lederberg. 1952. Genetic exchange in Salmonella. *Journal of Bacteriology*, **64**: 697–9.

Zuckerkandl, E. and L. Pauling. 1965. Molecules as documents of evolutionary history. *Journal of Theoretical Biology*, **8**: 357–66.

INDEX

Page numbers in **bold** refer to tables, and in *italics* refer to boxes and illustrations.